studies in physical and theoretical chemistry 86

THERMAL DECOMPOSITION OF IONIC SOLIDS

studies in physical and theoretical chemistry

Recent titles in this series

54 **Math/Chem/Comp 1987** edited by R.C. Lacher
55 **Semiconductor Electrodes** edited by H.O. Finklea
56 **Computational Chemistry** by M.D. Johnston, Jr.
57 **Positron and Positronium Chemistry** edited by D.M. Schrader and Y.C. Jean
58 ***Ab Initio* Calculation of the Structures and Properties of Molecules** by C.E. Dykstra
59 **Physical Adsorption on Heterogeneous Solids** by M. Jaroniec and R. Madey
60 **Ignition of Solids** by V.N. Vilyunov and V.E. Zarko
61 **Nuclear Measurements in Industry** by S. Rózsa
62 **Quantum Chemistry: Basic Aspects, Actual Trends** edited by R. Carbó
63 **Math/Chem/Comp 1988** edited by A. Graovac
64 **Valence Bond Theory and Chemical Structure** edited by D.J. Klein and N. Trinajstić
65 **Structure and Reactivity in Reverse Micelles** edited by M.P. Pileni
66 **Applications of Time-Resolved Optical Spectroscopy** by V. Brückner, K.-H. Feller and U.-W. Grummt
67 **Magnetic Resonance and Related Phenomena** edited by J. Stankowski, N. Pilewski, S.K. Hoffmann and S. Idziak
68 **Atomic and Molecular Clusters** edited by E.R. Bernstein
69 **Structure and Properties of Molecular Crystals** edited by M. Pierrot
70 **Self-consistent Field: Theory and Applications** edited by R. Carbó and M. Klobukowski
71 **Modelling of Molecular Structures and Properties** edited by J.-L. Rivail
72 **Nuclear Magnetic Resonance: Principles and Theory** by R. Kitamaru
73 **Artificial Intelligence in Chemistry: Structure Elucidation and Simulation of Organic Reactions** by Z. Hippe
74 **Spectroscopy and Relaxation of Molecular Liquids** edited by D. Steele and J. Yarwood
75 **Molecular Design: Chemical Structure Generation from the Properties of Pure Organic Compounds** by A.L. Horvath
76 **Coordination and Transport Properties of Macrocyclic Compounds in Solution** by B.G. Cox and H. Schneider
77 **Computational Chemistry: Structure, Interactions and Reactivity. Part A** edited by S. Fraga
Computational Chemistry: Structure, Interactions and Reactivity. Part B edited by S. Fraga
78 **Electron and Proton Transfer in Chemistry and Biology** edited by A. Müller, H. Ratajczak, W. Junge and E. Diemann
79 **Structure and Dynamics of Solutions** edited by H. Ohtaki and H. Yamatera
80 **Theoretical Treatment of Liquids and Liquid Mixtures** by C. Hoheisel
81 **Mössbauer Studies of Surface Layers** by G.N. Belozerski
82 **Dynamics of Excited Molecules** edited by Kozo Kuchitsu
83 **Structure, Fluctuation and Relaxation in Solutions** edited by H. Nomura, F. Kawaizuma and J. Yarwood
84 **Solid State NMR of Polymers** edited by I. Ando and T. Asakura
85 **Vibrational Spectra: Principles and Applications with Emphasis on Optical Activity** by P.L. Polavarapu

studies in physical and theoretical chemistry 86

THERMAL DECOMPOSITION OF IONIC SOLIDS

Andrew K. GALWEY* and **Michael E. BROWN**
Chemistry Department
Rhodes University
Grahamstown
6140 South Africa

**Retired from the School of Chemistry, Queen's University of Belfast*
Present address: 18 Viewfort Park Dunmurry, Belfast BT17 9JY,
Northern Ireland

1999

ELSEVIER
Amsterdam - Lausanne - New York - Oxford - Shannon - Singapore - Tokyo

ELSEVIER SCIENCE B.V.
Sara Burgerhartstraat 25
P.O. Box 211, 1000 AE Amsterdam, The Netherlands

First edition 1999

Library of Congress Cataloging in Publication Data
A catalog record from the Library of Congress has been applied for.

ISBN: 0-444-82437-5

∞ The paper used in this publication meets the requirements of ANSI/NISO Z39.48-1992 (Permanence of Paper).

Printed in The Netherlands.

THE INTEREST IN DECOMPOSING SOLIDS IS
NOT IN THE DESTRUCTION OF THE REACTANT
BUT IN THE DETERMINATION OF THE
CHEMICAL PROCESSES BY WHICH THE COMPONENTS
OF THE CRYSTAL ARE TRANSFORMED INTO THE
PRODUCTS.

PREFACE

There is no doubt that the extensive but dispersed literature concerned with decompositions of solids is in need of review. The number of comprehensive surveys of this important, active and well-defined subject area that have been published in the last fifty years (the effective life-time of the topic) is remarkably small. Even reviews of more restricted aspects, such as the thermal properties of particular groups of reactants, are surprisingly few in number.

In the immediate past there has been little progress towards understanding the factors that control thermal stability and determine the mechanisms of specific reactions of solids. The absence of any agreed basis for the classification of different types of these reactions means that new contributions cannot be related to information already available.

In all areas of science it is essential to reconsider periodically the theoretical concepts that are accepted as the unifying framework for the interpretation of measured data and the classification of experimental observations. The recognition of behavioural similarities for the decompositions of different substances is a central feature in the formulation of reaction mechanisms and in developing theory. Over the years, the range of crystalline reactants for which observations are available, has increased greatly. However, many thermal studies of solids do not report sufficient information, in depth or type, to enable significant progress to be made in comparing and classifying these rate processes.

Our account presents some personal perceptions of the concepts and the models that have found the most successful applications in the recent literature, with the aim of initiating a debate intended to contribute towards advancement of the theory of the subject. There are many citations of our own publications. These naturally emphasize the conclusions for which we feel the greatest personal confidence. References are made to older works when the chemical concepts are more satisfactorily described, or the discussions show greater insights than in the more modern articles. The need for greater emphasis on *chemical*, rather than *mathematical*, representations of behaviour (molecular models for mechanistic proposals, comparisons of properties with reference to chemical compositions, etc.) is a main theme of this book.

The first part of the book (Chapters 1 to 6) is concerned with theoretical aspects of the subject. The second part (Chapters 7 to 17) surveys groups of reactions classified by similarities of chemical composition. The final Chapter (18) reviews the subject by discussing unifying features identified as significant and proposes possible directions for future progress.

This book is expected to be of greatest value and interest to scientists concerned with the chemical properties and reactions of solids, including chemists, physicists, pharmacists, material scientists, crystallographers, metallurgists and others. This wide coverage of the literature dealing with thermal reactions of solids will be of value to both academic and industrial researchers by reviewing the current status of the theory of the subject. It could also provide a useful starting point for the exploitation of crystalline materials in practical and industrial applications. The contents should also be relevant to a wide variety of researchers, including, for example, those concerned with the stabilities of polymers and composite materials, the processing of minerals, the shelf-lives of pharmaceuticals, etc.

ACKNOWLEDGEMENTS

The authors acknowledge with gratitude the wordprocessing skills of Mrs M. Kent and Mrs B. Tarr in the preparation of the text. The encouragement and support of our wives, Kathleen Galwey and Cindy Brown, has been invaluable.

The progress of writing this book was notably accelerated by the award of Visiting Lectureships (at critical stages) to AKG by Rhodes University and to MEB by the Queen's University of Belfast.

AKG
MEB

November, 1998

ABOUT THE AUTHORS

Andrew Galwey's university career in research, teaching and administration was entirely at the Queen's University of Belfast, Northern Ireland. He was appointed as Assistant Lecturer in the Department of Chemistry in 1957 and retired 38 years later as Reader in Physical Chemistry and Head of Teaching in the School of Chemistry. Dr Galwey's PhD was awarded in 1958 by the University of London for research carried out in the Royal College of Science, Imperial College, under the supervision of Dr (later Professor) P.W.M. Jacobs. In 1973 the degree of DSc (London University) was awarded for a thesis, "Some Reactions of Solids", based on his published research.

Dr Galwey has spent periods of study leave in England, Italy and South Africa and has presented Invited Lectures at many International Conferences. His interests in solid state chemistry continue, in spite of his retirement, and include applications in environmental protection, the geology of mineral formation and heterogeneous catalysis.

Michael Brown's career follows a similar pattern of long service, at Rhodes University, Grahamstown, South Africa. He was appointed a Junior Lecturer in the Department of Chemistry in 1962 and is at present Professor of Physical Chemistry. He obtained his PhD in 1966 for research done under the supervision of Professor E.G. Prout. The collaboration between Galwey and Brown began in 1971, when Dr Brown spent a sabbatical year at the Queen's University of Belfast, and continues still. Dr Brown has also spent periods of study leave in the Cavendish Laboratory in Cambridge and the Group Technical Centre of ICI Explosives in Ardeer, Scotland. His research interests include the reactions of pyrotechnic compositions.

Both authors can trace their academic ancestry to the Garner School in Bristol that made such a seminal contribution in early research concerned with the chemistry of solids, particularly decompositions. They have worked in continuous collaboration for more than 25 years, both in Belfast and in Grahamstown, publishing many articles under joint authorship in a wide range of journals and at international conferences. During this co-operative research effort, of unusual duration and extent, they have not only developed areas of common interest but have both retained complementary research activities in different areas. Together they have been able to cover a wider range of topics than either could have individually.

THERMAL DECOMPOSITION OF IONIC SOLIDS

SUMMARY CONTENTS

PART A: THEORY AND BACKGROUND (Chapters 1 to 6)

PART B: THERMAL DECOMPOSITIONS OF SELECTED IONIC SOLIDS (Chapters 7 to 17)

THERMAL DECOMPOSITION OF IONIC SOLIDS

DETAILED CONTENTS

PART A: THEORY AND BACKGROUND (Chapters 1 to 6)

PART B: THERMAL DECOMPOSITIONS OF SELECTED IONIC SOLIDS (Chapters 7 to 17)

PART A: THEORY AND BACKGROUND

Chapter 1

INTRODUCTION

1.1. DECOMPOSITION OF SOLIDS

Information about the thermal stability of solid materials of all kinds is of great practical and technological importance and much of the knowledge that has been assembled is of an empirical nature. The necessary background for interpreting this information draws heavily on the interdisciplinary areas between chemistry, physics and branches of engineering, so that the topic is seldom treated at any depth, if at all, in general textbooks concerned with these individual fields. Even in textbooks on chemical kinetics, the rates and mechanisms of the thermal decompositions of solids are given less attention than other heterogeneous reactions. An exception is the recent text by House [1]. Several more specialized books have appeared dealing with solid state chemistry in general [2-14] and Young's book [15], published in 1966, dealt with some aspects of the decomposition of solids.

The present book describes and discusses the current status of the subject, emphasizing the fundamental concepts and illustrating the most successful applications of these ideas through selected examples. The early Chapters (2 to 6) are concerned with the theoretical background to the subject, while the later Chapters (7 to17) describe the behaviour exhibited by selected representative groups of reactants. The final Chapter (18) reviews the current status of the subject, discusses problems outstanding and considers possible future developments.

In this introductory Chapter, the background necessary for the consideration of the decomposition of solids is presented, and references are made to more detailed treatments of relevant aspects.

1.1.1. Heterogeneous and homogeneous reactions

One of the reasons that the decomposition of solids has attracted less interest than it deserves, is almost certainly that the kinetic behaviour of these heterogeneous rate processes [14,16-18] is subject to a different set of controls from those operative in the more widely described and discussed homogeneous reactions. The most obvious difference is that the term *concentration* is usually inapplicable to reactions of solids, where the heterogeneous behaviour arises because material at the boundary surfaces and in other defective regions of the reactant is more susceptible to chemical change than identical material remote from such zones. Reactions of crystalline solids frequently occur preferentially within a thin zone of reactant-product contact, the *reaction interface*, which advances into undecomposed reactant. Kinetic characteristics are determined, at least partially, by geometric variations of this active interface, rather than by the amount or the concentration of undecomposed reactant available. Additional factors which have no counterparts in homogeneous reactions, and may influence the rates at which solids decompose, include melting, sintering, nonstoichiometry, impurities and crystal defects. These factors are discussed in greater detail below.

1.1.2. The literature

There are few reviews of this large and important subject, although the extensive and sustained interest in the field is demonstrated by the large and continuing flow of original research papers. These papers are often in interdisciplinary journals, which adds to reviewers' difficulties. In addition to the general references mentioned above [1-18], there is a comprehensive review [19] covering the field up to 1978 which specifically emphasizes the kinetics of reactions of solids.

1.2. THE SOLID STATE

1.2.1. The perfect solid

In discussions of the properties of real gases it is helpful to start from the concept of an "ideal gas". Similarly, in considering the solid state, the starting point is the concept of the "perfect solid". In a perfect solid at zero kelvin the constituent units, whether they be atoms, molecules or ions, form a completely and perfectly ordered three-dimensional array. (The *crystal lattice* is defined as the points of triple intersection of three families of equidistant parallel planes.) This structural arrangement is the result of the interaction of the bonding forces amongst the units (and the zero-point energy). When the constituent unit is an atom with spherically symmetrical bonding forces, the stable disposition is that of closest-packed spheres. At higher temperatures, thermal energy results in increased vibration and rotation of the constituents, and it is their average positions which then coincide with the lattice points. These motions, although random, are limited in extent. Each constituent remains in the vicinity of its original site. Long-range order is maintained and the assemblage is rigid. The concept of the perfect crystalline solid provides the reference framework against which the various types of imperfection envisaged in real solid materials can be represented. Perfection and imperfection are both important in the present context because the overwhelming proportion of the material comprising most crystals is believed to approximate closely to the ideal state. Many of the physical properties of a crystal are, however, controlled by the atypical and sometimes relatively low concentrations of zones of imperfection. These are believed to exhibit a disproportionately large influence on chemical reactivity.

1.2.2. Thermal events

As a solid is heated, the amplitudes of the vibrations of the lattice constituents are increased and eventually a temperature will be reached where one (or more) of the following changes will occur.

(i) **Phase transition** [20,21]. A new arrangement of constituents may become stable thereby accommodating a different pattern of movement of the constituents at the higher temperature. For example, many metal perchlorates undergo a phase transition from a rhombohedral to a cubic structure [22]. This structural modification allows increased movement of the tetrahedral $C\ell O_4^-$ ion at higher temperatures.

(ii) **Melting** [23]. When sufficient energy becomes available, the forces of attraction between constituents become insufficient to maintain the ordered arrangement of the solid and the system relaxes to the more disordered arrangement of constituents in a liquid. At temperatures just above the melting point, the density of the liquid is usually comparable with that of the solid from which it was formed, indicating a relatively small change in the packing of the constituents. The densities of most liquids then decrease fairly rapidly with increasing temperature, representing increasing disorder. For some complex molecules the change from solid to liquid may occur in stages. The structures of intermediate order are known as *liquid crystals* [23].

(iii) **Sublimation**. When the bonding forces in the solid are very rapidly overwhelmed by the kinetic energy of the constituents, direct transition to the disordered arrangement of a gas may occur, without the intervention of a liquid phase. Melting and sublimation are discussed in more detail in Sections 2.4.3 and 2.4.5.

(iv) **Thermal decomposition**. When the bonding forces *within* constituent molecules or ions of the solid are weaker than those *between* the atoms constituting these units, increasing the temperature may result in bond redistribution and the formation of products chemically different from the reactant. Such chemical processes are referred to as *thermal decomposition* (or *crystolysis*). The kinetics and mechanisms of reactions involving such bond redistributions form the major topic of this book.

These processes, which are often reversible, are represented symbolically in Table 1.1 below. It is not unusual for melting or sublimation to occur concurrently with thermal decomposition.

Table 1.1. Thermal events on heating a single solid, A(s).

A (s,structure 1)	→	A (s,structure 2)	Phase transition
A (s)	→	A (ℓ)	Melting
A (s)	→	A (g)	Sublimation
A (s)	→	B (s) + gas or gases	Thermal decomposition

The differences in structural order and in mobility of the constituents explain the general observation that, during thermal decompositions which are accompanied by melting, the reaction usually proceeds more rapidly in the liquid than in the crystalline state. Fusion diminishes the intracrystalline attractive forces which may confer stability on each lattice constituent. The greater disorder in the liquid also enables a reactant to adopt the reaction configuration more easily.

1.2.3. Glasses

When liquids composed of complex molecules or ions (e.g. sucrose or silicates) are cooled rapidly a glass may be formed. A glass resembles a solid in many of its physical properties, e.g. rigidity, but differs in that the constituents do not show the regular (lattice) arrangement of a crystalline solid. Bonding between constituents is strong, but there is no long-range order, so that such materials can be regarded as supercooled liquids. The vitreous or glassy state is distinct from both liquid and solid states, but possesses some of the properties of each. On heating, glasses soften over a significant temperature interval to form a viscous liquid, but do not usually decompose.

1.2.4. Liquid crystals

Crystalline solids composed of highly asymmetric molecules may undergo stepwise increases in disorder as the temperature is increased. Such transitions are reversible and occur at a fixed characteristic temperature [23]. Movement in specific crystallographic directions may become possible, resulting in a liquid crystal in which structure and properties are anisotropic.

1.3. CLASSIFICATION OF SOLIDS

One convenient way of classifying solids is according to the dominant bonding forces between the constituents in the crystal, i.e. as *molecular, covalent, ionic or metallic* crystals. Hydrogen bonding may, in addition, contribute significantly to crystal stability and is often important in hydrates.

1.3.1. Molecular crystals

In molecular crystals the identity of each individual molecule is preserved. The van der Waals attractive forces between molecules, which give the solid its coherence, are weak compared with the (usually covalent) bonds between the atoms comprising each molecule. Intermolecular bonding is a consequence of the interactions between (i) permanent dipole moments, (ii) the induction of a dipole in one molecule due to the presence of a dipole in a neighbour and (iii) London dispersion dipoles arising as a result of temporary displacements of centres of positive and negative charge during vibration. The dimensions and structures of individual molecules in such solids differ in only minor respects from those found for the same molecules in the liquid or the gaseous states. The distances of closest approach between molecules are larger than intramolecular bond distances and, as a consequence of the weak intermolecular bonding, such solids tend to be soft and have low melting points, e.g. iodine, benzene, etc.

On heating, most molecular solids melt without chemical changes of the constituents. In a restricted range of compounds, characterized by their considerable internal strains and consequent instability, or containing several reactive groups, as in many explosives, appreciable decomposition may occur at temperatures below the melting point. Decomposition (or pyrolysis) of reactants in which intramolecular bonds are thermally less stable than intermolecular bonds, usually yields non-volatile products chemically resembling the reactant. Decomposition may be accelerated by solution formation or eutectic fusion. Such behaviour occurs during decompositions of a number of (strained or high energy) organic compounds, including certain peroxides, nitrocompounds, etc.

1.3.2. Covalent crystals

Each constituent atom of a covalent crystal is linked to its neighbours through directed covalent bonds. The crystal structure is determined by the spatial dispositions of these bonds. Because primary valence forces are involved, such solids are hard and have high melting points, e.g. diamond, silicon carbide, etc. Relatively few entirely covalent solids have been studied at elevated temperatures and it is, therefore, premature to comment on their decomposition characteristics.

1.3.3. Ionic crystals

In an ionic crystal the structure adopted represents the most efficient packing of ions of opposite charge subject to the additional influence of thermal energy. Electrostatic forces are non-directional and the structure adopted is determined by three main controls: (i) the relative numbers of cations and anions, (ii) the relative sizes of positive and negative ions and (iii) ionic shapes. Additional species, including hydrated or otherwise coordinated ions together with molecules of water (or solvent) of crystallization, may be incorporated in the crystal structure. Most of the reactions described and discussed in subsequent chapters are decompositions of ionic solids, because such rate processes are most amenable to quantitative kinetic measurements. Supplementary studies of spectra, electrical conductivity, etc., often provide important supporting evidence in the formulation of decomposition mechanisms.

Thermal decompositions of ionic solids may involve the release of a ligand (as in dehydration), or the breakdown of an inherently unstable lattice constituent (usually occurring at a reaction interface and sometimes promoted by the product phase), or a chemical interaction between two lattice constituents. In some solids two or more chemical changes may occur consecutively, e.g. dehydration often precedes decomposition or, more rarely, concurrently.

1.3.4. Metals

The structures of metals and alloys are usually close-packed arrangements of similar sized spheres. Each atom has a high coordination number. In addition to bonding to nearest neighbours, the valence electrons from each atom have freedom to migrate in an applied field so that metals are good conductors of electricity, e.g. copper, iron, lead, etc. Two complementary theoretical descriptions of metals have been given. The *valence bond* approach is concerned with the short distance interatomic forces between neighbouring atoms. The *band theory* considers long-range electronic properties, extending across the crystal as a whole. The application of both approaches to the properties of solids is described more fully below.

There have been few detailed studies of the thermal decompositions of metallic and related solids (including interstitial and intermetallic compounds) and generalizations cannot yet be made. There are some indications that, on heating, the smaller (i.e. the non-metallic) component in hydrides, nitrides, carbides etc., may diffuse to the metallic phase boundary and be expelled. The controlling factor may either be the ease of diffusion of this sometimes interstitial constituent (e.g. H,N,C, etc.) through the (host) reactant lattice, or it may be the ease of removal (desorption) of such species at the phase boundary, perhaps after reaction, e.g. $2H \rightarrow H_2$. The latter process is similar to the contributory steps in some heterogeneous catalytic reactions. Interface development may also participate in chemical changes involving metallic reactants.

1.3.5. Solids containing more than a single bond type

The classes above refer to solids in which there is a single dominant bond type. In many crystalline substances the structural elements are less simple, and the following additional factors may contribute to cohesive forces within the crystal.

(i) There may be two, or more, different types of interatomic bond. This group would include the majority of the vast number of coordination compounds and many silicates.

(ii) Individual bond types may be intermediate in character between the limiting types of linkage mentioned above.

(iii) Additional structural linkages may participate in determining lattice type and stability. Hydrogen bonding and coordinate bonds are examples.

Aspects of the structures of solids which arise through these considerations are illustrated by reference to the following types of crystalline materials.

Graphite provides an example which illustrates both (i) and (ii) above. There are two distinct bond types and one of these is intermediate between two of the classes above. Within any single sheet, parallel to the cleavage face, carbon atoms are interlinked in the hexagonal array by strong, directed covalent bonds. Between adjoining parallel, stacked sheets, the weaker cohesion is provided by attractive forces having character between that of van der Waals bonding and a metallic bond.

In sheet, layer and many three-dimensional silicates the linkages between the atoms constituting such extended anions are strong, directed and largely covalent bonds (Si-O), while the cations are electrostatically attracted to the surrounding oxygens of the condensed anion through non-directed ionic bonds. The dispositions of these different bonding forces, (i) above, explain the cleavage and other physical properties of such minerals.

Many ionic crystals contain water and/or other groups coordinated with a charged lattice constituent. In complex salts, ligands are often associated with the cation, the anion, or both, through coordination, resulting in the presence in the solid of several types of bonding.

During the thermal decomposition of a solid, bonds of different types may be ruptured and/or reorganized. For example, loss of water from a crystal containing a hydrated cation involves the rupture of coordinate bonds and the associated hydrogen bonds. As a result of the change of cation size, shape and possibly bonding properties, the product will generally have a different crystal structure and any reaction mechanism proposed must take both bonding and structural transformations into account. When the structure of the product maintains a definite geometrical relationship with the structure of the reactant, the reaction is said to be *topotactic*.

Another possible way of classifying solids is according to their electrical properties as conductors, semiconductors or insulators. This aspect of the solid state is discussed briefly in Section 1.5. below.

1.4. CRYSTAL IMPERFECTIONS [24-27]

The perfect crystal is regarded as a regular and infinite, three-dimensional arrangement of the constituent atoms, ions or molecules. This structure may be represented mathematically as a lattice and each lattice point is occupied by the correct atom, ion or molecule. Real crystals are envisaged as coherent aggregates of small volumes (about 10 nm edge cubes) of almost perfect crystal, each bonded to its neighbours by localised zones of imperfection, known as *grain boundaries*. The total crystal is composed of a mosaic of almost perfect lattice domains, each slightly misaligned with respect to its neighbours. Even the most nearly perfect regions deviate from the theoretical ideal in a number of respects. The main types of imperfection are listed in Table 1.2. These may, of course, occur in various combinations and in larger or smaller assemblages. Hannay [27] gives a review of the literature on crystal defects. Because of their importance in solid decomposition mechanisms, brief descriptions of each type of defect and aspects of their probable mode of participation in such reactions are given below.

In the neighbourhood of structural imperfections, the normal bonding of crystal constituents is distorted and this generally leads to increased reactivity in these regions. Reactions thus tend to be initiated in the vicinity of defects, and the reaction interface, identified as the zone within which many chemical changes occur in the solid state, can be regarded as a mobile complex imperfection which advances into the adjoining unchanged reactant. The imperfections listed in Table 1.2., singly or in combination, have been identified or suggested as participants in initiation or the development of interfaces in decompositions of one or another solid.

1.4.1. Phonons

Above 0 K all crystal constituents vibrate about their mean lattice positions. Quanta of thermal energy are referred to as *phonons*. On heating, amplitudes of atomic vibrations are increased, the crystal expands and, at appropriate temperatures, phase transformations, melting, sublimation or chemical reactions will occur. Enhanced mobility, in addition to increasing the tendency towards breaking bonds, either within or between lattice components (decomposition or phase changes, respectively), may also increase the ease of diffusion of all species present in the crystal.

1.4.2. Electrons and positive holes

The valence electrons in metals have considerable freedom of movement (see Section 1.3.4). This contrasts with ionic crystals where the bonding electrons are normally localized within the ions, and with covalent and molecular crystals where electrons are localized in the constituent bonds.

Table 1.2. Types of imperfection which may occur in real crystals, representing deviations from the ideal lattice.

Predominantly electronic imperfections
1. PHONONS (vibrations of atoms about mean positions).
2. EXCITONS (energetic combinations of electrons and positive holes).
3. ELECTRONS and POSITIVE HOLES ("excess" or "missing" electrons).

Imperfections localized in the vicinity of a small number of lattice sites
4. POINT DEFECTS (Schottky and Frenkel defects, unoccupied lattice sites, misplaced units).
5. NON-STOICHIOMETRY (excess of one structural constituent).
6. IMPURITIES (foreign constituents accepted into the structure).

Extended imperfections
7. DISLOCATIONS (edge and screw: imperfections in lattice alignments).
8. SURFACES (including crystal faces, edges, corners, steps, etc.).

The presence of an "excess" electron at a localized site within a crystal corresponds to a cation of decreased charge (or, more unusually, an anion of increased charge), e.g. M^{2+} ions in a compound containing mainly M^{3+} ions (and X^- anions):

$M^{3+}\ 3X^-\ M^{3+}\ 3X^-\ M^{3+}\ 3X^-$ normal crystal

$M^{3+}\ 3X^-\ \mathbf{M^{2+}}\ 3X^-\ M^{3+}\ 3X^-$ "excess" electron

The electron can move by a "hopping" process from one ion to a neighbour:

$$\mathbf{M^{2+}}\ 3X^-\ M^{3+}\ 3X^-\ M^{3+}\ 3X^- \rightarrow M^{3+}\ 3X^-\ \mathbf{M^{2+}}\ 3X^-\ M^{3+}\ 3X^-$$

Similarly, a positive hole, i.e. a deficit of an electron, can be represented as:

$M^{2+}\ 2X^-\ \mathbf{M^{3+}}\ 2X^-\ M^{2+}\ 2X^-\ M^{2+}\ 2X^-$

Movement of the positive hole in one direction is equivalent to electron transfer in the opposite direction:

$M^{2+}\,2X^-\,\mathbf{M^{3+}}\,2X^-\,M^{2+}\,2X^-\,M^{2+}\,2X^- \rightarrow M^{2+}\,2X^-\,M^{2+}\,2X^-\,\mathbf{M^{3+}}\,2X^-\,M^{2+}\,2X^-$

positive hole p^+ or h →

electron (e^-) ←

Electrons and positive holes are present in equal numbers, thereby maintaining the overall charge neutrality of the crystal, but are separated at suitable distances, represented diagrammatically as follows:

$M^{2+}\,2X^-\,\mathbf{M^{3+}}\,2X^-\,M^{2+}\,2X^-\,M^{2+}\,2X^-\,\mathbf{M^{+}}\,2X^-$

p^+ or h $\qquad\qquad\qquad\qquad e^-$

or

$M^{2+}\,2X^-\,\mathbf{M^{3+}}\,2X^-\,M^{2+}\,2X^-\,M^{2+}\,X^-\,\mathbf{X^{2-}}\,M^{2+}\,2X^-$

p^+ or h $\qquad\qquad\qquad\qquad e^-$

1.4.3. Excitons

Combination of an electron and a positive hole annihilates both defects. Maintenance of an energetic association between an electron and a positive hole, for sufficient time to permit migration of the associated pair to occur through the crystal, gives rise to a defect called an *exciton*. Further discussion of these defects is given in Section 1.5.1. which considers the movements of electrons in solids.

1.4.4. Point defects

In atomic or molecular solids, common types of point defects are the absence of an atom or molecule from its expected position at a regular lattice site (a *vacancy*), or the presence of an atom or molecule in a position which is not on the regular lattice (an *interstitial*). In ionic solids, these point defects occur in two main combinations. These are *Schottky defects*, in which there are equal numbers of cation and anion vacancies within the crystal, and *Frenkel defects*, in which there are cation vacancies associated with an equal number of "missing" cations located at non-lattice, interstitial positions. Both are illustrated in Figure 1.1. Point defects are also found in association with altervalent impurities, dislocations, etc., and combinations of vacancies with electrons or positive holes give rise to various types of colour centres (see below).

Point defects are important in conferring mobility on crystal components. They also increase the entropy of a crystal and there is always a small equilibrium concentration present that can be calculated by thermodynamic methods [24].

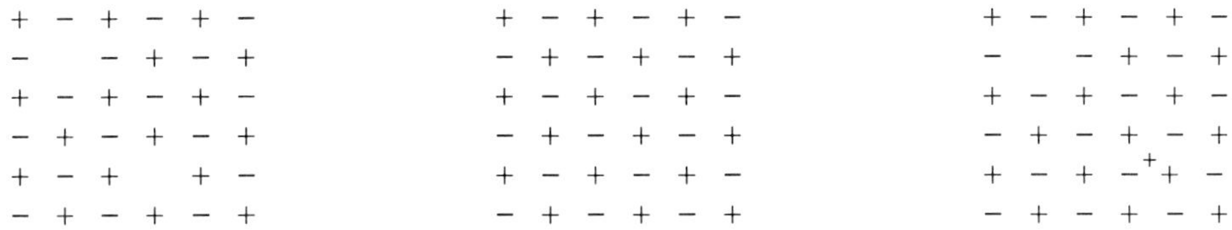

Schottky defect	Perfect crystal	Frenkel defect
Equal numbers of cation and anion vacancies	Every lattice site correctly filled	Cation vacancy accompanied by a distant cation in an interstitial position

Figure 1.1. Schematic representation of point defects in an ionic crystal. Imperfections occur in pairs to preserve the overall charge balance.

1.4.5. Non-stoichiometry

In a non-stoichiometric compound the additive is an excess of a constituent already present in the normal crystal. The most extensively investigated non-stoichiometric compounds are the Group I metal halides and metal chalcogenides. On heating potassium chloride, for example, in potassium vapour, excess metal is incorporated in the crystal as excess K^+ ions together with electrons and anion vacancies ($V_{C\ell-}$):

$$K + KC\ell \rightarrow KC\ell + V_{C\ell-} + e^-$$

The combination of the vacancy with the electron gives an *F-centre* (colour centre).

An anion vacancy has an overall positive charge and thus readily traps an electron. A crystal containing such imperfections develops a blue colour due to spectral absorption having a maximum at wavelengths between 600 and 700 pm. The intensity of absorption is directly proportional to the stoichiometric excess of Group I metal incorporated and to the decrease in salt density resulting from the concentration of anion vacancies generated.

Non-stoichiometric oxides may contain either an excess of metal (e.g. zinc oxide, $Zn_{1+\delta}O$) or a deficiency of metal (e.g. nickel oxide, $Ni_{1-\delta}O$). Some metal oxides (e.g. titanium oxide) may be prepared as either excess-metal or metal-deficit oxides. Excess metal is incorporated in the structure as interstitial cations together with the electrons removed through ionization. In metal-deficit or excess-oxygen oxides, excess oxygen is incorporated in the structure by formation of cation vacancies and positive holes, which may be achieved by oxidation of cations, e.g.

$$NiO(s) + \tfrac{1}{2}O_2(ads) \rightarrow NiO(s) + V_{Ni}^{2+} + 2Ni^{3+}$$

1.4.6. Impurities

Impurities, i.e. components chemically different from the constituents which form the perfect crystal structure, may be accommodated in the structure, in similar ways to the above, provided that the atoms, ions or molecules of the additive are chemically compatible with those of the host phase. This usually means that their size, shape and bonding properties are similar to those of the constituents of the host material. Incorporation is a specific process. Some binary, or more complex, combinations have only restricted ranges of mutual solubility, while other mixtures are capable of forming the complete range of intermediate *solid solutions* from 100%A to 100%B.

The incorporation of altervalent impurity ions in an ionic solid will usually require the concurrent generation of vacancies to maintain the electrostatic charge balance, e.g. each Ca^{2+} ion in $KC\ell$ or each Cd^{2+} ion in $AgC\ell$ necessitates the formation of one cation vacancy. Alternatively, in transition-metal compounds the charge anomaly can be accommodated by a change of valency in the host cation, e.g. Li_2O can be accommodated in NiO through formation of positive holes, i.e. Ni^{3+} ions.

An impurity located at a lattice site introduces strain that, in general, tends to reduce stability. Impurities may become mobile at elevated temperature and may accumulate within the distorted material that constitutes dislocations, internal surfaces, etc. (see below), or even as included or separate particles.

Unintentional and perhaps unrecognized incorporation of impurities may occur during crystal preparation. The addition of a controlled quantity of a selected additive is termed *doping*, and specific systems of this type are discussed in later chapters. Many studies of the influences of additives on decompositions have been reported. Fewer investigations have been concerned with the influences of compositional variations over extensive ranges on the chemical reactivities of solid solutions.

An impurity may participate in the electron-transfer steps involved in decompositions of ionic compounds, and the influences of electron donors or electron acceptors have been measured. Additives may also influence bond redistribution processes, by acting as a catalyst capable of promoting the necessary slow chemical step.

1.4.7. Dislocations (or line defects)

A dislocation is an imperfection which extends across many lattice units and represents a distortion of the lattice framework. Two distinct types of dislocation have been recognized. An *edge dislocation* (Figure 1.2.(a)) can be conveniently portrayed as the insertion of a plane of atoms part of the way through the crystal, requiring distortion of the adjoining planes, which bend inwards within a zone of strain beyond the "additional" plane to accommodate the standard spacing. A *screw*

(a)

(b)

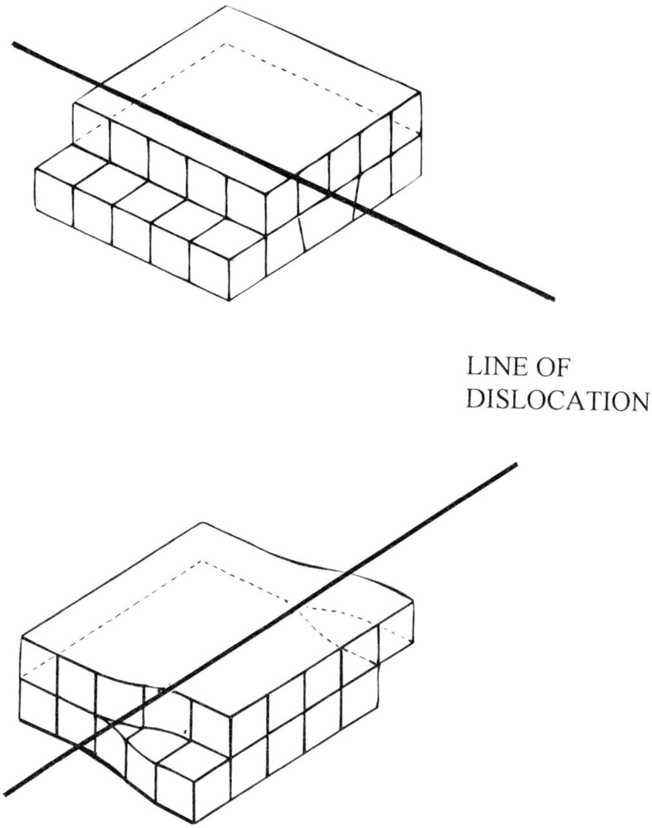

Figure 1.2. Dislocations in a primitive cubic lattice.
(a) Edge dislocation.
(b) Screw dislocation.

dislocation (Figure 1.2.(b)) can be regarded as a unit lattice displacement within a lattice plane, along a line of atoms terminating within the crystal, in a direction vertical to that of the lattice plane. A line traced on the crystal plane, encircling the displacement termination, traces a helix. Dislocation lines of either type cannot terminate within a crystal, but must extend to a surface, or form a loop, and complex imperfections can branch.

The intensity of a dislocation, which is the vector addition of individual simple imperfections within a selected region, can be quantitatively expressed by the *Burgers vector*. This is the difference in distance and direction between a path traced by moving from one structural constituent to the next around the zone of the real crystal of interest and an exactly comparable path traced in a region of perfect lattice. The Burgers vector of an edge dislocation is normal to the line of dislocation, while that of the screw dislocation is parallel to the line of dislocation.

Dislocations are high-energy imperfections and equilibrium concentrations predicted by thermodynamic calculations are very small. The significantly greater concentrations found in real crystals are ascribed to their initial generation within very small seed crystals and important participation in subsequent growth processes. Addition of material at the surface discontinuity associated with a screw dislocation results in rapid development, compared with the slower overall growth on a perfect surface where the nucleation of a new plane of atoms is difficult. Growth of crystals ultimately results in the appearance of a complex internal system of interlinking dislocations (the *Frank net*) arising from preferential growth at step edges associated with these imperfections, and the intergrowth of neighbouring crystals which are not perfectly aligned.

Material located within the distorted region comprising the dislocation core is relatively less stabilized by the structural bonding forces of the crystal than that in the regular structure and, consequently, is more reactive than identical constituents within regions of perfect lattice. Two factors contribute to this enhancement of reactivity: the elastic strain energy may decrease the activation energy required for a bond redistribution step, and the stereochemical constraints on reaction may be modified or decreased within the anomalous region of crystal. The onset of a slow process, such as dissolution in a poor solvent, or low temperature sublimation, usually results in preferential removal of material close to a dislocation, thus generating *etch pits* in the immediate vicinity of sites at the surface terminations of dislocation lines. The shapes and orientations of etch pits are controlled by the underlying crystal structure. The development of etch pits has been exploited as a useful technique for the detection of dislocations and for characterizing their concentrations and dispositions. Such information can then be compared with the sites of nucleation of decomposition, which may coincide with these sites of preferred reaction. Decomposition is believed often to commence at dislocation termination sites at crystal surfaces.

The disordered structure within dislocations may also contain other defects and be a zone of preferential concentration of impurities where the strain associated with the presence of each anomalous constituent is decreased. These also provide the paths for low-temperature ionic migration. Intergranular material, between blocks of relatively perfect material, contains relatively high dislocation densities and also impurities. These have been recognized generally as zones of preferred onset of chemical reactions.

1.4.8. Surfaces

A surface can be described as a crystal imperfection, in the sense that the perfect lattice is represented as extending indefinitely, whereas all real crystals are terminated by usually well-defined boundaries. The constituents of each boundary layer experience a force field which is different from that found within the bulk, in that the neighbouring constituents with which it interacts, are distributed on one side only. The lattice spacing for the outer layer may thus be slightly greater than that within the crystal. The reactivity of surface components may be appreciably greater than that of similar entitities stabilized within a symmetrical force field.

Volatile products of a reaction, once generated, can escape readily from surface sites. Additional local deformation of crystal regularity occurs in the vicinity of a point of surface termination of a dislocation, in the presence of impurities, or at zones of more complex crystal imperfection or damage. All these factors tend to increase the ease of onset of reaction. Crystal surfaces, particularly in the vicinity of specific sites of imperfection, are often identified as zones of initiation of reaction (*nucleation*).

Different exposed crystal surface planes have different relative reactivities. The flat growth surfaces of solution-grown crystals tend to retain small irregularities, due to uneven development and accumulation of impurities during removal of last traces of solvent. (Surfaces may also be damaged during handling.)

The outer faces of grown crystals tend, in some systems, to be more reactive than faces exposed by cleavage, though the influence of a possible difference in properties of the crystallographic surface exposed must also be considered. At edges and corners of crystals and at steps on the surface, the lack of symmetry in the structure, and hence the reactivity, is generally greater than for the surface constituents as a whole, but the amount of material involved is much less.

1.5. MOBILITY OF DEFECTS IN SOLIDS

During chemical changes proceeding in the solid state, movement of charge and/or of material is involved and the transport processes depend strongly upon the presence, and hence movement, of defects of various kinds.

1.5.1. Electrons and positive holes

Movement of electrons in metals occurs readily and metals are thus good conductors of electricity. The simplest model of a metallic crystal regards the metal atoms as being completely ionised, with the ions so formed embedded in an "electron gas". The crystal constitutes a potential well, within which the electrons are accommodated, and there is a large rise in potential at the crystal boundaries. This barrier can, however, be overcome at high temperatures, when electrons leave the surface with an energy distribution conforming to the Maxwell-Boltzmann distribution law. The *valence bond* theory of solids is concerned with individual bonds between neighbouring atoms and is most successful in predicting bond lengths. Because the number of valence electrons present in a metal is insufficient to occupy the available bond orbitals completely, the metal is regarded as a resonance hybrid of the very many structures possible, for example:

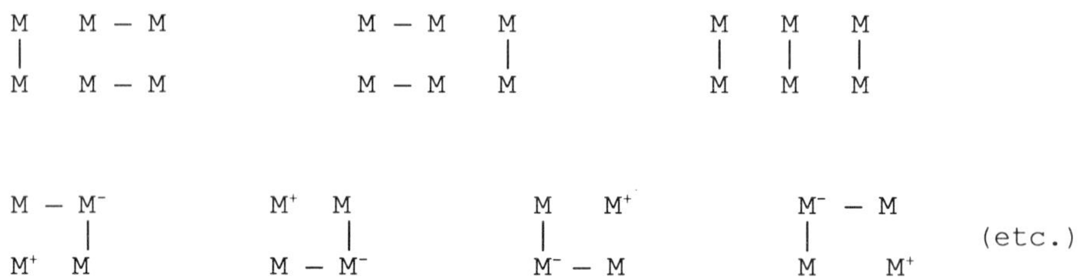

(M - M covalent bond; M^+, M^- metal ions; | "free" electron).

For metal crystals of appreciable size, the number of possible contributions to the alternative bond distributions is extremely large and this results in a considerable contribution to crystal stability from resonance energy. Because orbitals are incompletely occupied, application of an electric field results in electron migration.

Solids containing different electronic structures, such that all available orbitals are doubly occupied, are insulators because free migration of electrons is not possible. This is the situation in covalent crystals. Orbital overlap with the production of covalent shared, but incompletely filled, electron bonds does not occur in molecular or ionic crystals, which are therefore not electronic conductors.

The molecular orbital alternative to the valence bond theory, the *band* or *zone theory of solids*, considers properties of less-localized electrons free to move through a periodic potential field arising from the positive-ion cores. It is more successful in predicting bulk properties of crystals such as thermal and electrical conductivities. Both models lead to an electronic structure for solids in which the discrete permitted energy levels for electrons in isolated atoms are replaced by *bands* of allowed energy levels, separated by gaps of forbidden energies. Each atom within the solid contributes one level, capable of accommodating two electrons, to a band that extends throughout the crystal as a whole. In a crystal of any appreciable size (N atoms, say, where N is of the order of 10^{22}) the levels are so closely spaced that they may be regarded as continuous bands of allowed energies. The development of such a structure is represented schematically in Figure 1.3. Only the outermost electrons in each atom or ion interact sufficiently to form bands of significant width. The widths of these bands and of the energy gaps between bands are characteristic of each particular solid and refer to the equilibrium internuclear separations.

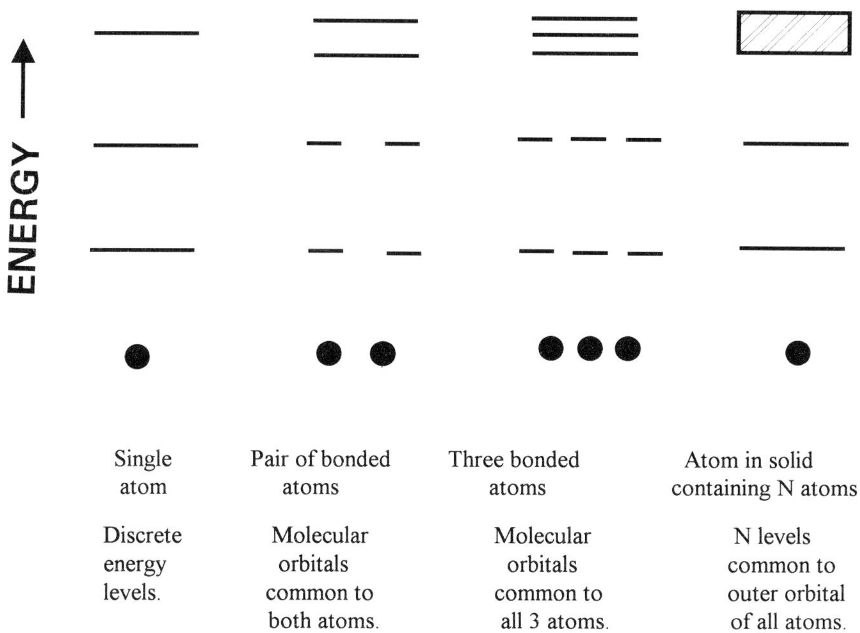

Figure 1.3. Schematic representation of the development of an energy band through interaction of the outer atomic orbital. Inner electron levels are not significantly altered.

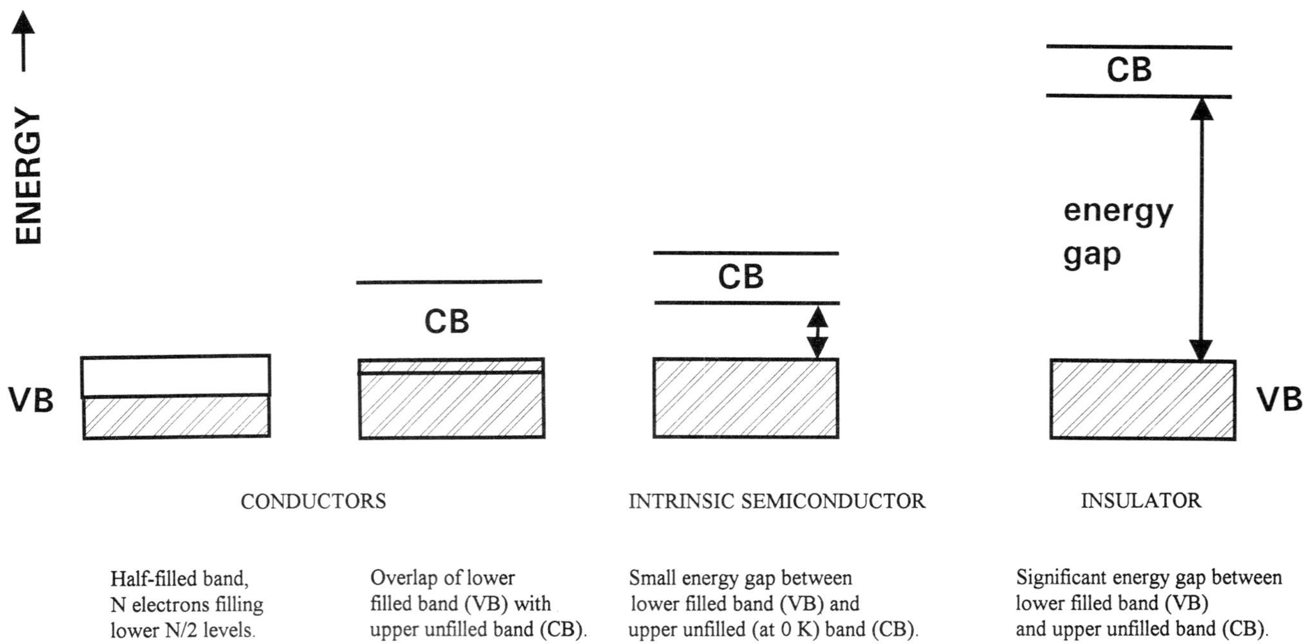

Figure 1.4. Schematic representation of distribution of electrons in the energy bands of conductors, intrinsic semiconductors and insulators (inner levels have been omitted for clarity). VB - valence band; CB - conduction band.

The extent of filling of the highest occupied energy band determines the electrical properties of the solid, see Figure 1.4., each level is capable of accommodating two electrons (of opposite spin). If this band is incompletely filled, an electron may readily migrate through the crystal by gaining small increments of energy under the influence of an applied field. Such a solid is a *conductor*. If the band is completely filled (at 0 K), the width of the gap between this band (the *valence band*, VB) and the next empty band (the *conduction band*, CB) is important. If this energy gap is large (several eV), electrons are not readily promoted to higher energy levels and the solid is an *insulator*. If the gap is small, at temperatures above 0 K some electrons are thermally excited into the conduction band and the solid shows a low electrical conductivity, which increases with increasing temperature. This is an *intrinsic semiconductor*.

In a crystal containing N atoms, each band is capable of accommodating 2N electrons. If each atom contributes one valence electron, the valence band will be incompletely filled and the substance will be a conductor, e.g. Group I metals. Similarly, trivalent elements such as $A\ell$ will produce incompletely-filled valence bands. Divalent elements, such as the Group II metals, would then be expected to produce completely-filled valence bands. Calculations show, however, that the VB and the CB overlap for these elements, giving rise to a composite, incompletely-filled band and the elements are conductors. Tetravalent elements, e.g. C (diamond), Si and Ge, are either insulators or intrinsic semiconductors, depending upon the width of the energy gap between the VB and the CB.

Semi-conductors have conducting properties between those of insulators and conductors. The band structures of conductors, intrinsic semiconductors and insulators are represented schematically in Figure 1.4. The widths and the separations of the bands are dependent upon the internuclear spacings of the constituents, so that the band structure may be modified in the vicinity of the surfaces of the crystal, by the occurrence of surface reactions, or by interactions with gases, liquids or other solids.

Electrical-conducting properties may be induced in an insulator, or enhanced in an intrinsic semiconductor, by incorporation of suitable impurities within the crystal during preparation. Impurities may include a non-stoichiometric excess or deficit of one or other of the crystal constituents as discussed earlier. At low concentrations, impurity atoms or ions will have electronic structures similar to those of isolated atoms or ions (with allowance for some interaction with the matrix). If the energy levels for the valence electrons of these impurity atoms or ions lie within the energy gap between the VB and the CB of the host insulator or semiconductor, they will modify the energy requirements for promotion of electrons from one band to the other. If the impurity levels are initially occupied by electrons, these electrons will require less energy to be promoted to the CB (Figure 1.5.(b)) and once the electrons are promoted, either thermally or by an applied electric field, the substance will behave as a semiconductor. If the levels are unoccupied, some electrons may be promoted from the filled VB to the impurity levels (Figure 1.5.(c)), thereby generating positive holes in the VB. The solid then has the properties of a semiconductor in that some transport of current is possible via these positive holes. Insulators, suitably "doped" in this way are called *extrinsic* semiconductors. Those in which the majority of charge carriers are electrons are classed as *n-type* semiconductors, e.g. excess-metal ZnO. Those in which the majority of charge carriers may be represented as positive holes are classed as *p-type* semiconductors, e.g. metal-deficit NiO.

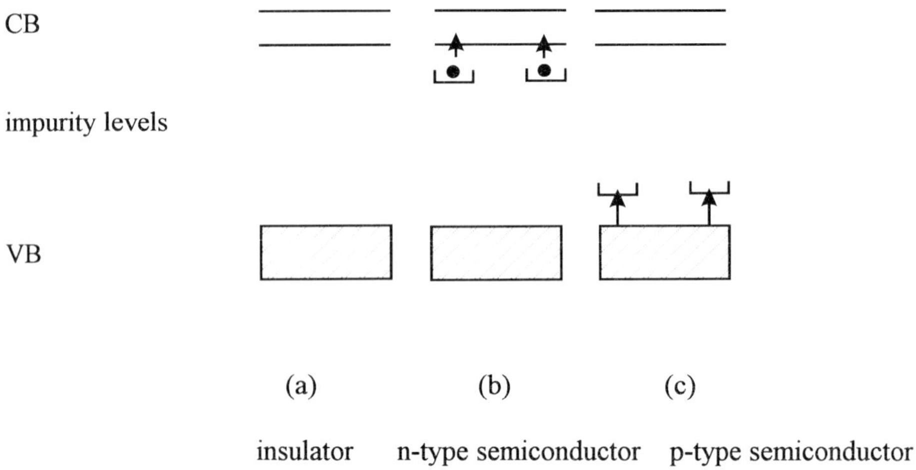

Figure 1.5. Extrinsic semiconductors containing impurities that interact with the band structure of the host material.

At high concentrations of impurities, the band structure of the host material will inevitably be distorted. In general, the conductivity of a metal decreases as the temperature is raised because thermal vibration of the constituents decreases lattice order and hence the ease of movement of electrons. In contrast, thermal promotion of electrons in a semiconductor increases the number of charge carriers, thereby increasing the conductivity with temperature.

Absorption of radiation in the far ultraviolet by an insulator, may be sufficient to promote an electron from the full VB to the empty CB, wherein it is free to migrate. This is observed as *photoconductivity*. At slightly lower energies, the electron may be promoted to an exciton level (e.g. the first unoccupied level of the metal atom) below the conduction band. The exciton is uncharged and undissociated and in this form can migrate through the lattice, but with no net transfer of electrons and, hence, there is no conductivity. The energy associated with the excitons may, however, move to and be transferred at reaction interfaces, or other extended imperfections, and excitons have been identified as probable participants in the mechanisms of many solid state reactions. Photoconductivity can also result from promotion of electrons from, or to, impurity levels. Light absorption associated with such transitions (Figure 1.6.) occurs at wavelengths greater than the absorption edge. In molecular crystals, intermolecular separations are too great to permit electronic

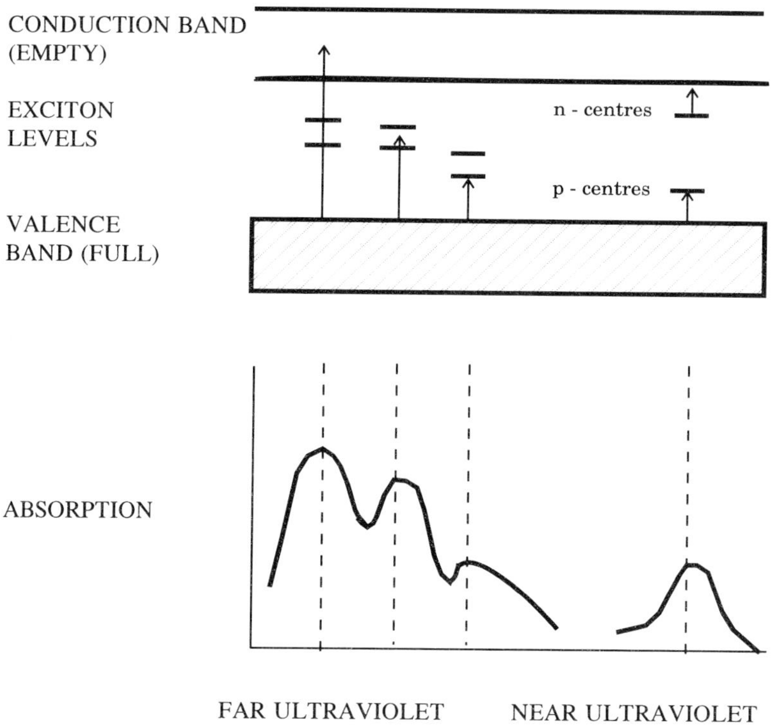

Figure 1.6. Schematic representation of electron promotion processes in ionic crystals. At the shortest wavelengths, electrons are promoted to the conduction band and photoconductivity is observed. Exciton formation occurs on promotion to levels below the conduction band, this energetic combination of electron and positive hole does not conduct electricity but may participate in decomposition processes. Electrons or positive holes may also be generated from impurities.

motions, except in specialized compounds where there is already extensive delocalization of bonding, such as graphite, for which the extended interplanar resonance systems possess some metallic character.

1.5.2. Mobility of point defects

For any atom, ion or molecule to move from one site to another in the crystal structure, it must have a site into which it can move (e.g. a vacancy, interstitial site or other defect) and it must overcome the appropriate potential energy barrier that opposes such migration, i.e. the activation energy for movement must be available by phonon transfer from the thermal vibrations of neighbours.

Three mechanisms of diffusion predominate in ionic solids:

(i) *a vacancy mechanism*, where the diffusing species occupies a normal site and moves into an adjacent vacant site (Figure 1.7. (a));

(ii) *an interstitial mechanism*, where the diffusing species occupies an interstitial position and migrates by moving directly from one interstitial position to another (Figure 1.7. (b)); and

(iii) *an interstitialcy mechanism*, where the diffusing species, originally in an interstitial position, moves into a normal site and displaces the original occupant into an interstitial position (Figure 1.7. (c)). Direct interchange of positions or concerted ring mechanisms are less likely (and more difficult to demonstrate) though such models have been discussed in the literature.

In the absence of a concentration gradient, diffusion of any species in a solid will approximate to a random-walk type of process. Diffusion under the influence of a concentration gradient is governed by Fick's laws:

1st Law: $J = -D(\delta C/\delta x)$

2nd Law: $\delta C/\delta t = D(\delta^2 C/\delta x^2)$

where J is the diffusion flux, D the diffusion coefficient, $\delta C/\delta x$ the concentration gradient in the x direction, and t is time.

The diffusion coefficients, D, at temperature T, for either interstitial ions or vacancies [28,29] are of the form:

$$D = \upsilon a^2 \exp(-E_a/\mathbf{k_B}T)$$

where υ is the frequency of vibration of the defect, a is the distance between adjacent defect positions and $\mathbf{k_B}$ is the Boltzmann constant. E_a is the activation energy for diffusion and its value decreases with thermal expansion of the crystal.

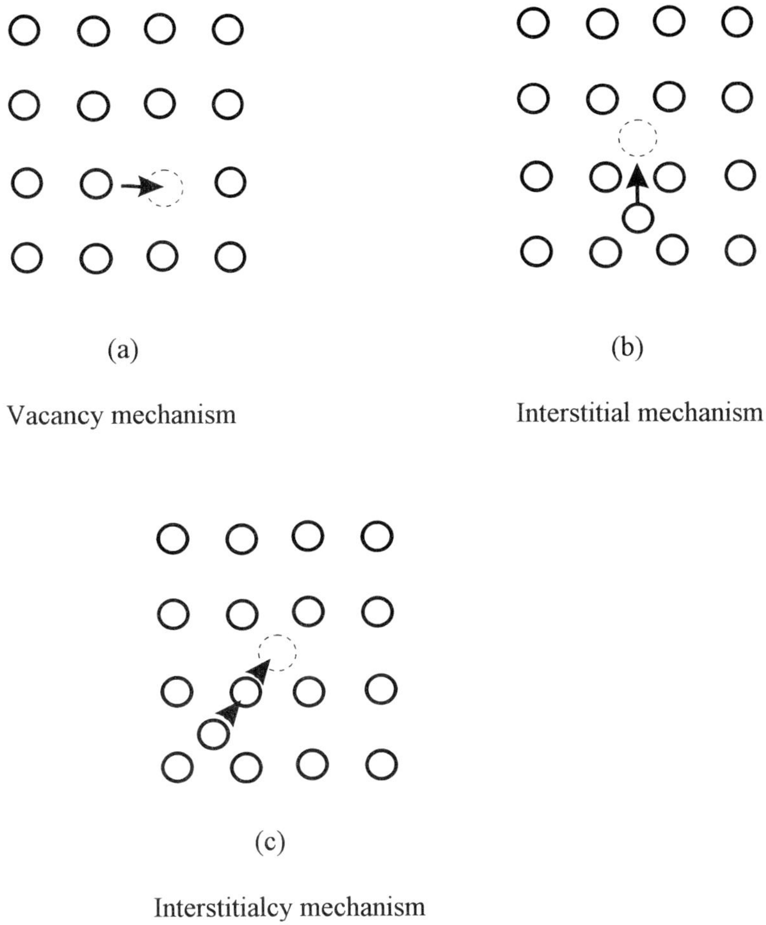

Figure 1.7. Mechanisms for migration of point defects.

The potential energy barrier is also lowered in the direction of an applied electric field, so that migration of interstitials or vacancies in the direction of the field is favoured. The electrical conductivity, σ, due to the migration of an interstitial or a vacancy, is related to the diffusion coefficient, D, for that defect by the Einstein-Nernst relation:

$\sigma/D = n\,e^2/\mathbf{k_B}T$ (n = number of defects, e = electronic charge).

The conductivities of different preparations of the same ionic substance tend to be irreproducible at low temperatures, because the amount of material migrating in regions of extended imperfection, such as dislocations, grain boundaries and surfaces, exceeds that migrating in the bulk of the crystal. Imperfection structures are crystal-specific and can differ considerably between materials apparently prepared identically. The activation energy for migration in the more perfect regions of the crystal is much higher than in the defective regions. As the temperature increases and approaches the melting point of the crystal, the rate of migration in the bulk may approach that in the defective regions. Different preparations of the same compound will then behave similarly.

For strongly ionic solids where cation and anion sizes are comparable, e.g. NaCℓ, KBr, etc., Schottky defects will predominate and both transport numbers t_+ and t_- are greater than zero ($t_e = 0$, no current is carried by electron migration). When the size of the cation is considerably smaller than the anion, e.g. AgBr, AgCℓ, etc., Frenkel defects occur and the interstitial cations are the dominant current carriers ($t_+ \approx 1$).

Vacancies may participate in the nucleation step in thermal decomposition. The aggregation of vacancies precedes the initiation of structural reorganization. Removal of any lattice constituent, by decomposition or migration, is an important contributory factor in interface reactions. The availability of vacancies, perhaps in combination with other imperfections, provides the space which may be required during bond redistribution steps.

1.5.3. Mobility of extended defects

On heating, constituents of the crystal within the strained zone of a line dislocation can be displaced relatively readily and the dislocations can thus migrate to surfaces where they are eliminated, or those with opposing Burgers vectors may cancel each other. Such reduction in dislocation concentration is termed *annealing*. Cold working (distortion or strain imposition) of a crystal increases its dislocation density. Impurities may retard the movement of dislocations through the crystal (e.g. carbon in steel).

The reactivity of surface material depends, to some extent, on its crystallographic environment. The mobility of a species may vary from one set of crystal faces to another. Individual crystal constituents tend to coordinate with a maximum number of nearest neighbours. Coordination on surfaces is necessarily unsymmetrical and isolated constituents tend to migrate to sites such as step edges where stability can be enhanced by increased coordination. Surface mobility is thus often appreciable at temperatures well below the melting point. Such behaviour is intermediate between solid and liquid states and is important both in sintering and as a transfer mechanism in solid-solid reactions. Migration across surfaces may be complicated by concurrent adsorption and desorption processes. The surface is thus a zone of preferential initiation of reaction and is an important factor in determining the generation and the subsequent geometry of interface development.

Rapidly grown crystals, or the products of fast high temperature reactions, tend to be disordered. The high surface-to-volume ratio results in enhanced reactivity that diminishes on recrystallization.

1.6. A SIMPLIFIED MODEL OF THE DECOMPOSITION PROCESS

Decomposition of a single pure solid substance A may be represented as:

$$A(s) \rightarrow B(s) + C(g)$$

In any real crystal of A there will be imperfect regions within which the constituents are more reactive than in the bulk of the solid. At a suitably high temperature, decomposition is initiated by the appropriate bond redistribution process within these imperfect regions. These regions have been associated, in different substances, with one or more of the imperfections discussed in Section 1.4. The initially isolated aggregates of product, still embedded in a matrix of original reactant, are usually sufficiently mobile (Section 1.5.) to reorganize into *nuclei* of product phase B. Mechanisms of these processes, and the stage of reorganization at which a nucleus may be regarded as having been established, will be discussed in more detail in later chapters. Nucleation does, however, result in the formation of a particle usually regarded as a crystal of product solid, part of the surface of which remains in chemical contact with the crystal from which it was formed. The enhanced reactivity associated with the reactant/product interface is a fundamental assumption accepted in discussions of solid state decomposition mechanisms. Locally promoted reaction may be a consequence either of the strain introduced by distortion of the structure, or the autocatalytic activity of the product formed, or both. Further decomposition may then result from either the formation of further nuclei, or, more significantly, through the *growth* of the existing nuclei by addition of further product across the reactant/product interface. Growth represents progressive advance of the interface into unchanged reactant solid.

Whether the nucleation or the growth step will exert the greater influence on the overall chemical change will depend upon the relative activation energies. Growth of nuclei is often anisotropic leading to different shapes of nuclei (see Figure 1.8., which shows nuclei of product formed during the dehydration of crystal hydrates [30,31]). Growth cannot continue indefinitely and nuclei eventually impinge on each other so that advance ceases along areas of contact ("overlap"). Growth may also cease if contact is lost between product and the remaining reactant. Nucleation usually, but not invariably, occurs in the surface region and this may be followed by

Figure 1.8. Micrographs of product nuclei formed on dehydration of alums. For more detailed explanations see reference [30]. (Reproduced with permission of the Royal Society, London.)

rapid coverage of the surface with product. Thereafter the reactant/product interface advances into the crystal, following a geometric course determined by the shape of the original reactant particle and the dispositions of the active growth nuclei.

This simplified model of the decomposition process is provided here to introduce the more detailed accounts given in the later chapters. The relatively simple processes now accepted as being involved in decompositions have provided a model for developing kinetic principles applicable to a wider range of solid state reactions.

REFERENCES

1. J.E. House, Principles of Chemical Kinetics, Wm.C. Brown Publishers, Dubuque, USA, 1997.
2. W.E. Garner (Ed.), Chemistry of the Solid State, Butterworths, London, 1955.
3. H. Schmalzried, Solid State Reactions, Verlag Chemie, Weinheim, 2nd Edn, 1981.
4. N.B. Hannay (Ed.), Treatise on Solid State Chemistry, Vol.4, Reactivity of Solids, Plenum, New York, 1976.
5. J.A. Hedvall, Solid State Chemistry, Elsevier, Amsterdam, 1966.
6. P.P. Budnikov and A.M. Ginstling, Principles of Solid State Chemistry, (Translated by K. Shaw), MacLaren, London, 1968.
7. A.K. Galwey, Chemistry of Solids, Chapman and Hall, London, 1967.
8. N.B. Hannay, Solid-State Chemistry, Prentice-Hall, New Jersey, 1967.
9. V.V. Boldyrev, M.Bulens and B. Delmon, The Control of the Reactivity of Solids, Elsevier, Amsterdam, 1979.
10. A.R. West, Solid State Chemistry and its Applications, Wiley, Chichester, 1984.
11. A.K. Cheetham and P. Day (Eds), Solid-State Chemistry Techniques, Clarendon, Oxford, 1986.
12. J. Šesták, Thermophysical Properties of Solids, Comprehensive Analytical Chemistry, Vol XII, Part D, (Ed. G. Svehla), Elsevier, Amsterdam, 1984.
13. H. Schmalzried, Chemical Kinetics of Solids, VCH Publishers, New York, 1995.
14. V.V. Boldyrev (Ed.), Reactivity of Solids: Past, Present and Future, IUPAC, Blackwell Science, Oxford, 1996.
15. D.A. Young, Decomposition of Solids, Pergamon Press, Oxford, 1966.
16. B. Delmon, Introduction a la Cinétique Hétérogène, Technip, Paris, 1969.
17. P. Barret, Cinétique Hétérogène, Gauthier-Villars, Paris, 1973.

18. P. Barret (Ed.), Reaction Kinetics in Heterogeneous Chemical Systems, Elsevier, Amsterdam, 1975.
19. M.E. Brown, D. Dollimore and A.K. Galwey, Reactions in the Solid State, Comprehensive Chemical Kinetics, Vol.22, (Eds C.H. Bamford and C.F.H. Tipper), Elsevier, Amsterdam, 1980.
20. J.W. Christian, Transformations in Metals and Alloys, Pergamon, Oxford, 1965; Vol. 1, 2nd Edn, 1975.
21. C.N.R. Rao and K.J. Rao, Phase Transitions in Solids, McGraw-Hill, New York, 1978.
22. F. Solymosi, Structure and Stability of Salts of Halogen Oxyacids in the Solid Phase, Wiley, London, 1977, p.123.
23. A.R. Ubbelohde, Melting and Crystal Structure, Clarendon, Oxford, 1965; The Molten State of Matter, Interscience, New York, 1978.
24. A.L.G. Rees, Chemistry of the Defect Solid State, Methuen, London, 1954.
25. A. Kelly and G.W. Groves, Crystallography and Crystal Defects, Longman, London, 1970.
26. A.M. Stoneham, Theory of Defects in Solids, Oxford University Press, Oxford, 1975.
27. N.B. Hannay (Ed.), Treatise on Solid State Chemistry, Vol.2, Defects in Solids, Plenum, New York, 1974.
28. N.F. Mott and R.W. Gurney, Electronic Processes in Ionic Crystals, Oxford University Press, Oxford, 1940, p.33.
29. F.S. Stone in W.E. Garner (Ed.), Chemistry of the Solid State, Butterworths, London, 1955, Chap.2.
30. A.K. Galwey, R. Spinicci and G.G.T. Guarini, Proc. R. Soc. (London), A378 (1981) 477.
31. H. Tanaka, N. Koga and A.K. Galwey, J. Chem. Educ., 72 (1995) 251.

Chapter 2

STOICHIOMETRY AND EXTENT OF DECOMPOSITION

2.1. INTRODUCTION

There are several possible reasons for undertaking a study of the thermal decomposition of a solid substance. These may range from the practical necessity of knowing the limit of the temperature range over which a material will remain stable for a proposed application, to theoretical considerations directed towards understanding the processes involved at the molecular level during decomposition. The formulation of detailed reaction mechanisms is the eventual aim of all kinetic studies, but the investment of time and effort required to unravel all of the details may not be possible or easily justified. Effort spent in determining a reaction mechanism may often, however, be repaid by the control gained over the properties of the solid product(s) of decomposition. For example, these solids are often extremely active as heterogeneous catalysts for other reactions and their properties may be controlled by appropriate variation of the conditions used for the decomposition of the original reactant.

The experimental stages in the study of a decomposition of a solid are as follows. Some or all of these stages may be omitted depending on the facilities available and the overall objectives of the work, though such omissions necessarily restrict the value and reliability of the conclusions obtained:

(i) reactant preparation,
(ii) reactant characterization,
(iii) determination of the stoichiometry of reaction, including characterization of the compositions, structures, topotactic relationships and textures of the solid product(s) (initial products may differ from those resulting from concurrent or consecutive secondary reactions),
(iv) determination of the thermochemistry of reaction,
(v) kinetic measurements (which may be isothermal and/or non-isothermal), including data capture (yield, time, temperatures), storage and analysis,

(vi) identification of intermediates, which may include measuring their concentration-time relationships and investigation of properties of these possible participants,
(vii) complementary studies, including microscopic observation of textural changes during and at completion of reaction, the possible participation of melting, etc., and
(viii) effects of varying decomposition conditions and/or pretreatment of the reactant (e.g. irradiation, crushing, ageing, etc.).

In the following Chapters (3 to 6), the main features of these stages are outlined and examples are given of studies which illustrate the successful use of appropriate experimental techniques, together with the necessary background theory. Applications of these principles and techniques are described in later Chapters (7 to 17), where the decompositions of specific groups of compounds are discussed in greater detail.

2.2. THE FRACTIONAL DECOMPOSITION (α)

The conventional method of measuring the amount of decomposition which has taken place is by the dimensionless *fractional decomposition*, α, which ranges from 0.00 for the pure reactant, to 1.00 signifying complete decomposition. On the assumption that both solid and gaseous products maintain a constant composition, α at any time during the decomposition may be measured directly from:

(i) the *mass-loss* at that time relative to the overall mass-loss when decomposition is complete, so $\alpha = (m_o - m_t)/(m_o - m_f)$ (where m_o is the initial mass of reactant and m_f the mass of the solid residue at $t = \infty$); or
(ii) the *pressure* of accumulated gaseous products in an initially evacuated constant volume apparatus, then $\alpha = (p_t - p_o)/(p_f - p_o)$ (where p_o is the pressure in the system at the start of reaction and p_f is the pressure when decomposition is complete).

Analogous expressions are applicable to other experimentally measured parameters related to extents of reaction, e.g. diffraction peak intensity in X-ray measurements, heat evolution in DSC, etc. When product compositions vary during the course of a reaction (or with temperature in non-isothermal work) care must be exercised to ensure that the parameter used to measure α is quantitatively defined with reference to the chemical change being investigated. If a decomposition takes place in several distinguishable stages it is essential to determine separate sets of α - time data for each stage. Values of α are usually recorded as a function of time at constant temperature in conventional kinetic studies, or as a function of temperature, usually in a linear heating programme, in thermoanalytical studies.

A technique, developed by Rouquerol [1] and known as constant rate thermogravimetry requires more sophisticated controllers to adjust the furnace temperature in such a way as to maintain a constant preselected value of $d\alpha/dt$ over the major portion of the reaction. This increased complexity of the apparatus results in a simplification in the data treatment.

2.3. THE REACTANT

Probably the most widely used methods of reactant preparation are by crystallization, or by precipitation from solution. Sometimes use can be made of a silica-gel crystal growth technique [2], especially when the compound formed is a sparingly-soluble precipitate from aqueous solution.

High temperature methods used in synthetic solid state chemistry [3] such as solid-solid reactions, crystallization from melts and vapour-phase transport are not appropriate for the preparation of compounds of low thermal stability.

Even if the reactant is used as supplied by a manufacturer, it is important that as much information as possible should be collected on its method of preparation and its treatment (e.g. grinding, exposure to radiation, ageing during exposure to water vapour or oxygen, etc.) in the time that has elapsed since preparation. Some or all of these factors may have a profound influence on the reactivity of the sample [4]. The consequences of an unknown history may be expunged by suitable recrystallization of the sample.

Single crystals of a sufficiently large size for a decomposition study are often difficult to obtain [5]. Studies using individual well-defined and relatively perfect crystals can, in principle, provide deeper insight into the factors controlling reaction than is possible through use of less well-defined reactants, such as powders, with a range of particle sizes. Work with a single crystal enables the relative reactivities of different crystal faces to be compared [6]. Such ideal reactants differ considerably from the type of material usually used in industrial processes where reactants are often fine powders. Particle size and crystal perfection are often important parameters in controlling the reactivity of a solid reactant sample, so that single crystals are preferable for mechanistic studies.

Many crystals grown from aqueous solution are hydrates (see Chapter 7). Dehydration often precedes loss of other ligands or decomposition of other constituent ions. If water loss is the reaction to be studied, the possibility that partial dehydration may have occurred during storage must be investigated. Information on the water content of the reactant and the importance of any intermediate steps in dehydration must be considered. The conditions during dehydration can influence the reactivity of the anhydrous salt, or the lower hydrate, obtained. The product formed during dehydration in vacuum is often amorphous or microcrystalline, while in higher water vapour pressures the product may be crystalline. This solid product may then be the reactant for a solid state decomposition study. In such a study it is often assumed that the initial water loss step is completed rapidly at the higher decomposition temperature. It is important, however, to consider possible influences on the decomposition of textural changes in the original reactant, of water retention and of hydrolysis processes.

Impurities in the reactant, even when present in very small amounts, may participate in, and hence alter, both the rates and mechanisms of decomposition reactions. It is thus necessary to know what impurities are present, their concentrations and spatial distributions. Sometimes the effects of specific additives, introduced as solid solution (doping), may be studied to give information on the reaction mechanism. Although not always provided, a full elemental analysis of the reactant is an essential prerequisite for the study of the decomposition. There is discussion of futher aspects of reactant characterization in Chapter 6.

2.4. CHANGES PRECEDING OR ACCOMPANYING DECOMPOSITION

2.4.1. Introduction

Heating of a crystalline solid increases the amplitude of the vibrations (phonons) of the constituent atoms. In the absence of other changes, this must ultimately lead to melting, with the loss of long-range order, but the retention of some short-range structure. However, before this catastrophic breakdown of crystal order, there may be locally enhanced activity associated with the several types of imperfections mentioned in Chapter 1.4. The activation of electrons and positive holes results in increased electronic mobility, and conductivity, in the solid. Point defects permit the movement of material, including impurities, throughout the more perfect regions of the crystal. Such mobility occurs most particularly in zones of extended imperfection where dislocations may migrate. These, as regions of instability, can be eliminated by coalescence or displacement to surfaces due to growth of the more perfect regions of crystals (*annealing*). Surfaces also become more reactive as the melting point is approached. There may be direct volatilization of material (*sublimation*) or superficial migration to establish contacts between previously separate particles with overall diminution of surface area (*sintering*).

Vibrations of the crystal components may destabilise a lattice structure above a particular temperature, but this does not necessarily result in melting if an alternative phase is stable. Normally the form stable at the higher temperature, generated by a reversible recrystallisation, is less ordered. An example is the orthorhombic to cubic transformation of the alkali metal perchlorates [7] which is associated with increased rotational freedom of the perchlorate anion. Studies of phase transformations are of value in advancing understanding of solid state decompositions.

Alternatively, or concurrently with processes of the above type, chemical changes of crystal components can occur. Such events can again be associated with structural imperfections, most notably reactant/product interfaces or crystal surfaces (regarded as extended and complicated crystal imperfections). Breakdown of individual ions may occur at temperatures below those of recrystallization or melting. These include reactions such as the loss of H_2O from hydrated ions in hydrates, the breakdown of unstable ions such as N_3^-, MnO_4^-, etc, or interactions involving more than a single ion, such as those in the decomposition of $NH_4C\ell O_4$.

The occurrence of a phase transition in the solid, or melting, recrystallisation, sintering or sublimation during a solid state decomposition introduces complexities into the formulation of a reaction mechanism. The process of heating a solid may result in changes to the imperfection structure so that, at the reaction temperature, the properties of the reactant may be different from those determined for the original preparation. Important changes result from a precursor chemical step, as occurs, for example, when dehydration of a prepared hydrated reactant precedes decomposition. The rapid initial structural changes, which may include recrystallization, following water loss, results in modifications of reactant particle sizes, together with changes in imperfection concentrations and distributions. The fundamental nature of the retexturing that occurs during such precursor reactions is often not discussed in the interpretation of kinetic data.

2.4.2. Solid phase transitions (including recrystallization)

Recrystallization of a solid may occur for a variety of reasons. The growth of zones of more perfect crystal structure at elevated temperatures proceeds through the elimination of regions of local distortion, e.g. dislocations, grain boundaries, etc., without change in chemical composition or lattice configuration [8]. Where two phases are present, the larger and more perfect particles of discontinuous phase may grow at the expense of the smaller, so reducing the interfacial (and possibly other forms of) energy [9,10].

Recrystallization of a solid may also produce a new lattice configuration. Many solids (inorganic or organic, covalent or ionic) can assume different crystal structures (*polymorphs*) dependent upon their method of preparation and subsequent treatment. Decompositions of polymorphic solids may be preceded by changes from one crystal structure to another (*phase transition* or *transformation*) [9,11-16]. The strictest definition of a phase transition precludes any change of composition, but the structural changes will generally produce intermediate solids with properties which may differ considerably from those associated with the original solid reactant. The higher temperature phase usually has the more open structure, with higher symmetry than the lower temperature phase. The higher temperature structure allows increased freedom of motion of one (or more) lattice component, e.g. the ability of nonspherical molecules (or ions) to rotate [7]. The increase in entropy is comparable with the melting process, though the long-range order is not destroyed.

Buerger [11] has classified transitions as either *reconstructive* or *displacive*. Reconstructive transitions involve a major reorganization of the crystal structure (e.g. graphite ⇄ diamond), while displacive transitions involve bond distortion rather than bond breaking (e.g. quartz ⇄ tridymite ⇄ cristobalite). These differences are illustrated in Figure 2.1.

Buerger's more detailed classification scheme [11] refers to changes in the interactions in the (first) coordination sphere; interactions with next nearest neighbours (the second coordination sphere) and order ⇄ disorder transitions (which, in turn, may be substitutional, e.g. random distributions of Li^+ and Fe^{3+} over the octahedral sites in fcc $LiFeO_2$, or orientational, e.g. the ferroelectric ⇄ paraelectric transition in $NH_4H_2PO_4$).

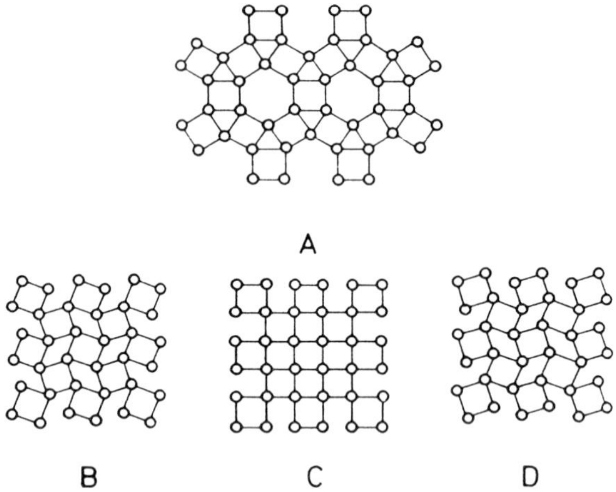

Figure 2.1.
Reconstructive and displacive transitions. A transition from structure A to any of the other structures illustrated would be a reconstructive transition, involving major reorganization of the bonding, while transitions between B, C and D are displacive transitions, involving bond distortion rather than bond breaking [11].

Ehrenfest's classification (see [11]) into *first-order* and *second-order* transitions is based on thermodynamic criteria. First-order transitions have discontinuities in the first derivatives of the Gibbs energy with respect to temperature (= entropy) and pressure (= volume), while second order transitions have discontinuities in the second derivatives of the Gibbs energy (= heat capacity, thermal expansion and compressibility). DSC (or DTA) may be used to measure the enthalpy changes and transition temperatures associated with first-order transitions, while shifts in the DSC baseline arising from changes in heat capacity of the sample are used to detect and measure second-order transitions.

The rates at which phase transitions occur vary greatly. At one extreme are the transitions in minerals occurring only on geological timescales, while at the other extreme are very rapid and readily reversible transitions [17]. West [11] has given a schematic diagram illustrating the dependence of the rates of a first-order transition (I $\rightleftarrows$ II) on temperature (see Figure 2.2.). At temperatures close to the transition temperature T_c (region A) the rates in either direction are slow because $\Delta G \approx 0$. At temperatures further from T_c, the rates increase. In region B the I → II transition rate increases rapidly. In region C the II → I rate increases and may reach a maximum at T_M. If it is possible to supercool the high temperature form II to below T_M, then the rate of the II → I transition decreases in region D.

Most reconstructive transitions take place by processes of formation and growth of nuclei of the product phase and the kinetics are often described by the model developed by Johnson, Mehl, Avrami and Erofeev (see Chapter 3). These reactions are, therefore, identical with other decompositions and represent simple chemical changes. Studies are often relatively difficult and greater interest has been directed towards the experimentally more accessible reactions that yield gaseous products.

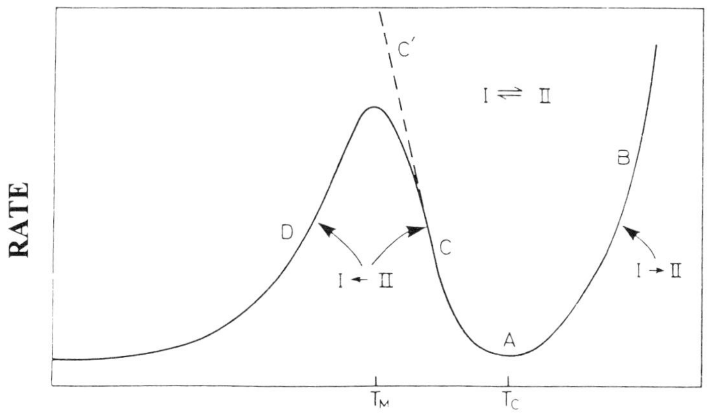

Figure 2.2.
The dependence of the rate of a first-order transition (I → II) on temperature [11].

An example of an intensively studied set of polymorphs whose decompositions are of great theoretical and practical importance (see Chapter 12) is $CaCO_3$ which may exist (in order of decreasing thermodynamic stability) as calcite, aragonite or vaterite [18]. Vaterite can be prepared by precipitation from aqueous solutions under carefully controlled conditions. A DTA curve for the vaterite → calcite transition is shown in Figure 2.3. The transition is exothermic (ΔH = -34.3 J g^{-1}) with onset at 704 K. Isothermal extent of conversion against time curves were described [18] by the Johnson, Mehl, Avrami, Erofeev model (see Chapter 3) with $n = 2$. The measured Arrhenius parameters were E_a = 210 kJ mol^{-1} and A = 1.15 x 10^{14} min^{-1}. The decomposition of vaterite and its concurrent transformation to calcite under various conditions were compared [18] with the decomposition of calcite under the same conditions (see Chapter 12).

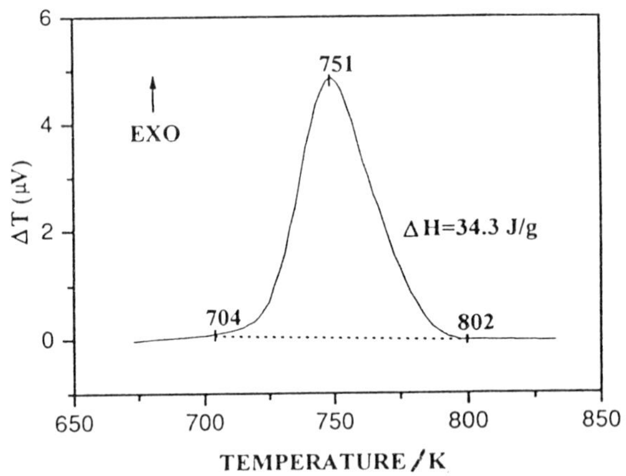

Figure 2.3.
DTA curve for the vaterite → calcite transition at a heating rate of 10 K min^{-1} in flowing argon (reproduced from [18] with permission).

2.4.3. Melting

Theoretical explanations which have been advanced to account for the decrease in order occurring at the temperature of fusion of a crystalline solid include an increase in the amplitude of thermal vibrations so that the stabilizing forces of the crystal are overcome, and/or that there is a marked increase in the concentrations of lattice defects (vacancies) or dislocations. Within a few degrees of the melting point, the physical properties of many solids begin to deviate significantly from their values at lower temperatures. These effects are clearly interconnected so that the individual contributions are not readily distinguished. Amplitudes of vibrations of atoms are increased in the vicinity of a vacancy, and dislocation multiplication must also influence the concentration of vacancies.

Ubbelohde [19] has emphasized the sparsity of data on the properties of solids and liquids in the vicinity of the melting point. Fusion is accompanied by a loss of structural order, which always contains a positional contribution and sometimes also contains an orientational contribution, so that the entropy increase may include more than a single term. In some solids the increasing disorder, resulting from the temperature rise, may involve one or more phase transformations below the melting point.

An important feature of the fusion process is the general inability of solids to superheat and this observation is usually discussed in theoretical treatments. While a few exceptions to this generalization are known [8], it is generally agreed that almost all solids start to melt upon the attainment of the appropriate temperature, unlike the reverse process where supercooling of most liquids can be demonstrated under appropriate conditions. Christian [13], discussing melting as a nucleation and growth process, suggests that the surface may be its own nucleating agent and lattice defects or impurities retained in such regions may aid the formation of the melt. Theoretical treatments of melting must consider the interactions between two phases (solid and liquid) which differ in their degrees of ordering, thus theories based on properties of a single phase (the solid) alone, have been criticized as being unrealistic.

The Lindemann [20] theory of melting (1910) considers the consequences of increasing temperature on the vibrational amplitudes of the atoms in a solid and suggests that, when the displacements occurring in such motion reach a particular fraction (about 10%) of the lattice spacing, the mutual interactions of such vibrations result in an overall loss of crystal stability. Various developments of this approach have been attempted including consideration of the anharmonicity associated with increased vibrational amplitudes and the distributions of energy among the constituent bonds of multiatomic molecules. The Lennard-Jones-Devonshire [21] theory (1939) considers the energy required to interchange lattice constituents between site and interstitial positions. The energy requirement for each such transposition is decreased when many occur within a small volume, and the result can be regarded as the production of a quasi-crystalline liquid. Both these approaches are critically discussed, with appropriate supporting evidence, in Ubbelohde's books [19].

Kuhlmann-Wilsdorf [22] (1965) provided a new theoretical treatment of the role of dislocations in melting. She concluded that fusion occurred when the free energy for formation of glide dislocation cores became negative. There is a well-defined temperature above which dislocations are freely generated. Thus proliferation of such cores tends to fill the crystal to capacity and, with the accompanying absorption of latent heat, this results in the inability of the material to withstand a shearing stress. Consideration has also been given to the concentrations of vacancies near the melting point [23,24], and the roles of increasing vibration [23,25] and expansion of the lattice [26,27].

Coker *et al.* [28] have studied the fusion of tetra-*n*-amyl ammonium thiocyanate and identified the greatest structural change as occurring in the phase transition that precedes melting. The solid and molten states of the salt are believed to possess generally similar structures in which the significant difference is the ability of the hydrocarbon group to kink and unkink after fusion has occurred. The electrical conductivity of the solid increased as the melting point was approached. Allnatt and Sime [29] similarly found an anomalously large increase in the electrical conductivity of sodium chloride within about 4 K of the melting point. From comparisons between results for pure, strained and Sr^{2+} doped crystals, it was concluded that fusion is not a consequence of either accidental impurities or dislocations.

Clark *et al.* [30] considered premelting transitions in the context of a general theory of disorder in plastic crystals. Entropy changes were attributed to contributions from two sources: increasing rotation of the molecular constituents of the crystal and the expansion of the lattice, against the van der Waals forces, necessary to accommodate this increased disorder. The observed changes were collated with molecular shapes.

As with other solid state properties, the increased relative contribution from surface energy in very small (micrometre size and below) crystals influences the melting point [8,31,32]. Experiments with metals (Pb, Bi, Sn) have shown unusually enhanced [32] diffusion and movement of small crystals in the vicinity of the melting point. The changes in fusion behaviour of lead [31] are not attributable to a size effect alone, but the phase change proceeds through a liquid skin mechanism. The behaviour of bismuth [8] is different, being time dependent, and may arise through a delay in nucleation of crystals.

2.4.4. Sintering

When closely packed particles are heated to temperatures around the Tammann temperature [33], which is approximately $0.5T_m$ (where T_m is the melting temperature), the particles join at the points of contact (*sinter*) and the intervening pores formed gradually disappear with time. In addition to macroscopic measurements of surface area and pore volume, the progress of sintering can be measured [34] from (i) the size of the "neck" that forms between particles (Figure 2.4.(a)) and (ii) the extent to which the particle centres approach each other (the "shrinkage" of the sample, Figure 2.4.(b)).

(a)

(b)

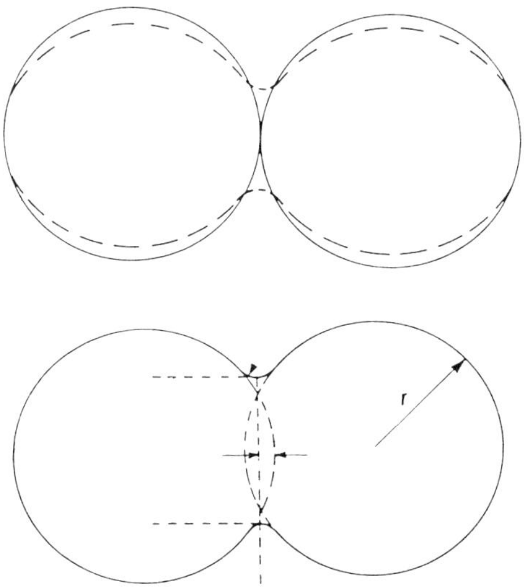

Figure 2.4.
Neck formation between spherical particles: (a) without shrinkage but with a decrease in particle radius, r; (b) with shrinkage [34].

The driving forces, transport mechanisms and kinetics of sintering have been extensively studied [34,35]. Searcy and Beruto [36] have reviewed the theories of both isothermal and nonisothermal (i.e. in a temperature gradient) sintering of crystalline solids. Early theoretical models generally ignored the effects of anisotropy and assumed that temperature gradients provided no significant driving force for sintering. The simplest case treated [36] is of particles containing only a single atomic species and vacancies. Classical thermodynamics attributes the increased reactivities of small particles to surface tension effects (liquids), or stress (solids) [37].

The Kelvin equation [34] relates the excess Gibbs energy, ΔG, of a small spherical particle to its radius, r:

$$\Delta G = 2\gamma_{sv}\,\Omega / r$$

where γ_{sv} is the solid-vapour surface energy, and Ω is the molar volume/the Avogadro constant. The curvature of the surface at the neck between particles is opposite in sign to that of the spherical particles and, hence, a decrease in Gibbs energy is predicted if matter is transported from the convex surface of the sphere to the concave surface of the neck. One mechanism for such transport is by evaporation and recondensation at the neck.

Searcy and Beruto [36] explain how, in crystalline solids, the minimization of particle free energy may be achieved through redistribution of vacancies. Because the vacancy concentration decreases as the particle size decreases, the vapour pressures of small crystals increase.

The rate determining step in the sintering of crystalline solids has often been assumed to be diffusion [33,37,38], whether it be via surfaces, grain boundaries or within the bulk solid structure, but an alternative possibility is the rate at which new crystal layers are nucleated or grow. Analyses have been given [36] of the kinetics of vacancy formation or annihilation at surfaces. The "neck regions" between particles are favoured because less strongly bonded atoms from corners, edges, kinks, etc., can transfer there and be held more strongly by bonding across the interface. The shapes of the boundaries formed during grain growth are then determined by the relative rates of nucleation and growth of new crystal layers.

Very detailed and thorough studies [36] of the isothermal sintering, in the presence of water vapour, of CaO and MgO samples, obtained by decomposition of the corresponding carbonates, showed that both pore and grain growth occurred. The disappearance of closed pores in preference to open pores supported the mechanism described above for layer growth.

The effects of temperature gradients are explained [36] in terms of movement of vacancies from higher to lower temperature regions which then enhances atom-vacancy exchanges in the lower temperature regions. Unless temperature gradients are sufficient to allow matter to be transported to the coldest parts during the time frame considered, these gradients will cause differential densification or coarsening of the surfaces.

2.4.5. Sublimation

During sublimation, constituents of the solid are directly transferred to the gas phase without the intervention of a liquid phase. Gilles [39] has provided six "principles of vaporization reactions". (i) All substances vaporize, but in different modes (see below) and at different rates depending upon the temperature and the environment. (ii) Vaporization processes may be represented by chemical reactions. (iii) Thermodynamic factors determine the maximum extent of vaporization. (iv) Kinetic factors determine the actual processes and the predicted extent of vaporization may not be reached. (v) The possibility of formation of solid-solutions should be considered, as should (vi) the reactivities of all constituents (e.g. residual gases, sample containers, etc.) of the systems under examination.

Somorjai and Lester [40] have reviewed the literature concerned with the sublimation of solids, including an analysis of the sequence of chemical steps required to convert crystal components into the observed volatile species. They emphasize that investigations under conditions of equilibrium between the solid and vapour provide thermodynamic information, but that kinetic information has to be obtained under conditions far from equilibrium and, frequently, by evaporation into a vacuum. The temperature dependence of the vacuum evaporation rate is used to estimate the activation energy of vaporization. Problems in formulating mechanisms and in identifying the rate limiting step are discussed [40].

For one face of a monatomic single crystal, the flux into vacuum [40] is:

$$J_A / (\text{mol cm}^{-2} \text{ s}^{-1}) = k_v \ [\text{A}]_s \qquad (2.1)$$

where k_v is the rate coefficient and $[\text{A}]_s$ is the surface concentration of A atoms. The values of k_v and $[\text{A}]_s$ will vary from one crystal face to another.

The maximum possible rate of vaporization of species A (molar mass, M) from the surface, J_{max}, at a given temperature is the rate which would be attained at dynamic equilibrium:

$$J_{max} = p_A \, (2\pi MRT)^{-1/2}$$

where p_A is the *equilibrium* vapour pressure of A. If the rate of vaporization into vacuum is less than J_{max}, an *evaporation coefficient*, α_v, is introduced, where:

$$\alpha_v = J_A(T) \, / \, J_{max}(T)$$

so that:

$$J_A = \alpha_v \, p_A \, (2\pi MRT)^{-1/2} \qquad (2.2)$$

known as the *Hertz-Knudsen-Langmuir* (HKL) equation [40]. α_v serves as a measure of the deviation of experimental conditions from those at true equilibrium.

Somorjai and Lester [40] detail some of the problems likely to be encountered in vaporization measurements. These include: (i) the effects of variations of J_A with crystal surface, which are particularly important in the use of polycrystalline samples; (ii) the self-cooling resulting from the endothermic vaporization, which may cause temperature gradients in the sample, especially at high fluxes; and (iii) the complications caused by vapour-vapour collisions when measurements are made in a significant partial pressure of vapour.

Gole [41] has distinguished the following modes of vaporization. (i) Simple solids which show no unusual barriers to evaporation and condensation (their vaporization and condensation coefficients are close to unity); e.g. monatomic elements such as Zn. (ii) "Retarded" vaporization of molecular and ionic solids (measured vaporization coefficients are very much smaller than those predicted from equilibrium vapour pressures). This group is further sub-divided into (a) vaporization without dissociation (e.g. alkali halides); (b) vaporization with dissociation (e.g. $ZnO(s) \rightarrow Zn(g) + \frac{1}{2}O_2(g)$); and (c) "incongruent" vaporization in which a condensed phase and a vapour are formed (e.g. $BN(s) \rightarrow B(s) + \frac{1}{2}N_2(g)$).

Somorjai and Lester [40] point out that the HKL equation (2.2) "is entirely inadequate, however, to describe the kinetics of vaporization", because the equation applies to a surface in which all the atoms are available for desorption and are bound by the same vaporization energy. Real surfaces are heterogeneous and different sites have different binding energies. Evaporation occurs more readily from dislocation edges or kink sites on surface terraces, possibly following migration onto planar crystal faces. (This behaviour enables dislocations to be revealed through the formation of etch pits resulting from a small amount of initial sublimation which occurs preferentially from these least stable zones of imperfection.)

Vaues of α_v may be very much less than unity and be temperature dependent. Somorjai and Lester [40] comment that "all the kinetic information is contained in the evaporation coefficient and its variation with conditions of vaporization", and they recommend the avoidance of the use of α_v in describing the rates of evaporation of solids under non-equilibrium conditions. The rate of sublimation is dependent on the attainment of sufficient energy by suitably disposed surface molecules (possibly accompanied by electron or proton transfer in ionic solids). The overall rate of reactant removal is sensitive to the presence of impurities at the surface. The reverse reaction may be significant if the volatile material is not immediately removed from the vicinity of the reactant particles. Arrhenius parameters measured for sublimation processes may include a term which represents a temperature dependent concentration of surface intermediates [42]. The observation that measured evaporation rates are lower than those estimated from equilibrium vapour pressures suggests that the kinetics may be determined by a surface dissociation that precedes evaporation. This view is supported by evidence that, in selected systems, specific additives can considerably promote evaporation rates. For example [40], the evaporation rate of red phosphorous between 550 and 675 K was found to be increased by three orders of magnitude by the presence of thallium.

The observed vaporization rates from heterogeneous surfaces have been explained [40] in terms of a stepwise model. For example, the rate equations for the three-step process at a metal surface (see Figure 2.5.):

$$\text{A (ledge)} \underset{k_2}{\overset{k_1}{\rightleftarrows}} \text{A (adsorbed)}$$

$$\text{A (adsorbed)} \underset{k_4}{\overset{k_3}{\rightleftarrows}} \text{A (g)}$$

$$\text{A (ledge)} \underset{k_6}{\overset{k_5}{\rightleftarrows}} \text{A (g)}$$

where only atoms at ledge sites (c) and adsorption sites (d) participate, have been derived [40]. The rate constants, except k_4 and k_6, include activation energies which depend upon the strengths of the bonds broken in the different bonding situations. The E_a value associated with k_5 is large, so that vaporization of adsorbed atoms will dominate the evaporation process. The rate of supply of adsorbed atoms may, however, be limited by surface diffusion from ledges and it can be predicted [40] that, for such a limitation, the evaporation rate can be as low as J_{max} /3. Values lower than this are taken to indicate direct vaporization from kink or ledge sites.

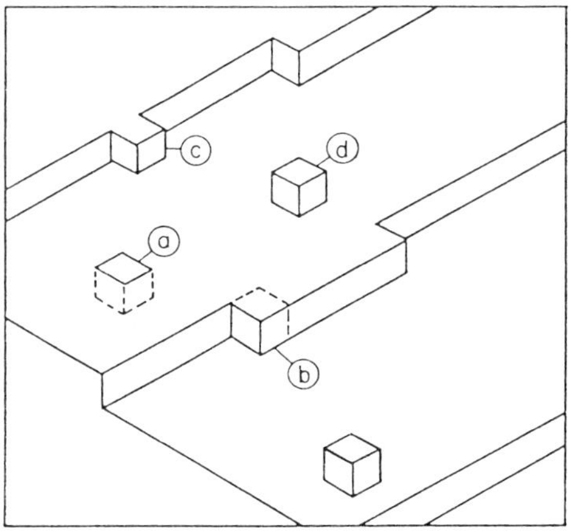

Figure 2.5.
Model of a metal surface showing atoms at (a) surface sites; (b) kink sites; (c) ledge sites, and (d) adsorption sites [40].

Reproduced from G.A Somerjai and J.E. Lester, in Progress in Solid State Chemistry, Vol.4 (1967), Pergamon Press, with permission.

The review by Somorjai and Lester [40] includes a detailed consideration of the mechanisms of evaporation of cadmium sulfide between 920 and 1070 K:

$$CdS(s) \longrightarrow Cd(g) + \tfrac{1}{2}S_2(g)$$

The effect of light in increasing the rate of volatilization was ascribed to charge transfer. The presence of excess cadmium in nonstoichiometric crystals decreased sublimation rates, as was also found for doping with copper. From a quantitative consideration of the carefully measured kinetic data (discussed in greater detail in [40]) it is concluded that the reaction mechanism is as follows:

$$CdS(surface) + 2p \rightleftarrows S(surface) + V_S^+(surface) + n$$

$$\uparrow\downarrow$$

$$\tfrac{1}{2}S_2(surface) \rightleftarrows \tfrac{1}{2}S_2(g)$$

$$CdS(surface) + 2n \rightleftarrows Cd(surface) + V_{Cd}^-(surface) + p$$

$$\uparrow\downarrow$$

$$Cd(g)$$

where V signifies a vacancy at a cation or anion site, and p and n are positive holes and electrons, respectively. Thus several distinct rate processes contribute to the overall chemical change. Electron transfer, together with redistributions in bonding (e.g. $2S \rightleftarrows S_2$), results in chemical changes accompanying sublimation. Product thus forms on a surface that itself is undergoing reorganization, including the generation of imperfections. The above example shows the potential complexity of desorption processes. More detailed investigations, such as those described in [40] for CdS, are required to characterize the chemistry of a wider range of reactants in what might be regarded as the simplest of all solid-surface rate processes.

2.4.6. Sublimation and thermal decomposition

Sublimation may be regarded as a form of decomposition of a solid. L'vov and Novichikhin [43,44] have interpreted the measured rates of product formation for a number of decompositions by regarding them as sublimation-type reactions. Decomposition of compound S is regarded as forming the gaseous products A, B, C, etc.:

$$S(s) \longrightarrow aA(g) + bB(g) + cC(g) \qquad (2.3)$$

The flux of product A (J_A) can be expressed in terms of its partial pressure (p_A)

corresponding to the hypothetical equilibrium of reaction (2.3) through the Hertz-Knudsen-Langmuir equation (2.2) above:

$$J_A = p_A / (2\pi MRT)^{1/2}$$

$$= a\gamma[\, F^{1/\upsilon}(2\pi MRT)^{1/2}]^{-1} \exp(\Delta S_T^o / \upsilon R) \exp(- \Delta H_T^o / \upsilon RT) \qquad (2.4)$$

where a is the coefficient of reactant A in equation (2.3); $\gamma = 1.013 \times 10^5$ Pa atm^{-1}; and $F = a^a\, b^b\, c^c$ .....

$\upsilon = a + b + c + ...$

$M = M_A^{a/\upsilon} \times M_B^{b/\upsilon} \times M_c^{c/\upsilon} \times ......$

ΔS_T^o and ΔH_T^o are the entropy and enthalpy changes for reaction (2.3), and the further assumption is made that the apparent activation energy:

$$E_a = \Delta H_T^o / \upsilon.$$

Additional refinements to the theory are provided [43] to correct for evaporation from a surface rather than an effusion cell and the small influence of temperature on the pre-exponential factor.

L'vov and Novichikhin [43,44] applied this model to the analysis of thermal decomposition data for oxysalts ($AgNO_3$, $BaSO_4$, $BaCO_3$), nitrides (BN, Be_3N_2, Mg_3N_2, AℓN, GaN), an oxide (ZnO), a sulfide (CdS), a selenide (CdSe), $NH_4C\ell$ and the elements P, As and Sb. Calculated values of E_a for the assumed reactions (in the range between 172 and 532 kJ mol^{-1}) agreed quite well, on the whole, with their experimental values, and they noted that the approach, although in a preliminary form that "requires checking and refinement", was adequate for the interpretation of the low values of the evaporation coefficient, α_v, (10^{-7} to 10^{-1}) in all the cases considered.

Values of α_v were used to find the reaction stoichiometry that gave acceptable agreement of magnitudes of evaporation rates and E_a. For example, it was concluded that AℓN decomposition at 1700 K yields 60% of the nitrogen in the form of free atoms:

$$A\ell N(s) \rightarrow A\ell(g) + 0.6N(g) + 0.2\, N_2(g) \text{ rather than } (\rightarrow A\ell(g) + 0.5N_2(g)).$$

The value of these conclusions would be much greater if confirmed by analytical measurements of the yields of the individual *primary* product species evolved from the reacting surfaces.

The direct application of equation (2.4) to sublimation rates [43,44] bears a formal similarity to the Polanyi-Wigner treatment of interface reactions (Chapter 3). The number of product molecules formed in unit time is assumed to be directly proportional to the frequency of activation within a population of energetically identical precursors at the reactant surface. This may be an acceptable model where a surface step is followed by rapid evaporation so that reaction is regarded as the first-order volatilization of species within a homogeneous boundary array. Somorjai and Lester [40] regard this sublimation model to be an oversimplification. Sublimation is, in principle, simpler than the interface advance process considered in the Polanyi-Wigner treatment. The absence of a solid product in sublimation excludes the complicating contributions arising from reversibility within the post-interface layer [45]. Data referring to the forward reaction only are regarded as providing information concerning the chemical step in salt breakdown. At low $p(CO_2)$ (less than $10^{-2}\ p_{eq}$) in the $CaCO_3$ decomposition [46], for example, the controlling step in CO_2 evolution is believed to be either condensed phase diffusion of CO_2 (migrations of CO_3^{2-} and O^{2-} ions in different directions) or the breakdown of $CO_3^{2-} \rightarrow CO_2 + O^{2-}$. This reaction is considered in greater detail in Chapter 12.

Surface reactions, however, are *not* necessarily controlled by a single rate determining process but may proceed through a series of consecutive, interlinked steps [40]. This behaviour is not, in general, represented by the transition state theory as developed for application to single-step homogeneous processes involving random energetic encounters in space. The surface of a solid provides a quite different reaction environment and it may be necessary to develop an alternative reaction rate theory to accommodate the following features that do not apply to interactions between individual and freely moving gaseous or solute reactants.

(i) The volatilization of products during sublimation may be achieved only through a sequence of several consecutive steps, including the following: (a) transfer of one or more crystal constituents from a lattice site to an adsorbed state, perhaps accompanied by vacancy formation; (b) electron transfer and/or bond reorganization within and between the precursors to desorption; and (c) desorption of the volatilized species. These processes may require several steps each and may be interlinked.

(ii) The crystal face on which these changes are occurring is inhomogeneous and undergoes dynamic reorganization during material losses. The chemical properties of individual surface constituents of the reactant are believed [40] to depend on location and the numbers of neighbours to which each is bonded at different crystal sites: faces, dislocations, step edges, etc. Thus, unlike homogeneous reactions, reactivities of chemically similar entities can differ and vary with position and time. The interactions between crystal components and adsorbed material are sufficiently close to require consideration as a single chemical system.

(iii) Close packing on surfaces may promote multiple encounters (a strong "cage effect") leading to tri- or tetramolecular processes which are believed not to contribute significantly in homogeneous rate processes. Surface species may participate in chemical changes that do not yield a volatile product: such parallel but undetected reactions cannot be considered in the theory.

In summary, there is evidence [40] that the controls of surface reactions differ fundamentally from those applicable to homogeneous rate processes. Factors determining the rate of product formation, through the interactions of various adsorbed species retained within active zones across inhomogeneous crystal faces, may be expected to differ from those applicable to the results of random encounters of individual identical reactant species in a gaseous or solution environment. The surprising feature is that the theory developed to apply to homogeneous reactions has been so successful for so long in explaining the behaviour of reactions involving solids. Nevertheless the factors that control the rates of reactions at crystal faces are still not thoroughly understood. For many surface reactions the identities and concentrations of the essential precursors to product formation are not known. The significance of Arrhenius parameters measured for reactions of solids has been debated (see Chapter 4).

The success of studies such as those for CdS sublimation [40] may be used to advance understanding of the mechanisms and controls of these reactions so that the principles may be applied to a wider range of heterogeneous reactions, such as the thermal decompositions of solids, other heterogeneous catalytic reactions, etc.

2.4.7. Sublimation and heterogeneous catalysis

The recognition that sublimations of solids proceed through a series of distinct chemical steps draws attention to the similarities between such processes and the mechanisms of heterogeneous catalytic reactions. In heterogeneous catalysis, rates of surface steps are often expressed by concentration terms for adsorbed species that are formally analogous to those used in homogeneous catalysis [47]. For many reactions of interest, however, the identities and numbers of the essential adsorbed participants are not known. Mechanistic conclusions have to be based on indirect evidence through the assumption that chemical changes proceed within a two-dimensional layer subject to kinetic laws similar to those applicable to gases. Constituents from the solid may also participate within the active zone, but the area of solid surface actively contributing to product formation is often not known. Aspects of this model that merit further consideration include the following.

(i) Many of the fundamental kinetic studies of reactions on surfaces have used relatively perfect crystal faces, low temperatures and low concentrations of adsorbed species. Under conditions of greatest catalytic activity, at higher temperatures and with saturated surfaces, the chemical changes occurring, and their controls, may be

quite different. Such reactions can be accompanied by retexturing of the solid surface, demonstrating the involvement, in the active layer, of components from the underlying solid [48]. Sublimation may involve similar surface rearrangements.
(ii) Chemisorbed species can be strongly retained at surfaces at low temperatures and constituents may be evolved at higher temperatures only after further chemical changes [49].
(iii) During reactions involving interactions of adsorbed intermediates, the surface concentrations of precursors to product formation may vary with temperature which will influence the apparent magnitude of E_a [50]. Different contributions from this effect within a series of comparable reactions proceeding within similar temperature intervals can lead to compensation behaviour [51].

Many of the kinetic characteristics of heterogeneous reactions can be satisfactorily represented by rate equations similar in form to those developed through studies of homogeneous processes. However, the identities, properties, concentrations (etc.) of the surface participants in heterogeneous processes of all types may be difficult to characterize. The experimentally accessible and relatively less complicated sublimation reactions offer a route to establishing a theoretical representation and increasing understanding of aspects of surface processes that may apply to a wider range of reactions, perhaps including chemical changes proceeding on catalyst surfaces and at interfaces.

A transition state model satisfactorily represents [47] the observed fit of rate data for surface reactions to the Arrhenius equation. The distribution function for the most energetic quanta within a solid, if assumed to extend into the surface layer, has been shown [52] to have the same form as the Maxwell-Boltzmann equation. It cannot be accepted, however, that the value of E_a measured for the overall change can be identified with a single step energy barrier to reaction where interlinked and/or consecutive processes participate. Much more information is required to interpret the significance of the magnitudes of A and E_a for most surface and interface processes. If reaction proceeds to product in a single step, perhaps through breakdown of an oxyanion [46], then behaviour may be represented by the model discussed by L'vov *et al.* [43,44]. Where there are several consecutive contributory steps, more detailed investigations are always required to elucidate the mechanism.

2.4.8. Topochemistry and topotactic reactions

Günter and Oswald [53] have provided a working definition of the term *topotactic chemical reaction*. The defining characteristics are that (i) the product is formed in one or several equivalent crystallographic orientations with respect to the reactant crystal, and (ii) the chemical change can proceed throughout the entire volume of the reactant. If the solid product is oriented in a small number of non-equivalent directions, reaction is regarded as being a number of different, concurrent processes.

Boldyrev [54] has discussed the terminology in detail and classified topochemical reactions on the basis of the number of structural elements of the reactant that are conserved in the product. For example, three-dimensional close-packing of anions is retained in the reactions [53]:

α-FeOOH $\rightarrow$ α-Fe_2O_3 ; α-AlOOH $\rightarrow$ α-Al_2O_3; and Fe_3O_4 $\rightarrow$ γ-Fe_2O_3.

Two-dimensional conservation of structure occurs when the relative positions of dense-packed layers of ions are displaced with respect to each other in the reactions:

α-Fe_2O_3 $\rightarrow$ Fe_3O_4; CoOOH $\rightarrow$ Co_3O_4; and $Ni(OH)_2$ $\rightarrow$ NiO.

One-dimensional conservation of structure occurs where the reactant and product both have chain-type structures, as in the following reactions:

$Cu(OH)_2$ $\rightarrow$ CuO; $CaSO_4.2H_2O$ $\rightarrow$ $CaSO_4$; and $MoO_3.H_2O$ $\rightarrow$ MoO_3.

The essential characteristic of *topotaxy* is thus that the product is oriented crystallographically with respect to one or more structural elements of the reactant. This characteristic has considerable value in the formulation of reaction mechanisms because evidence of topotaxy can be used to identify or suggest the atoms in the reactant unit cell that are displaced or removed during the reaction. It can also be assumed that other atoms will be relatively unmoved. Maintained linkage across the interface leads to retained orientational relationships between the reactant and product. By assuming minimum displacements and reorganization of retained atoms, the chemical steps involved in the conversion of reactant to product may be able to be identified [55-57].

The topotactic preparation of crystalline solids has been reviewed by Oswald and Günter [58] and further aspects of topochemical reactions have been discussed by Prodan [59].

2.4.9. Changes on heating a crystalline solid

The types of changes which can occur on heating a crystalline solid are summarized in Table 2.1. Not all may be identifiable for any particular solid and some of the changes may overlap. {D} in the Table indicates that the process may be regarded as a *decomposition* if significant bond reorganization or rupture amongst or between crystal constituents is involved. For example, the melting of a molecular crystal involves only redistribution of van der Waals forces and thus is not regarded as a decomposition. In contrast, the breakdown of the permanganate ion in $KMnO_4$ is definitely a decomposition. Arrows indicate whether the parameter mentioned increases (↑) or decreases (↓), while (↕) means that changes may involve either increases or decreases.

2.5. THERMODYNAMIC FEASIBILITY OF DECOMPOSITION

The simplest decompositions can be represented as:

$$A(s) \rightarrow B(s) + C(g) \qquad (2.5)$$

Many reactions of this type are assumed to take place in a single stage, with no side reactions, other than the possible occurrence of the reverse process (see below). Three well-studied examples are:

$$CaCO_3(s) \rightarrow CaO(s) + CO_2(g)$$

$$4MnO_2(s) \rightarrow 2Mn_2O_3(s) + O_2(g)$$

$$CuSO_4.5H_2O(s) \rightarrow CuSO_4.3H_2O(s) + 2H_2O(g)$$

Decompositions are frequently studied isothermally and under constant pressure. The sign of the change in Gibbs energy, ΔG, for the reaction will thus determine the spontaneity of the decomposition for the temperature and pressure conditions applying. From the general relationship $\Delta G = \Delta H - T\,\Delta S$ for the decomposition, the entropy change, ΔS, will almost inevitably be positive, so that the sign of ΔG will be determined largely by the sign of the enthalpy change, ΔH. If the values of ΔS and ΔH are assumed not to vary markedly with temperature, exothermic decompositions will be thermodynamically feasible at all temperatures (and hence are irreversible) and will be limited only by kinetic factors (see Chapter 3). Endothermic decompositions will only be thermodynamically feasible at temperatures greater than $T = \Delta H/\Delta S$. Below this temperature the reverse reaction, i.e. recombination, is expected.

After an initial particle of the solid product phase B has appeared, the activities of both the solid reactant, A, and the solid product, B, are assumed to remain unaltered throughout the course of the decomposition (until no A remains).

The equilibrium constant, K_a, for a reversible reaction of the type given in equation (2.5), at temperature T, is:

$$K_a = [B(s)]\,[C(g)]/[A(s)]$$

Table 2.1. Changes occurring on heating a crystalline solid

	Numbers of imperfections	Structural order	Surface reactivity	Surface area	Particle size	Decomposition
ANNEALING Decrease of internal disorder	↓	↑ (small)	↓	-	-	-
SINTERING Decrease of surface area	↓	↑	↓	↓	↑	-
SUBLIMATION Volatilization from surfaces	-	-	↕	↕	↓	{D}
MELTING Decrease in long-range order	↑	↓	?	↓	-	{D}
PHASE TRANSFORMATION New crystal structure (← recrystallization, from LOSS OF STRUCTURAL CONSTITUENT)	↑	↓	↕	↕	↕	{D}
LOSS OF STRUCTURAL CONSTITUENT → recrystallization (see PHASE TRANSFORMATION)						
LOSS OF STRUCTURAL CONSTITUENT → amorphous; DECOMPOSITION → amorphous	↑	↓	↑	↑	↓	{D}
DECOMPOSITION → recrystallization	↑	↓	↑	↑	↑	{D}

{D} indicates that the process may be regarded as a *decomposition*.
Arrows indicate whether the parameter mentioned increases (↑) or decreases (↓).
↕ means that changes may involve either increases or decreases of the parameter.

where the square brackets denote activities. Because [A(s)] = [B(s)] = 1,

$$K_a = [C(g)] = f_{C(g)} \approx p_{C(g)} = p_{eq}$$

where $f_{C(g)}$ is the fugacity of the gaseous product C. The fugacity can be approximately represented by the pressure, $p_{C(g)}$, which, for convenience, when a single gas only is present, is written as p_{eq}, the equilibrium dissociation pressure of C at temperature T.

The relationship between the standard state, designated by the symbol $^{\ominus}$ (all activities of reactants and products being equal to unity), and the equilibrium state is:

$$\Delta G^{\ominus} = - RT \ln K_a = - RT \ln(p_{eq}/\text{bar}) \tag{2.6}$$

When the prevailing pressure of product C is less than the equilibrium pressure at temperature T, the decomposition is feasible, but at higher pressures the reaction cannot proceed. If the decomposition is carried out in a flowing stream of carrier gas, or in a gas mixture with the partial pressure of C maintained below the equilibrium value at the reaction temperature, the decomposition should proceed to completion (although the operation of kinetic or geometrical factors, or the occurrence of secondary reactions, may decrease the rate or prevent this in practice).

Differentiation of the relationship (derived from equation (2.6)):

$$\ln K_a = - \Delta G^{\ominus}/RT = - \Delta H^{\ominus}/RT + \Delta S^{\ominus}/R$$

with respect to temperature, leads to the van't Hoff equation:

$$\text{d}\ln K_a/\text{d}T = \Delta H^{\ominus}/RT^2$$

For the decomposition, $K_a = p_{eq}$, and hence:

$$\text{d}\ln p_{eq}/\text{d}T = \Delta H^{\ominus}/RT^2$$

If $\Delta H^{\ominus}$ is virtually independent of temperature over the interval T_1 to T_2, then

$$\ln(p_2/p_1) = \Delta H^{\ominus} (T_2 - T_1) / RT_1T_2$$

or $\ln(p)$ is proportional to T^{-1}.

This approximate relationship for reversible endothermic decompositions has been used to estimate average values of the standard enthalpy of decomposition, ΔH°, from experimental equilibrium pressure measurements across a suitable temperature range.

Extension to more complicated reactions requires appropriate modification of the mathematical expression, e.g. for the Mond process:

$Ni(s) + 4CO(g) \rightarrow Ni(CO)_4(g)$

$K_p = \{p(Ni(CO)_4)\} / \{p(CO)\}^4$

so that:

$\ln[\{p(Ni(CO)_4)\} / \{p(CO)\}^4]$ is proportional to T^{-1}.

For exothermic solid-solid reactions involving no gases, reaction should proceed to completion provided that reactant/product interfaces maintain physical contact and that diffusion of reactants through the solid product does not become too slow.

2.6. THERMODYNAMIC PREDICTION OF DECOMPOSITION PRODUCTS

2.6.1. Introduction

Thermodynamic data can be used to predict the final (equilibrium) products of many decompositions. From considerations of stoichiometry only, two or more alternative primary products can often be expected. For example, decompositions of metal carboxylates (described in greater detail in Chapter 16) can lead to the formation of metal, metal oxide, or metal carbonate, as residual products, depending upon the electropositive character of the cation, together with the appropriate oxides of carbon (see equations (2.7 to 2.9) below). Thermodynamic predictions can be useful in mechanistic studies because evidence that observed *primary* reaction products differ from those identified as being most stable can be used in formulating the reaction mechanism. The principles used are outlined below for two of the systems where this approach has been most sucessful: the decompositions of metal oxalates and of metal perchlorates. For these substances the thermodynamic data are available and the products of decomposition are relatively simple. In similar thermodynamic analyses for other solid reactants, however, the following potential difficulties may be encountered. More than a single crystallographic form of a solid product may exist, so appropriate thermodynamic data are required. The energy of a solid may be modified significantly by the total surface area and by crystal imperfections. Movement of gases, either volatile products or gaseous reactants, away from or towards the reaction zone may be inhibited by a layer of solid product, so that the atmosphere prevailing may be difficult to characterize. In some reactions the solid products promote secondary reactions between primary products.

2.6.2. Decompositions of metal carboxylates

Decompositions of metal carboxylates (see Chapter 16) can lead to the formation of metal, metal oxide, or metal carbonate, as residual products:

$$M(COO)_2(s) \rightarrow M(s) + 2CO_2(g) \qquad (2.7)$$

$$M(COO)_2(s) \rightarrow MO(s) + CO(g) + CO_2(g) \qquad (2.8)$$

$$M(COO)_2(s) \rightarrow MCO_3(s) + CO(g) \qquad (2.9)$$

Decompositions of other metal carboxylates, such as acetates and other anions containing hydrocarbon radicals, give a more complicated range of volatile products and the solid product may also contain the metal carbide, identified as being formed by secondary reactions. It is useful to be able to predict, on thermodynamic grounds, the probable solid product which, in turn, may provide information about the gaseous products to be expected.

Decompositions of metal formates may be represented as:

$$M(HCOO)_2(s) \rightarrow M(s) + CO(g) + CO_2(g) + H_2O(g)$$

$$M(HCOO)_2(s) \rightarrow MO(s) + 2CO(g) + H_2O(g)$$

Secondary reactions may occur between the primary gaseous products and formulation of mechanisms requires consideration of the possible participation of processes such as the water-gas equilibrium:

$$H_2O(g) + CO(g) \rightleftarrows H_2(g) + CO_2(g)$$

For example, Fahrenfort *et al.* [42] considered the possible role of the intermediate formation of nickel formate in the nickel catalyzed decomposition of formic acid. They showed that when the volatile reaction products were rapidly removed, the carbon monoxide concentration was greater than that expected from the water-gas equilibrium. When the gaseous products remained in contact with the residual solid, containing catalytically active nickel metal, the $[CO]/[CO_2]$ ratio was that expected from the water-gas equilibrium. The composition of the *primary* products thus differed from the thermodynamic prediction. Such observations are important in formulating reaction mechanisms involving surface processes. Furthermore, correlation of the structures of primary products, as opposed to those expected at equilibrium, with the structures of reactant components can be used as evidence in identifying the bond-breaking step controlling the reaction rate.

Standard changes in the Gibbs energy for a proposed reaction at 298 K may be calculated from tables of standard Gibbs energies of formation, if a value for the reactant is available:

$$\Delta G^{\circ}_{\text{reaction}} = \Sigma(\Delta G^{\circ}_{\text{f,products}}) - \Delta G^{\circ}_{\text{f,reactant}}$$

The thermodynamically favoured path may thus be predicted for the particular conditions of reaction. To obtain values of $\Delta G^{\circ}_{\text{reaction}}$ at the decomposition temperature, use may be made of the data provided in graphical form by Ellingham [60] ("Ellingham Diagrams") for the variation of $\Delta G^{\circ}_{\text{f}}$ with temperature for various oxides, based on the assumption that $\Delta H^{\circ}_{\text{f}}$ and $\Delta S^{\circ}_{\text{f}}$ are independent of temperature. Dollimore [61] has illustrated the use of Ellingham diagrams in predicting the thermodynamic products of metal oxalate decompositions, as follows. The equilibrium constants for reactions (2.7) and (2.8) above are, respectively:

$$K_1 = [\text{M}]\,[\text{CO}_2]^2\,/[\text{MC}_2\text{O}_4]$$

and

$$K_2 = [\text{MO}]\,[\text{CO}]\,[\text{CO}_2]/[\text{MC}_2\text{O}_4]$$

The ratio of these two equilibrium constants:

$$K_2/K_1 = [\text{MO}]\,[\text{CO}]/[\text{M}]\,[\text{CO}_2] = K_3$$

is the equilibrium constant for the reaction:

$$\text{M(s)} + \text{CO}_2\text{(g)} \rightleftarrows \text{MO(s)} + \text{CO(g)} \qquad (2.10)$$

The problem thus becomes that of establishing which is the spontaneous direction of reaction (2.10) for the particular metal M under consideration. Reaction (2.10) can, in turn, be regarded as the difference of the reactions:

$$M(s) + \tfrac{1}{2}O_2(g) \rightleftarrows MO(s) \qquad (2.11)$$

and

$$CO(g) + \tfrac{1}{2}O_2(g) \rightleftarrows CO_2(g) \qquad (2.12)$$

so that

$$\Delta G^{\circ}_{(2.10)} = \Delta G^{\circ}_{(2.11)} - \Delta G^{\circ}_{(2.12)}$$

The Ellingham diagrams for reaction (2.12) and three examples of reaction (2.11) with M = Ni, Zn and Aℓ are shown in Figure 2.6. Note that the numerical values of ΔG°_{f} are calculated for 1 mole of O_2. It can be seen that the plot for the oxidation of nickel metal lies above that for the oxidation of CO at all temperatures, so that $\Delta G^{\circ}_{(2.10)}$ will be positive for all temperatures and Ni metal is thus the expected product. In contrast, the plot for the oxidation of zinc lies below that for the oxidation of CO, so that $\Delta G^{\circ}_{(2.10)}$ will be negative for all temperatures and hence ZnO will be the expected product. More comprehensive Ellingham diagrams [60] show that the main products of the decompositions of the carboxylates of copper, lead, nickel, cobalt and cadmium, in inert atmospheres or vacuum, are expected to be the metals, while zinc, manganese and tin carboxylates should give metal oxides. For more electropositive cations the residual product is expected to be a carbonate (equation (2.9)) and behaviour will be controlled by the stability of this solid product.

The Gibbs energy plot for the oxidation of iron:

$$Fe(s) + \tfrac{1}{2}O_2(g) \rightleftarrows FeO(s)$$

lies below that for the oxidation of CO up to 1000 K, where it crosses over, but it is only 17 kJ mol^{-1} above it at the decomposition temperature of iron(II) oxalate (about 600 K). The product of this decomposition in inert atmospheres is found to be a mixture of metal and metal oxides, with the composition being very dependent upon the experimental conditions, especially the partial pressures of the gaseous products present during reaction [62,63].

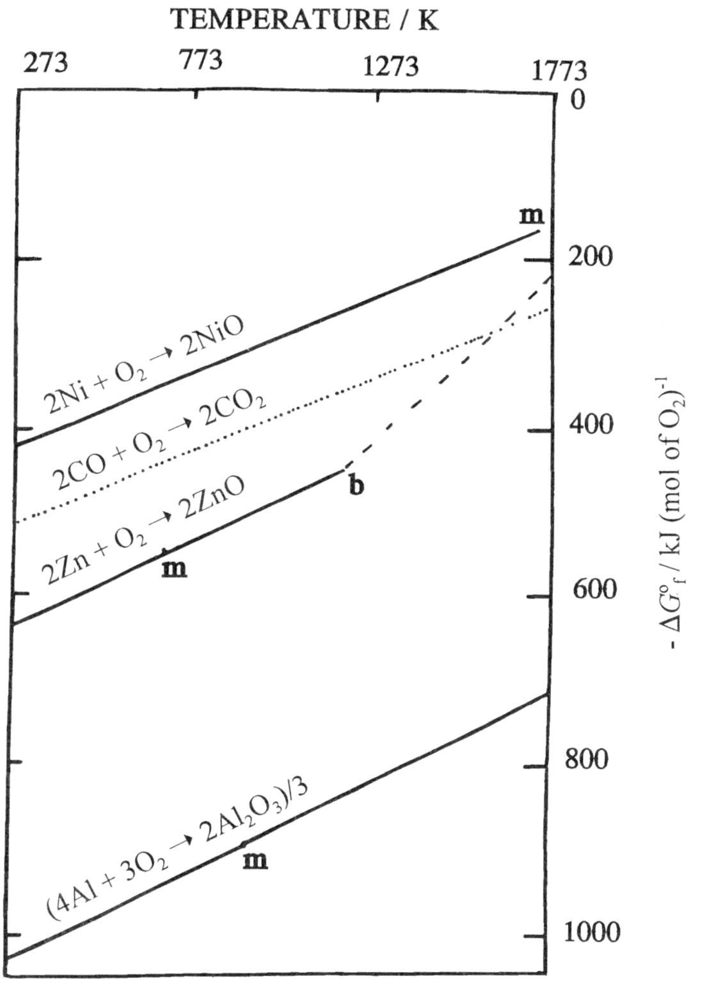

Figure 2.6.
The Ellingham diagram for the reactions:

$M(s) + \frac{1}{2}O_2(g) \rightleftarrows MO(s)$ (2.11)

(with M = Ni, Zn and Aℓ)and

$CO(g) + \frac{1}{2}O_2(g) \rightleftarrows CO_2(g)$ (2.12)

(m indicates melting and b boiling temperatures).

For trivalent metal oxalates the most probable reactions are either:

$$M_2(C_2O_4)_3(s) \rightleftarrows 2M(s) + 6CO_2(g)$$

or

$$M_2(C_2O_4)_3(s) \rightleftarrows M_2O_3(s) + 3CO_2(g) + 3CO(g)$$

(although iron(III) oxalate gives iron(II) oxalate as an intermediate in the overall decomposition [62]). The treatment for the decompositions to oxide is similar to that above, except that ΔG° for the reaction:

$$\{4M(s) + 3O_2(g) \rightleftarrows 2M_2O_3(s)\}/3$$

must be compared with that for the oxidation of CO (reaction (2.12)). The Gibbs function plot for the oxidation of $A\ell$ (Figure 2.6.) lies far below that for CO, so that ΔG° for the reaction:

$$2A\ell(s) + 3CO_2(g) \rightleftarrows A\ell_2O_3(s) + 3CO(g)$$

is negative, and hence the reaction is thermodynamically feasible, at all temperatures. Chromium(III) oxalate also gives the oxide Cr_2O_3 as product, but antimony(III) oxalate gives the metal.

Andrew [64], in a review of the pitfalls encountered in applying thermodynamic data to inorganic chemical processes, uses as an example the reaction:

$$2CO(g) \xrightarrow[\text{catalyst}]{\text{Ni}} CO_2(g) + C(s)$$

and points out that the value of the standard Gibbs energy of formation for C(s) depends upon the temperature of deposition of the carbon. Above 870 K the deposit differs little from graphite. At lower temperatures (670 K) the deposit is, however, almost amorphous so that the magnitude of the equilibrium "constant", $\{p(CO_2)\}/\{p(CO)\}^2$, is approximately a tenth of that appropriate for graphite. Similar problems are to be expected when dealing with other finely divided and defective solid decomposition products. The role of the atmosphere in influencing the decompositions of oxysalts has been discussed by Dollimore [65].

2.6.3. Decompositions of metal perchlorates

Markowitz [66] has discussed the prediction of the products of the thermal decomposition of metal perchlorates. The major final solid products may be either the chloride or the oxide:

$$M(C\ell O_4)_n(s) \rightarrow MC\ell_n(s) + 2nO_2(g) \tag{2.13}$$

$$M(C\ell O_4)_n(s) \rightarrow \tfrac{1}{2}M_2O_n(s) + \tfrac{1}{2}nC\ell_2(g) + 1.75nO_2(g) \tag{2.14}$$

There may be simultaneous or preliminary dehydration processes for hydrated salts. Group I metal perchlorates, as well as calcium, barium and silver perchlorates [59] are represented by reaction (2.13). Aluminium and iron(III) perchlorates are represented by reaction (2.14). The solid product of decomposition of magnesium perchlorate is a mixture of magnesium oxide and magnesium chloride.

The relative values of the standard changes in Gibbs energy, $\Delta G^{\ominus}$, for reactions (2.13) and (2.14) will depend upon the values of $\Delta G^{\ominus}_{formation}$ of the solid chloride and the equivalent amount of the solid oxide. Decomposition thus tends to proceed in the direction yielding the solid product which has the lower value of $\Delta G^{\ominus}_{formation}$. The values of $\Delta G^{\ominus}_{formation}$ for $MgC\ell_2$ and MgO at 298 K are very similar (- 591.8 and - 569.4 kJ mol^{-1}, respectively) so that a mixture of solid products is formed. Reversible interconversion of these two products is possible through the reaction:

$$MgO(s) + C\ell_2(g) \rightleftarrows MgC\ell_2(s) + \tfrac{1}{2}O_2(g)$$

In principle, if it were possible to control accurately the ratio of the partial pressures, $\{p(O_2)\}^{1/2}/\{p(C\ell_2)\}$, the solid reaction product could be predetermined, but in practice the prevailing pressures at the reactant/product interface may differ appreciably from the controlled values.

Markowitz [66] also notes that the entropy changes for reactions (2.13) and (2.14) are similar for the majority of metal perchlorates. Hence the differences in $\Delta G^{\ominus}$ for reactions (2.13) and (2.14) arise from the differences in $\Delta H^{\ominus}$ for these two reactions. As discussed above, this difference will then be determined by the differences in the enthalpies of formation of equivalent amounts of the chloride and the oxide. Such values are often more accessible than values of $\Delta G^{\ominus}_{formation}$. The validity of this approach is confirmed [66] by a consideration of available data for the decomposition products of metal perchlorates.

2.7. REVERSIBLE SOLID STATE REACTIONS

Many solid state reactions are reversible. Examples include the endothermic dehydrations and carbonate dissociations [67] such as:

$$Ca(OH)_2(s) \rightleftarrows CaO(s) + H_2O(g)$$

$$CaCO_3(s) \rightleftarrows CaO(s) + CO_2(g)$$

Enthalpies of dissociation may be determined from measurements of the variation of the equilibrium pressure of the gaseous product with reaction temperature [60]. In the study of the kinetics of these reactions (Chapters 8 and 12) consideration must be given to the possible influence of the reverse reaction on the rate measurements [68]. Kinetic parameters should be measured at very low pressures of H_2O or CO_2 in the reaction vessel [45]. At higher pressures, equilibria may be established within the pores of the solid product. This is given as the explanation for the frequent observation that the value of the enthalpy of dissociation of a particular carbonate is close to the value of the activation energy measured for evolution of CO_2 [45].

2.8. EXPERIMENTAL DETERMINATION OF THE STOICHIOMETRY AND EXTENT OF REACTION

2.8.1. Introduction

Reaction kinetics is concerned with rates of chemical transformations and thus experimental measurements must be of parameters which can be accurately related to changes in extent of a defined and characterized rate process, or set of rate processes. Yield-time data can only be significant if the stoichiometry of the reaction of interest is established, including analytical confirmation of the identities of the products, as well as any possible reaction intermediates.

The experimental methods selected for use will depend on the complexity of the reaction being studied. Decompositions that yield a single gaseous product are most easily studied and investigations of reactions of this type have contributed most to the development of this subject. A wide variety of experimental methods can be used, including measurements of mass loss, gas evolution, heat absorption or evolution, X-ray diffraction changes, etc. Even in such relatively simple reactions, allowance may have to be made for reactions with gases (such as water or oxygen) in the surroundings, including allowance for the reversibility of the decomposition itself.

Decompositions which yield *no* volatile products, including phase transformations, may be the simplest reactions in principle, but experimental methods are more limited (e.g. heat absorption or evolution, X-ray diffraction) and interpretations of results are less straightforward. The occurrence of phase transformations, including recrystallization of amorphous reactants or products, and small but significant contributions from hydrolysis by residual water, may easily be overlooked.

When decomposition is evidently not simple, with two or more processes being observed to overlap, sufficient information must be obtained to enable the contributions from each of the participating reactions, including secondary reactions between products, to be distinguished. Use of mass spectrometry or Fourier transform infrared spectroscopy coupled with thermogravimetry (see below) allows the evolution of individual gaseous species to be measured as functions of time or temperature.

Conventional kinetic studies are carried out under *isothermal* conditions, but the kinetic analyses based on experiments during which the reactant is subjected to a *programmed temperature* variation have provided additional interesting possibilities.

Whatever technique is chosen from those available [69], the reliability of the observations and conclusions can always be increased by complementary and supplementary measurements devised by the researcher for the problem in hand.

2.8.2. Measurements of accumulated gas pressure

Many of the early studies of the kinetics of decomposition of solids were carried out by measuring the pressure of product gases, accumulated in an initially evacuated constant-volume apparatus, as a function of time at constant temperature. A wide range of pressure sensors is available. The sample size, apparatus volume and gauge range have to be within compatible limits. Knowledge of the volume of the system is necessary for stoichiometric calculations. Limited product analyses can be achieved by the use of suitable refrigerant traps to condense particular gaseous products. The responses of the gauge used to the individual and combined constituents of gas mixtures have to be determined. Yields may have to be corrected for non-ideality of the gases, possible adsorption processes [70] and, as the total pressure increases, of contributions from reverse reactions.

2.8.3. Evolved gas analysis

The above method can be extended and selectivity increased by use of any analytical technique capable of distinguishing and measuring yields of individual products. Most of the routine methods of gas analysis [71-75] have been used to determine the compositions of gaseous products from solid state decompositions. The sophistication of the analysis may range from a single determination of the amount and composition of the accumulated gas at the end of reaction, to the continuous or frequent monitoring of individual gaseous products with temperature or with time. These profiles can be used for the determination of kinetic information.

When the gaseous products are simple and can be readily predicted from the composition of the reactant, e.g. nitrogen from metal azides or oxygen from metal permanganates, it may be possible to determine the overall stoichiometry from simple gas-law calculations from pressures of gas collected under calibrated conditions. Even such systems, however, may yield atomic or ionic, rather than molecular, species as intermediates [76] and these require more specific methods of detection.

In all gas analysis systems, precautions have to be taken against, or allowance made for, the condensation of vapours in the sampling system and/or the occurrence of secondary reactions between gaseous primary products, sometimes catalyzed by the surfaces of the reaction vessel or sampling system.

Many commercial thermobalances (see below) are supplied with evolved gas analysis (EGA) facilities, usually either a coupled quadrupole mass spectrometer (QMS) or a Fourier transform infrared (FTIR) spectrometer [74,75]. Mass spectrometry (MS) is versatile, fast and conclusive in its results, but instruments are expensive and at least part of the system has to operate under high vacuum. The high sensitivity of MS makes it a suitable technique for measuring rates of product evolution during the early stages of reaction of small samples, possibly even single crystals. Measurements during these early stages can provide information on the chemical steps participating in the processes of nucleation and of early growth. The occurrence of solid-phase transitions or melting may also release gas or solvent molecules trapped during crystallization, (emanation thermal analysis (ETA) [77] and thermoparticulate analysis (TPA) [78]).

Where a limited range of gaseous products is to be expected, e.g. the decomposition of metal sulfates or metal carbonates, detectors for specific gases may be incorporated into the experimental system [79].

Analysis by gas chromatography (GC), although it involves relatively inexpensive equipment, has several disadvantages in that monitoring of products cannot be continuous, because time has to be allowed for the component of the gaseous mixture with the longest retention time to be eluted from the column before the next sample is introduced. GC peaks have to be identified through comparison with samples of known gases or gas mixtures. Quantitative analysis is more difficult.

2.8.4. Thermogravimetry

Thermogravimetry (TG), one of the major techniques of thermal analysis [80-83], measures changes in mass during either programmed temperature or isothermal heating of the sample. Programmed temperature measurements provide a rapid means of determining the temperature range within which the reactant decomposition proceeds at a measurable rate and of determining whether decomposition occurs in one or more distinct steps. The overall mass-loss measured during decomposition, or the mass-losses after each of the separable steps, provide immediate information concerning the stoichiometry of the reaction. For kinetic measurements, the equipment is often used in the isothermal mode, although it is possible (see Chapter 5) to obtain kinetic information from programmed-temperature experiments. TG and EGA (see above) are often coupled to provide a means of characterization of the decomposition products.

The simplest balance mechanism, readily adaptable to decomposition studies, is that based on measurement of the extension of a quartz spiral spring. Quartz crystal microbalances can be used for very sensitive measurements [84] (e.g. on single crystals) but the temperature range is limited.

The material and the size and shape of the sample container, as well as the disposition of the sample within it, may all influence the rate of decomposition. TG equipment is well suited to studies of the influences of controlled pressures of gases (including volatile products) on reaction rates.

Complementary measurements of magnetic transitions using *thermomagnetometry* (which requires a thermobalance and an applied magnetic field) can provide useful information on the oxidation states of intermediates formed during decomposition of appropriate samples.

2.8.5. Differential scanning calorimetry (DSC) or differential thermal analysis (DTA)

Two other major techniques of thermal analysis, DSC and DTA, may be used to measure quantitatively the progress of reaction [80-83,85] by measuring the heat absorbed or released during reaction. These techniques can also detect other heat changes arising as a result of melting or solid-solid transitions and may also be coupled with EGA (see above). The DSC signal, $\mathrm{d}q/\mathrm{d}t$, where q is the energy transferred to or from the sample relative to the inert reference material, is usually assumed to provide a measure of the reaction rate, $\mathrm{d}\alpha/\mathrm{d}t$, and is used in methods of kinetic analysis (Chapter 5). Valuable information on possible changes of mechanism may be obtained by comparison of the extents of reaction calculated from different experimental techniques, e.g. TG and DSC or DTA.

2.8.6. Microscopic examinations

Direct examination, using optical or electron microscopy, of the textural changes that occur during decompositions, can provide conclusive evidence on the nature and rates of processes such as the formation, disposition and growth of nuclei; the changing geometry of a reaction interface; the occurrence of melting, etc. Ideally such measurements require a heated sample stage in a controlled atmosphere but, in practice, samples of partially decomposed reactant (at a range of known α values) and the residual solid obtained after complete decomposition, are examined on the assumption that quenching to the observation temperature has not caused significant textural changes. Care must be exercised to avoid deterioration of samples between preparation and examination.

Fracture of partially decomposed crystals before examination [86] can reveal internal details, possibly more representative of the sample as a whole than the information obtained from the relatively limited amount of surface material. Replication (using polymer film) can be used to examine surfaces that would be unstable in the microscope and can reveal internal crack or pore structures in suitable materials [86].

2.8.7. X-ray diffraction methods

The measurements of mass-loss, together with some knowledge of the chemical characteristics of the reactant and of the thermodynamic factors applicable (Section 2.6 above), can only suggest or eliminate possibilities for the constituents which are retained in the solid product(s). It is essential to identify the phases present in the residue more precisely, usually by X-ray diffraction or spectroscopic methods. Problems arise if the product is non-crystalline and/or strongly absorbing of radiation such as ultraviolet or infrared. Such examinations can be used to confirm the identities of both the reactant and the product phases and should be extended to include samples reacted to a range of α values to determine whether any crystalline intermediates participate. The possibility that reaction may be topotactic (see Section 2.4.8. above) should also be considered.

In addition to the difficulties of possible lack of crystallinity, there is nearly always the problem that the solid phase(s) examined are those existing after the system has been quenched to room temperature (unless sophisticated simultaneous heating and examination systems are available [87-89]). The residues from many solid state decompositions are finely-divided, reactive and may even be pyrophoric. If such residues are exposed to the laboratory atmosphere, they will usually differ considerably from the product formed in the reaction environment.

The results of analyses of both gaseous and solid product(s) need to be related to all the analytical information available on the reactant to ensure a consistent overall elemental balance when reporting the reaction stoichiometry.

2.9. TEMPERATURE MEASUREMENT, CONTROL AND CALIBRATION

The reliability and precision of all kinetic data collected depend sensitively on the constancy of the temperature within reactant sample and the reaction vessel that contains it. The control of the temperature of the reaction vessel by the furnace must include due allowance for the heat evolved or absorbed by the sample during the decomposition process (see below). Numerous designs of furnaces have been used.

An ingenious arrangement, suitable for intermediate temperatures in glass systems, is the double-boiler system suggested by Prout and Herley [90]. For high temperatures (above about 1200 K) special and expensive construction materials are required.

Perturbation of the sample temperature occurs on introduction of the reactant together with any container into the heated zone and due consideration of the time and magnitude of these changes must be incorporated into kinetic analyses. The endothermic contribution from an initial dehydration reaction may delay the onset of a subsequent decomposition which itself may either liberate or absorb heat.

Self-heating, i.e. the generation of temperatures within the bulk of the sample that are significantly greater than the controlled reaction environment, is particularly important in the very exothermic decompositions of explosives [91,92].

The heat balance equation in a reacting solid is:

rate of heat accumulation in crystal	=	rate of heat production by reaction	-	rate of heat loss by radiation	-	rate of heat loss by conduction	-	rate of heat loss by convection

The last term (convection) is dependent upon the gaseous atmosphere in which decomposition takes place and will approach zero if the gaseous products are removed by continuous pumping. It is usually assumed that, during decomposition of single crystals, the crystal is effectively isolated from its support by the product gases released, making the conduction contribution zero, but for powdered samples packed in metal decomposition vessels, conduction may be significant. While the crystal undergoes exothermic decomposition, the temperature of the sample, T, may rise appreciably above that of the furnace, T_f. When T is greater than T_f, the sample will lose heat to the surroundings. If it is assumed that reaction is confined to the surface of the sample (area, s), then the rate of heat production by decomposition is:

$$dQ_1/dt = s\, Q\, A \exp(-E_a/RT)$$

where $Q = \Delta H$, the enthalpy of decomposition, and A and E_a are the Arrhenius parameters for the reaction. The rate of loss of heat by radiation is:

$$dQ_2/dt = s\, \epsilon\, \sigma (T^4 - T_f^4)$$

where ϵ is the emissivity of the surface and σ is Stefan's constant (5.670×10^{-8} J m^{-2} K^{-4} s^{-1}). Heat losses due to convection of the surrounding gas are dependent upon the gas pressure, the temperature difference ($T - T_f$), the surface exposed to the gas and the coefficient of heat transfer, h (units: J m^{-2} s^{-1} K^{-1}).

2.10. THERMOCHEMICAL MEASUREMENTS

Using differential scanning calorimetry (DSC) (or, less directly, differential thermal analysis (DTA)) (see Section 2.8.5., above) it is possible to measure several of the thermodynamic properties of solids and of solid state reactions. The DSC response is directly proportional to the heat capacity, C_p, of the sample, so that by use of a calibrant it is possible to obtain values of this fundamental thermodynamic property, at a particular temperature, or as an average over a specified temperature range. Other thermodynamic properties are readily derived from such measurements:

$$\Delta H_{(T1-T2)} = \int_{T_1}^{T_2} C_p \, dT$$

and

$$\Delta S_{(T1-T2)} = \int_{T_1}^{T_2} (C_p/T) dT = \int_{T_1}^{T_2} C_p \, d\ln T$$

Using these relationships, the enthalpy and/or entropy changes associated with the occurrence of phase transitions, melting, sublimation or decomposition of the sample can be determined. Such processes generally result in considerable changes in the heat capacity of the sample, which appear on the DSC trace as marked deviations from the extrapolated baseline.

The values of the enthalpy changes for the processes that have occurred on heating, are obtained by measurement of the area under the response curves. The calibration factor for the instrument concerned is usually determined by melting a weighed sample of a pure metal of known enthalpy of fusion, and comparing the response area obtained with that observed for the sample under similar conditions. The major advantage of DSC over DTA is that the calibration factor is virtually independent of temperature. For DTA, the variation of the calibration factor with temperature has to be determined. Computer-controlled instruments usually incorporate this correction in the software or hardware. Changes in heat capacity make the determination of the baseline for integration more difficult. Brennan *et al.* [93] have suggested procedures for constructing the baseline. The enthalpy values obtained are average values over the temperature range from peak onset to reestablishment of baseline, and this temperature interval is usually considerably removed from the usual reference temperature of 298 K used for most tabulated data. Corrections, which are usually small, can be made using the Kirchhoff equation:

$$\Delta H_{T2} = \Delta H_{T1} + \int_{T_1}^{T_2} \Delta C_p \, dT$$

if heat capacity values are available or can be determined.

Measurements of the entropy change, ΔS, accompanying a thermal event in the sample, requires determination of the C_p values and hence evaluation of the integral:

$$\int_{T_1}^{T_2} (C_p / T)\, \mathrm{d}T$$

Such measurements have proved to be of considerable value [94] in studying the ordering or disordering accompanying phase transitions in polymers and in coordination compounds. The glass transition [94] is not accompanied by an enthalpy change and appears as a discontinuity in the heat capacity of the sample, as recorded by the baseline trace of the DSC.

In formulating mechanisms of solid state decompositions it is important to consider the possible occurrence of melting (Section 2.4.3., above). Melting usually enhances the rate of decomposition because the reactant constituents are no longer subject to stabilizing crystal forces. Overall melting is usually readily detected, but small, localized contributions from melting are often difficult to confirm or demonstrate to be absent. Examination of DSC or DTA responses may provide some information.

2.11. COMPARISON OF THERMOCHEMICAL AND KINETIC PARAMETERS

The magnitude of the enthalpy change, ΔH, measured for the process which has taken place in the sample, can be readily converted, from the original sample mass, to a nominal mole of reactant. Such enthalpy changes (after correction to the standard temperature, as above) can be compared with tabulated values, or may be used to calculate previously unknown enthalpies of formation. It may prove of interest to compare the molar enthalpy of decomposition with the apparent activation energy, E_a, obtained from kinetic measurements (Chapter 5). The significance of the activation energy for solid state reactions is a matter of continuing controversy (see Chapter 4) and requires an extension of the theory. However, formally the unit expressing this parameter contains a nominal mole which, presumably, refers to the amount of material involved in the high energy configuration which appears to be widely regarded as the equivalent of the activated complex in homogeneous reactions [52,95]. Enthalpy measurements for endothermic decompositions and E_a values can then be compared on the conventional energy profile representing the course of reaction (Figure 2.7.). The activated state is assumed to be identical for the forward and reverse processes. This generalization may apply to homogeneous reactions, but need not hold for the decompositions of solids, where the precursors to change may arise within the different crystal structures of the reactant and product. Little is known about these possible differences.

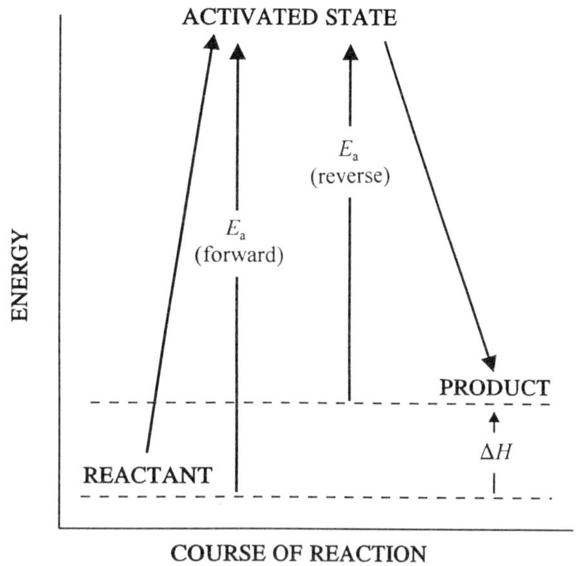

Figure 2.7.
Schematic energy profile during the course of an endothermic reversible reaction.

The minimum value of E_a for an endothermic process is equal to ΔH. This corresponds to an activation energy of zero for the reverse reaction. The evaporation of a liquid is an example of such a process and the decompositions of some carbonates are believed to show this behaviour. If, however, the measured value of the activation energy turns out to be *less* than the calculated value of ΔH, the possibilities are that the mechanism of decomposition is complex and that ΔH does not refer to the rate determining step, or that the "mole" referred to in the scaling of ΔH does not correspond to a mole of material in the activated state. By adjustment of the value of ΔH to refer to some smaller amount of substance, using the (admittedly limited) criterion that E_a must be greater than, or perhaps equal to ΔH, an estimate may be obtained of the amount of substance corresponding to a mole of material in the activated state. For example, if the enthalpy of removal of a bidentate ligand from a coordination compound, calculated per mole of ligand, yields a value between one and two times higher than the measured activation energy, then the activated state might involve breaking of a single ligand coordinate bond. Such conclusions are of course speculative but receive some support from spectroscopic measurements (Chapter 6).

2.12. CONCLUSIONS

Conclusions based on the widest possible variety of experimental observations provide the most reliable mechanistic interpretations. The necessity for quantitative characterisation of the reactant, the products and any intermediates has been emphasised. This characterisation includes the chemical composition, impurities, imperfections, crystal sizes, lattice structures and physical textures of all participating phases and the determination of their roles in the reactions investigated. Any study directed towards determining the detailed chemical mechanism of a particular reaction must establish the stoichiometry and identify the *initial* products, which may differ from those at equilibrium. These products may be compared with those predicted from thermodynamic considerations.

The choice made from the variety of experimental techniques described above and in Chapters 7 to 17, dealing with individual reactants, requires that the parameter(s) measured are quantitatively related to the extent of a defined reaction (or set of reactions). Use of two (or more) complementary techniques, as well as microscopic observations, increase the reliability of the reaction mechanism which is formulated by reconciling *all* the experimental observations available.

REFERENCES

1. J. Rouquerol, Thermochim. Acta, 144 (1989) 209.
2. A.R. West, Solid State Chemistry and its Applications, Wiley, Chichester, 1984, Chap.2.
3. E.S. Halberstadt and H.K.Henisch, J. Cryst. Growth, 3,4 (1968) 363; H.K. Henisch, J. Dennis and J.I. Hanoka, J. Phys. Chem. Solids, 26 (1965) 493.
4. V.V. Boldyrev, M. Bulens and B.Delmon, The Control of the Reactivity of Solids, Elsevier, Amsterdam, 1979.
5. I.D. Begg, P.J. Halfpenny, R.M. Hooper, R.S. Narang, K.J. Roberts and J.N. Sherwood, Proc. R. Soc. (London), A386 (1983) 431; Phil. Mag., A46 (1982) 559; J. Mater. Sci., 18 (1983) 81.
6. E.K. Powell and A.W. Searcy, Metall. Trans., 11B (1980) 427; W.E. Garner and H.V. Pike, J. Chem. Soc., (1937) 1565; W.E. Garner, Chemistry of the Solid State, Butterworths, London, 1955, Chap.8.
7. F. Solymosi, Structure and Stability of Salts of Halogen Oxyacids in the Solid Phase, Wiley, London, 1977, p.196, 123-4.
8. S.J. Peppiatt, Proc. R. Soc. (London), A345 (1975) 401.
9. A.K. Ardell, Mechanisms of Phase Transformations in Crystalline Solids, Institute of Metals, Monograph and Report Series, No. 33 (1969) 111.

10. R.V. Day and J. Barford, Nature (London), 217 (1968) 1145.
11. A.R. West, Solid State Chemistry and its Applications, Wiley, Chichester, 1984, Chap.12; M.J. Buerger, Fortschr. Miner., 39 (1961) 9; Soviet Physics, Crystallography, 16 (1972) 959.
12. C.N.R. Rao and J. Gopalakrishnan, New Directions in Solid State Chemistry, Cambridge University Press, Cambridge,1986, Chap.4.
13. J.W. Christian, Transformations in Metals and Alloys, Pergamon, Oxford, 1965.
14. C.N.R. Rao and K.J. Rao, Phase Transitions in Solids, McGraw-Hill, New York, 1978.
15. J. Burke, The Kinetics of Phase Transformations in Metals, Pergamon, Oxford, 1965.
16. L. Nagel and M. O'Keefe, International Review of Science, Inorganic Chemistry, Vol.10, (Ed. L.E.J. Roberts) Series 1, Chap.1, Butterworths, London, 1972.
17. N.H. Hartshorne and M.H. Roberts, J. Chem. Soc., (1951) 1097.
18. M. Maciejewski, H.-R. Oswald and A. Reller, Thermochim. Acta, 234 (1994) 315.
19. A.R. Ubbelohde, Melting and Crystal Structure, Clarendon, Oxford, 1965; The Molten State of Matter, Wiley, Chichester, 1978.
20. F.A. Lindemann, Z. Phys., 11 (1910) 609.
21. J.E. Lennard-Jones and A.F. Devonshire, Proc. R. Soc. (London), A169 (1939) 317; A170 (1939) 464.
22. D. Kuhlmann-Wilsdorf, Phys. Rev., 140 (1965) 1599.
23. V.A. Slyusarev and A.I. Burakhovich, see: Chem. Abs., 80 (1974) 31228 m.
24. T. Gorecki, Z. Metallk., 65 (1974) 426.
25. K. Honda, Prog. Theor. Phys., 52 (1974) 385.
26. H. Mori, H. Okamoto and S. Isa, Physica, 73 (1974) 237.
27. T. Riste, Anharmonic Lattices, Structural Transitions and Melting, Noordhoft, Leiden, The Netherlands, 1974.
28. T.G. Coker, B. Wunderlich and G.J. Janz, Trans. Faraday Soc., 65 (1969) 3361.
29. A.R. Allnatt and S.J. Sime, Trans. Faraday Soc., 67 (1971) 674.
30. T. Clark, M.A. McKervey, H. Mackle and J.J. Rooney, J. Chem. Soc., Faraday Trans. I, 70 (1974) 1279.
31. S.J. Peppiatt and J.R. Sambles, Proc. R. Soc. (London), A345 (1975) 387.
32. M. Blackman, N.D. Lisgarten, S.J. Peppiatt and M.S. Rahman, Nature (London), 258 (1975) 139.
33. A.J.E. Welch, Chemistry of the Solid State (Ed. W.E. Garner), Butterworths, London, 1955, p.307.

34. J.E. Burke, Chemical and Mechanical Behaviour of Inorganic Materials (Eds A.W. Searcy, D.V. Ragone and U. Colombo), Wiley-Interscience, New York, 1970.
35. J.E. Burke and J.H. Rosolowski, Treatise on Solid State Chemistry (Ed. N.B. Hannay), Plenum, New York, 1976, Vol.4, Chap.10.
36. A.W. Searcy and D. Beruto, Sci. Ceram., 14 (1988) 1.
37. H. Schmalzried, Solid State Reactions, Verlag Chemie, Weinheim, 1981, p.205.
38. W.D. Kingery, H.H. Bowen and D.R. Uhlmann, Introduction to Ceramics, Wiley, New York, 2nd Edn, 1976, Chap.10.
39. P.W. Gilles, J. Amer. Ceram. Soc., 58 (1975) 279.
40. G.A. Somorjai and J.E. Lester, Prog. in Solid State Chem., 4 (1967) 1; Adv. High Temp. Chem., 2 (1969) 203.
41. J.L. Gole, Ann. Rev. Phys. Chem., 27 (1976) 525.
42. J. Fahrenfort, L.L. van Reyen and W.M.H. Sachtler, The Mechanism of Heterogeneous Catalysis, (Ed. J.H. de Boer), Elsevier, Amsterdam, 1960, p.23.
43. B.V. L'vov and A.V. Novichikhin, Thermochim. Acta, 290 (1997) 239; 315 (1998) 135,145; B.V. L'vov, Spectrochim. Acta B, 52 (1997) 1.
44. B.V. L'vov *et al.* ; Spectrochim.Acta, B, 36 (1981) 397; Zh.Anal.Khim., 36 (1981) 2085; 39 (1984) 221, 1206; 40 (1985) 792.
45. D. Beruto and A.W. Searcy, J. Chem. Soc., Faraday Trans.I, 70 (1974) 2145.
46. T. Darroudl and A.W. Searcy, J. Phys. Chem., 85 (1981) 3971.
47. G.C. Bond, Catalysis by Metals, Academic, London, 1962.
48. A.K. Galwey, P. Gray, J.F. Griffiths and S.M. Hasko, Nature (London), 313 (1985) 668.
49. A.K. Galwey, Proc. R. Soc. (London), A271 (1963) 218; Geochim. Cosmochim. Acta, 36 (1972) 1115.
50. A.K. Galwey, Adv. Catal., 26 (1977) 247.
51. A.K. Galwey, Thermochim. Acta, 294 (1997) 205; A.K. Galwey and M.E. Brown, Thermochim. Acta, 300 (1997) 107.
52. A.K. Galwey and M.E. Brown, Proc. R. Soc. (London), A450 (1995) 501.
53. J.R. Günter and H.-R. Oswald, Bull. Inst. Chem. Res. Kyoto Univ., 53 (1975) 249.
54. V.V. Boldyrev, React. Solids, 8 (1990) 231.
55. H.-R. Oswald, Thermal Analysis, Proc. 6th Int. Conf., 1 (1980) 1.
56. H.-R. Oswald and J.R. Günter, React. Solids, Proc. 10th Int. Conf., (1984) 101.

57. A. Reller and H.-R. Oswald, J. Solid State Chem., 62 (1986) 306; Pure Appl. Chem., 61 (1989) 1323.
58. H.-R. Oswald and J.R. Günter, Crystal Growth and Materials, North-Holland, Amsterdam, 1977, p.416.
59. E.A. Prodan, J. Thermal Anal., 32 (1987) 1635; React. Solids, 4 (1988) 363.
60. H.J.T. Ellingham, J. Soc. Chem. Ind., 63 (1944) 125.
61. D. Dollimore, J. Chem. Soc., (1963) 2617.
62. M.A. Mohamed and A.K. Galwey, Thermochim. Acta, 213 (1993) 269, 279.
63. D. Broadbent, D. Dollimore and J. Dollimore, J. Chem. Soc. A, (1967) 451.
64. S.P.S.Andrew, The Industrial Use of Thermochemical Data, Chem. Soc. Spec. Tech. Publ. No. 34, (Ed. T.I. Barry), Chem. Soc., London, 1980, p.91.
65. D. Dollimore, J. Thermal Anal., 11 (1977) 185.
66. M.M. Markowitz, J. Inorg. Nucl. Chem., 25 (1963) 407.
67. P.E. Halstead and A.E. Moore, J. Chem. Soc., (1957) 3873.
68. S. Rajam and A.K. Galwey, J. Chem. Soc., Faraday Trans. I, 78 (1982) 2553.
69. M.E. Brown, D. Dollimore and A.K. Galwey, Reactions in the Solid State, Comprehensive Chemical Kinetics, Vol.22, (Eds C.H. Bamford and C.F.H. Tipper), Elsevier, Amsterdam, 1980, Chap.2.
70. M.G. Burnett, A.K. Galwey and C. Lawther, J. Chem. Soc., Faraday Trans., 92 (1996) 4310.
71. W. Lodding, Gas Effluent Analysis, Edward Arnold, London, 1967.
72. H. Hachenburg, Industrial Gas Chromatographic Trace Analysis, Heyden, London, 1973.
73. B. Thompson, Fundamentals of Gas Analysis by Gas Chromatography, Varian, Palo Alto, 1977.
74. P.J. Haines, Thermal Methods of Analysis, Blackie, London, 1995, Chap.5.
75. S.B. Warrington, Thermal Analysis - Techniques and Applications, (Ed. E.L. Charsley and S.B. Warrington), Royal Society of Chemistry, Cambridge, 1992, p.84.
76. B.V. L'vov, Thermochim. Acta, 291 (1997) 179.
77. V. Balek and J. Tölgyessy, Emanation Thermal Analysis, Wilson & Wilson's Comprehensive Analytical Chemistry, (Ed. G. Svehla), Vol. XII, Part C, Elsevier, Amsterdam, 1984.
78. E. Kaiserberger (Ed.), Special Issue: Coupling Thermal Analysis and Gas Analysis Methods, Thermochim. Acta, 295 (1997) p.1-186.
79. D.J. Morgan, J. Thermal Anal., 12 (1977) 245.
80. M.E. Brown, Introduction to Thermal Analysis, Chapman and Hall, London, 1988.

81. W.W. Wendlandt, Thermal Analysis, 3rd Edn., Wiley, New York, 1986.
82. P.J. Haines, Thermal Methods of Analysis, Blackie, London, 1995.
83. E.L. Charsley and S.B. Warrington, Thermal Analysis - Techniques and Applications, Royal Society of Chemistry, Cambridge, 1992.
84. V.B. Okhotnikov and N.Z. Lyakhov, J. Solid State Chem., 53 (1984) 161.
85. G. Höhne, W. Hemminger and H.-J. Flammersheim, Differential Scanning Calorimetry, Springer, Berlin, 1996.
86. A.K. Galwey and M.A. Mohamed, Thermochim. Acta, 121 (1987) 97.
87. N. Gérard, G. Watelle-Marion and A. Thrierr-Sorel, Bull. Soc. Chim. Fr., (1968) 4367; J.-C. Mutin, G. Watelle-Marion, Y. Dusausoy and J. Protas, Bull. Soc. Chim. Fr., (1972) 4498.
88. W. Engel, N. Eisenreich, M. Hermann and V. Kolarik, J. Thermal Anal., 49 (1997) 1025.
89. I. Hatta, H. Takahashi, S. Matuoka and Y. Amemiya, Thermochim. Acta, 253 (1995) 149.
90. E.G. Prout and P.J. Herley, J. Chem. Ed., 37 (1960) 643.
91. A.R. Ubbelohde, Chemistry of the Solid State, (Ed. W.E. Garner), Butterworths, London, 1955, Chap.11.
92. M.E. Brown, Thermochim. Acta, 148 (1989) 521.
93. W.P. Brennan, B. Miller and J.C. Whitwell, Ind. Eng. Chem. Fundam., 8 (1969) 314.
94. B. Wunderlich, Thermal Analysis, Academic, Boston, 1990.
95. P.D. Garn, Thermochim, Acta, 135 (1988) 71; 160 (1990) 135.

Chapter 3

KINETIC MODELS FOR SOLID STATE REACTIONS

3.1. INTRODUCTION

Factors that have been identified as controlling the rates of most solid state reactions, alone or in combination, include the following.
(i) *Chemical reactions.* These are the bond redistribution steps by which the chemical constituents of the reactant are transformed into those of the product phase. These changes are often represented as electronic activation processes occurring at a reactant/product interface. This is an advancing zone of enhanced reactivity within which chemical reactions may be accompanied by structural and textural changes.
(ii) *Reaction geometry.* Chemical changes at the interface result in its systematic advance into the unchanged reactant. The reaction rate is generally assumed to be directly proportional to the aggregate areas of all contributing interfaces. Hence product yields vary with time according to the changing geometry of these interfaces as reaction proceeds.
(iii) *Diffusion.* Reaction rates may be influenced by the transportation of participants to or from the reaction interface.
Other types of rate control have been proposed, usually with reference to particular reactions. Some of these are mentioned in the later chapters concerned with the thermal behaviour of specific compounds.

Several kinetic models, based on quantitative consideration of the three factors above, have been developed and are discussed in this Chapter. Valuable reviews of the development of such models representing the way in which the extent of reaction, α, may be expected to vary with time, at constant temperature, have been given by Jacobs and Tompkins [1], Young [2], Delmon [3], Barret [4], Brown *et al.* [5], Sesták [6], Schmalzried [7] and, most recently, Jacobs [8].

The geometric models of solid state reactions are based upon the processes of *nucleation* and *growth* of product nuclei by interface advance. These processes are discussed individually in the next section, followed by a description of the ways in which these contributions are combined to give rate equations for the overall progress of reaction.

3.2. NUCLEATION AND GROWTH OF NUCLEI

Nucleation can be defined as "the initial establishment of a new and discrete product particle within the solid reactant" [9]. Two types of chemical change may be involved: (i) the chemical transformation of one or more constituents (e.g., ions or molecules) of the reactant into the constituents of the product; and (ii) recrystallization of reacted material to the lattice structure of the product phase. The result of nucleation, perhaps requiring several steps, is the generation of the active reactant/product interface at which reaction occurs preferentially during its subsequent advance into unchanged reactant as the nucleus grows.

Growth is thus the maintained reaction within the interface between the reactant and product phases. The zone within which chemical changes occur may be regarded as an advancing energetic wave, while components of the the reactant and of the residue undergo relatively minor displacements. Reasons for the enhanced reactivity at advancing interfaces are discussed below.

Jacobs and Tompkins [1,8] have discussed the energetics of nucleation. The product species formed initially will be in crystal lattice positions very close to where these were located in the original reactant. At a later stage, sufficient product species will have accumulated to allow a localized transformation to the normal structure of the solid product. If this critical accumulation is not achieved, the reverse of the initial decomposition step may occur with reformation of reactant. Galwey and Laverty [9] discuss the stages involved in the development of a nucleus and identify the *growth nucleus* as being generated when reacted material recrystallises into the product phase.

Recent studies of dehydration reactions (see Chapter 7) have shown that changes across reactant surfaces, identified with water loss, precede the appearance of growth nuclei. Nucleation may occur, therefore, within this modified boundary layer and, in some hydrates, a rapid onset of nucleation can follow exposure of partially dehydrated material to water vapour [10].

Identification of the chemical changes that contribute to nucleation is a matter of considerable experimental difficulty. The sites of locally enhanced reactivity at which nucleation occurs are changed beyond recognition by the formation of the product particle. Characterization of these changes presents a considerable challenge to the laboratory techniques available, particularly for a site embedded within a larger crystal, the boundary layer of which may have already undergone chemical or structural modifications [10].

Because reactions are often initiated at only a relatively small number of *nucleation sites* [9] it is concluded that these are zones of the reactant crystal which possess significantly enhanced local reactivity. The first stage in their development may involve restricted changes within the reactant phase leading to the generation of an *incipient nucleus*. Proceeding from this precursor to the *germ nucleus* is expected to require an investment of energy. This is regarded as the difficult step and perhaps includes recrystallization of reacted material to generate the product phase. Thereafter, the interface of the *growth nucleus* that appears advances by preferred chemical change at the reactant-product contact.

The relative ease of nucleation exerts an important influence on the kinetics of the overall reaction. When the nucleation step occurs infrequently, sometimes ascribed to a large activation energy, a sigmoid yield-time curve is observed. In contrast, if nucleation is rapid, all surfaces of the reactant are rapidly converted to product early in the reaction and the interface so formed advances inwards. Yield-time curves are then predominantly or entirely deceleratory.

3.3. KINETICS OF NUCLEATION

Nuclei are generated predominantly, perhaps exclusively, at or near crystal surfaces and the numbers participating at different times have been measured for some decompositions using microscopy. Several models for nucleus generation have been developed [1-6,8]. These include:

(i) single-step nucleation,
(ii) instantaneous nucleation,
(iii) linear nucleation,
(iv) multi-step nucleation, and
(v) branching nucleation.

Single-step nucleation requires the assumption (which is difficult to demonstrate for most reactions of interest) that the generation of a single molecule (atom, ion-pair, etc.) of product constitutes the establishment of a growth nucleus. The kinetics of single-step nucleation are based on the assumption that there are N_0 potential nucleus forming sites having equal probability of being converted into nuclei. The rate of nucleation among these equivalent sites is then a random, first-order rate process:

$$dN/dt = k_N (N_o - N)$$

where N growth nuclei are present at time t and k_N is the rate constant for nucleation. Integration with ($N = 0$, $t = 0$) leads to:

$$N = N_o[1 - \exp(- k_N t)]$$

and differentiation again gives the *exponential law of nucleation*:

$$dN/dt = k_N N_o \exp(- k_N t)$$

Instantaneous nucleation occurs for this model when k_N is *large*, so that all possible nuclei are formed at the onset of reaction: $N = N_o$ at virtually zero time and no further nuclei are generated during the subsequent reaction. When k_N is *small*, the rate of nucleation is approximately constant as reaction proceeds, because the number of sites, $(N_o - N)$, undergoes little change. This is known as the *linear law of nucleation*:

$$N = k_N N_o t$$

Multi-step nucleation. A different approach was provided by Bagdassarian [11] and developed by Allnatt and Jacobs [8,12]. It was assumed that several distinct steps are required to generate a growth nucleus. The steps could be either successive processes in the immediate vicinity of a particular initiated site, or the interactions of several mobile energetic participants (excitons, interstitial ions, etc.), with the product species presumably immobilized at the nucleation site.

Bagdassarian [11] considered the kinetics of the sequence of steps through which a *germ* nucleus accommodates p atoms (ions or molecules) of product and, on the attainment of this number, was converted to a *growth* nucleus. The rate coefficients for the addition of individual atoms up to the number, p, were assumed to be equal:

$$k_0 = k_1 = k_2 = ... = k_{p-1} = k_I$$

For the continued growth of assemblages of more than p atoms the rate coefficients were again equal, but greater in value, so that:

$$k_p = k_{p+1} = k_{p+2} = ... = k_g \text{ and } k_g > k_I$$

This results in an acceleratory nucleation process known as the *power law of nucleation* of the form:

$$dN/dt = C \eta t^{\eta - 1}$$

where C is a constant and η is an integer (see below).

A more general derivation of this model was provided by Allnatt and Jacobs [8,12]. The differential equation which expresses the number of germ nuclei, N_i, which have received i atoms is:

$dN_i/dt = k_{i-1} N_{i-1}(t) - k_i N_i(t)$

The limiting conditions which apply are:

$N_i = 0$ for $i < 0$; $N_i = 0$ for $i > 0$ for time $t = 0$;

$N_i = N_0[0]$ for $i = 0$ for time $t = 0$;

$N_i = 0$ for $i = 0$ for time $t = \infty$.

Integration, using the Laplace transformation and taking due consideration of the limits, gives:

$$N_i(t) = N_0[0]\, k_0\, k_1\, k_2 \ldots k_{i-1} \left\{ \left(\sum_{l=0}^{l=i} \exp(-k_l t) \right) / \left(\pi_{j=0}^{j=i} (k_j - k_l) \right) \right\} \quad (j \neq l)$$

If $k_i t << 1$, the simplifications which result are, for one step:

$N_1(t) = N_0[0]\, k_0\, t = K_1\, t$

for two steps:

$N_2(t) = N_0[0]\, k_0\, k_1\, t^2 = K_2\, t^2$

for η steps:

$N_\eta(t) = N_0[0]\, k_0 \ldots k_\eta\, t^\eta = K_\eta\, t^\eta$ and $dN_\eta/dt = K\, t^{\eta-1}$

These are the various forms of the *power law of nucleation.*

Branching nucleation. This process is based on the assumption that energy released from an exothermic reaction step may be capable of directly activating reaction in a neighbouring species, as in homogeneous chain reactions where short-lived, highly reactive intermediates are formed. It is accepted, however, that energy chains cannot be sustained in solid state reactions other than under conditions of burning or explosion. Energy will be dissipated to crystal vibrations more rapidly than it can be passed on to a neighbouring reactant constituent.

An alternative mechanistic representation [13] of autocatalytic behaviour [5,8,14] which results in a rate equation formally similar to the chain branching model, assumes that stresses produced by reaction cause cracking of the reactant crystal that increases the surface available for nucleation. These equations are of the form:

$$\mathrm{d}N/\mathrm{d}t = k_{\mathrm{N}} N_{\mathrm{o}} + k_{\mathrm{B}} N - k_{\mathrm{T}} N$$

where k_{N}, k_{B} and k_{T} are the rate coefficients for nucleation (≡ chain initiation), branching and termination of growth nuclei, respectively. These qualitatively represent crack divergence, leading to maintained nucleation on the new surfaces exposed, and loss of reactant to product formed. Replacement of $k_{\mathrm{B}} - k_{\mathrm{T}}$ by k_2 gives:

$$\mathrm{d}N/\mathrm{d}t = k_{\mathrm{N}} N_0 + k_2 N$$

which, on integration between the limits 0 and t and 0 and N, gives:

$$N = (k_{\mathrm{N}} N_0 / k_2)\{\exp(k_2 t) - 1\}$$

and the dependence of nucleation rate on t is:

$$\mathrm{d}N/\mathrm{d}t = k_{\mathrm{N}} N_0 \exp(k_2 t)$$

Branching nucleation forms the starting point for a derivation of the Prout-Tompkins reaction model [13] (see below). The main rate equations for nucleation are summarized in Table 3.1. and illustrated in Figure 3.1.

Kinetics of nucleation may be studied by microscopic examinations of surfaces from which numbers of growth nuclei/unit area (N) are counted after known reaction times. Photographs of the surface may be taken at appropriate intervals and used to obtain the quantitative information. The nucleation law is sometimes also inferred from the overall kinetic behaviour. Precise fits to rate laws (including nucleation) require large numbers of accurate observations to enable reliable statistical analyses of the data to be made and it is frequently difficult to obtain sufficient numbers of reliable (N,t) values. The rate equations in Table 3.1. represent, to an acceptable approximation, all types of behaviour that can realistically be expected [10].

These relationships provide a sufficient foundation for kinetic analyses where the nucleation step, although the essential precursor to all subsequent changes, contributes only a single term to the overall composite kinetic expression.

Table 3.1.
Nucleation rate laws for solid state decompositions

Rate law	Nucleation rate $dN/dt =$	Number of nuclei present at time t $N =$	η value used in rate equation
Exponential	$k_N (N_o - N)$ $k_N N_o \exp(- k_N t)$	$N_o [1 - \exp(- k_N t)]$	1
Linear	$k_N N_o$	$k_N N_o t$	1
Instantaneous	∞	N_o	0
Power	$C \eta t^{\eta - 1}$	$C t^{\eta}$	> 1 (usually 2 or 3)
Branching	$k_N N_o \exp(k_2 t)$	$(k_N N_o / k_2)(\exp(k_2 t) - 1)$	-

N_o = initial total number of potential nucleation sites; N = number of growth nuclei present at time t; (dN/dt) = rate of nucleation, k_N = nucleation rate constant; k_B = branching rate constant; k_T = termination rate constant: $k_2 = k_B - k_T$. C is a constant.
η = number of steps involved in the formation of a nucleus.

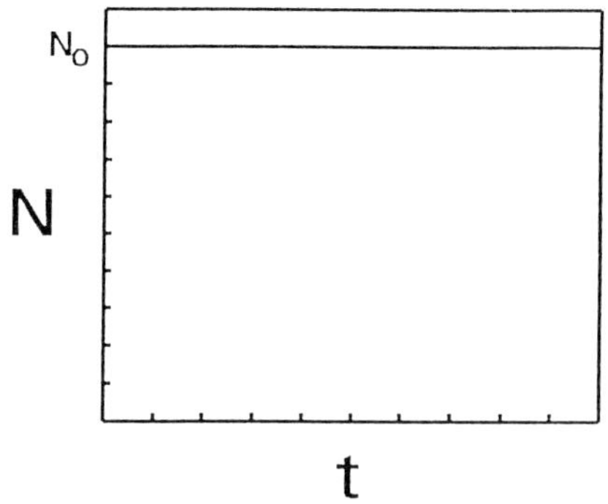

(a) Instantaneous nucleation: all possible nucleation sites are converted to growth nuclei early in the reaction.

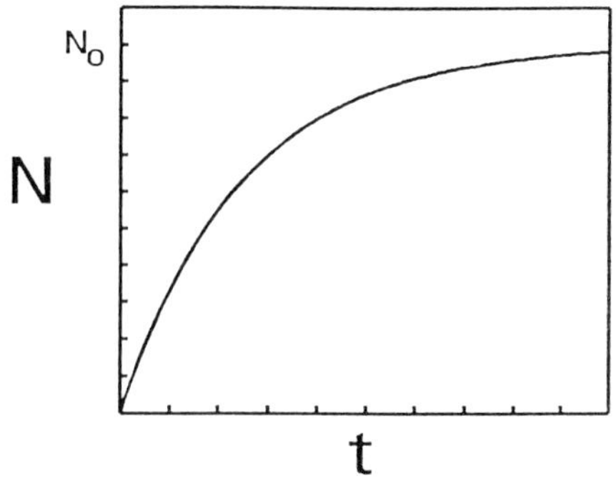

(b) Exponential nucleation: first order increase in number of sites activated.

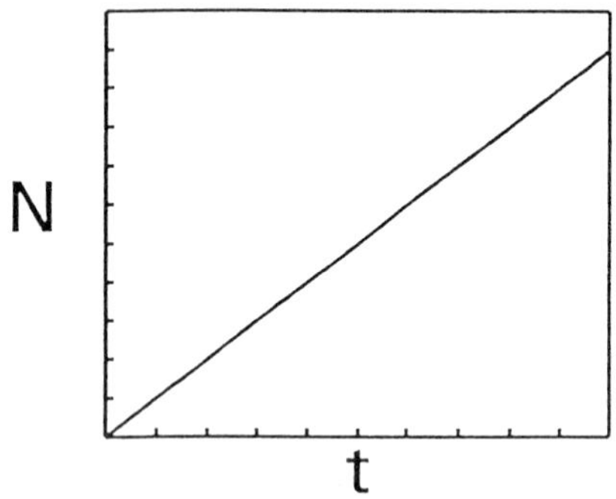

(c) Linear nucleation: constant rate of nucleation (in practice rate will diminish as area of unreacted surface decreases).

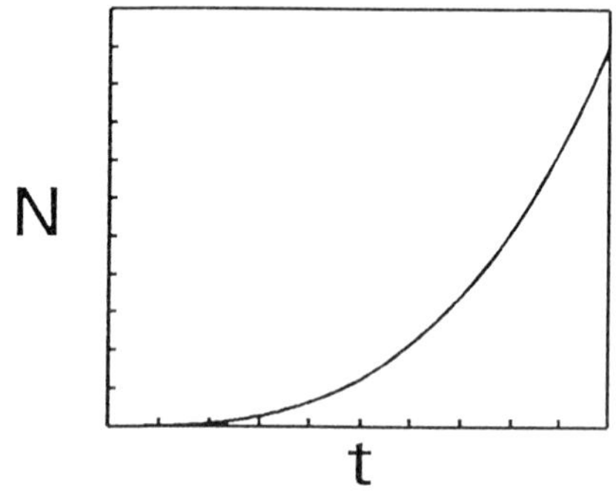

(d) Power or acceleratory nucleation (again the rate will diminish as area of unreacted surface decreases).

Figure 3.1.
Idealized representations of the variations with time of numbers of growth nuclei (N) present at time, t, for four of the nucleation rate equations given in Table 3.1.

3.4. KINETICS OF GROWTH

3.4.1. Growth of nuclei

Nuclei can be identified under the microscope because of the textural changes that accompany reaction. Decomposition is often accompanied by loss of volatile products, together with the appearance of cracks because the residual products do not fully occupy the volume of the former reactant material. Recrystallization yields small particles with crystallite textures that scatter light and the opaque intranuclear (product) material is recognizable. The advancing boundary surface of each nucleus is the reaction interface.

A characteristic feature of the kinetics of most solid decompositions is that *rates of interface advance are constant* [1,2,5,8]. This is expressed by:

$$\mathrm{d}r/\mathrm{d}t = k_G(t - t_o)$$

where r is a linear dimension of the nucleus at time t and k_G is the rate constant for growth of a nucleus formed at time t_o, after the commencement of reaction.

Rates of growth of small nuclei (often at the lower limits of microscopic detection) may be different from those subsequently attained. Low rates of initial growth have been ascribed to the instability of very small nuclei, but instances of relatively more rapid early growth of small nuclei are also known [10]. However, it is generally agreed that, after initial deviations, all nuclei within a particular reactant sample grow at the same rate.

Growth of nuclei in three dimensions ($\lambda = 3$, where λ is the *number of dimensions* term for use in kinetic analysis) means that the interface expands outwards at an equal (constant) rate in all directions, generating hemispheres of product and an increasing area of active reactant/product contact. Such unrestricted freedom of expansion is not always possible, however, and nucleus growth may be effectively confined to two dimensions ($\lambda = 2$) in solids possessing a layer structure, or where the reactant is in the form of thin crystals [15]. Two-dimensional growth may also arise when many closely-spaced nuclei are generated in the form of linear arrays. These grow initially in three dimensions which, however, soon leads to coalesence between near neighbours and generation of a cylindrical reaction zone where the increase in diameter with time fits the square law ($\lambda = 3 \rightarrow \lambda = 2$). This pattern of behaviour has been proposed for nickel oxalate decomposition [16]. Growth in a single dimension ($\lambda = 1$) may occur where severe restrictions on nucleus development are imposed by the crystal structure or the sample morphology, for example in reactions of fibre-like crystals.

Three main constraints tend to decrease the rate of product formation through nucleus growth. These are *coalesence* of nuclei, *ingestion* of sites potentially capable of generating nuclei, and crystal *boundaries*, and are explained below.

3.4.2. Unrestricted growth

If the above constraints on growth are initially disregarded, a nucleus formed at time t_j occupies a volume $v(t, t_j)$ where:

$$v(t, t_j) = \sigma[r(t, t_j)]^{\lambda}$$

and λ is the number of dimensions in which the nucleus grows (1, 2 or 3), σ is the shape factor (e.g. $4\pi/3$ for a spherical nucleus and 8 for a cubic nucleus) and the radius of the nucleus at time t is:

$$\mathrm{r}(t, t_j) = \int_{t_j}^{t} \mathrm{F}(t)\, \mathrm{d}t$$

where $\mathrm{F}(t)$ is the rate equation for growth. It is generally assumed that the rate of linear advance of the reactant/product interface is constant, hence:

$$\mathrm{r}(t, t_j) = k_G\,(t - t_j)$$

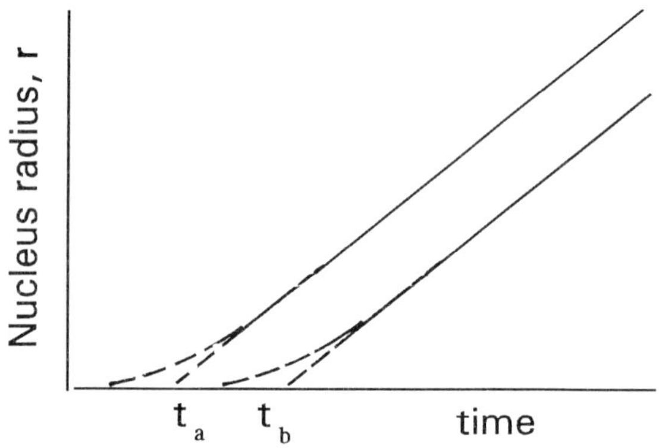

Figure 3.2.
The rate of interface advance during nucleus growth is constant for most soild state decompositions for which measurements have been made. The rates of growth for the two nuclei shown here are equal. It is inferred, by extrapolation, that these were formed at times t_a and t_b but, as implied by the dotted lines, growth may have been slower during the very early stages (at or below the limits of microscopic resolution).

The assumption that the rate of increase of the radius (or other linear dimension)is constant is a simplification (see Figure 3.2.) because growth rates often vary with crystallographic direction. Consequently, in many reactions the nuclei exhibit characteristic shapes [1]. Hemispherical nuclei, indicative of equal rates of growth in all directions, are found, for example, in the decomposition of barium azide [17] and in the dehydration of chrome alum [10,18-20]. The early work of Garner and his collaborators [20] on the dehydrations of large, relatively perfect crystals of alums and other hydrates (see Chapter 7) included microscopic examinations of the textural changes, such as crack dispositions, which accompanied nucleus growth. Measurements of rates of nucleation and of growth provided direct comparisons with the product yields predicted by geometric kinetic models [1]. Detailed characterization of nucleus development is, however, sometimes difficult or even impossible, and only the advance of an identifiable reactant/product interfacial zone may be observable.

3.4.3. Restrictions on growth of nuclei

Two factors are identified as limiting nucleus development, by preventing unrestricted expansion, and thereby curtailing the acceleratory process. These are the *coalescence* of nuclei and the *ingestion* of potential nucleus forming sites.

(i) Coalescence refers to the loss of interface which occurs when the reactive zones of two or more growing nuclei meet. This is sometimes incorrectly described as "overlap of nuclei", by reference to their projected shape, had growth continued in the absence of interference. The limits to growth of individual nuclei are set by the spacing of the original nucleation sites.

(ii) Ingestion refers to the elimination of a site at which another nucleus could otherwise have developed by the growth of an existing nucleus. Potential nucleation sites which never develop as independent nuclei are sometimes referred to as "phantom" nuclei, since their growth (imagined in the theoretical treatment) yields no real product. Growth is also limited by the dimensions of the crystal in which the nucleus is developing: product formation cannot extend beyond the reactant boundaries. The sizes and shapes of the reactant crystallites influence kinetic behaviour in ways that are discussed further below.

During reaction, the number of potential nucleus-forming sites decreases from the initial value N_0, at $t = 0$, to $N_1(t)$ at time t, because $N(t)$ sites have been activated to form nuclei and $N_2(t)$ sites have been eliminated by ingestion, so that:

$$N_1(t) = N_0 - N(t) - N_2(t)$$

$$- \mathrm{d}N_1 = \mathrm{d}N + \mathrm{d}N_2$$

Substituting $dN = k_N N_1 dt$ and $dN_2 = [N_1/(1 - \alpha)] d\alpha$ and integrating, the rate equation for nucleation is [1]:

$$dN/dt = k_N N_0 (1 - \alpha) \exp(- k_N t)$$

This is the exponential law (Table 3.1.) modified to allow for loss of potential reaction sites as reaction proceeds.

In a more general analysis, Allnatt and Jacobs [8,12] removed the restriction of single-step nucleation.

3.5. COMBINATION OF KINETICS OF NUCLEATION AND GROWTH

In interface advance reactions, the total volume of product at time t, $V(t)$, is given by the volume summation to include all individuals of the set of growth nuclei. Due allowance must be made for the different starting times at which each participating nucleus (linear dimension, $r(t, t_j)$) begins to grow. In principle this may be expressed as an integration for any combination of a rate equation for nucleation (Table 3.1.) together with a rate equation for growth :

$$V(t) = \sigma \, {}_{o}\!\int^{t} [\, {}_{ti}\!\int^{t} r(t, t_j)]^{\lambda} \, [dN/dt]_{t = tj} \, dt \qquad (3.1)$$

where σ is a shape factor for growth nuclei. $V(t)$ may be related to the fractional extent of reaction at time t, $\alpha = V(t)/V_f$. The final volume, V_f, corresponds to completion of reaction.

In principle, equation (3.1) may be integrated for any combination of rate laws for nucleation and for growth to give a rate expression of the form $g(\alpha) = kt$. In practice, this cannot be performed generally because there is no functional relationship between the nucleation and the growth terms.

The simplest kinetic expression based on the geometric model of interface advance is the *power law*. The rate of interface advance in three dimensions, at all surfaces of N instantaneously formed growth nuclei, is assumed to be unrestricted and constant. The total volume of product in the N hemispherical product assemblages after time t (each of radius $k_G t$) is then $V = (2\pi N/3)(k_G t)^3$. The exponent is determined by the number of dimensions in which growth proceeds. If nucleation continues during reaction, the acceleratory character is increased by the value of η given in Table 3.1. The power law is more generally expressed as $\alpha^{1/n} = kt$, where $n = \eta + \lambda$ (representing the nucleation law and the number of dimensions in which nuclei grow). This acceleratory equation can be applicable only to the early stages of reaction when the assumption of unrestricted nucleus growth is acceptable.

Coalesence of neighbouring nuclei decreases the acceleratory contribution, leading to the deceleratory behaviour characteristic of the later stages of reaction. Hemispherical nuclei, indicative of equal rates of growth in all directions, are found, for example, in the decomposition of barium azide [17] and in the dehydration of chrome alum [10,19,20]. Other reactions develop nuclei of different shapes characteristic of preferred advance in certain crystallographic directions.

The fractional decomposition, on the assumption that nuclei grow *without interference*, α', is obtained from $\alpha'(t) = V(t)/V_f$ where V_f is the volume of product formed when decomposition is complete.

If nucleus growth proceeds in λ dimensions at a constant rate of advance, but the effects of overlap are disregarded, and is accompanied by nucleation according to the power law, the resulting rate equation has the form [1,5,8]:

$$\alpha = (kt)^n$$

where $n = \eta + \lambda$, and η is the exponent in the nucleation law. A correction, t_0, to the time scale may be necessary (where there is an induction period, heat-up time, etc.):

$$\alpha = k^n(t - t_0)^n$$

For random nucleation according to the exponential law (Table 3.1.), followed by "normal" growth [1,8] in three dimensions, and $k_N t$ small:

$$\alpha \approx (\sigma N_0/4)(k_G / k_N)^3 (k_N t)^4$$

i.e. $n = 4$, $\eta = 1$ and $\lambda = 3$.

For k_N large (i.e. virtually instantaneous nucleation) and three-dimensional growth:

$$\alpha \approx B(k_N t)^3$$

i.e. $\eta = 0$ and $\lambda = 3$.

As discussed above, these rate equations make no allowance for the restrictions on growth of the nuclei. It is necessary to relate the unrestricted fractional decomposition, α', to the true value, α. A general but complicated solution to the problem has been provided [1,8]. For the simpler case of three-dimensional growth of randomly-distributed nuclei on large crystals, Avrami [21] has shown that α and

α' may be related by $d\alpha = d\alpha'(1-\alpha)$ which satisfies the requirements that when $\alpha = 0$, $d\alpha/dt = d\alpha'/dt$, and when $\alpha = 1$, $d\alpha/dt = 0$, but $d\alpha'/dt$ is finite. Hence:

$$\int_0^{\alpha} (1-\alpha)^{-1}\, d\alpha = \int_0^{\alpha'} d\alpha' \qquad \text{and} \qquad -\ln(1-\alpha) = \alpha'$$

so that power law expressions of the form $\alpha' = k^n(t-t_0)^n$ can be modified to:

$$-\ln(1-\alpha) = k^n(t-t_o)^n \quad \text{or} \quad [-\ln(1-\alpha)]^{1/n} = k(t-t_o) \qquad (3.2)$$

$n = \eta + \lambda$, is based on the nucleation rate law (η) and number of dimensions in which nuclei grow (λ).

Erofeev's approach [5,8,22], with contributions from Kholmogorov and Bel'kevich [23,24], was more general than that of Avrami and was based on the probability of the reaction step occurring in a particular time interval. The rate of reaction is:

$$d\alpha/dt = P\exp\left(-\int_0^t P\, dt\right) = P(1-\alpha)$$

where $P\, dt$ is the mean probability that an individual molecule will react in time interval dt. For an isothermal unimolecular reaction, $P = K$ = constant, so that:

$$\alpha = 1 - \exp(-Kt)$$

In solid-state reactions, the rate of nucleation may be given by either of the expressions: dN/dt = const. or $dN/dt = t^n$ const. For both expressions, at time t the probability ($P\, dt$) is proportional to the total volume of the spherical layers at the peripheries of nuclei which originated at time t_j. The radii of the spheres at the inner and outer boundaries of these layers are:

$$r = k_r(t-t_j)$$

and

$$r + dr = k_r(t + dt - t_j)$$

Thus

$$P\, dt = dt\int_0^t 4\pi k_r^3(t-t_j)^2\,(dN/dt)\, dr$$

Identifying P with α and substituting the two nucleation laws mentioned above, leads to the expressions:

$\alpha = 1 - \exp(-Kt^4)$ (when dN/dt = constant)

and

$\alpha = 1 - \exp(-Kt^{\eta+3})$ (when $dN/dt = t^{\eta-1}$)

Where a reaction proceeds through the development of cylindrical nuclei, initiated at edges or surface cracks in the reactant solid, the expression is:

$\alpha = 1 - \exp(-Kt^3) = 1 - \exp(-k^3t^3)$

and for flat nuclei:

$\alpha = 1 - \exp(-Kt^2) = 1 - \exp(-k^2t^2)$

The Erofeev treatment, therefore, yields the rate equation:

$-\ln(1 - \alpha) = Kt^n = k^nt^n$

Mampel's approach [25] to the derivation of kinetic expressions for nucleation and growth processes, with allowance for coalesence of neighbouring nuclei, also leads to a form of equation (3.2) through consideration of geometrical models of random placement of circular discs or hemispheres on a surface. This analysis, which also includes the effects of the particle size of the reactant, is not amenable to a general solution, but the following special cases are of interest. (i) When α is small, $\alpha = k^4t^4$ (a form of the power law). (ii) When the reactant particles are small, there is effectively a single nucleus generated on each crystallite and $\ln(1 - \alpha) = C - kt$ (where C and k are constants). (iii) When reactant particles are large, the form of the equation resembles the contracting volume equation (see below).

Written in the general form:

$$[-\ln(1 - \alpha)]^{1/n} = k(t - t_0) \qquad (n = \eta + \lambda) \qquad (3.2)$$

Equation (3.2) is often referred to as the *Avrami-Erofeev* (A-E) equation, or more fittingly, on account of the substantial contributions from other workers, especially Johnson and Mehl [26] in the field of metallurgy, as the *Johnson-Mehl-Avrami-Erofeev-Kholmogorov* (JMAEK) equation. The values of n obtained from kinetic analysis of the data for the overall reaction are interpreted [27,28] in terms of the nucleation rate equation and the shapes of the growth nuclei as indicated in Table 3.2.

Allowance for the effects of particle size on the overall rate of decomposition, complicates the integration of equation (3.1). Only limited aspects of this problem are considered in the Mampel approach [25]. Some qualitative guidelines on the effect of particle size on a reaction proceeding by a nucleation and growth process are as follows. For large crystals, the maximum slope of the α-time curve is expected to increase as the reactant is sub-divided, because a larger number of nuclei can be formed on the greater surface available per unit mass of solid. With further reduction in crystallite size, however, the volume of product which is formed as a consequence of each relatively difficult nucleation step is restricted to the volume of the particle nucleated. When crystallite dimensions are sufficiently small, the rate of reaction is controlled by nucleation in an assemblage of similar reactant particles and the first-order expression applies.

In a further development of the probability treatment used by Erofeev, Kolar-Anic *et al.* [29,30] and Dorko *et al.* [31] have pointed out that equation (3.2) can be rearranged and compared with a form of the Weibull distribution [32,33]:

$$f(t) = 1 - \exp[-(t - t_o)/K^{-1}]^n$$

where K^{-1}, n and t_o are the scale, shape and location factors of the distribution, respectively.

Johnson and Kotz [33] discuss in detail the Weibull and other distributions which find application when conditions of strict randomness of the exponential distribution are not satisfied. Gittus [34] has discussed some situations in which the Weibull distribution may be expected to find application, including nucleation and growth processes in alloy transformations, and Dorko *et al.* [31] have used the Weibull distribution function for the kinetic analyses of reactions in which decomposition is accompanied by melting. Using this approach, it was found possible to determine the separate solid phase and liquid phase contributions to the overall decomposition. Taplin [35] questioned the value of the approach developed by Kolar-Anic *et al.* [29,30] and pointed out that this establishes no connection between these equations and the distributions of energy or of reactivity in the system. The objective is not to generate (possibly empirical) rate equations capable of describing the relationships between the extent of reaction and time, but to provide a physical and geometric model of the reaction system where the parameters involved have quantitative significance [33].

Table 3.2.
Values of n in the Avrami - Erofeev (JMAEK) equation [27,28]

$$- \ln (1 - \alpha) = k^n (t - t_0)^n \tag{3.2}$$

n	Nucleation rate law (Table 3.1.)	η	λ
> 4	power	2 or 3	3
4	linear	1	3
3 to 4	exponential	0 to 1	3
3	instantaneous	0	3
	linear	1	2
2.5	exponential	0 to 1	2
2	instantaneous	0	2
	linear	1	1
1	instantaneous (iv)	0	1

Notes:
(i) Values of $\eta > 1$ do not appear to have been described in reactions for which $\lambda = 2$ or 1, although such behaviour is in principle possible.
(ii) Fractional values of n can be interpreted either as diffusion controlled growth or may arise from distributions of sizes or shapes of reactant particles.
(iii) The alternative interpretations mentioned for particular values of n cannot be distinguished from kinetic evidence. A distinction may be possible based on information obtained from microscopic observations of the shapes and distribution of sizes of nuclei.
(iv) This is the first-order rate equation.

3.6. GEOMETRICAL MODELS

3.6.1. Introduction

In decompositions of solids where the initiation of reaction results in the rapid development of a large number of closely-spaced growth nuclei on all surfaces, or on specific crystallographic faces, the overall kinetics of reaction are determined by the geometry of advance of a coherent reaction interface from these boundaries towards the centres of the reactant particles [1-5,8,36-40]. This type of behaviour can occur where either (i) the Gibbs energy for nucleus formation is comparable with that of subsequent growth, or (ii) the development of nuclei is anisotropic, so that there is initial rapid and preferential advance of the reaction interface across the surfaces to generate rapidly a comprehensive product layer covering all reactant faces [41].

In reactions of this type, the induction period, if any, may be too short to permit detection and, during this time, there is virtually instantaneous and dense nucleation across all active surfaces. The maximum reaction rate is attained at a low α and, thereafter, the α-time curve is deceleratory. There is, thus, no sharp distinction between such kinetic behaviour and the later stages of the nucleation and growth processes discussed above. In some early work [42], the influence of slow nucleation and an acceleratory period was removed by artificial initiation of reaction (nucleation) across all surfaces, so that the kinetic analysis was simplified to the consideration of a process advancing inwards from all faces of a crystal of known size and geometry.

3.6.2. Rapid dense nucleation on all surfaces

If reaction is initiated rapidly and extensively across all faces of a cube [1,2] of edge a, then, after a short time, close spacing of nuclei results in the rapid (low α) generation of a coherent reaction zone that, in the absence of diffusion effects (considered below), advances inwards at a constant rate (k_G). The progress of this reaction is described by [1-5]:

$$\alpha = [a^3 - (a - 2k_G t)^3]/a^3 \quad \text{or} \quad 1 - (1 - \alpha)^{1/3} = (2k_G/a)\, t \qquad (3.3)$$

Equation (3.3) is referred to as the *contracting volume* (*cube* or *sphere*) equation and is the simplest example of a more general family of kinetic expressions [1-5,43-45], which describe different rates of interface advance in different crystallographic directions and variations in dimensions and shapes of crystallites [3]. For reaction in a rectangular particle of edge dimensions, a, b and c, into which the rates of interface advance are k_a, k_b and k_c, respectively, the appropriate rate equation is:

$$\alpha = [abc - (a - 2k_a t)(b - 2k_b t)(c - 2k_c t)]/(abc)$$

(the factor 2 arises because reaction commences on the two opposite sides of each crystal and two interfaces advance towards the centre).

When a $\approx$ b >> c and $k_a = k_b = k_c$, the relative decreases in the dimensions of the particle of unreacted solid are small in the a ($\gg k_a t$) and b($\gg k_b t$) directions, so that the reaction is zero order, $\alpha \approx (2k_c/c)\, t$. The rate of such a reaction in thin discs or plate-like particles of reactant is thus effectively constant, $d\alpha/dt = k$, because the rate of interface advance into the very much smaller c dimension is constant and the influence of the contraction at the edges is small [46]. Similar consideration of a reaction readily initiated at all surfaces of a cylinder or column-shaped prism of reactant (a >> b ≈ c and $k_a = k_b = k_c$) shows that:

$$1 - (1 - \alpha)^{1/2} = (2k_G/b)\, t \quad (n = 2)$$

Rate equations for other reactant shapes may be derived from consideration of the geometry of the original crystal, or crystallites, of reactant [47,48]. Allowances can be made for contributions to the overall kinetic behaviour from known distributions of reactant particle-sizes by using summations for an appropriate range of (measured) dimensions.

3.6.3. Rapid dense nucleation restricted to specific crystal surfaces

When nucleation occurs on certain crystal surfaces only, rate equations based on geometric considerations may be derived as outlined above, with the restriction that the advance of interfaces into the bulk of the reactant particle proceeds inwards from those crystal surfaces upon which nucleation occurs. When the interface moves inwards from the edges only of a disc or plate-like particle:

$$1 - (1 - \alpha)^{1/2} = kt \qquad (n = 2)$$

This is known as the *contracting area* (*disc, cylinder or rectangle*) equation. Rate equations have been derived which include allowance for the angles between the active crystal faces [43,44].

The contracting geometry equation expressed in its most general form is:

$$1 - (1 - \alpha)^{1/n} = kt$$

where n is the number of dimensions in which the interface advances. In the *contracting volume* equation, $n = 3$; for the *contracting area* equation, $n = 2$; and, when there is linear advance of the interface in a single direction, $n = 1$, the equation becomes $\alpha = kt$, i.e., the reaction proceeds at a constant rate, which is the zero-order rate equation.

Rates of advance of the interfaces for this set of kinetic expressions are constant. Inclusion of a diffusion limitation term, which results in a progressive decrease in the rate of interface advance, is considered in Section 3.8. below.

3.6.4. Other geometric models

Nucleation according to the exponential law:

$$dN/dt = k_N N_0 \exp(- k_N t)$$

is a continuing process, so that reaction in those crystallites which were nucleated first may be completed before reaction is initiated in other particles. Allowance for this termination of interface advance, resulting from the finite size of reactant particles,where nucleation is a slow process, may be incorporated into the geometric analysis [1,5].

Topley and Hume [36] developed a model based on the rapid initial formation of (on average) a single nucleus on the surface of each particle of reactant, represented as a sphere of radius, a. The advancing reaction interface penetrates the reactant at an equal rate in all inward directions ($k_G = dr/dt$) and the volume of material reacted at time t is that volume of a sphere, having its centre at the site of surface nucleation and of radius $k_G\, t$, which falls within the reactant. The fractional reaction, which is the zone of interpenetrating spheres, at time t is:

$$\alpha = 0.5\ (k_G\, t/a)^3\ [1 - (3k_G\, t/8a)]$$

The resulting α-time curves are sigmoid with maximum rate at $\alpha = 16/27$ (= 0.593). The assumption that there is a single nucleus per crystallite is generally unrealistic, but the model has been applied [36] to the dehydration of $CaCO_3.6H_2O$, within the restricted range ($0.3 < \alpha < 0.7$).

3.7. MODELS BASED ON AUTOCATALYSIS

The rate equations considered under this heading were originally formulated on the concept of "nucleus branching" (Table 3.1.), analogous to homogeneous chain reactions [5,8]. Occurrence of "branching", without restriction on multiplicative development or allowance for termination, would lead to an exponential dependence of α on t:

$$\alpha = k \exp\ (k_B\, t)$$

where k_B is the rate coefficient for nucleus branching. Acceleratory behaviour cannot be maintained indefinitely in any real reaction and allowance for termination can be made by including a term $k_T(\alpha)$, which is a measure of the probability of termination which increases with α, so that:

$dN/dt = k_N N_o + N[k_B - k_T(\alpha)]$

where (α) expresses a dependence upon α. When k_N is large, all available initiation sites (N_o) are rapidly exhausted. When k_N is very small, $N[k_B - k_T(\alpha)] >> k_N N_o$, and:

$dN/dt = N[k_B - k_T(\alpha)]$

If it is assumed that the rate of reaction ($d\alpha/dt$) is proportional to the number of participating nuclei (N), integration of this expression requires knowledge of the relationship between k_B, $k_T(\alpha)$ and α. This model has been re-examined recently in considerable detail by Jacobs [8].

$$d\alpha/dt = (k_B - k_T\ \alpha)\alpha \qquad (3.4)$$

At the point of inflection (α_i, t_i), from equation (3.4):

$d^2\alpha/dt^2 = (k_B - 2k_T\ \alpha_i) = 0$

and $k_B = 2k_T\ \alpha_i$

Assuming [8] that the ratio $k_B/\ k_T$ is independent of α:

$d\alpha/dt = k_B\alpha\,(1 - \alpha\,/\,2\alpha_i)$

for $\alpha \geq \alpha_o$, where α_o is the extent of initial decomposition that occurs before the time t_o when chain branching becomes dominant. Hence:

$$\ln\,[\alpha\,/(1 - \alpha\,/\,2\alpha_i)] - \ln\,[\alpha_o\,/(1 - \alpha_o\,/\,2\alpha_i)] = k_B(t - t_o) \qquad (3.5)$$

which Jacobs [8] has termed the generalized Prout-Tompkins (GPT) equation.

Prout and Tompkins [13] considered the particular case of sigmoid α-time curves, with the point of inflection α_i at 0.50. Equation (3.5) then becomes:

$\ln\,[\alpha\,/(1 - \alpha)] - \ln\,[\alpha_o\,/(1 - \alpha_o)] = k_B(t - t_o)$

which has the same form as equation (3.6):

$$\ln[\alpha/(1 - \alpha)] = k_B t + c \qquad (3.6)$$

which is referred to as the *Prout-Tompkins* equation or, sometimes, the *Austin-*

Rickett equation [49]. The sigmoid form of the α-time curve expressed by equation (3.6) is similar in shape to those derived from the Avrami-Erofeev equation.

Jacobs [8] has discussed the assumptions made in deriving the GPT equation (3.5): (i) that the probability of termination is $k_T\ \alpha$, and (ii) that the ratio $k_B/\ k_T$ is independent of α. The need for proper consideration of the inflection point (and the initial decomposition α_o) was illustrated by re-analysis of the data for the decomposition of $AgMnO_4$ in terms of the GPT equation [8]. Slight adjustment of α_i from the 0.500 of the original PT equation to 0.466 for the GPT equation gave a remarkable improvement in fit of the data for whole crystals, while the data for ground crystals required $\alpha_i = 0.213$.

If k_B varies inversely with time (i.e., there is a decrease in the effectiveness of branching as reaction proceeds) it follows that [50]:

$$\ln\,[\alpha/(1 - \alpha)] = k_B \ln t + c$$

which is known as the *modified Prout-Tompkins* equation. The energy chain theory on which this model was originally based is now regarded as unacceptable, because energy quanta released from reaction are expected to be dispersed as thermal energy and not specifically transferred to potential reactant species.

Product generation at developing nuclei can promote particle disintegration, exposing more surfaces on which reaction is then initiated leading to further decomposition of crystallites. This autocatalytic model was clearly and graphically described by Prout and Tompkins [13]. However, observational support for the participation of crack development could not be obtained from a detailed, high-magnification examination of the surface textures of partially reacted $KMnO_4$ crystals [51]. An alternative explanation of this autocatalytic behaviour may be found [52] in the acceleration of reaction proceeding in an increasing volume of melt.

3.8. DIFFUSION MODELS

3.8.1. Diffusion processes

Diffusion processes participate in, and may even control the rates of many types of chemical reactions involving solids. For example, (i) in decompositions of solids the rate may be controlled by the diffusive escape of a volatile product, and (ii) in reactions between solids a solid product may be formed between the reactants [5], so that progress is maintained by transport across the barrier layer (Figure 3.3.).

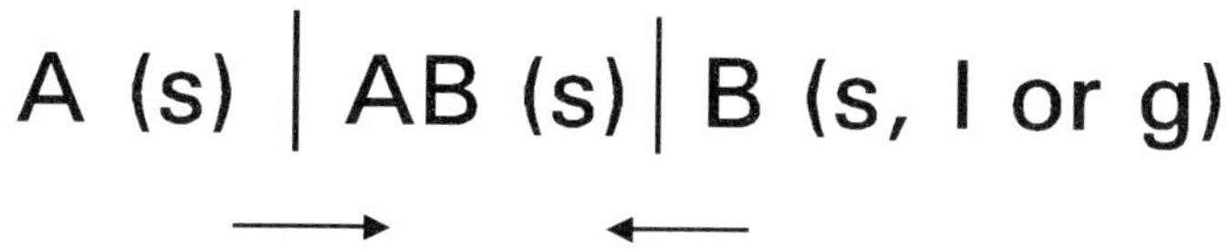

Figure 3.3.
Continuation of reaction requires maintained transport of constituents of A and/or B, across the coherent and adherent layer of product AB. The rate of reaction diminishes progressively with increase in thickness of the barrier layer (AB).

Rate equations require terms expressing the contributions from both diffusion control and the geometric changes resulting from interface advance. For some reactions, diffusion may be slower than the chemical processes occurring at the reaction interface and hence control the overall rate of decomposition. Cations are usually more mobile within the barrier layer than the larger anions. The anions maintain the immobile crystal structure within which the chemical changes occur.

A complication that may arise in kinetic studies is that two or more crystalline products may be generated and the rate may be controlled by diffusion across both of these contiguous phases, represented schematically by:

$$AX/BX \rightarrow AX/ABX_2/BX \rightarrow AX/A_2BX_3/ABX_2/BX$$

where A and B represent the migrating cations and X is the immobile (larger) anion. Diffusion coefficients of A and B in the phases ABX_2 and A_2BX_3 may be different and due allowances for their individual contributions must be incorporated into the kinetic analysis of such systems.

In the decompositions of some particularly stable crystalline materials, the reactant structure does not undergo recrystallization or disintegration, although there may be modification of lattice parameters following the loss of a small stable molecule, such as H_2O or NH_3, from the reactant phase. Such molecules diffuse outwards between structural components that are sufficiently stable to survive unmodified. Reaction rates are controlled by Fick's laws, ease of movement being determined by the dimensions of the intracrystalline channels. The participation of diffusion control is often recognized by the appearance of the characteristic $t^{1/2}$ term in the rate equation established. Theoretical aspects of diffusion control have been discussed by Okhotnikov *et al.* [53-57].

Transport processes may be dominant in reactions where the reactant structure does not undergo immediate major reorganization. Reactions of this type are often characterized by anisotropy of interface advance in layered structures where migrations are exclusively in the interlayer planes. Examples of diffusion controlled reactions have been reported for the dehydrations of some crystalline hydrates, including $CaSO_4.2H_2O$ [54,58] and $CaHPO_4.2H_2O$ [59].

3.8.2. Rate equations for diffusion

The simplest rate equation, which applies when the reaction zone has a constant area and the rate of product formation decreases in direct proportion to the thickness of the product barrier layer, is:

$$\alpha = (k_D t)^{1/2} \tag{3.7}$$

known as the *parabolic law* [5-7,60,61].

Variations in behaviour become important when diffusion in or across the barrier layer is inhomogeneous as a consequence of cracking or due to the development of more than a single layer of product. Such behaviour may sometimes be satisfactorily represented by the alternative empirical rate equations:

$$\alpha = k_1 \ln (k_2 t + k_3) \qquad \text{(the } \textit{logarithmic law}\text{)}$$

$$\alpha = k_1 t + k_2 \qquad \text{(the } \textit{linear law}\text{)}$$

3.8.3. Contributions from both diffusion and geometric controls

For a diffusion-limited reaction proceeding in spherical particles (radius a) the rate expression is obtained by combining the parabolic law, equation (3.7), with the contracting volume equation (3.3) (with $n = 3$) to give:

$$[1 - (1 - \alpha)^{1/3}]^2 = kt/a^2 \tag{3.8}$$

usually referred to as the *Jander* equation [62]. If allowance is made for differences in the molar volumes of reactant and of product (ratio, z^{-1}) then Carter [63] and Valensi [64] showed that equation (3.8) becomes:

$$[1 + (z - 1)\alpha]^{2/3} + (z - 1)(1 - \alpha)^{2/3} = z + 2(1 - z)kt/a^2 \tag{3.9}$$

When z approaches unity, equation (3.9) reduces to the *Ginstling-Brounshtein* equation [65]:

$$1 - 2\alpha/3 - (1 - \alpha)^{2/3} = kt/a^2$$

If the rate of reaction with diffusion control is also proportional to the amount of reactant present based on the contracting sphere model, then:

$$[(1 - \alpha)^{-1/3} - 1]^2 = kt/a^2$$

The *Dunwald-Wagner* equation [66], based on Fick's second law of diffusion into or out of a sphere of radius a is:

$$\ln[6/(\pi^2(1 - \alpha))] = \pi^2 Dt/a^2$$

where D is the diffusion coefficient of the migrating species.

The rate equation for diffusion-controlled reaction in two dimensions, for example of a cylindrical particle, radius a, is [67]:

$$(1 - \alpha)\ln(1 - \alpha) + \alpha = k_1 t/a^2$$

Incorporation of a diffusion term in nucleation and growth reaction models has been proposed by Hulbert [68]. Interface advance is assumed to fit the parabolic law and is proportional to $(Dt)^{1/2}$, but the nucleation step is uninhibited. The overall rate expressions have the same form as the Avrami-Erofeev equation:

$$-\ln(1 - \alpha) = (kt)^m$$

but the exponent $m = \eta + \lambda/2$. Diffusion control reduces the relative importance of the acceleratory process because, for constant η and λ, n must always be greater than m. Measurement of m cannot be used to identify the kinetic model unambiguously because there are alternative, acceptable combinations of η and λ, see Table 3.2. Even the existence of diffusion control is not necessarily demonstrated from the value of the exponent, because common values of m and n may result from diffusion- or interface-limited reaction models. There has been no unambiguous demonstration that a reaction of this type, in which there is diffusion control within growing nuclei while nucleation continues occurs in any particular rate process. The model seems to be inherently improbable because the growing nuclei cannot be spherical, so that surface expansion would be expected to dominate, spreading nuclei across surfaces and resulting in contracting volume geometry and a fit to the Jander equation.

Hulbert [68] has also discussed the consequences of the relatively large concentrations of lattice imperfections, including, perhaps, metastable phases and structural deformations, which may be present at the commencement of reaction but later decrease in concentration and importance. If it is assumed [69] that the rate of removal of defects is inversely proportional to time (the Tammann treatment) and this effect is incorporated in the Valensi approach, equation (3.9) is modified by replacement of t by $\ln t$.

3.9. MODELS BASED ON 'ORDER OF REACTION'

The kinetics of many solid state reactions have been reported as being satisfactorily represented by the first-order rate equation [70] (which is also one form of the Avrami-Erofeev equation ($n = 1$)). Such kinetic behaviour may be expected in decompositions of fine powders if particle nucleation occurs on a random basis and growth does not advance beyond the individual crystallite nucleated.

Reactions in melts, vitreous materials, polymers, etc. can justifiably be analyzed by equations based on a concentration dependence of rate. Some reactions proceeding in vitreous reactant phases have been shown to conform to second or even third-order rate equations. Progressive melting of a solid reactant during decomposition results in acceleratory behaviour [52,71-73] and comprehensive melting before dehydration was observed to result in an approximately constant rate of water evolution [74,75].

The kinetics of a first-order reaction are very similar to those represented by the contracting volume equation [70], except in the final stages of reaction when α approaches 1.00. In measurements of reactivity, or in comparisons of properties of similar substances, the first-order expression can sometimes be used as a convenient empirical measurement of rate. The assumption of first-order behaviour is often made in the kinetic analyses of programmed temperature experiments (see Chapter 5). The software supplied with many commercial instruments often provides only order-based equations for kinetic analysis of data, whereas other equations more obviously applicable to solids, such as those given here, are not tested.

3.10. ALLOWANCE FOR PARTICLE-SIZE EFFECTS

The derivations of the rate equations given above are based on reaction models that include consideration of the changes in geometry of the interface during its advance within a single crystal of simple shape, usually a cube or a sphere. This is clearly an idealized representation of the general reactant which will more usually be composed of several crystals and, for powders, very many crystallites having a range of sizes and shapes. Variations in average particle-sizes and size distributions can affect the observed kinetic behaviour [76-81].

The role of reactant particle dimensions on the kinetics of nucleation and growth processes is more difficult to predict. Here the dominant controlling factor is whether or not the advancing interface can progress from one nucleated particle to a neighbouring unreacted crystallite which it touches [53].

Attempts have been made to allow for the influence of particle-size distributions on kinetic behaviour [76-83], but most usually it is assumed (perhaps implicitly) that all reactant particles behave similarly. Allowances for size effects, which are difficult to quantify, are most often based on numerical integration across an assumed or measured distribution of particle dimensions [82,83].

Hutchinson *et al.* [82] have shown that the decomposition of each individual particle can be regarded as fitting the contracting volume equation (3.3), written in the form:

$$\alpha_p(t, a) = 1 - [1 - (t - t_b)/(\rho a/k)]^3$$

where ρ is the sample density, k/ρ the constant rate of interface advance and t_b the time at which the particle of radius a begins to react. It was found that $t_b = Cr^{1/3}$ where C is the induction period constant. The mass fractions, F(a), of particles of radius a were assumed to form a normal distribution. The total fractional decomposition at time t, is then:

$$\alpha(t) = \int_0^{A'} (a)\alpha_p(t,a)\,\mathrm{d}a$$

$$= \int_0^{A_1} \mathrm{F}(a)\,\mathrm{d}a + \int_{A_1}^{A_2} \alpha_p(t,a)\mathrm{F}(a)\,\mathrm{d}a$$

The first term allows for particles of radius $\leq A_1$, which are completely decomposed at time t, and the second term includes particles which are only partially reacted at the same time. The expression for $\alpha(t)$ must be solved [82] separately, by numerical integration, for each value of t.

McIlvried and Massoth [83], using a similar approach for both the contracting volume and diffusion-controlled models, with normal and log-normal particle size distributions, calculated generalized plots of α against reduced time t (defined by $t = kt/\mu$) for various values of the standard deviation of the distribution, σ (log-normal distribution) or the dispersion ratio, σ/μ (normal distribution with mean particle radius, μ). The effects of particle-size distributions on the values calculated for α were small, influences being larger for diffusion-controlled models than for the contracting volume reaction. At short times, the calculated values of α for reaction of a range of particle sizes under diffusion control were larger than those calculated for a uniform particle-size, owing to the relatively more rapid reaction of the small particles. Conversely, at later stages, α values calculated for a range of particle sizes were smaller than for a uniform system, because the larger particles react slowly. For a given particle-size distribution, with known values of σ and μ, the experimental values of α may be plotted against log t and matched against theoretical curves [83] to determine the rate coefficient, k.

There are several other effects which result in deviations by real systems from the idealized models described above. (i) Subsidiary interfaces may develop resulting in a zone, rather than a surface, of reaction. (ii) The volume of product will generally be different from that of the reactant from which it was derived, and thus the effective reaction interface may not extend across the whole surface of the nucleus. This can result in particle disintegration. (iii) In reversible reactions, a volatile product may be adsorbed on the surface of the residual phase locally inhibiting reaction and hence the observed rate of product formation is less than that expected for the amount of reactant that has decomposed. (iv) Diffusion control may become significant in reversible reactions.

Allowance for these effects could, in principle, be incorporated in any kinetic model, but the integration invariably becomes complicated and additional constants (unmeasured) may be introduced, bringing uncertainty to the interpretation.

3.11. THE MOST IMPORTANT RATE EQUATIONS USED IN SOLID STATE REACTIONS

The rate equations which have found most widespread application to solid state reactions are summarized in Table 3.3. Other functions can be found in the literature. The expressions are grouped according to the shape of the *isothermal* α-time curves as acceleratory, sigmoid or deceleratory. The deceleratory group is further subdivided according to the controlling factor assumed in the derivation, as geometrical, diffusion or reaction order.

General rate equations have been proposed [84-86] to cover those listed in Table 3.3. Either or both are often referred to as the *Sesták - Berggren* equation:

$$d\alpha/dt = k\,\alpha^m(1 - \alpha)^n$$

$$d\alpha/dt = k\,\alpha^m(1 - \alpha)^n[-\ln(1 - \alpha)]^p$$

Sesták [87] has suggested the term "*accomodation coefficient*" for any additional terms, other than $(1 - \alpha)^n$,which are needed to modify equations based on reaction order (RO) for application to heterogeneous systems. (See also Koga [88]).

TABLE 3.3.

THE MOST IMPORTANT RATE EQUATIONS USED IN KINETIC ANALYSES OF SOLID STATE REACTIONS

	$g(\alpha) = k(t' - t_o) = kt$	$f(\alpha) = (1/k)\,(d\alpha/dt)$	$\alpha = h(t)$
1. Acceleratory α-time			
Pn power law	$\alpha^{1/n}$	$n(\alpha)^{(n-1)/n}$	
E1 exponential law	$\ln \alpha$	α	
2. Sigmoid α-time			
A2 Avrami-Erofeev	$[-\ln(1-\alpha)]^{1/2}$	$2(1-\alpha)[-\ln(1-\alpha)]^{1/2}$	
A3 Avrami-Erofeev	$[-\ln(1-\alpha)]^{1/3}$	$3(1-\alpha)[-\ln(1-\alpha)]^{2/3}$	
A4 Avrami-Erofeev	$[-\ln(1-\alpha)]^{1/4}$	$4(1-\alpha)[-\ln(1-\alpha)]^{3/4}$	
An Avrami-Erofeev	$[-\ln(1-\alpha)]^{1/n}$	$n(1-\alpha)[-\ln(1-\alpha)]^{(n-1)/n}$	$1 - \exp[-(kt)^n]$
B1 Prout-Tompkins	$\ln[\alpha/(1-\alpha)]$	$\alpha(1-\alpha)$	$\{1 + [\exp(kt)]^{-1}\}^{-1}$

TABLE 3.3. (contd)

	$g(\alpha) = k(t' - t_o) = kt$	$f(\alpha) = (1/k)\ (d\alpha/dt)$	$\alpha = h(t)$
3. Deceleratory α-time			
3.1 Geometrical models			
R2 contracting area	$1 - (1 - \alpha)^{1/2}$	$2(1 - \alpha)^{1/2}$	$1 - (1 - kt)^2$
R3 contracting volume	$1 - (1 - \alpha)^{1/3}$	$3(1 - \alpha)^{2/3}$	$1 - (1 - kt)^3$
3.2 Diffusion models			
D1 one-dimensional	α^2	$1/2\alpha$	$(kt)^{1/2}$
D2 two-dimensional	$(1 - \alpha)\ln(1 - \alpha) + \alpha$	$[- \ln(1 - \alpha)]^{-1}$	
D3 three-dimensional	$[1 - (1 - \alpha)^{1/3}]^2$	$3/2(1 - \alpha)^{2/3}[1 - (1 - \alpha)^{1/3}]^{-1}$	$1 - [1 - (kt)^{1/2}]^3$
D4 Ginstling-Brounshtein	$1 - (2\alpha/3) - (1 - \alpha)^{2/3}$	$3/2[(1 - \alpha)^{-1/3} - 1]^{-1}$	

TABLE 3.3. (contd)

	$g(\alpha) = k(t' - t_o) = kt$	$f(\alpha) = (1/k)\,(d\alpha/dt)$	$\alpha = h(t)$
3.3 'Order of reaction' models			
F0 zero order	α	1	kt
F1 first order	$-\ln(1 - \alpha)$	$1 - \alpha$	$1 - \exp(-kt)$
F2 second order	$[1/(1 - \alpha)] - 1$	$(1 - \alpha)^2$	$1 - (kt + 1)^{-1}$
F3 third order	$[1/(1 - \alpha)^2] - 1$	$(1 - \alpha)^3$	$1 - (kt + 1)^{-1/2}$

NOTES:

(1) Rate of reaction, $d\alpha/dt = k\,f(\alpha)$; $g(\alpha)$ is the integrated function of α that is linear with time and $h(t)$ is the function of t that is linear with α.

(2) Rate coefficients, k are different in each expression and times, t, are assumed to have been corrected for any induction period, t_0.

(3) The units of k are always expressed as $(\text{time})^{-1}$. In those equations incorporating an exponent, the correct notation is $\alpha = k^n t^n$ and NOT $\alpha = k\,t^n$.

3.12. ISOTHERMAL YIELD-TIME CURVES

The most characteristic and commonly observed shapes of *experimental isothermal* α-time and ($d\alpha/dt$)- time curves found for decompositions of solids are schematically represented in the composite diagrams in Figure 3.4. Different segments of the most general curve (applicable to particular reactions) are delimited by the axes A to E. The significance of each of these segments is explained further below.

The isothermal α - time plots corresponding to the rate equations in Table 3.3. (calculated on the arbitrary basis that $\alpha = 0.98$ at $t = 100$ min) are illustrated in Figure 3.5. and some of the relative rate [= $(d\alpha/dt)/(d\alpha/dt)_{max}$] - time plots are shown in Figure 3.6. (The relative rate - time plots for the third-order rate equation and the diffusion models are too deceleratory for comparison and have been omitted.)

The most general form of the experimental α-time relationship for decompositions of solids is shown in Figure 3.4., **Curve A**. This includes an initial, relatively rapid small evolution of gas, that is often identified as the decomposition of less stable surface material or impurities. The small contribution (typically 1-3% of the total product) from this separate rate process should be subtracted before kinetic analysis of the subsequent main reaction. Onset of the dominant rate process may be deferred during an *induction period* that is required for the generation of growth nuclei. Once growth nuclei have been established, the rate increases progressively during the *acceleratory period* to the maximum rate after which the *deceleratory period* continues to the completion of reaction at F. Many such reactions have an approximately linear reaction rate extending significantly on either side of the point of inflection. Examples of this general reaction type include the decompositions of nickel oxalate [16] and silver malonate [89].

Curve B lacks the initial process, but an induction period is required for the appearance of growth nuclei before onset of the sigmoid α-time curve. Examples include the decompositions of NH_4ClO_4 [90] and $KMnO_4$ [51].

Curve C is characterized by the rapid onset of an acceleratory reaction and no induction period is found. Any initial small evolution of gas will probably be obscured by onset of the main reaction. Such behaviour is found in the decompositions of iron(II) and iron(III) oxalates [91].

Curves D and E are difficult to distinguish. There may be a short acceleratory stage before onset of the dominant or exclusively deceleratory rate process. Examples of behaviour of this type include the dehydrations of $Li_2SO_4.H_2O$ [70,92], lignite [93] and $Ca(OH)_2$ [94].

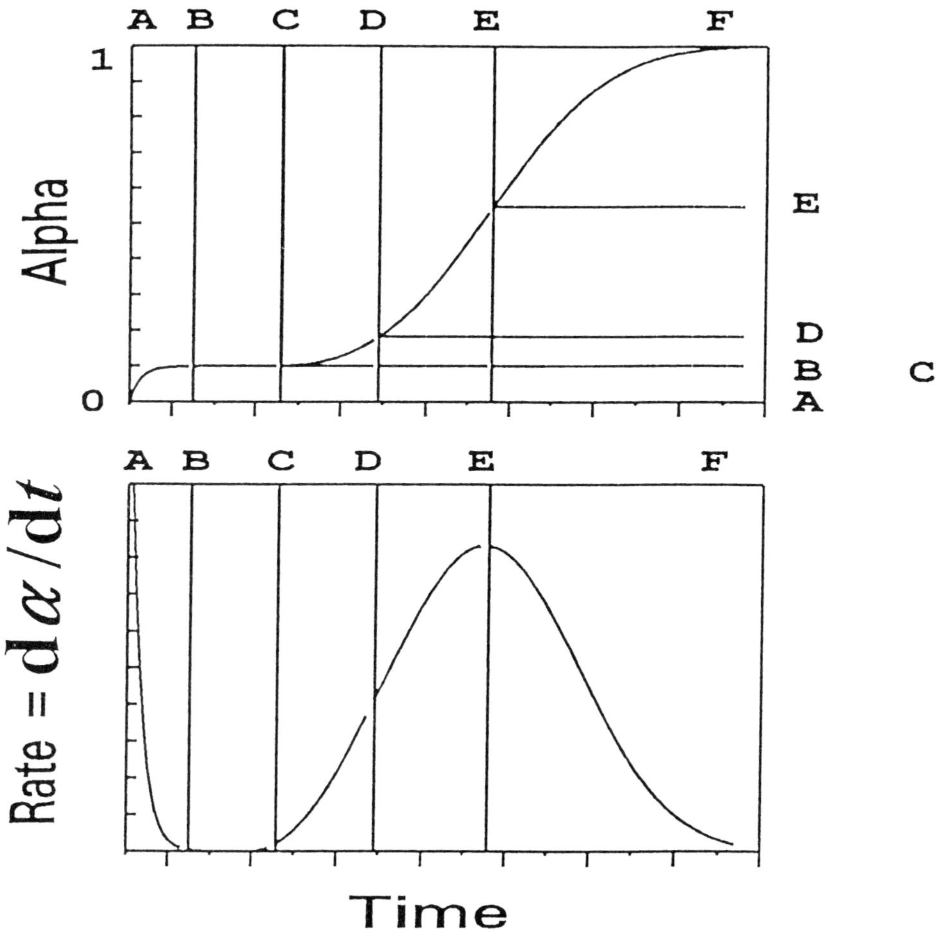

Figure 3.4. Various typical shapes of isothermal α-time and $(d\alpha/dt)$ - time curves observed for the decompositions of solids. The alternative sets of axes A - E represent the different types of behaviour, characteristic of various reactants, that are discussed in the text.

Point F refers to the completion of reaction. In processes involving advance of an interface, product formation will cease when the migrating interfaces have swept throughout the reactant volume and interfaces will have annihilated each other on

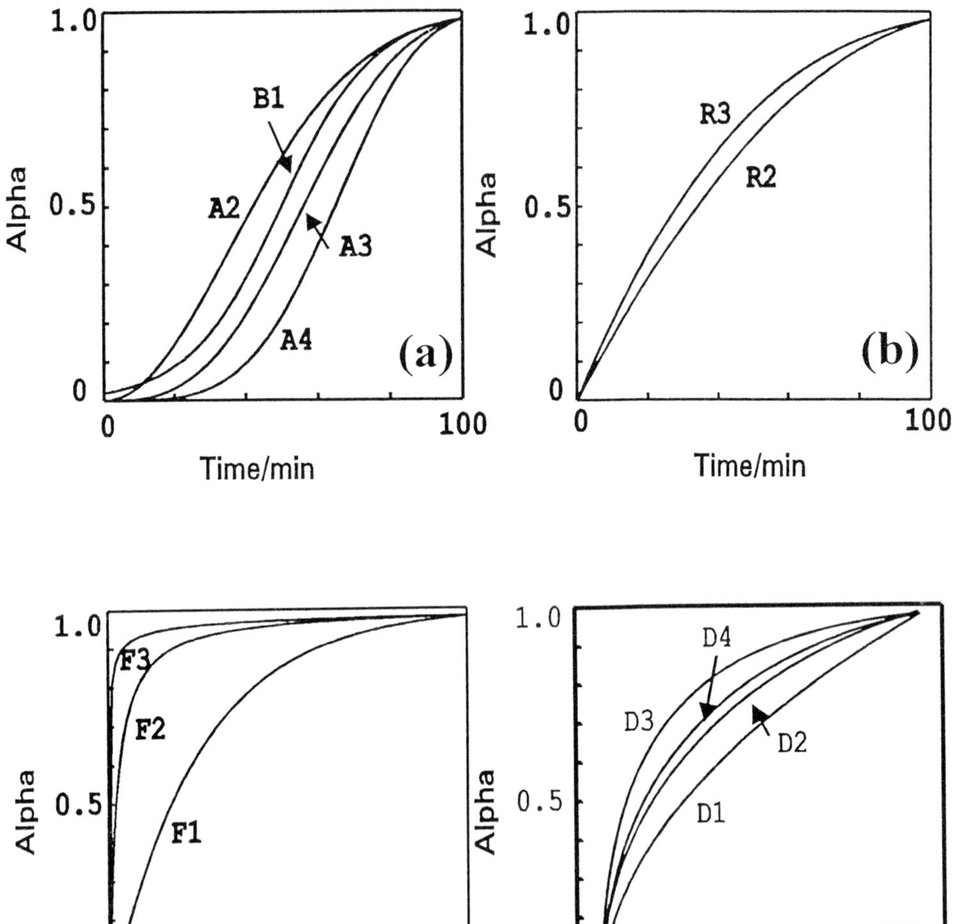

Figure 3.5.
Isothermal α - time plots corresponding to the rate equations in Table 3.3. (calculated on the arbitrary basis that $\alpha = 0.98$ at $t = 100$ min.): (a) sigmoid models; (b) geometrical models; (c) reaction order (RO) models; and (d) diffusion models.

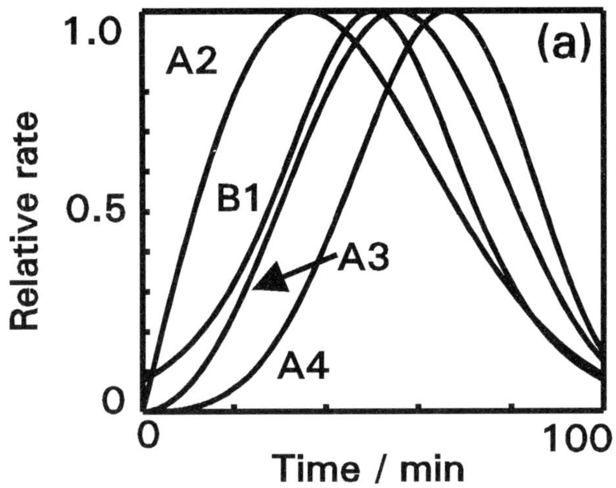

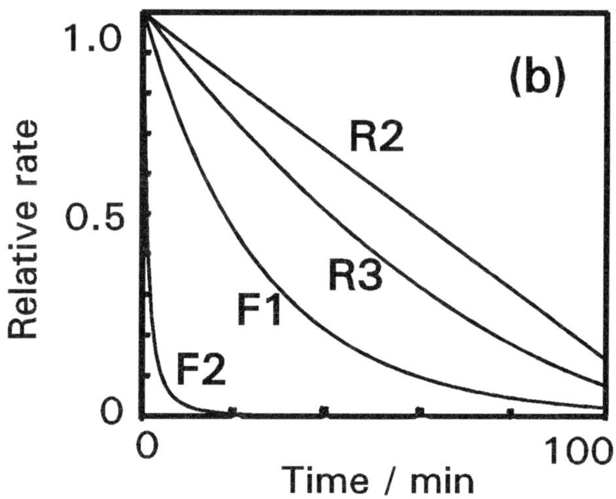

Figure 3.6.
Some of the isothermal relative rate [= $(d\alpha/dt)/(d\alpha/dt)_{max}$] - time plots corresponding to the rate equations in Table 3.3.: (a) sigmoid models; (b) geometrical and reaction order (RO) models. (NOTE: (i) the relative rate - time plots for the third-order rate equation and the diffusion models are too deceleratory for useful comparison; and (ii) calculation on an arbitrary basis, e.g. that α = 0.98 at t = 100 min., results in plots of relative rates against α, in place of time, having similar shapes).

meeting at the particle centres. This means that the termination of reaction will be relatively abrupt, as shown in Figure 3.4. If, however, kinetic behaviour is controlled by concentration terms, as in homogeneous rate processes, the approach to completion will be much slower and no sharp cessation is found.

In addition to these rate processes, a number of kinetic studies reported in the literature include α-time curves that show significant differences from the pattern represented in Figure 3.4. Examples of predominantly zero-order behaviour have been described [5] and some decompositions proceed to completion in more than a single stage as was found for the dehydrations of some hydrates, e.g. $CuSO_4.5H_2O$ [95], and the decompositions of several copper(II) carboxylates [96].

Yield-temperature curves, typically obtained from programmed temperature experiments using the techniques of thermal analysis, are illustrated in Chapter 5, after consideration of the effects of temperature on the rates of solid state reactions in Chapter 4.

3.13. CONCLUSIONS

Microscopic observations of the textural modifications of the surfaces of solids undergoing decomposition have identified localized zones of product (nuclei) which grow in size as reaction continues. These observations of chemical changes occurring preferentially within an advancing reactant/product interface zone led to the geometrical model, representing the kinetics of such reactions, described in this Chapter. Enhanced reactivity in particular zones of the reactant and in the vicinity of the reactant/product interface do not have parallel concepts in homogeneous kinetics and, hence, the set of rate equations (Table 3.3.) applicable to crystolysis [89] reactions has been developed. This set has been remarkably successful in representing yield-time data for numerous different rate processes (see Chapters 7 to 17). In spite of this success however, the possibility of recognizing additional patterns of behaviour should not be overlooked. The existence of this set may, in fact, have inhibited the development of alternative models. On the other hand, models providing excellent mathematical descriptions of experimental results are of limited value if they contain parameters which are not physically interpretable [35].

The demonstration that yield-time data for a particular rate process are satisfactorily represented by a rate equation from the set in Table 3.3. is a useful initial result (as in all kinetic analyses), but formulation of a reaction model requires more support than a statistical comparison with the fit of the same data to other rate equations. Some important aspects of kinetic analysis are summarized below.

(i) *Reaction geometry and kinetic model.* The fit of a rate equation is often interpreted as support for a particular geometric pattern of generation and advance of the reactant/product interface. Such an interpretation may not always be warranted because alternative models may exist, for example, melt formation can lead to acceleratory behaviour [52,71-73] with kinetics apparently similar to nucleation and growth processes. These ambiguities can often be resolved and proposals confirmed or rejected by microscopic examinations. The fit of a diffusion-control model would, similarly, require identification of the diffusing species to be of value in formulating a reaction mechanism. Identification of a reaction-order model is often an approximation, without detailed mechanistic significance, and further experimentation under different conditions may provide more interpretable data.

(ii) *Uncertainties in rate equations.* The rate equations in Table 3.3. may have been derived through the use of approximations and/or assumptions that are not applicable to the reactant in question. For example, few real samples consist of particles of uniform shape and size. In addition, reactant crystals generally have inhomogeneous distributions of imperfections that may contribute to the initiation and development of reaction. Disintegration of the reactant, particularly in the final stages of decomposition, can influence reaction kinetics, as can contributions from reactant self-heating or self-cooling. Detailed modelling of such complications usually requires introduction of additional parameters, often of unknown magnitude.

(iii) *Theoretical background.* Recent work on hydrates has shown [10] that reactions at surfaces are more complicated than has been generally assumed. The concept of the interface as a thin reaction zone of molecular dimensions is also being reconsidered [94,97,98].

In conclusion, a careful balance has to be maintained between the mathematical aspects of modelling and the molecular processes which the models are attempting to represent [99]. Progress in mathematical techniques and theory has far outstripped insights into the controls of reactivity and understanding of the bond redistribution processes involved in crystolysis [89] reactions.

REFERENCES

1. P.W.M. Jacobs and F.C. Tompkins, Chemistry of the Solid State, (Ed. W.E. Garner), Butterworths, London, 1955, Chap. 7; P.W.M. Jacobs, Mater. Sci. Res., 4 (1969) 37.
2. D.A. Young, Decomposition of Solids, Pergamon, Oxford, 1966.
3. B. Delmon, Introduction a la Cinétique Hétérogène, Technip, Paris, 1969.
4. P. Barret, Cinétique Hétérogène, Gauthier-Villars, Paris, 1973.

5. M.E. Brown, D. Dollimore and A.K. Galwey, Reactions in the Solid State, Comprehensive Chemical Kinetics, (Eds C.H. Bamford and C.F.H. Tipper), Vol. 22, Elsevier, Amsterdam, 1980.
6. J. Sesták, Thermophysical Properties of Solids, Comprehensive Analytical Chemistry, Vol. XIID, Elsevier, Amsterdam, 1984.
7. H. Schmalzried, Solid State Reactions, 2nd Edn, Verlag Chemie, Weinheim, 1981; Chemical Kinetics of Solids, VCH, Weinheim, 1995.
8. P.W.M. Jacobs, J. Phys. Chem., B, 101 (1997) 10086.
9. A.K. Galwey and G.M. Laverty, Solid State Ionics, 38 (1990) 155.
10. A.K. Galwey, R. Spinicci and G.G.T. Guarini, Proc. R. Soc. (London), A378 (1981) 477.
11. C. Bagdassarian, Acta Physicochim. URSS, 20 (1945) 441.
12. A.R. Allnatt and P.W.M. Jacobs, Can. J. Chem., 46 (1988) 111.
13. E.G. Prout and F.C. Tompkins, Trans. Faraday Soc., 40 (1944) 488.
14. M.E. Brown, Thermochim. Acta, 300 (1997) 93.
15. A.K. Galwey, D.M. Jamieson and M.E. Brown, J. Phys. Chem., 78 (1974) 2664.
16. D.A. Dominey, H. Morley and D.A. Young, Trans. Faraday Soc., 61 (1965) 1246.
17. A. Wischin, Proc. R. Soc. (London), A172 (1939) 314.
18. J.A. Cooper and W.E. Garner, Trans. Faraday Soc., 32 (1936) 1739.
19. J.A. Cooper and W.E. Garner, Proc. R. Soc. (London), A174 (1940) 487.
20. W.E. Garner and T.J. Jennings, Proc. R. Soc. (London), A224 (1954) 460.
21. M. Avrami, J. Chem. Phys., 7 (1939) 1103; 8 (1940) 212; 9 (1941) 177.
22. B.V. Erofeev, C.R. Dokl. Akad. Sci. URSS, 52 (1946) 511.
23. P.I. Bel'kevich, Vestn. Akad. Nauk Beloruss, SSR, 4 (1952) 120.
24. P.I. Bel'kevich, and B.V. Erofeev, Vestn. Akad. Nauk Beloruss, SSR, 4 (1952) 127.
25. K.L. Mampel, Z. Phys. Chem. A, 187 (1940) 43, 235.
26. W.A. Johnson and R.F. Mehl, Trans. AIME, 135 (1939) 416.
27. J.W. Christian, The Theory of Transformations in Metals and Alloys, Pergamon, Oxford, 2nd Edn, 1975, p.489.
28. J. Burke, The Kinetics of Phase Transformations in Metals, Pergamon, Oxford, 1965, p.192.
29. Lj. Kolar-Anic, S. Veljikovic, S. Kapor and B. Bubljevic, J. Chem. Phys., 63 (1975) 663.
30. Lj. Kolar-Anic and S. Veljikovic, J. Chem. Phys., 63 (1975) 669.
31. E.A. Dorko, W. Bryant and T.L. Regulinski, Anal. Calorim., 3 (1974) 505.
32. W. Weibull, J. Appl. Mech., 18 (1951) 293.

33. N.L. Johnson and S. Kotz, Distributions in Statistics: Continuous Univariate Distributions, Wiley, New York, 1970.
34. J.H. Gittus, Appl. Statistics, 16 (1967) 46.
35. J.H. Taplin, J. Chem. Phys., 68 (1978) 3325.
36. B. Topley and J. Hume, Proc. R. Soc. (London), A120 (1928) 210.
37. R.S. Bradley, J. Colvin and J. Hume, Proc. R. Soc. (London), A137 (1932) 531; Phil. Mag., 14 (1932) 1102.
38. B. Topley, Proc. R. Soc. (London), A136 (1932) 413.
39. J. Hume and J. Colvin, Proc. R. Soc. (London), A132 (1932) 548.
40. P. Barret, R. de Hartoulari and R. Perrat, C.R. Acad. Sci., C, 248 (1959) 2862.
41. M.E. Brown, B. Delmon, A.K. Galwey and M.J. McGinn, J. Chim. Phys., 75 (1978) 147.
42. W.E. Garner and M.G. Tanner, J.Chem. Soc., (1930) 47.
43. J. Hume and J. Colvin, Proc. R. Soc. (London), A125 (1929) 635.
44. W.D. Spencer and B. Topley, J. Chem. Soc., (1929) 2633.
45. S.J. Gregg and R.I. Razouk, J. Chem. Soc., (1949) 536.
46. R.J. Acheson and A.K. Galwey, J. Chem. Soc. A, (1968) 942.
47. T.A. Clarke and J.M. Thomas, J. Chem. Soc. A, (1969) 2227, 2230.
48. K.C. Sole, M.B. Moolman and M.E. Brown, J. Chem. Soc., Faraday Trans., 86 (1990) 525.
49. J.B. Austin and R.L. Rickett, Met. Technol. N.Y., (1938) T.P. No. 964; Trans. AIME, 135 (1939) 396.
50. E.G. Prout and F.C. Tompkins, Trans. Faraday Soc., 42 (1946) 482.
51. M.E. Brown, A.K. Galwey, M.A. Mohamed and H. Tanaka, Thermochim. Acta, 235 (1994) 255.
52. N.J. Carr and A.K. Galwey, Proc. R. Soc. (London), A404 (1986) 101.
53. A.K. Galwey, G.M. Laverty, N.A. Baranov and V.B. Okhotnikov, Phil. Trans. R. Soc. (London), A347 (1994) 139, 157.
54. V.B. Okhotnikov, S.E. Petrov, B.I. Yakobson and N.Z. Lyakhov, React. Solids, 2 (1987) 359.
55. V.B. Okhotnikov and I.P. Babicheva, React. Kinet. Catal. Lett., 37 (1988) 417.
56. V.B. Okhotnikov, I.P. Babicheva, A.V. Musicantov and T.N. Aleksandrova, React. Solids, 7 (1989) 273.
57. V.B. Okhotnikov, N.A. Simakova and B.I. Kidyarov, React. Kinet. Catal. Lett., 39 (1989) 345.
58. M.C. Ball and L.S. Norwood, J. Chem. Soc. A, (1969) 1633; (1970) 1476.
59. M.C. Ball and M.J. Casson, J. Chem. Soc., Dalton Trans., (1973) 34.

60. T.B. Grimley, Chemistry of the Solid State, (Ed. W.E. Garner), Butterworths, London, 1955, Chap.14.
61. A.J.E. Welch, Chemistry of the Solid State, (Ed. W.E. Garner), Butterworths, London, 1955, Chap.12.
62. W. Jander, Z. Anorg. Allg. Chem., 163 (1927) 1; Angew. Chem., 41 (1928) 79.
63. R.E. Carter, J. Chem. Phys., 35 (1961) 1137, 2010.
64. G. Valensi, C.R. Acad. Sci., C, 202 (1936) 309; J. Chem. Phys., 47 (1950) 489.
65. A.M. Ginstling and B.I. Brounshtein, Zh. Prikl. Khim., 23 (1950) 1327.
66. B. Serin and R.T. Ellickson, J. Chem. Phys., 9 (1941) 742.
67. J.B. Holt, J.B. Cutler and M.E. Wadsworth, J. Amer. Ceram. Soc., 45 (1962) 133.
68. S.F. Hulbert, J. Br. Ceram. Soc., 6 (1969) 11.
69. C. Kroger and G. Ziegler, Glastech. Ber., 26 (1953) 346; 27 (1954) 199.
70. M.E. Brown, A.K. Galwey and A. Li Wan Po, Thermochim. Acta, 203 (1992) 221; 220 (1993) 131.
71. C.E.H. Bawn, Chemistry of the Solid State, (Ed. W.E. Garner), Butterworths, London, 1955, Chap.10.
72. A.K. Galwey and G.M. Laverty, Proc. R. Soc. (London), A440 (1993) 77.
73. A.K. Galwey, L. Pöppl and S. Rajam, J. Chem. Soc., Faraday Trans. I, 79 (1983) 2143.
74. S.D. Bhattamisra, G.M. Laverty, N.A. Baranov, V.B. Okhotnikov and A.K. Galwey, Phil.Trans. R. Soc. (London), A341 (1992) 479.
75. A.K. Galwey, G.M. Laverty, V.B. Okhotnikov and J. O'Neill, J. Thermal Anal., 38 (1992) 421.
76. L.L. Bircumshaw and B.H. Newman, Proc. R. Soc. (London), A227 (1954) 115, 228.
77. P.G. Fox, J. Solid State Chem., 2 (1970) 491.
78. W.E. Garner and A.J. Gomm, J. Chem. Soc., (1931) 2123.
79. K.J. Gallagher, Reactivity of Solids, (Ed. G.-M. Schwab), Elsevier, Amsterdam, 1965, p.192.
80. S. Miyagi, J. Japn. Ceram. Soc., 59 (1951) 132.
81. H. Sasaki, J. Amer. Ceram. Soc., 47 (1964) 512.
82. R.W. Hutchinson, S. Kleinberg and F.P. Stein, J. Phys. Chem., 77 (1973) 870.
83. H.G. McIlvried and F.E. Massoth, Ind. Eng. Chem. Fundam., 12 (1973) 225.
84. J. Sesták and G. Berggren, Thermochim. Acta, 3 (1971) 1.

85. W-L. Ng, Aust. J. Chem., 28 (1975) 1169.
86. J. Málek and J.M. Criado, Thermochim.Acta, 175 (1991) 305.
87. J. Sesták, J. Thermal Anal., 36 (1990) 1997.
88. N. Koga, Thermochim.Acta, 258 (1995) 145.
89. A.K. Galwey and M.A. Mohamed, J. Chem. Soc., Faraday Trans. I, 81 (1985) 2503.
90. A.K. Galwey and P.W.M. Jacobs, Proc. R. Soc. (London), A254 (1960) 455.
91. A.K. Galwey and M.A. Mohamed, Thermochim. Acta, 213 (1993) 269, 279.
92. A.K. Galwey, N. Koga and H. Tanaka, J. Chem. Soc., Faraday Trans. I, 86 (1990) 531.
93. M.E. Brady, M.G. Burnett and A.K. Galwey, J. Chem. Soc., Faraday Trans. I, 86 (1990) 1573.
94. A.K. Galwey and G.M. Laverty, Thermochim. Acta, 228 (1993) 359.
95. W.L. Ng, C.C. Ho and S.K. Ng, J. Inorg. Nucl. Chem., 34 (1978) 459.
96. A.K. Galwey and M.A. Mohamed, Thermochim. Acta, 239 (1994) 211.
97. V.V. Boldyrev, Y.A. Gapanov, N.Z. Lyakhov, A.A. Politov, B.P. Tolochko, T.P. Shakhtshneider and M.A. Sheromov, Nucl. Inst. Method. Phys. Res., A261 (1987) 192.
98. G.G.T. Guarini, J. Thermal Anal., 41 (1994) 287.
99. A. Korobov, Heterogeneous Chem. Rev., 3 (1996) 477.

Chapter 4

THE INFLUENCE OF TEMPERATURE ON REACTION RATE

4.1. INTRODUCTION

Temperature exerts a strong influence on the rates of most solid state reactions. In theoretical discussions of the temperature dependencies of rates of reactions involving solids, the terminology developed for homogeneous rate processes has often been used without explicit modification to describe heterogeneous processes.

Differences between the models representing homogeneous and heterogeneous reactions will be discussed below, but first the similarities in behaviour will be outlined. In reactions of both types, reaction rates generally increase markedly with temperature. Rate coefficients almost invariably fit one or other form of the Arrhenius equation:

$$k = A \exp(- E_a/RT) \quad \text{or} \quad k = A'T^n \exp(- E_a/RT) \tag{4.1}$$

Where E_a, the *activation energy*, and A, the *pre-exponential* or *frequency factor*, are referred to as the Arrhenius parameters. E_a is often identified as the energy barrier which must be surmounted during the bond redistribution steps required to change reactants into products. A is identified with the frequency of occurrence of the high-energy reaction configuration, first represented as a molecular encounter (a collision) and later as a specific vibration in the reaction coordinate. Clearly, reference to a collision between reactant species is untenable where the reactant is immobilized in the lattice of a solid or the transformation proceeds at a narrow interface between two crystalline phases. The absolute reaction rate theory, in which the activated complex is regarded as a high energy atomic configuration, has been applied to other aspects of the behaviour of solids, e.g. diffusion in crystals, and is discussed in Section 4.5. below.

The original identification by Hood [1] of a linear relationship between log k and $1/T$ was empirical. Later the relationship was given some theoretical justification by van 't Hoff [2] who expressed the influence of temperature on equilibrium constants (K_e) by an equation of similar form:

$\ln K_e = -(\Delta H/RT) + \text{constant}$

Arrhenius [3] and Harcourt and Essen [4] developed the equation towards the form now universally referred to as the Arrhenius equation (4.1).

4.2. VARIABLES OTHER THAN TEMPERATURE

Temperature is, of course, only one of a number of variables which may influence the rates of chemical changes in crystals. Other possible variables include α, pressure of volatile product (most significant in reversible reactions), reactant pressure (in gas-solid reactions), etc. Thus the overall rate equation applicable to the decomposition of a solid may be a function of several variables [5]:

Reaction rate, $d\alpha/dt = F(\alpha, T, \ldots\ldots\ldots\ldots\ldots, X)$

It is usual to assume [5] that the contributions to rate control can be separated into independent functions of the individual variables in the form:

$\text{Rate} = F_1(\alpha)\, F_2(T) \ldots\ldots\ldots\; F_n(X)$

The influence of temperature on reaction rate can then be studied with the contributions from all the other controlling variables held constant. The standard method is to carry out several isothermal experiments at a series of different temperatures and express each set of observations in the form:

$\text{Rate} = k\, f(\alpha)$ or $g(\alpha) = k(t - t_0)$

The relationship between k and T is then usually found to be satisfactorily expressed by the linear relation between $\ln k$ and $1/T$, i.e, the Arrhenius equation (4.1). The Arrhenius parameters, A and E_a, provide a convenient and widely accepted method for the quantitative reporting of kinetic data that also permits comparisons of reactivities.

4.3. ALTERNATIVES TO THE ARRHENIUS EQUATION

The importance of the Arrhenius relationship in chemical kinetics has been emphasized by Laidler [6] in an impressive survey of the historical development of ideas relating to the application of equation (4.1) to diverse rate processes. He points out that, for many experimental studies, the data may exhibit acceptable alternative linear relationships, e.g., between $\ln k$ and T, and between $\ln k$ and $\ln T$.

This is because, over the usual limited experimental temperature interval, plots of sets of values of $1/T$, T and $\ln T$ against each other are approximately linear, as illustrated in Figure 4.1. The accuracy of most experimentally measured rate coefficients is usually insufficient to confirm or exclude a fit to one of these alternative functions. Relationships incorporating additional constants have been suggested, such as:

$$\ln k = a + b/T + c \ln T + d\,T$$

but these are described by Laidler [6] as "theoretically sterile", in that the constants do not lead to deeper understanding of the mechanisms of chemical reactions.

Flynn [7] has reviewed the forms of several possible k, T functional relationships but does not discuss the theoretical implications of the equations listed.

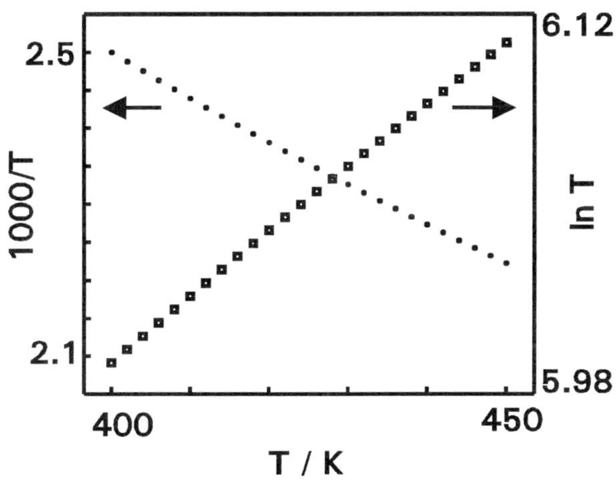

Figure 4.1.
Scaled plots of ($\ln T$)/3 and $1000/T$ against T to show the approximate linearity of the functions over the limited temperature intervals for which experimental data are usually obtainable and from which Arrhenius parameters are calculated.

4.4. DETERMINATION OF THE MAGNITUDES OF THE ARRHENIUS PARAMETERS, E_a and A

4.4.1. Arrhenius plots

The experimental methods used to determine the values of a set of rate coefficients (k_1, k_2, ..., k_n) at a series of temperatures (T_1, T_2, ... , T_n), either from a set of isothermal experiments, or from one or more programmed temperature experiments, are discussed in Chapter 5. The chemical reaction(s) to which the kinetic measurements apply should be clearly identified. Values of ln k_i are then plotted against $1/T_i$ in what is referred to as an Arrhenius plot. Linear regression (with suitable weighting for the ln function [8]) yields values for the slope (E_a/R) and the intercept (ln A) with their standard errors.

Arrhenius plots should be examined carefully because plots may be curved or show discontinuities in slope. The temperatures at which such discontinuities occur may be significant in interpreting behaviour of the reactant, e.g., there may be a phase change of the reactant within the temperature range being investigated.

Data should be collected across the widest temperature interval that is experimentally accessible to obtain the most reliable A and E_a values. The reproducibilty of rate coefficients for decomposition should be assessed at or near both ends of the temperature interval. Reported magnitudes of A and E_a should include confidence limits, estimated by the criteria mentioned, together with the temperature and α ranges to which they apply. These are not always to be found in literature reports.

When determining Arrhenius parameters, the possibility that the reliabilities of k values may vary systematically with temperature should be considered. This requires careful analysis of data for the most rapid reactions where the number of α or yield-time readings may be limited by the characteristics of the detection system. More significant, however, are the probable contributions from reactant self-heating or self-cooling on the measured reaction rates in reactant particles where the reaction enthalpy and heat flow can result in deviation from the reaction environment temperature. While such temperature dependent errors may result in variation of the slope and deviation from linearity of the Arrhenius plot, detailed studies of the effect do not appear to have been made. A discussion of some aspects of the phenomenon has been given [9]. At lower temperatures, conditions closer to the isothermal reaction are more likely to prevail. The slowest rate processes that can be investigated depend on the equipment available. Automatic data collection obviously extends the lower temperature range that is accessible to study and such developments should increase the reliability of reported magnitudes of Arrhenius parameters.

In studying any particular reaction it is desirable to demonstrate that the shape of the α-time curve does not change across the temperature range investigated. This may be judged from the α-time ranges over which plots of g(α) against time are linear, or by comparisons of plots of α against reduced-time at different temperatures. Changes in reaction kinetics and mechanism are sometimes found within the temperature interval investigated. For example, the decomposition of CoOOH ($\rightarrow Co_3O_4$) [10] below about 570 K is a nucleation and growth process with E_a = 214 kJ mol^{-1}, whereas above this temperature there is phase boundary control and E_a is significantly lower at 101 kJ mol^{-1}.

4.4.2. The rate coefficient, *k*

Care should be taken in defining the procedure for calculating values of k from the experimental data. There is always the possibility that the apparent k is a compound term containing several individual rate coefficients for separable processes (such as nucleation and growth). It is important that the dimensions of k (and hence of A) should be (time)$^{-1}$. For example, the power law (Table 3.3.) should be written as $\alpha^{1/n} = kt$ and not as $\alpha = k't^n$. Similarly the Avrami-Erofeev equation (An) is $[-\ln(1 - \alpha)]^{1/n} = kt$. The use of k' in place of k in the Arrhenius equation will produce an apparent activation energy E_a' which is n times the conventional activation energy E_a, obtained using k.

Incorporation of scaling factors, which may arise from measurement of a quantity proportional to k rather than k itself, so that $k'' = ak$, will lead to displacements in the y-intercepts, so that $\ln A = \ln A'' - \ln a$.

Benson [11] has discussed the requirements for accurate measurement of rate coefficients, k, for reactions and the uncertainties introduced into the values of E_a and A through the uncertainties in k and the temperature T for each isothermal experiment. He concludes that T must be known to about 0.03% (0.2 K at 600 K) to limit the error in k to 1.0%. Control of reaction temperatures to better than 1 K is difficult above 600 K and the resulting errors are at least 5% in k and about 10% in E_a.

4.4.3. Quantity dimensions

The slope of the Arrhenius plot has units of (temperature)$^{-1}$ and activation energies are obtained in units of (J mol^{-1}) by dividing the measured slope by the gas constant, R in J K^{-1} mol^{-1}. There is a difficulty, however, in interpreting the significance of the term "mole" in solid state reactions. In some reversible reactions, the enthalpy change (ΔH) is approximately equal to E_a, because the activation energy for the reverse reaction is small or approaching zero. Therefore, if an independently measured ΔH value is available (from DSC or DTA data), and is referred to a mole of reactant, an estimation of what could constitute "mole of activated complex" can be made.

4.4.4. Preliminary processes

The main decomposition reaction in a number of solids is preceded by an initial process [12], giving a small yield of product, typically 1 to 2%. This has often been identified as the breakdown of relatively less stable surface material, impurities or precursor reactions during the induction period. Data from the two separate processes should be analyzed independently to obtain A and E_a values for the initial process. Alternatively, the activation energy for the induction period may be estimated from Arrhenius-type plots of ln t_d against $1/T$ where t_d is the induction period. This may be specified as the time required to initiate the main reaction (Figure 3.4., curve A) and can be measured using one of several criteria: the time required to reach $\alpha = 0.01$ (or some other suitable value); the extrapolated intercept on the time axis of the plot of g(α) against time; the onset of acceleratory behaviour, etc. E_a values for such initial reactions are sometimes smaller than those for the main reaction, consistent with the view that this is the decomposition of a less-stable surface layer. Aspects of the role of outer surfaces in the decompositions of crystalline solids have been discussed by Guarini *et al.* [13].

4.4.5. Reliability of measured E_a values

Although determination of the activation energy is a major objective of most kinetic studies of solid state decompositions, the accuracy of such reported E_a values is frequently difficult to assess. Reproducibility of measurements is not always good and relatively few values have been confirmed in independent studies. E_a values are often reported to several significant figures, without provision of realistic esitmates of the uncertainties in the measurements. Arrhenius plots are generally assumed to be linear and few tests are made for possible deviations. This assumption is fundamental to most methods of non-isothermal kinetic analysis (see Chapter 5) where the consequences of deviation (or mechanism change) are often ignored. The precision achievable in measurement of E_a values for reactions of solids is unlikely to match that routinely achieved in homogeneous rate processes.

We distinguish three classes of E_a values reported for solid state decompositions.
(i) For irreversible reactions, values of E_a are largely independent of prevailing conditions, as in the decompositions of $NH_4C\ell O_4$, $KMnO_4$, NiC_2O_4, where values reported by different workers are in good agreement.
(ii) For reversible decompositions, where careful experimental measurement has excluded any contribution from the reverse process, values of E_a may have mechanistic significance.
(iii) For reactions which are kinetically sensitive to reaction conditions, E_a values represent only the overall temperature coefficient for reaction under the set of conditions used and probably do not have any further (or fundamental) significance. Many of the values of E_a given in the literature refer to reactions of this type, where there is the possibility of reversibility and this and other factors influencing kinetic behaviour have not been characterized.

Few tests of the reproducibility of E_a values for decompositions of types (i) and (ii) are to be found in the literature. Interdependent variations of A and E_a for processes classified as type (iii) may give rise to one form of compensation behaviour described below.

Both increased accuracy of measured E_a values for solid state decompositions and realistic assessments of uncertainty limits are needed to provide data suitable for the theoretical interpretation of the significance of the activation energy.

4.5. THE POLANYI-WIGNER EQUATION

An early attempt to explain Arrhenius-type behaviour in a solid state decomposition reaction was made by Polanyi and Wigner [14] who derived the expression:

$$\mathrm{d}x/\mathrm{d}t = (2\,\nu E_a/RT)\, x_0 \exp(-\,E_a/RT) \qquad (4.2)$$

The linear rate of advance of the interface, $\mathrm{d}x/\mathrm{d}t$, is expressed in terms of x_0, the incremental advance from unit reaction, and ν a vibration frequency. The term $2E_a/RT$ arises because the critical energy required can be achieved through three degrees of vibrational freedom. This model was developed through a transition state treatment where rate control was identified as the dissociation of a particular bond, strength E_a, within the advancing decomposition interface, followed by evaporation of the surface-held species. This theory of reaction rates was discussed in some detail by Bradley [15] and later successfully applied by Topley [16] to the loss of water from copper sulfate pentahydrate. (It is emphasized that discussions refer specifically to the *linear* rate of interface advance. The systematic rate changes arising from reaction geometry are accommodated in the rate equation $g(\alpha) = kt$.)

Young [17] discussed the Polanyi-Wigner equation in terms of the thermodynamics of the activation process ($\Delta S^‡$ and $\Delta H^‡$). Equation (4.2) may be written as:

$$\mathrm{d}x/\mathrm{d}t = (k_B T/h) x_0 \exp(\Delta S^‡/R) \exp(-\Delta H^‡/RT)$$

for an irreversible reaction (where k_B is the Boltzmann constant and h the Planck constant). It can be seen from this expression that the expected values for ν in equation (4.2) are calculated on the assumption that the entropy of activation ($\Delta S^‡$) is zero. Values of A calculated for reactions which did not fit the Polanyi-Wigner treatment, the so-called "abnormal" reactions [18], correspond to values of $\Delta S^‡$ of from 200 to 500 J K^{-1} mol^{-1}.

The magnitude of ν, the vibrational frequency of a crystal constituent at a surface or interface, is expected, from equation (4.2), to be of the order of 10^{13} s^{-1}. Values of the pre-exponential factor, A, calculated from experimental results have often been compared with this expected frequency and reactions have been classified (e.g., [18]) as "normal" when agreement is within a factor of 10 to 10^2, together with an E_a value close to the enthalpy of dissociation. When both A and E_a values are appreciably larger than is consistent with the above criteria, the reactions are described as "abnormal". The use of these terms has been criticized [19] as unacceptable because this assumes the universal applicability of the Polanyi-Wigner description to reactions of this type. Such a conclusion is premature because important aspects of the conditions obtaining at reaction interfaces and the controls determining reactivities have not yet been acceptably established.

A comparative survey of the available A and E_a values reported for decompositions of solids [19] showed a slight predominance of A values between 10^{11} and 10^{13} s^{-1}. This observation was not sufficiently strong to confirm the general applicability of the Polanyi-Wigner treatment. Similarly, the magnitudes of activation energies showed a spread, mainly between $100 < E_a < 230$ kJ mol^{-1}, with no obviously dominant value.

4.6. TRANSITION-STATE DESCRIPTION

Shannon [20] assumed that atoms or molecules at the reaction interface can have vibrational, and possibly also torsional and rotational, degrees of freedom with a Boltzmann distribution of energies. He also assumed that the rate determining step is decomposition and not any prior or subsequent process. Using transition-state terminology, the rate coefficient, k_r, for the reaction:

$$A(s) \rightarrow [A(s)]^{\ddagger} \rightarrow \text{products}$$

can be written as:

$$k_r = (k_B T/h)\,(Q^{\ddagger}/Q)\,\exp(-E_0/RT) \tag{4.3}$$

where $Q^{\ddagger}$ is the partition function for the activated complex $[A(s)]^{\ddagger}$, with the contribution from motion along the reaction coordinate removed, and Q is the partition function for the reactant.

Calculation of Q for the reactant is reasonably straightforward, but the assignment of a value to $Q^{\ddagger}$ requires assumptions about the structure of the activated complex. Shannon [20] used published kinetic data for 29 reactions to calculate the ratio $(Q^{\ddagger}/Q)$ in equation (4.3), and three patterns of behaviour were considered with (i) $(Q^{\ddagger}/Q) < 1$, (ii) $(Q^{\ddagger}/Q) \sim 1$ and (iii) $(Q^{\ddagger}/Q) > 1$. Two mechanisms of decomposition were distinguished [20]: (i) direct loss of gaseous product from the interface (dissociations of $CaCO_3$ and $MgCO_3$), and (ii) formation of mobile adsorbed material followed by desorption (applicable to some dehydrations where water is the mobile species and $Q^{\ddagger}/Q > 1$).

Cordes [21] extended the theoretical analysis to consider intracrystalline reactions including bimolecular processes. The treatment assumed that the energy of a molecule was not altered when a nearest neighbour became an activated complex, and that the only influence of the product species is to define an interface. In qualitative terms, low values of A were associated with strongly held ("tight") surface complexes and higher values with more mobile complexes.

4.7. THE SIGNIFICANCE OF THE ACTIVATION ENERGY

Many interpretations of the Arrhenius parameters have been proposed and discussed in the literature. Anderson [22] regarded A and E_a as having empirical rather than theoretical significance. Both must be reported to define adequately the reaction rate. Redfern [23] has suggested that the activation energy is the *average* excess energy that a reactant molecule must possess to react. The reaction dynamics treatment [24] identifies E_a as the difference between the energy of the molecules undergoing reaction and the overall average energy. Bertrand *et al.* [25] regard E_a as a composite function, comprised of contributions from numerous parameters, including the deviation from equilibrium and the thermal gradient.

Vyazovkin and Linnert [26] also regard the thermal decomposition of a solid as a multistep process. They report a quantitative analysis of the systematic variations of values of E_a with α using theoretically calculated data for two model processes: (i) a simple reversible reaction and (ii) a reversible reaction followed by an irreversible step. It was shown that the dependence of E_a on α for measurements of the dehydrations of a number of crystalline hydrates was consistent with the latter model, (ii).

Garn [5,27] believes that no discrete activated state is generated in decompositions of solids. This conclusion is based on the observed wide variations in magnitudes of calculated activation energies. Vibrational interactions transfer energy rapidly within a crystalline solid so that no substantial difference from the average energy can be achieved or sustained. The statistical distributions upon which the Arrhenius equation is based (including the Boltzmann function) are thus not applicable to solids and consequently this use of the Arrhenius equation lacks a theoretical foundation.

4.8. STRUCTURE OF THE ACTIVE REACTION ZONE AND INTERFACE CHEMISTRY

4.8.1. The rate controlling step

The application of absolute reaction rate theory to a chemical change at an interface is only useful if the calculations refer to an identified, or at least reliably inferred, model of the controlling bond redistribution step. This is a problem, because it is particularly difficult to characterize the structures of the immediate precursors to reaction in many solid state rate processes of interest. The activated species are inaccessible to direct characterization because they are usually located between reactant and product phases. The total amount of reacting material present within this layer, often of molecular dimensions, is small and irreversible chemical and textural changes may accompany opening of such specialized structures for examination or analysis. Moreover, the presence of metallic and/or opaque, ill-crystallized product phases may prevent or impede the experimental recognition of participating intermediates or essential textural features.

The Polanyi-Wigner treatment is based on the simplest model that can be envisaged for interface phenomena. Real interfaces, involving a wider range of bond redistributions than the simple dissociation of a coordinate link to water, often accompanied by restructuring, are undoubtedly more complicated. In discussing the kinetics of reactions involving effectively immobilized crystal components, it is appropriate to mention that there may be significant contributions from mechanisms that have vanishingly small probabilities in homogeneous processes. Trimolecular steps, or steps of even higher molecularity, may occur between static entities, suitably positioned. Sequences of consecutive reactions ("surface chains") may become effective routes to chemical change when the first interaction is not followed by immediate (post-collisional) separation of the participants. Temperature dependent changes in the effective concentrations of the precursors to the essential step will contribute to the magnitude of the apparent activation energy [28] which then cannot be identified with a single step rate process.

4.8.2. Types of interface structures

Examples of the diversity of possible interface structures are as follows.

Topotactic interfaces. Primary valence forces may link closely juxtaposed, or perhaps coherent, reactant and product phases so that the crystalline product retains the orientation and some structural features of the reactant [29,30]. The interface, with thickness of molecular dimensions, is defined by the discontinuity of structure and bonding within which reactivity is locally enhanced by the strain field.

Product catalysis. If the residual solid is a catalyst for breakdown of a reactant constituent, decomposition may occur within chemisorbed material at the product surface side of the interface. This model has been discussed with reference to the decompositions of some metal carboxylates where the metal is identified as a known (or expected) heterogeneous catalyst for the breakdown of the carboxylic acid concerned. These reactions may proceed through the adsorbed anion, for example in the decompositions of nickel formate [31,32] and silver malonate [33]. Anions chemisorbed on the metal surface decompose in the asymmetric bonding field, accompanied or followed by cation transfer to maintain charge balance.

Thick interfacial zone. Some interfaces give evidence of an extended structure, for example Boldyrev *et al.*, [34] show that there are water losses from within a zone up to about 150 μm thickness during the dehydration of $Li_2SO_4.H_2O$. In other dehydrations, water loss occurs in advance of the recrystallization interface as discussed for the reactions of $NiSO_4.6H_2O$ [35] and of $Ca(OH)_2$ [36].

Intervening molten phase. In several decompositions, formerly regarded as proceeding in the solid phase, microscopic evidence has shown the intervention of liquid, within which it seems possible that breakdown processes can proceed more rapidly [37]. These include the decompositions of $(NH_4)_2Cr_2O_7$ [38], NH_4ClO_4 [39] and copper(II) malonate [40]. Liquefaction during reaction, without comprehensive melting, can be difficult to recognize [40], but the kinetic behaviour is expected to be controlled by parameters characteristic of homogeneous reactions.

Excluding this last interface model (where reaction proceeds in a liquid) the reactant-product contacts described involve close interactions, often coherent bonding, between crystalline phases. Even if both solids are relatively perfect, the juxtaposition surface is a plane of bond structure discontinuity between the different substances. In most real systems it is to be expected that there will be some poorly crystallized material within which there will be locally enhanced freedom of vibrational or other movements capable of enhancing the probability of chemical reaction. Such zones may be less ordered than a crystal but more ordered and much less mobile than a liquid. This must be accompanied by local distortion of the band structure and the following discussion considers the distribution of energy within the interfacial zone [41]. A proposed representation of the electron accommodating levels across the phase junction is shown in diagrammatic form in Figure 4.2. The band structure is shown (dotted) together with interface levels ascribed to imperfections associated with the local bond distributions.

4.8.3. Energy distributions

The rate controlling steps in solid state decompositions are often identified [42] either as electron transfer or as bond rupture. The appropriate energy distribution functions applicable to electronic energy are based on Fermi-Dirac statistics while Bose-Einstein statistics apply to phonons. The relevant forms of these energy distributions for application to solid state reactions are as follows.

Electronic energy. The probability (F) that a discrete and non-degenerative energy state E_e is occupied by an electron is expressed [43] by Fermi-Dirac statistics as:

$$F = (1 + \exp [(E_e - E_f)/k_B T])^{-1}$$

where E_f is the Fermi level. When the magnitude of E_e exceeds that of E_f by several times the magnitude of $k_B T$ ($\approx$ 2.5 kJ mol^{-1} at 300 K and 5 kJ mol^{-1} at 600 K), this approximates to [43]:

$$F \approx \exp [-(E_e - E_f)/k_B T] \qquad (4.4)$$

which has the same form as the Maxwell-Boltzmann function.

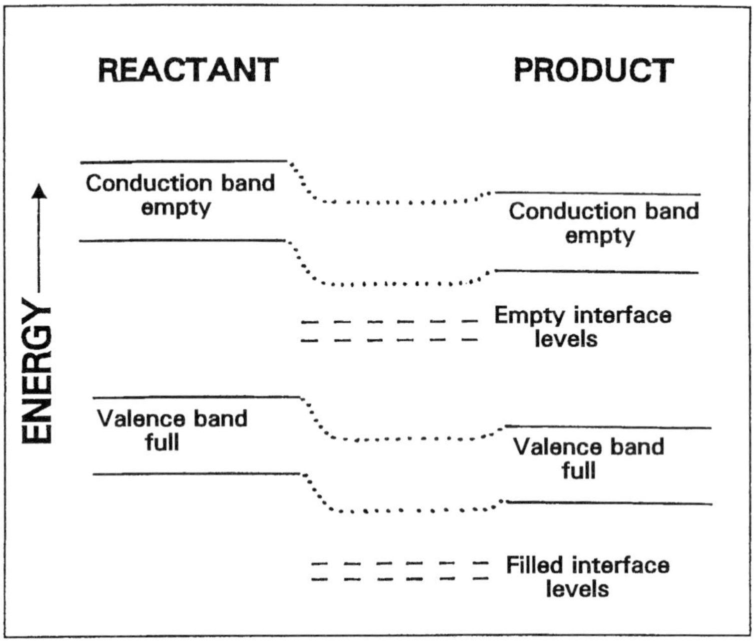

Figure 4.2.
A schematic representation of the electron accommodating interface levels across the reactant-product contact zone.

Phonon energy. If the random-walk migrations of mobile phonons [44] extend into surface-held species and vibration there is rate limiting for the reaction (as in the Polanyi-Wigner treatment) then the energy distribution function, n(ω), can be expressed through Bose-Einstein statistics as:

$$n(\omega) = [\exp(h\omega/2\pi k_B T)-1]^{-1}$$

where ω is the angular frequency of the lattice mode. When $h\omega$ is significantly greater than $2\pi\ k_B T$ the function decreases exponentially [45]:

$$n(\omega) \approx \exp(-h\omega/2\pi k_B T) \qquad (4.5)$$

an expression which is again analogous to the Maxwell-Boltzmann function.

In equations (4.4) and (4.5) above it is shown that the distribution functions for electron energy and for phonon energy approximate to the form of the Maxwell-Boltzmann expression at the highest range of the energy spread. Decompositions are usually (intentionally) investigated under temperature conditions such that bond activation is a relatively rare phenomenon and thus associated with that small fraction of the (electron or phonon) energy distribution that is satisfactorily represented by the Maxwell-Boltzmann expression. Activation then results in occupancy of those interface levels regarded as localized sites which, when energized, become precursor states from which bond reorganizations may follow.

This model of interface energetics assumes the participation of electron accommodating levels of a type hitherto not characterized. Obtaining evidence for the existence of such interface levels is expected to be experimentally difficult but, in favourable systems, absorptions might be detecable by spectroscopy at wavelengths just below the absorption edge (Figure 1.6). The model is, however, analogous to the familiar representation of impurity levels extensively used in the theory of semi-conductors [46]. It also is similar to the surface sites discussed in consideration of chemisorption [47]. If the reaction interface is portrayed as a complex bonding discontinuity, or as an array of imperfections, interface acceptor levels are unsatisfied orbitals and/or sites associated with the reactive species that participate in the chemical change.

Activation of *interface levels* by the most energetic electron or phonon quanta according to distributions expressed by equations (4.4) or (4.5) provides a theoretical explanation of the fit of the Arrhenius equation to solid state decompositions proceeding at an interface [41].

4.9. THE COMPENSATION EFFECT

4.9.1. Definition

Compensation behaviour is defined as the existence of a relationship between the values of A and E_a for a group of related rate processes such that:

$$\ln A = a + bE_a \quad (4.6)$$

'Compensation' refers to the behaviour pattern in which a rise in E_a (which will decrease the rate of a reaction at any particular temperature) is partially or completely offset by an increase in A [28]. From equation (4.6), a temperature, T_i ($=(bR)^{-1}$) exists [48], at which all reactions of the set expressed by equation (4.6) proceed at equal rates. Alternative descriptions for the phenomenon are *isokinetic behaviour* or the *θ-rule*, where $\theta = T_i$ is the temperature of equal reaction rates.

Compensation behaviour has been noted for a wide variety of diverse chemical and physical phenomena, in addition to some solid state decompositions which are reviewed below. The effect is perhaps best known from heterogeneous catalysis [28,48] and is also found [28] for homogeneous reactions in solution, electrode processes, viscosity, electrical conductivity, etc. The numerous reports in the literature and wide occurrence of the effect have led to a wide debate about the theoretical significance of isokinetic phenomena. Some problems concerning the magnitudes of experimental errors and the importance of defining unambiguously what is meant by the compensation effect have recently been stressed [49]. The topic remains controversial because no resolution of conflicting attitudes has yet been achieved. Many workers regard the effect as an artefact and certainly no generally acceptable theoretical explanation has so far been provided. Koga [50] has recently reviewed the significance of compensation behaviour to solid state reactions analyzed by thermal methods and his bibliography lists many relevant publications. The probability of mutual dependence of Arrhenius parameters is discussed with reference to: (i) the temperature interval of the reaction, (ii) α, (iii) f(α) and (iv) the isokinetic hypothesis. The possibility that the compensation parameter b (equation (4.6))is 'even more directly correlated with the strength of the bond to be broken than the activation energy' is discussed by Zsakó *et al.* [51].

For rate processes in which the Arrhenius parameters are independent of reaction conditions, it may be possible to interpret the magnitudes of A and E_a to provide insights into the chemical step that controls the reaction rates. However, for a number of reversible dissociations (such as $CaCO_3$, $Ca(OH)_2$, $Li_2SO_4H_2O$, etc.) compensation behaviour has been found in the pattern of kinetic data measured for the same reaction proceeding under different experimental conditions. These observations have been ascribed to the influence of procedural variables such as sample masses, pressure, particle sizes, etc., that affect the ease of heat transfer in the sample and the release of volatile products. The various measured values of A and E_a cannot then be associated with a particular rate controlling step. Galwey and Brown [52] point out that few studies have been specifically directed towards studying compensation phenomena. However, many instances of compensation behaviour have been recognized as empirical correlations applicable to kinetic data to kinetic data reported for groups of related rate processes. A possible qualitative explanation for this is that the same reaction, studied under different conditions, proceeds within almost identical temperature intervals, fulfilling or approximating to the isokinetic condition. Thus there is a common reactivity level (*k* value) which means that the variations of *A* and E_a must conform to the compensation pattern (equation 4.6) [48].

4.9.2. Common features

Compensation effects apply to sets of reactions incorporating some common feature [28]. No compensation trend was discernable within a large group of diverse solid state decompositions [19]. Defining the rate processes that constitute a recognizable set is thus a problem that needs attention. At least two types of compensation sets can be recognized in solid state decompositions: (i) decompositions of groups of different but chemically related reactants, and (ii) decomposition of a single reactant under different conditions.

4.9.3. Decompositions of chemically related but different compounds

The compensation phenomena considered here apply to apparent *A* and E_a values measured for similar reactions in groups of chemically related compounds. The controlling steps in such breakdowns may be closely comparable, and occur within the same type of crystallographic environment and/or have other common mechanistic features.

As examples of series of related reactions, compensation effects have been described [53] for the thermal decompositions of [CoX_2 (aromatic amine)$_2$] type complexes (7 reactions) and also for a series of cobalt (III) and chromium (III) complexes (22 compounds studied in which two compensation trends were identified). Later work [54] examined the dehydrations and deamminations of dioximine complexes (two compensation trends identified), and [$Co(NCS)_2$(ammine)$_2$]-type complexes (three compensation trends identified). The systems involving larger entropy changes required less energy for activation [53]. Separate compensation plots for the dehydrations and the decompositions of eleven alkali and alkaline-earth metal dithionates were described by Zsakó *et al.* [55].

Vyazovkin and Linnert [56] argue that kinetic data, *A* and E_a, values obtained on the assumption of a one-step reaction may be incorrect because the possibility that thermal decompositions proceed by multistep processes has been ignored. This potential error can be avoided by using isoconversional methods to calculate Arrhenius parameters as a function of α. A real isokinetic relationship in a multistep process can be identified from the dependence of T_i, and its confidence limits, on α. The contribution from the second reaction step is negligible at the start of chemical change and thereafter rises as α increases.

4.9.4. Decomposition of a single reactant under different conditions

Several instances of compensation behaviour have been described, typically for endothermic, reversible decompositions. Kinetic characteristics are sensitive to reaction conditions, most notably the effects of the presence of the volatile product and heat flow to and within the reaction zones undergoing self-cooling. The following are examples of reactions studied in detail.

Calcium carbonate. For calcite dissociation it has been emphasized [57] that it is essential to remove all contributions from the reverse reaction to obtain reliable kinetic data. At low pressures (below about 10^{-4} Torr), the value of E_a found [57] (205 kJ mol^{-1}) was appreciably greater than the dissociation enthalpy (178 kJ mol^{-1}). In many other studies, under less rigorously defined experimental conditions, values of $E_a \approx \Delta H$. It has, however, long been accepted [58] that kinetic parameters (rate equation $g(\alpha) = kt$, A and E_a) are influenced by reaction conditions. In particular, the apparent magnitudes of the Arrhenius parameters, reported by different workers, using varied experimental methods including rising temperature experiments, exhibit a remarkable spread, from A about 10^2 s^{-1} and E_a = 108 kJ mol^{-1} to exceptionally large apparent values of A (more than 10^{100} s^{-1}) and E_a greater than 1500 kJ mol^{-1} [59,60]. Clearly this spread of values cannot be associated with a single, simple chemical reaction.

The existence of compensation behaviour can be accounted for as follows. All samples of calcite undergo dissociation within approximately the same temperature interval, many kinetic studies include the range 950 to1000 K. The presence of CO_2 (product) may decrease reactivity and a delay in heat flow into the reactant will decrease the reaction temperature. Thus, under varied conditions, the reaction occurs close to a constant temperature. This is one of the conditions of isokinetic behaviour (groups of related reactions showing some variations of T_i within the set will nonetheless exhibit a well-defined compensation plot [61]). As already pointed out, values of A and E_a calculated for this reaction, studied under different conditions, show wide variation. This can be ascribed to temperature-dependent changes in the effective concentrations of reaction precursors, or in product removal [28] at the interface, and/or heat flow. The existence of the (close to) constant T_i for the set of reactions, for which the Arrhenius parameters include wide variations, requires (by inversion of the argument presented above) that the magnitudes of A and E_a are related by equation (4.6).

Recognizing this sensitivity of reaction rates to prevailing conditions, several studies have reported systematic measurements of the influences of procedural variables on kinetic parameters. Wilburn *et al.* [62] used TG and DTA (1 and 7 K min^{-1}) data to measure $CaCO_3$ decomposition rates and peak temperatures. DTA and DTG curves were shown to depend on sample mass, heating rate and the partial pressure of CO_2. (A generally similar pattern of behaviour was reported for the under rising temperature (5 K min^{-1}) conditions by Wei and Luo [64] showed that A and E_a values for dispersed samples ($A \approx 10^{14}$ s^{-1} and $E_a \approx 310$ kJ mol^{-1}) were greater than those for aggregated material ($A \approx 10^7$ s^{-1} and $E_a \approx 210$ kJ mol^{-1}). Decomposition times were longer in samples composed of larger particles. Magnitudes of apparent Arrhenius parameters increased with % CO_2 in the atmosphere from $E_a = 311$ kJ mol^{-1} and $A = 3 \times 10^{14}$ s^{-1} in nitrogen to $E_a = 1065$ kJ mol^{-1} and $A = 3 \times 10^{44}$ s^{-1} in 30% CO_2. Karagiannis *et al.*, [65] reported a systematic study of the effects of procedural variables for the same decomposition, using DTA and DTG measurements analyzed by a modified Coats-Redfern method. Quantitative trends in the changes of E_a values with variations of sample mass and heating rate are reported, based on the fit of rate data to a reaction order rate equation.

These largely qualitative trends, demonstrate the significant influences of reactant mass, particle size, partial pressure of CO_2, and other variables, on this reaction and are important in identifying factors that must be controlled in manufacturing techniques. The quantitative effects of CO_2 availability and self-cooling on the rates and chemistry of interface processes, could perhaps be determined by extending the Beruto and Searcy approach [57] to reactions in controlled CO_2 pressures [66] or at higher temperature where self-cooling contributions become significant.

Calcium hydroxide. The dissociation of calcium hydroxide (reversible and highly endothermic) exhibits many points of similarity with calcite decomposition and it is not surprising, therefore, that this reaction exhibits compensation [36]. The relationship between A and E_a fits equation (4.6) and, as with calcite, this may be the result of contributions from the reverse process.

Dehydration of lithium sulfate monohydrate. This reaction has recently been considered [67] but not adopted as a system for possible use in comparative kinetic studies. Various complementary experimental techniques [68] were used to investigate the same reaction. The Arrhenius parameters obtained exhibited compensation behaviour that was explained by kinetic influences similar to those described for calcite. The difficulties inherent in obtaining reproducible kinetic data make this reaction unsuitable for use as a model system in comparative experiments by different researchers.

4.10. REACTION MECHANISMS

References to the formulation of reaction mechanisms throughout this chapter have emphasized the possibility that the transition state theory of reaction kinetics *may* not be applicable to chemical changes proceeding in the solid state and crystolysis reactions in particular. For many of the rate processes of interest, little information is available concerning interface structures at the molecular scale. The reaction is available concerning interface structures at the molecular scale. The reaction environment in a solid is inherently more complicated than the (brief) transitional encounter and exchange of energy upon which transition state theory is based. In the crystalline environment, neighbouring species may exert appreciable influences on the controlling bond rearrangement steps. Repeated encounters and interactions with short-lived intermediates are possible and topotactic or other structural constraints may operate, leading to situations far removed from that of a "free" activated complex.

Only for very few reactions has the value of the activation energy been confirmed, within experimental error, by two (or more) independent studies. Examples of decompositions where this has been achieved, include $KMnO_4$, $RbMnO_4$ and $CsMnO_4$ [69] (E_a = 160 to 170 kJ mol^{-1}) and NH_4ClO_4 [39] (E_a about 140 kJ mol^{-1}). These decompositions are irreversible. Activation energies reported for reversible processes are much more dependent upon local conditions within the reaction zone. This is apparent from the compensation behaviour observed (as discussed above) for the dissociations of $CaCO_3$, $Ca(OH)_2$ and $Li_2SO_4.H_2O$. Thus carefully controlled experimental conditions are required [57,70] to obtain values of E_a which may be identified with the controlling forward step, uninfluenced by the reverse reaction.

In the absence of a more completely characterized model for the reaction interface, it is often assumed (implicitly) that the transition state theory is applicable. This assumption may have hampered the development of a better model. Two areas which show potential for refinement of the theory are detailed textural studies, where possible, of reaction zones and spectroscopic studies of electron energy distributions during reactions in solids [41].

Attempts are often made to interpret the experimentally determined values of E_a in terms of rupture of a particular bond [71], to provide support for a proposed mechanism [72], or to compare E_a with the dissociation enthalpy [73] of a reversible reaction [18]. L'vov and co-workers [74,75] (see more detail in Chapter 2) have recently explored the consequences of treating decomposition as complete sublimation [76] of the reactant followed by condensation of those components which yield the solid residue. They have compared E_a values with enthalpies of vaporization.

Many of the published E_a (and A) values do not adequately define the experimental conditions which applied, the reproducibility of the experiments and numerical estimates of the uncertainties in these values, or the reaction stoichiometry upon which their calculations are based.

REFERENCES

1. J.J. Hood, Phil. Mag., 6 (1878) 731; 20 (1885) 323.
2. H. van't Hoff, Vorlesungen ober Theoretische und Physikalische Chemie, Braunschweig, Berlin, 1903.
3. S. Arrhenius, Z. Phys. Chem. (Leipzig), 4 (1889) 226.
4. A.V. Harcourt and W. Essen, Phil. Trans. R. Soc. (London), A186 (1895) 187; 212 (1913) 187.
5. P.D. Garn, Thermochim. Acta, 135 (1988) 71; 160 (1990) 135.
6. K.J. Laidler, J. Chem. Educ., 49 (1972) 343; 61 (1984) 494.
7. J.H. Flynn, Thermochim. Acta, 300 (1997) 83.
8. P.R. Bevington, Data Reduction and Error Analysis for the Physical Sciences, McGraw-Hill, New York, 1969, p.180.
9. A.K. Galwey and P.W.M. Jacobs, Trans. Faraday Soc., 56 (1960) 581.
10. P. O'Brien and U. Patel, J. Chem. Soc., Dalton Trans., (1982) 1407.
11. S.W. Benson, The Foundations of Chemical Kinetics, McGraw-Hill, New York, 1960, p.91.
12. A.K. Galwey and M.E. Brown, J. Chem. Soc., Faraday Trans. I, 78 (1982) 411.
13. G.G.T. Guarini and L. Dei, Thermochim. Acta, 269/270 (1995) 79; M.E. Brown, A.K. Galwey and G.G.T. Guarini, J. Thermal Anal., 49 (1997) 1135.
14. M. Polanyi and E. Wigner, Z. Phys. Chem. Abt. A, 139 (1928) 439.
15. R.S. Bradley, Phil. Mag., 7 (1931) 290; J. Phys. Chem., 60 (1956) 1347.
16. B. Topley, Proc. R. Soc. (London), A136 (1932) 413.
17. D.A. Young, Decomposition of Solids, Pergamon, Oxford, 1966, p.43.
18. W.E. Garner, Chemistry of the Solid State, (Ed. W.E. Garner), Butterworths, London, 1955, Chap.8.
19. A.K. Galwey, Thermochim. Acta, 242 (1994) 259.
20. R.D. Shannon, Trans. Faraday Soc., 60 (1964) 1902.
21. H.F. Cordes, J. Phys. Chem., 72 (1968) 2185.
22. H.C. Anderson, Thermal Analysis, (Ed. H.G. McAdie), Chem. Inst. Canada, Toronto, 1967, p.37.
23. J.P. Redfern, Differential Thermal Analysis, (Ed. R.C. MacKenzie), Academic, New York, Vol.1, 1970, p.123.
24. M. Menzinger and R. Wolfgang, Angew. Chem., 8 (1969) 438.
25. G. Bertrand, M. Lallemant and G. Watelle, J. Thermal Anal., 13 (1978) 525.
26. S. Vyazovkin and W. Linnert, Int. J. Chem. Kinet., 27 (1995) 597.
27. P.D. Garn, J. Thermal Anal., 7 (1975) 475; 10 (1976) 99; 13 (1978) 581.

28. A.K. Galwey, Adv. Catal., 26 (1977) 247.
29. J.R. Günter and H.R. Oswald, Bull. Inst. Chem. Res. Kyoto Univ., 53 (1975) 249; H.R. Oswald, Thermal Analysis, (Ed. H.G. Wiedemann), Birkhaueser, Basel, 1980, Vol.1, p.1.
30. V.V. Boldyrev, React. Solids, 8 (1990) 231.
31. J. Fahrenfort, L.L. van Reyen and W.H.M. Sachtler, The Mechanism of Heterogeneous Catalysis, (Ed. J.H. de Boer), Elsevier, Amsterdam, 1960, p.23.
32. A.K. Galwey, M.J. McGinn and M.E. Brown, Reactivity of Solids, (Eds J.S. Anderson, M.W. Roberts and F.S. Stone), Bristol, Chapman and Hall, London, 1972, p.431.
33. A.K. Galwey and M.A. Mohamed, J. Chem. Soc., Faraday Trans.I, 81 (1985) 2503.
34. V.V. Boldyrev, Y.A. Gapanov, N.Z. Lyakhov, A.A. Politov, B.P. Tolochko, T.P. Shakhtshneider and M.A. Sheromov, Nucl. Inst. Method. Phys. Res., A261 (1987) 192.
35. G.G.T. Guarini, J. Thermal Anal., 41 (1994) 287.
36. A.K. Galwey and G.M. Laverty, Thermochim. Acta, 228 (1993) 359.
37. A.K. Galwey, J. Thermal Anal., 41 (1994) 267.
38. A.K. Galwey, L. Pöppl and S. Rajam, J. Chem. Soc., Faraday Trans. I, 79 (1983) 2143.
39. A.K. Galwey and M.A. Mohamed, Proc. R. Soc. (London), A396 (1984) 425.
40. N.J. Carr and A.K. Galwey, Proc. R. Soc. (London), A404 (1986) 101.
41. A.K. Galwey and M.E. Brown, Proc. R. Soc. (London), A450 (1995) 501.
42. M.E. Brown, D. Dollimore and A.K. Galwey, Reactions in the Solid State, Comprehensive Chemical Kinetics, (Eds C.H. Bamford and C.F.H. Tipper),Vol. 22, Elsevier, Amsterdam, 1980.
43. G.G. Roberts, Transfer and Storage of Energy by Molecules, The Solid State, (Eds G.M. Burnett, A.M. North and J.N. Sherwood),Vol.4, Wiley, London, 1974, p.153.
44. J.R. Hook and H.E. Hall, Solid State Physics, Wiley, Chichester, 2nd Edn, 1991, Chap.2.
45. R.J. Elliott and A.F. Gibson, Introduction to Solid State Physics and its Applications, Macmillan, London, 1974.
46. A.R. West, Solid State Chemistry, Wiley, Chichester, 1984, p.513; A.L.G. Rees, Chemistry of the Defect Solid State, Methuen, London, 1954.
47. J. Cunningham, Comprehensive Chemical Kinetics, (Eds C.H. Bamford and C.F.H. Tipper),Vol.19, Elsevier, Amsterdam, 1984, Chap.3, p.294, 305.
48. G.C. Bond, Catalysis by Metals, Academic Press, London, 1962, p.141.

49. M.P. Suárez, M. Palermo and C.M. Aldao, J. Thermal Anal., 41 (1994) 807.
50. N. Koga, Thermochim. Acta, 244 (1994) 1.
51. J. Zsakó, J. Horák and Cs. Várhelyi, J. Thermal Anal., 20 (1981) 435.
52. A.K. Galwey and M.E. Brown, Thermochim. Acta, 300 (1997) 107; A.K. Galwey, Thermochim. Acta, 294 (1997) 205.
53. J. Zsakó and Cs. Várhelyi, J. Thermal Anal., 7 (1975) 33; J. Zsakó, Cs. Várhelyi, G. Liptay and K. Szilágyi, J. Thermal Anal., 7 (1975) 41.
54. J. Zsakó, Cs. Várhelyi, G. Liptay and A. Borbély-Kuszmann, J. Thermal Anal., 31 (1986) 285;J. Zsakó, G. Liptay and Cs. Várhelyi, J. Thermal Anal., 31 (1986) 1027.
55. J. Zsakó, E. Brandt-Petrik, G. Liptay and Cs. Várhelyi, J. Thermal Anal., 12 (1977) 421.
56. S. Vyazovkin and W. Linnert, J. Solid State Chem., 114 (1995) 392.
57. D.Beruto and A.W. Searcy, J. Chem. Soc., Faraday Trans.I, 70 (1974) 2145.
58. A.L. Draper, Proc. 14th Robert A. Welch Found. Conf. Chem. Res., 14 (1970) 214.
59. J. Zsakó and H.E. Arz, J. Thermal Anal., 6 (1974) 651.
60. P.K. Gallagher and D.W. Johnson, Thermochim. Acta, 14 (1976) 255.
61. R.K. Agrawal, J. Thermal Anal., 31 (1986) 73.
62. F.W. Wilburn, J.H. Sharp, D.M. Tinsley and R.M. McIntosh, J. Thermal Anal., 37 (1991) 2003.
63. J.H. Sharp, F.W. Wilburn and R.M. McIntosh, J. Thermal Anal., 37 (1991) 2021.
64. H. Wei and Y. Luo, J. Thermal Anal., 45 (1995) 303.
65. G.N. Karagiannis, T.C. Vaimakis and A.T. Sdoukos, Thermochim. Acta, 262 (1995) 129.
66. T. Darroudl and A.W. Searcy, J. Phys. Chem., 85 (1981) 3971.
67. M.E. Brown, R.M. Flynn and J.H. Flynn, Thermochim. Acta, 256 (1995) 477.
68. M.E. Brown, A.K. Galwey and A. Li Wan Po, Thermochim. Acta, 203 (1992) 221; 220 (1993) 131.
69. M.E. Brown, A.K. Galwey, M.A. Mohamed and H. Tanaka, Thermochim. Acta, 235 (1994) 255.
70. T.B. Flanagan, J.W. Simons and P.M. Fichte, Chem. Commun., (1971) 370.
71. J.E. House and C. Flentge, Thermochim. Acta, 24 (1978) 117.
72. J. Ribas, A. Escuer and M. Montfort, Thermochim. Acta, 76 (1984) 201; Inorg. Chem., 24 (1985) 1874.
73. H. Tagawa and H. Saijo, Thermochim. Acta, 91 (1985) 67.

74. B.V. L'vov and A.V. Novichikhin, Thermochim. Acta, 290 (1997) 239; 315 (1998) 135,145; B.V. L'vov, Spectrochim. Acta, B, 52 (1997) 1.
75. B.V. L'vov *et al.* ; Spectrochim.Acta, B, 36 (1981) 397; Zh.Anal.Khim., 36 (1981) 2085; 39 (1984) 221, 1206; 40 (1985) 792.
76. P.J. Gardner and P. Pang, J. Chem. Soc., Faraday Trans. I, 84 (1988) 1879.

Chapter 5

ANALYSIS AND INTERPRETATION OF EXPERIMENTAL KINETIC MEASUREMENTS

5.1. INTRODUCTION

The main purposes of kinetic studies are: (i) to obtain information about the *reaction mechanism*, which may prove useful in modifying the course of that reaction or in predicting the behaviour of similar, as yet untested, reactions; and/or (ii) to measure *kinetic parameters*, which may allow behaviour to be interpolated or extrapolated to conditions of reaction different from those for which the measurements were made.

The experimental data used in such an analysis may be: (i) several sets of measurements of some property which has been demonstrated to be proportional to either the fractional extent of reaction, α, or the rate of reaction, $d\alpha/dt$, against time t, at a series of different *isothermal* temperatures; or (ii) similar measurements made during *programmed temperature* experiments, possibly at a series of different but constant heating rates.

5.2. THE EXPERIMENTAL RESULTS

Various ways of measuring the fractional extent of reaction, α, or the rate of reaction, $d\alpha/dt$, have been described in Chapter 3. Before a detailed kinetic analysis of any set of α, t (or α, T) results is attempted, it is useful to assess the validity of the results. The following factors require consideration. (i) It is essential to confirm that α has been meaningfully defined, related to reaction stoichiometry and is capable of accurate measurement. (Errors in the measurement of the final yield, e.g. p_f or m_f, may be particularly significant.) (ii) The main reaction may be preceded by a kinetically distinct initial surface process. (iii) The composition of the products may vary with the extent of reaction, α. (iv) Large enthalpy changes during reaction may cause local self-heating or self-cooling. (v) Rates of nucleation and/or growth may vary with the crystal surfaces involved and in different polymorphs of the same compound. (vi) The rate of reaction may depend on the surface area of reactant and may be affected by the pretreatment of the sample, e.g. by an initial dehydration step, pre-irradiation, grinding, ageing, adsorption, etc. (vii) If the gaseous product of a reversible reaction is not being effectively and completely removed from the neighbourhood of the sample, the apparent kinetic characteristics may be influenced so that there are variations in behaviour with prevailing conditions. (viii) The reactant may melt and/or sublime.

The *reproducibility* of the sets of α, t or α, T data under fixed conditions of isothermal temperature, or at a constant heating-rate, may be an indication of the validity of the experimental conditions, (vii) above. Results of successive similar experiments, using the same and different sample preparations, should be compared, including investigations of any influence of the sample mass on the rate data [1]. Such influences are often indications of the operation of factors (iv) self-heating or cooling, and/or (vii) hindered escape of gaseous product. While no clear criteria have been agreed as to what constitutes reproducible behaviour, the variations in α-time values between similar experiments can be measured and reported. Experiments using different reactant masses may be used to determine the extents and influences of self-heating (-cooling), reverse and secondary reactions, all of which are expected to increase with the amount of solid reacting. Kinetic characteristics may be markedly changed by reactant damage or pretreatment (e.g., an initial dehydration step, pre-irradiation, surface abrasion, ageing, grinding, pelleting, annealing etc.). Comparisons of reactivities of untreated reactant samples with those subjected to a particular treatment may provide information that can be used in the formulation of a reaction mechanism.

Some experimental techniques may produce relatively noisy α, t or α, T traces and it may be advisable to introduce some *smoothing* of the data, as no model is sufficiently sophisticated to allow for temporary small fluctuations. Any reluctance to resort to such a procedure can be weighed against the consideration that the extent of mathematical smoothing can be monitored, while the smooth experimental traces from other measurement techniques may contain an unknown amount and type of instrumental damping.

Thus data for isothermal kinetic analyses consist of sets of measured values of α, t; $(\mathrm{d}\alpha/\mathrm{d}t)$, t; $(\mathrm{d}\alpha/\mathrm{d}t)$, α, etc., that may be interconverted. The temperature (and accuracy limits of all data) for each experiment are also recorded. For programmed temperature (non-isothermal) experiments similar data are recorded, together with the important additional temperature-time relationship, so that (if required) values of α, t and T can be calculated for each measurement.

5.3. IDENTIFICATION OF THE RATE EQUATION THAT GIVES THE BEST REPRESENTATION OF THE EXPERIMENTAL DATA

One objective of a kinetic analysis is to identify which, if any, of the rate equations (in an appropriate form) from Table 3.3. provides the most acceptable description of the experimental α, t or α, T data. In deciding what constitutes an "*acceptable description*", there are at least two main aspects to be considered: (i) the purely mathematical "fit" of the experimental data to the relationship between α and t, $(d\alpha/dt)$ and t or $(d\alpha/dt)$ and α, required by the models listed in Table 3.3., together with the range of α across which this expression satisfactorily represents the data (whether the fit varies with temperature is also important) and (ii) the evidence in support of a kinetic model obtained by complementary techniques such as optical and electron microscopy, spectroscopy etc. (see Chapter 6).

5.4. METHODS OF KINETIC ANALYSIS OF ISOTHERMAL DATA

5.4.1. Introduction

The main approaches which have been used in kinetic analysis of isothermal data for decompositions and other reactions of solids are listed below and explained in greater detail in Sections 5.4.5. to 5.4.7.

(i) The linearity of plots of $g(\alpha)$ (from Table 3.3.) against time is determined.

(ii) Plots of α against measured values of reduced-time are compared with similar plots calculated for the rate equations in Table 3.3.

(iii) Plots of measured values of $(d\alpha/dt)$ against either α or t are compared with similar curves calculated for the rate equations in Table 3.3. [2].

(iv) The linearity of plots of $(d\alpha/dt)$ against $f(\alpha)$ (from Table 3.3.) is determined [3].

The decision as to which of the available rate equations provides the most acceptable fit to a given set of isothermal (α_i, t_i) data requires consideration of:

(i) the α-time ranges over which the models themselves can be most clearly distinguished [4];

(ii) the effects of experimental errors, including random scatter in the measured values of α, which may reduce distinguishability; and

(iii) the possibility that the reaction may be represented by more than a single rate equation (different equations are sometimes applicable across different α intervals and such composite kinetic interpretations require additional supporting evidence [5]).

There is no general agreement on the range of α over which the fit should be acceptable. Carter [6] stated that conformity should be observed over virtually the complete range $0.0 < \alpha < 1.0$. In contrast, many literature reports describe the applicability of different rate equations during comparatively limited α ranges and, hence, a sequence of models is proposed to apply during different stages of reaction. The acceleratory models can only apply during the initial part of a sigmoid α-t curve and another model is required to account for the deceleratory region of the curve.

The value of the geometric interpretations derived from macroscopic kinetic measurements is increased if they can be supported by complementary evidence such as microscopic observations.

Before considering the above methods of kinetic analysis of the overall reaction in detail, it is important to identify any particular problems which relate to the induction period, the acceleratory and the deceleratory regions (see Figure 3.4.).

5.4.2. The induction period

The delay, t_0, before the onset of the main reaction may include contributions from (i) the time, t_h, required for the sample to attain reaction temperature, (ii) a further delay, t_d, resulting from changes within the sample, e.g. water removal (endothermic and therefore self-cooling) from a hydrate, phase transitions, etc. and (iii) the time, t_i, required to complete the reaction steps preceding establishment of the main reaction, which is to be regarded as the true induction period. The values of t_h, t_i and t_d may vary differently with temperature so that the temperature dependence of $t_0(=t_h + t_d + t_i)$ may be complex and may be very different from that of the main reaction. The magnitude of t_0 can be measured from the intercept on the t-axis of a g(α)-time plot.

The existence of the induction period can introduce uncertainty into a reduced-time analysis (see below). The yield of product formed during a precursor (perhaps surface) reaction, having different kinetic properties from the subsequent decomposition, must be subtracted before kinetic analysis of the main reaction.

5.4.3. The acceleratory region

Kinetic analyses of solid state reactions have often concentrated on the initial stages of reaction, because the measurements obtained can sometimes be considered with reference to microscopic observations (Chapter 6). The power law and exponential law are often good descriptions of the early stages. Possible routes to the determination of the power law exponent are plots of $\alpha^{1/n}$ against time, or ln α against ln t. Corrections have to be made for the induction period, t_0, and any initial rapid evolution of gas, α_0. Plots of $\alpha^{1/n}$ against t were found [7,8] to be a most insensitive method for the determination of n, while the values of n from plots of log α against log t were up to 8% in error, the deviation being greatest at low α.

5.4.4. The deceleratory period

It is particularly difficult to correlate experimental data towards the end of the deceleratory region (α approaching unity) with the requirements of a particular rate equation [4]. This lack of sensitivity arises from the close similarities in the shapes of the α-time curves for some models (Figure 3.5.). The poor fit of kinetic equations in the final stages of reaction has been attributed to effects such as variations in particle-size, disintegration of crystallites, sintering, chemisorption of product gases on the solid residue (including the approach to an equilibrium), poisoning of reaction, onset of a further (different) reaction, etc. Some of the kinetic models also become undefined at long times, e.g. for the F1 equation $\alpha = 1.00$ can only be reached at infinite time. Errors in the determination of the final yield of products are significant. A random error of ± 5% can have [4] an appreciable effect (up to 20%) on the magnitude of the measured rate coefficient, k, and the shapes of α-time curves can change sufficiently to reduce the distinguishability of alternative kinetic models. The shapes of the latter parts of sigmoid curves are also similar to those of the deceleratory equations.

5.4.5. Testing of linearity of plots of g(α) against *t*

This method is the simplest and is well-suited to implementation in computer programs or spreadsheets [9]. Various standard statistical criteria, e.g. the correlation coefficient, r; the standard error of the slope of the regression line, s_b; or the standard error of the estimate of g(α) from t, s_{yx}, are used to quantify the deviation of a set of experimental points from the calculated regression line through them [10]. The inadequacies of r as an indicator of fit have been stressed [4,11]. The use of s_b is preferable in that its value is dependent upon the range of t used in the analysis.

The use of a single parameter to express the deviation of the data from the least-squares line does not, however, reveal whether deviations are systematic or approximately random. The magnitudes and directions of such deviations and their variations with α can, however, be useful [4] in identifying the most appropriate rate equation, and plots of residuals [$g(\alpha)_{EXPTL} - g(\alpha)_{PREDICTED}$], against time have been recommended. Each kinetic model within similar groups may then be associated with the alternative model with which it is most likely to be confused (Table 5.1). Note that high values of the correlation coefficient, r, are obtained on analysis using the incorrect expression, even over the wide range $0.05 < \alpha < 0.95$. When the range used in the analysis is shortened [4] to $0.20 < \alpha < 0.80$, distinguishability becomes even more difficult. A similar result is found if the wider α-range is used but a random error of up to 1% of the value of each α is introduced into the data set. The residual curves obtained on misapplication of an analyzing expression to each kinetic expression have been classified according to their shapes [4].

Once a satisfactory fit has been obtained for a rate equation, $g(\alpha) = k(t - t_0)$, the value of k and its standard error, s_b, may be determined from the slope of the plot. If the form of $g(\alpha)$ does not change with temperature, the values of k at a series of isothermal temperatures, T, can be used in a conventional Arrhenius plot (Chapter 4) to calculate values for E_a and A.

For those rate equations containing exponents, e.g the Avrami-Erofeev equations (An), it would appear that plots of $\ln[-\ln(1 - \alpha)]$ against $\ln(t - t_0)$ would provide the most direct method for the determination of the value of the exponent, n. Such plots are, however, notoriously insensitive and errors in t_0, together with contributions from initial rate processes, can influence the apparent value of n. Non-integer values of n have been reported [12].

Table 5.1.
Major problems of distinguishability in isothermal kinetic analysis [4]

Rate equation considered	Closest alternative analyzing expression	Correlation coefficient, r for "incorrect" analysis
Deceleratory group ($0.05 < \alpha < 0.95$)		
F1	R3	0.9902
R2	R3	0.9979
R3	R2	0.9979
Sigmoid group ($0.05 < \alpha < 0.95$)		
A2	A3	0.9960
A3	A4	0.9989
A4	A3	0.9989
B1	A4	0.9986
Deceleratory regions of sigmoid curves ($0.5 < \alpha < 0.97$)		
A2	R3	0.9997
A3	R3	0.9997
A4	R3	0.9993
B1	R3	0.9984

5.4.6. Reduced-time scales and plots of α against reduced-time

In the rate equation [$g(\alpha) = k\, t$] for the reaction investigated, $g(\alpha)$ is the kinetic model (see Table 3.3.), k is the rate coefficient which is a function of temperature (see Chapter 4), and t is the reaction time *after correction* for any induction period, t_0, (see Section 5.4.2.). The influence of temperature can be eliminated, provided that the kinetic model, $g(\alpha)$, does not change across the temperature range investigated, by introducing a reduced-time scale, t_{red}. This involves dividing each corrected time value, t, by the (corrected) time required to reach some reference value of α, often but not necessarily $\alpha = 0.50$. $t_{red} = t / t_{0.5}$. Thus, for example:

$$g(0.50) = k\, t_{0.50} \quad \text{and} \quad g(\alpha) = g(0.50)\,(t/t_{0.50}) = g(0.50)\, t_{red}$$

Plots of experimental values of α against reduced-time for isothermal experiments at different temperatures (and hence having different values of $t_{0.50}$) should then all fall on a single curve and this composite curve can be compared to the master curves for each kinetic model. Deviations (α_{THEOR} - α_{EXPTL}) for each point can be determined and comparisons of the magnitudes, and variations with α, of these differences can then be used to identify the rate equation giving the most acceptable kinetic fit to the data and its range of applicability. Any systematic change in kinetic behaviour with temperature is readily detected in the preparation of the composite curve. The master curves for each kinetic model can then be calculated using the rate equations from Table 3.3. in the form [$\alpha = h(t)$] with t being the reduced-time scale, t_{red}, and the rate coefficient being $g(0.50)$. Table 5.2. gives values of $g(\alpha)$ for some of the models from Table 3.3. at 0.10 intervals of α.

The choice of reference value of α is often selected as some value other than 0.50, mainly because the slope, $d\alpha/dt$, in the neighbourhood of $\alpha = 0.50$ can be large and hence the errors in identifying the time, $t_{0.50}$, at which $\alpha = 0.50$ may be significant. When there is an initial reaction, or uncertainty in the length of the induction period, it may be appropriate to use two common points for the scaling. One possibility [13] is $t_{red} = 0.00$ at $\alpha = 0.20$ and $t_{red} = 1.00$ at $\alpha = 0.90$ so $t_{red} = (t - t_{0.2})/(t_{0.9} - t_{0.2})$.

Several variations of the above procedure have been suggested by Jones *et al.* [7] to quantify the comparison of experimental results with the master curves. (i) The experimental value α_e at which $(t / t_{0.5})_e$ has the *same* value as that in the master data $(t / t_{0.5})_m$ can be plotted against the master value α_m. Where agreement between experiment and model is good, the plot of α_e against α_m will have slope of unity and pass through the origin. (ii) Similarly, plots of $(t / t_{0.5})_e$ against $(t / t_{0.5})_m$ for common values of α_e and α_m should have slopes of unity and pass through the origin if the model is a good description. (iii) Finally, a plot of values of t_e against $(t / t_{0.5})_m$ for common values of α_e and α_m should be a straight line of slope $(t_{0.5})_e$ through the origin. Advantages [7] of this method include direct evidence of deviations through variations from linearity of each plot. The analysis is also sensitive and applicable over the whole range $0 < \alpha < 1$.

Because $k = g(0.50)/t_{0.50}$ (at a constant temperature, T), it follows that $1/t_{0.50}$ (times corrected by subtraction of t_o) is proportional to k at temperature, T. The reciprocals of the time-scaling factors, $t_{0.50}$, (or at any other reference value of α) can thus be used as measures of the rate coefficient, and hence the temperature dependence of $t_{0.50}$ can be used to calculate the activation energy, E_a, of reaction without the necessity of identifying the kinetic model, $g(\alpha)$ [14]:

$$\ln (1/t_{0.50}) = - E_a/RT + \ln A - \ln g(0.50)$$

Table 5.2.
Values of $g(\alpha)$ (from Table 3.3.) calculated for various values of α

	g(α)								
α	0.1	0.2	0.3	0.4	0.5	0.6	0.7	0.8	0.9
model									
F1	0.1054	0.2231	0.3567	0.5108	0.6931	0.9163	1.2040	1.6094	2.3026
F2	0.1111	0.2500	0.4286	0.6667	1.0000	1.5000	2.3333	4.0000	9.0000
F3	0.2346	0.5625	1.0408	1.7778	3.0000	5.2500	10.1111	24.0	99.0
R2	0.0513	0.1056	0.1633	0.2254	0.2929	0.3675	0.4523	0.5528	0.6838
R3	0.0345	0.0717	0.1121	0.1566	0.2063	0.2632	0.3305	0.4152	0.5358
A2	0.0111	0.0498	0.1272	0.2609	0.4805	0.8396	1.4496	2.5903	5.3019
A3	0.0012	0.0111	0.0454	0.1333	0.3330	0.7693	1.7452	4.1689	12.208
A4	0.0001	0.0025	0.0162	0.0681	0.2308	0.7049	2.1012	6.7096	28.110
D1	0.0100	0.0400	0.0900	0.1600	0.2500	0.3600	0.4900	0.6400	0.8100
D2	0.0052	0.0215	0.0503	0.0935	0.1534	0.2335	0.3388	0.4781	0.6697
D3	0.0012	0.0051	0.0125	0.0245	0.0425	0.0692	0.1091	0.1721	0.2867
D4	0.0012	0.0050	0.0117	0.0221	0.0369	0.0573	0.0854	0.1249	0.1847

5.4.7. Differential methods

Several experimental techniques produce signals proportional to the rate of reaction, ($d\alpha/dt$), rather than to α itself. Even if the signal is proportional to α, numerical or electronic differentiation may be used to convert the α values to rate measurements. The assumption that the rate of heat evolution or absorption is proportional to the rate of reaction needs careful examination and confirmation. Among the possible causes of deviations from proportionality which can occur are the release of strain energy in crystals and/or melting of impurities.

Differential methods for kinetic analysis, proposed in the literature, include the following.

(i) Delmon [2] has suggested comparison of plots of ($d\alpha/dt$) against α, obtained from experiment, with master curves calculated for the various kinetic models which he has provided.

(ii) ($d\alpha/dt$) may be plotted against t_{red} [15], or the differences (α_{THEOR} - α_{EXPTL}) may be plotted against α.

(iii) The linearity of plots of experimentally measured rate ($d\alpha/dt$) values against the *derivative* functions of the solid state rate equations, $f(\alpha)$ in Table 3.3. may be tested [3].

Differential methods of kinetic analysis can provide better distinguishability amongst the available kinetic expressions, particularly for the sigmoid group of equations (A2 to A4 in Table 3.3.) and for the geometric processes (R2 and R3 in Table 3.3.).

For any rate equation of the form (Table 3.3.):

$$f(\alpha) = (1/k)(d\alpha/dt) = (1 - \alpha)^n$$

the apparent 'order of reaction', n, can be determined [16,17] from a plot of $\ln(d\alpha/dt)$ against $\ln(1 - \alpha)$. Knowledge is thus required of both ($d\alpha/dt$) and α. The values of n applicable to the R2 and R3 models (Table 3.3.) are 0.50 and 0.67, respectively.

5.5. NON-ISOTHERMAL KINETIC ANALYSIS

5.5.1. Introduction

As an alternative to the traditional, and arguably more fundamental, approach of extracting kinetic information from the results of several isothermal experiments at different temperatures, there has been great theoretical and practical interest in the problem of extracting similar information from one or more experiments during which the reaction temperature is changed according to some set programme (usually, but not necessarily [18-20], a linear increase of T with time). Rouquerol *et al.* [21,22] have developed a technique, referred to as constant rate thermal analysis (CRTA), in which the the sample is heated in such a way that reaction takes place at a constant rate. Ortega *et al.* [23] have extended this idea to control the temperature so that the reaction proceeds at a constantly increasing rate (acceleration). A *temperature-jump* or *step-wise* programme has also been suggested [24,25] in which, during a single experiment, the temperature is rapidly changed ("jumped") from one value to another. The reaction rates at the two (or more) temperatures are measured and used to calculate Arrhenius parameters for that particular α value.

It is impossible, within a reasonable space, to describe more than a few of the more generally accepted methods of data treatment. New and modified methods continue to appear in the literature and the complexity of some of the approaches make them difficult to classify.

As in isothermal kinetic studies, the analysis of non-isothermal kinetic (NIK) data is directed towards identifying the rate equation (Table 3.3.) and Arrhenius parameters which most satisfactorily describe the observations. Changes in kinetic characteristics during the course of a non-isothermal experiment are probably more easily overlooked than in a series of isothermal experiments. Possible changes include variations in the magnitudes of A, E_a and the form of the conversion function, g(α) or f(α). Such variations are often attributed to a change of reaction mechanism, occurring either at a particular value of α (which may also mean that α is not precisely defined in terms of an identified reaction stoichiometry), or over different temperature ranges. There may also be changes with temperature of the secondary reactions between products. The relative contributions from reversible, concurrent and consecutive reactions will also generally vary considerably with temperature.

Boldyreva [26] has criticized the use of the NIK approach for the determination of kinetic parameters and reaction mechanisms in the absence of more direct studies, and states that in certain technological situations, e.g. processes carried out under non-isothermal conditions, the rapidity with which the information is obtained and the similarities between laboratory and process conditions "may compensate for the absence of a physical meaning". Maciejewski [27] has also provided critical discussions of the usefulness of kinetic data for solid state reactions and has warned of the dangers of regarding measured kinetic parameters as being characteristic of the compound being studied, without reference to the experimental conditions used.

Isothermal and non-isothermal methods provide complementary routes to a common objective, the determination of the kinetic parameters for a selected reaction. Isothermal studies represent one limit of the numerous possibilities for temperature changes which can, in principle, be employed in non-isothermal studies. Both approaches have their advantages and their disadvantages and attempts to downplay or ignore the importance of either approach are unproductive.

Non-isothermal measurements may enable steps in a sequence of chemical processes to be distinguished in what might otherwise be incorrectly regarded as a single reaction.

5.5.2. Non-isothermal yield-temperature curves

The shapes of curves of α against T and $d\alpha/dT$ against T obtained from experiments using a constant rate of temperature increase (shown in Figures 5.1. and 5.2.), differ considerably from the curves of α against t and $d\alpha/dt$ against t under isothermal conditions (given in Chapter 3, Figures 3.5. and 3.6.) for the conversion functions listed in Table 3.3.

Those models based on an apparent order of reaction, n, (also including fractional values), i.e. F1, F2, F3, R2 and R3, Figure 5.1.(b), are difficult to distinguish at low values of α. Distinguishability improves for higher orders at higher values of α. The diffusion models, D1, D2, D3 and D4, give generally lower onset temperatures and flatter curves (representing extended temperature intervals) (Figure 5.1.(c)) than the nth-order group (Figure 5.1.(b)), while the Avrami-Erofeev (JMAEK) models, An, (Figure 5.1.(a)) have higher onset temperatures and steeper curves. Figure 5.2. shows the derivative curves corresponding to the integral curves given in Figure 5.1.

For a fixed model, e.g. R3, the influences of the other variables, the heating rate, β, the pre-exponential factor, A, and the activation energy, E_a, are shown in Figures 5.3., 5.4. and 5.5., respectively. Elder [28] has provided similar curves and Zsakó [29] has shown similar influences for the first-order (F1) model. Although the F1 model cannot be regarded as a realistic representation of many solid state decompositions, it has often been assumed to apply as an approximation. Figure 5.3. shows the regular effect on the theoretical R3 curve of doubling the heating rate in the range $\beta = 1$ to 16 K min^{-1}. Decreasing the pre-exponential factor by factors of ten in the range $A = 10^{17}$ to 10^{13} s^{-1} mainly affects the onset temperature and the acceleratory portion of the curves, Figure 5.4., with the remaining segments being almost parallel. Very similar behaviour is observed for curves in which the activation energy increases in steps of 5 kJ mol^{-1} as shown in Figure 5.5.

The overall shape of the thermal analysis curve is thus mainly determined by the kinetic model, while the position of this curve on the temperature axis is controlled by the values of E_a, A and, to a lesser extent, the heating rate β.

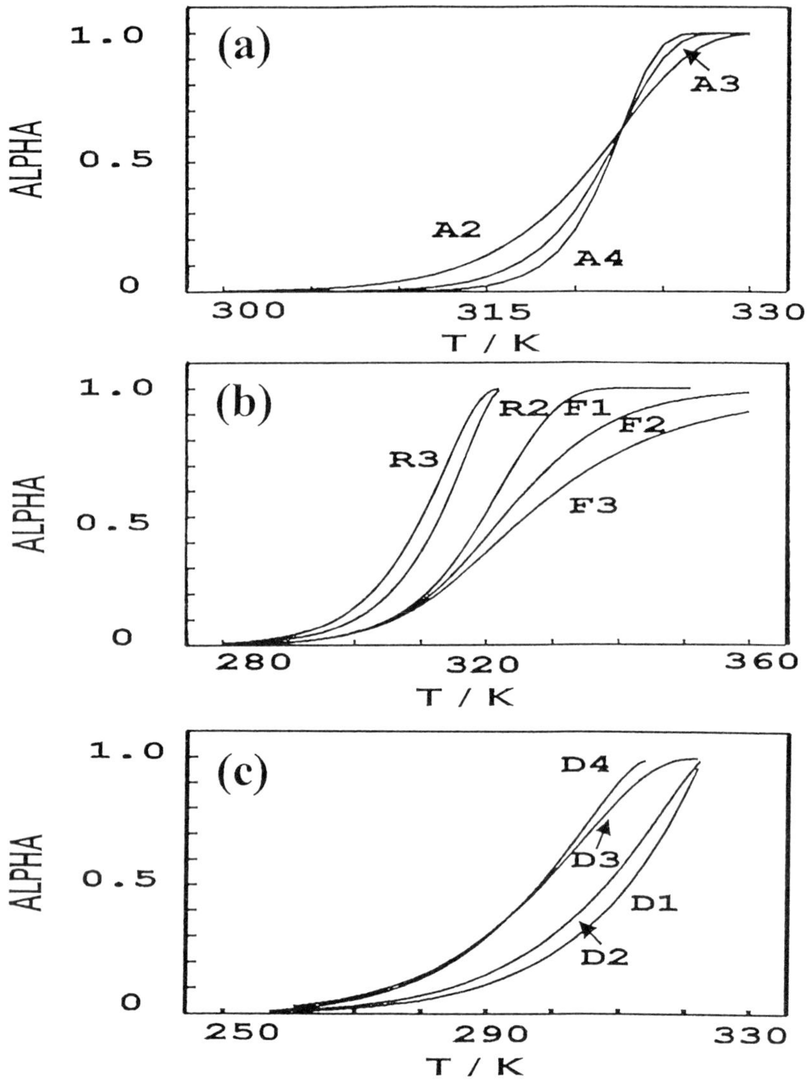

Figure 5.1.
Theoretical α, temperature curves for various kinetic models, with heating rate, $\beta = 1.0$ K min^{-1}; $A = 1.9 \times 10^{15}$ min^{-1} and $E_a = 100$ kJ mol^{-1}.
(a) the Avrami-Erofeev (JMAEK) models, An,
(b) models based on apparent order of reaction, n, and contracting geometry i.e. F1, F2, F3, R2 and R3, and
(c) diffusion models, D1, D2, D3 and D4.

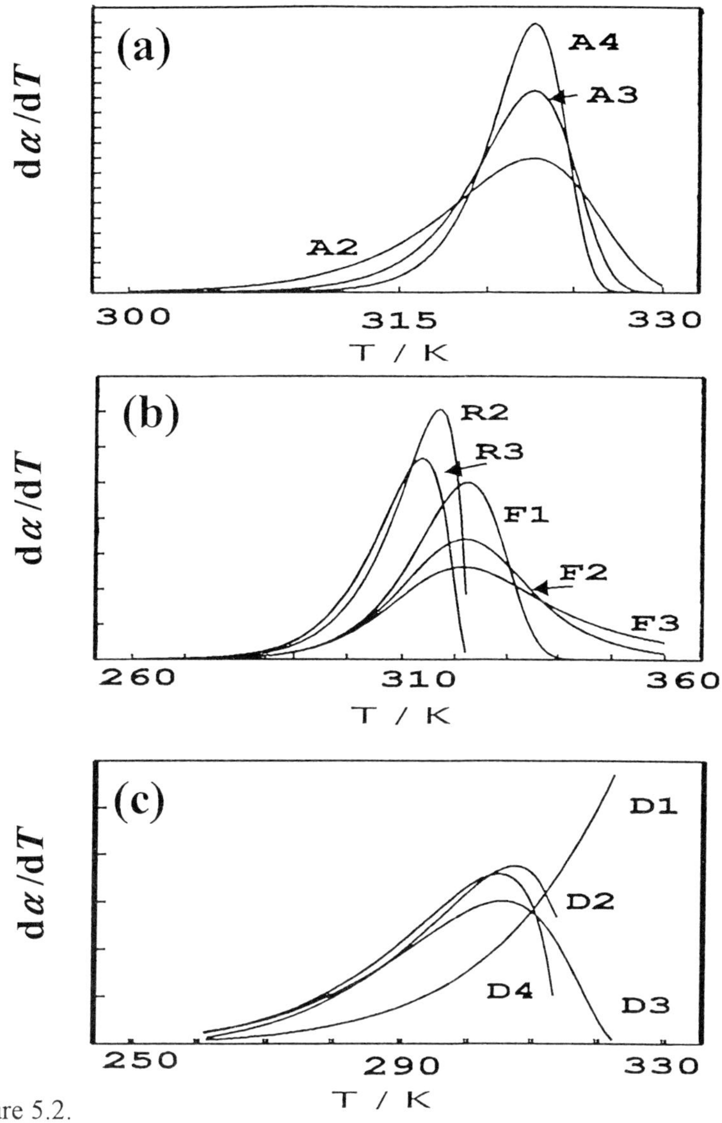

Figure 5.2.
Theoretical curves of dα/dT against temperature for various kinetic models, with heating rate, $\beta = 1.0$ K min^{-1}; $A = 1.9 \times 10^{15}$ min^{-1} and $E_a = 100$ kJ mol^{-1}.
(a) the Avrami-Erofeev (JMAEK) models, An,
(b) models based on apparent order of reaction, n, and contracting geometry i.e. F1, F2, F3, R2 and R3, and
(c) diffusion models, D1, D2, D3 and D4.

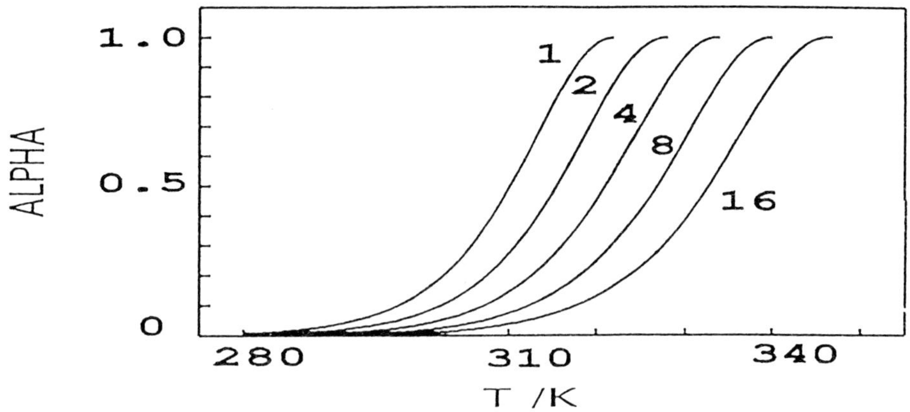

Figure 5.3.
The influence of the heating rate, β (1 to 16 K min^{-1}) on the α, T curve for the R3 model (A = 1.9 x 10^{15} min^{-1}; E_a = 100 kJ mol^{-1}).

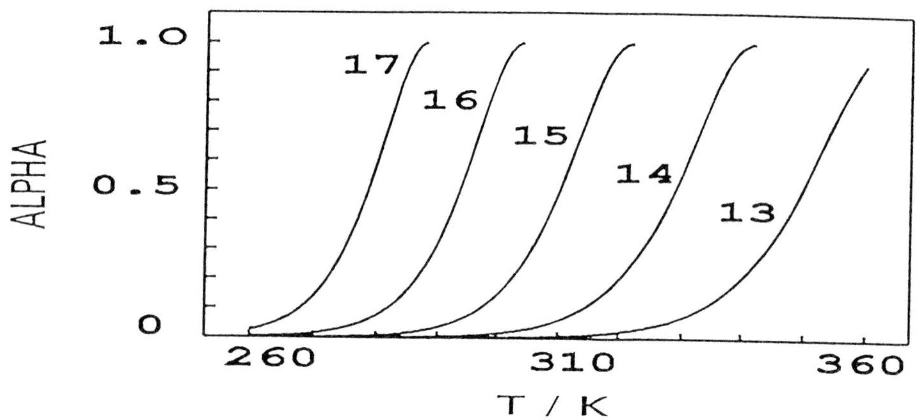

Figure 5.4.
The influence of the pre-exponential factor, A (1.9 x 10^{17} to 1.9 x 10^{13} min^{-1}) on the α, T curve for the R3 model (β = 1.0 K min^{-1}, E_a = 100 kJ mol^{-1}).

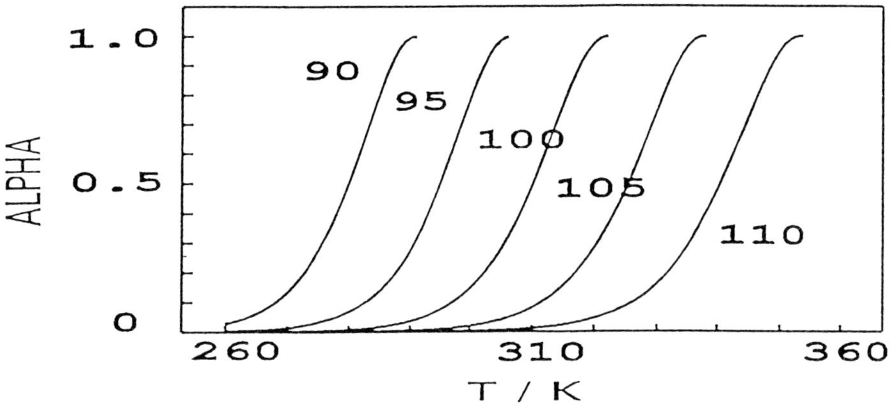

Figure 5.5. The influence of the activation energy, E_a (90 to 110 kJ mol^{-1}) on the α, T curve for the R3 model ($\beta = 1.0$ K min^{-1}, $A = 1.9 \times 10^{15}$ min^{-1}).

5.5.3. The "inverse kinetic problem (IKP)"

The usual starting point for the kinetic analysis of non-isothermal data is:

$$d\alpha/dT = (d\alpha/dt)(dt/dT) = (d\alpha/dt)(1/\beta) \tag{5.1}$$

where $\beta = (dT/dt)$, is the heating rate. Even this step has aroused continuing debate [30-32]. The Arrhenius equation (Chapter 4) is almost invariably assumed to be applicable [33] to the rate processes being studied, so that equation (5.1) may be expanded to:

$$d\alpha/dT = (1/\beta)(d\alpha/dt) = (A/\beta)\exp(-E_a/RT)\,f(\alpha) \tag{5.2}$$

where f(α) is a kinetic expression (see Table 3.3.).

A central problem of NIK [34] is the integration of the right-hand side of equation (5.2). Separating the variables leads to:

$$d\alpha/f(\alpha) = (A/\beta)\exp(-E_a/RT)\,dT \tag{5.3}$$

Integrating equation (5.3) between the limits, $\alpha = 0$ at $T = T_o$ and $\alpha = \alpha$ at $T = T$ gives:

$$_o\int^{\alpha} (f(\alpha))^{-1}\, d\alpha = {}_{To}\int^{T} (A/\beta) \exp(-E_a/RT)\, dT$$

$$g(\alpha) = {}_{To}\int^{T} (A/\beta) \exp(-E_a/RT)\, dT = {}_{o}\int^{T} (A/\beta) \exp(-E_a/RT)\, dT \qquad (5.4)$$

(because $_o\int^{To} (A/\beta) \exp(-E_a/RT)\, dT = 0$). Krís and Sesták [35] emphasize that the above separation of the variables α and T is not justified when significant self-heating or self-cooling occurs.

For any reaction under investigation, experimental measurements, obtained at a known heating rate, β, are converted to values of α and/or $(d\alpha/dt)$ at temperatures T. Often the analysis uses the rate equations familiar from homogeneous kinetics, so that $f(\alpha)$ is assumed to have the form $(1 - \alpha)^n$ and n, the apparent "reaction order" (RO), becomes the unknown.

Suggestions have been made [36-38] that a more general rate equation:

$$f(\alpha) = \alpha^m (1 - \alpha)^n [-\ln (1 - \alpha)]^p \qquad (5.5)$$

known as the Sesták-Berggren equation, should be used and values sought for m, n and p. Certain values of these indices correspond to the more familiar rate equations, e.g. if $m = n = 1$ and $p = 0$, the Prout-Tompkins or Austin-Rickett equation results (see Chapter 3).

If the rate expression and/or the Arrhenius parameters vary with α, i.e. the reaction mechanism changes, the kinetic analysis becomes very much more complicated and the results of such analyses become less reliable and less useful.

Whether rate measurements from *a single experiment* extending across a range of temperatures are, in principle, capable of providing a complete kinetic analysis (the conversion function, $f(\alpha)$ or $g(\alpha)$, and the magnitudes of E_a and A) has been debated. Agrawal [39] has discussed some of the problems of the *uniqueness* of the derived parameters. Problems occur whenever attempts are made to determine more than two parameters from a single curve. Use of kinetic expressions containing multiple α terms also leads to non-unique kinetic parameters, as does the existence of an apparent compensation effect. Criado *et al.* [40] have shown that the same *curve of α against T* can be generated using three different kinetic models with different Arrhenius parameters. Málek [41] has given an excellent account of the correlation between kinetic parameters and the kinetic models from which they are derived.

Vyazovkin and Lesnikovich have provided two interesting reviews [42,43] of methods and principles for the solution of the IKP. (These articles also serve as a very useful entry to the Russian literature). They emphasize that all inverse problems have ambiguous solutions. The ambiguity may arise from attempts to determine too many unknown constants from limited data, or appear when a set of experimental data can be alternatively described by different formal models and kinetic constants. The possible solving of the IKP using methods based on *generalized descriptions*, (or "synthesis" rather than "analysis") has also been discussed [43]. These methods are based upon the premise that different aspects of the real process may be best described by a synthesis of individual features of competing ideal models. Such a generalized description is the Sesták - Berggren equation (5.5) which, depending on the exponents, may represent the usual set of models (Table 3.3.). The number of parameters which can be determined depends upon the experimental data [44,45]. Usually this is limited to a maximum of two of the three exponents. Wyandt and Flanagan [46] describe the use of the Sesták - Berggren equation (5.5) in a non-isothermal kinetic method suggested by Zimmerman [47]. No assumptions are made about the mechanism of the solid state process being investigated. Values for m, n and p are calculated directly from the data, and then the most likely mechanistic model can be selected.

Some workers choose to describe their experimental results in terms of *partial* application of various models over different regions of time or temperature and others prefer *approximate* agreement with a single model (or very limited number of models) over as wide a range of α as possible.

In spite of the division of experimental approaches into isothermal and non-isothermal methods of kinetic analysis, Criado *et al.* [40] have pointed out that the alternative types of experimental conditions all represent approximations to the true sample temperature. Garn [48] has warned that factors such as heat transfer from the furnace to the sample, complicated by self-cooling or self-heating of the sample during reaction, and the influence of evolved product gases trapped in the vicinity of the sample on the rates of reversible reactions, all make increased contributions under non-isothermal conditions. Other difficulties encountered during kinetic analysis of the results of isothermal experiments, such as the fitting of data across as wide a range of α as possible and distinguishing amongst kinetic expressions with similar forms, are also magnified under programmed temperature conditions.

5.5.4. Classification of methods of NIK analysis

The methods used in the analysis of non-isothermal kinetic data can be classified as *derivative*, also referred to as *differential methods*, based on the use of equation

(5.2), or *integral methods*, based on the use of equation (5.4). Vyazovkin and Lesnikovich [42] suggested a classification, based on the method of calculation of the kinetic parameters, as either *"discriminatory"*, i.e the kinetic model f(α) or g(α) is identified, or *"non-discriminatory"*. Methods involving discrimination can be further sub-divided into methods of *"analysis"*, where a single model is sought to describe the experimental data, or methods of *"synthesis"*, where several models are combined to give a better description of the data. Another approach to classification could be based on the use of *linear* or *non-linear regression* during data manipulation.

Although values of A and E_a can be estimated from a *single* dynamic (i.e. α, T) experiment if a particular conversion function, f(α) or g(α), is assumed (by rather arbitrary trial-and-error comparisons) to apply, it is more usual to calculate kinetic parameters from two or more otherwise identical dynamic experiments at different heating rates (β).

5.5.5. Isoconversional methods

When more than one set of experimental results is available, the unknown form of the conversion function f(α) or g(α) may be eliminated by comparing measurements made at a common value of α under the two (or more) sets of different conditions. These *isoconversional* methods are thus model independent, or "non-discriminating" methods of estimating the Arrhenius parameters [14,42,43].

Isoconversional methods [43] rely on the general equation (see equation (5.2)):

$$\ln(\mathrm{d}\alpha/\mathrm{d}T)\beta = \ln(A/\ \mathrm{f}(\alpha)) - E_a/RT$$

The numerical values of E_a and of the combination (A/ f(α)) can be determined unambiguously, but the value of A depends upon the conversion function chosen, f(α). The treatment also assumes that the model f(α) remains the same over the whole range of α.

To avoid discarding potentially significant information, the parameters obtained from isoconversional methods may be used with the original data to determine the kinetic model [49], although this is often not done. Activation energies determined from isoconversional methods [43] are in good agreement with values from isothermal experiments.

5.5.6. Plots of α against reduced temperature

By analogy with the use of plots of α against reduced time in isothermal kinetic analysis to determine the appropriate conversion function, Meindl *et al.* [50] have suggested the use of plots of α against reduced temperature. Vyazovkin *et al.* [51] have compared these methods for the analysis of isothermal and of non-isothermal data. The shapes of plots of α against reduced temperature are less sensitive to the form of the kinetic model than are plots of α against reduced time. The isothermal and non-isothermal curves are not related, as has been suggested [52], simply by the heating rate, $\beta = \mathrm{d}T/\mathrm{d}t$, but the times, t, corresponding to an isothermal curve at $T = T_{iso}$ are obtained from measurements of α and T at heating rate β, by use of the relationship:

$$t = A \,{}_0\!\int^T \exp(-E_a/RT)\,\mathrm{d}T \,/\, [\beta A_{iso} \exp(-E_{iso}/RT_{iso})]$$

or the quantity, t^*, proportional to the time, where:

$$t^* = {}_0\!\int^T \exp(-E_a/RT)\,\mathrm{d}T$$

5.5.7. Examples of derivative (or differential) methods

Equation (5.2) can be written in a variety of forms, for example:

$$\ln[(\mathrm{d}\alpha/\mathrm{d}T)/\mathrm{f}(\alpha)] = \ln(A/\beta) - (E_a/R)(1/T) = y \qquad (5.6)$$

If assumptions are made about the form of f(α), then the slope and intercept of a plot of y against $(1/T)$ may be used to determine the values of E_a and A, respectively. Várhegyi [53] has pointed out that the sensitivity of the left-hand side of equation (5.6) is inversely proportional to $\mathrm{d}\alpha/\mathrm{d}t$. Thus, unless appropriate weighting factors are used, the derived kinetic parameters will be extremely sensitive to the data at the onset and the end of the α,T curve and have decreased sensitivity in the vicinity of the maximum $\mathrm{d}\alpha/\mathrm{d}T$, where most of the reaction proceeds.

Several different mathematical procedures have been suggested for extracting kinetic information from sets of α,T data obtained at different heating rates, β_i.

Friedman's method [54] is to plot $\ln(\mathrm{d}\alpha/\mathrm{d}t)_i$ against $1/T_i$, measured at the same value of α_i, from α,T curves at different heating rates β_i (or different isothermal reaction temperatures, T_i). The parallel lines obtained have slopes $= -E_a/R$ and different intercepts $= \ln[A\,\mathrm{f}(\alpha)_i]$. A value for A is obtained by extrapolation of a plot of the intercept against α_i to $\alpha_i = 0$.

Carroll and Manche [55], in a procedure otherwise identical with Friedman's above, suggested plotting $\ln[\beta_i\,(\mathrm{d}\alpha/\mathrm{d}T)_i]$ against $1/T_i$.

Flynn [56] suggested a further variation, namely to plot $T_i \ln(\mathrm{d}\alpha/\mathrm{d}t)_i$ against T_i for the same value of α_i at different heating rates β_i. The lines obtained have a common

intercept = $-E_a/R$, but different slopes = $\ln[A\ f(\alpha)_i]$.
Freeman and Carroll [57] assumed $f(\alpha) = (1 - \alpha)^n$ and considered incremental differences in $(d\alpha/dT)$, $(1 - \alpha)$ and $(1/T)$. This leads to the expression:

$$\Delta\ln(d\alpha/dT) = n\ \Delta\ln(1 - \alpha) - (E_a/R)\ \Delta(1/T)$$

which can be used to determine the value of E_a by plotting either:

$[\Delta\ln(d\alpha/dT)/\Delta\ln(1 - \alpha)]$ against $[\Delta(1/T)/\Delta\ln(1 - \alpha)]$

or:

$[\Delta\ln(d\alpha/dT)/\Delta(1/T)]$ against $[\Delta\ln(1 - \alpha)/\Delta(1/T)]$

Sesták *et al.* [58] have discussed some improvements to this method, including the extension to other rate equations. Criado *et al.* [59] have demonstrated that the Freeman and Carroll treatment does not allow an *n*th order rate equation to be distinguished successfully from other kinetic models. This was confirmed by Jerez [60] who pointed out the large errors involved in the regression procedure, and suggested a modification in the method used to calculate E_a and n involving the use of the point at which the rate is a maximum and the centre of gravity of the cluster of experimental points. Van Dooren and Muller [61] have examined the effects of sample mass, particle size and heating rate on the determination of n, E_a and A from DSC data using the Freeman and Carroll procedure.

Several approaches involve use of the *second derivative* of equation (5.6), or the version based on $f(\alpha) = (1 - \alpha)^n$, with respect to temperature [29,58] or with respect to α [29]. Differentiation of:

$$d\alpha/dT = (A/\beta)\exp(-E_a/RT)(1 - \alpha)^n \tag{5.7}$$

gives:

$$d^2\alpha/dT^2 = (d\alpha/dT)[(E_a/RT^2) - n(d\alpha/dT)/(1 - \alpha)]$$

and, because this derivative must be zero at the inflexion point of a curve of α against T, or the maximum of a curve of $d\alpha/dT$ against T, it follows that:

$$E_a/RT^2_{max} = (d\alpha/dT)_{max}\ n/(1 - \alpha_{max}) \tag{5.8}$$

from which E_a may be calculated if n is known and T_{max}, $(d\alpha/dT)_{max}$ and α_{max} have been measured [62].
Kissinger [63,64] used the combination of equations (5.7) and (5.8) which gives:

$$(A/\beta)\exp(-E_a/RT_{max})\, n(1 - \alpha_{max})^{n-1} = E_a/RT_{max}^2$$

Because $(1 - \alpha_{max})$ is a constant for a given value of n, a value for E_a may be obtained from the slope of a plot of $\ln(\beta/T^2_{max})$ against $1/T_{max}$ for a series of experiments at different heating rates, β_i. Augis and Bennett [65] modified the Kissinger treatment for use with the Avrami - Erofeev (or JMAEK) model (An). They plotted $\ln(\beta/(T_{max} - T_o))$ against $1/T_{max}$ where T_o is the initial temperature at the start of the heating programme, instead of $\ln(\beta/T^2_{max})$ against $1/T_{max}$. Elder [28] has generalized the Kissinger treatment to make it applicable to the full range of kinetic models. The generalized equation is:

$$\ln(\beta/T_{max}^{m+2}) = \ln(AR/E_a) + \ln L - E_a/RT_{max}$$

where m is the temperature exponent of the pre-exponential term in the modified Arrhenius equation (often taken as zero) and $L = -\ f(\alpha_{max})/(1 + mRT_{max}/E_a)$. This correction term was found to be relatively small, but helps in distinguishing between similar models. Llopiz *et al.* [66] have derived correction terms for the usual range of kinetic models. The values of E_a obtained were not very sensitive to the incorrect choice of model [28].
Ozawa [67,68] developed an integral method that is also applicable to derivative curves and is similar to the Kissinger method. The slope of a plot of $\ln \beta$ against $1/T_{max}$ is used to determine E_a. Van Dooren and Muller [61] found that both sample mass and particle size could influence the calculated magnitudes of the apparent kinetic parameters determined from DSC experiments using the methods of Kissinger and of Ozawa. The two methods gave similar values for E_a with a slightly lower precision for the Kissinger method. It was suggested [61] that temperatures at $\alpha = 0.5$ (half conversion) should be used in place of T_{max}. The applications of this approach have been reviewed by Tanaka [69].
Borchardt and Daniels [70,71] developed a method for DTA studies of homogeneous liquid-phase reactions, that involves the calculation of the rate coefficient, k, from the expression [72]:

$$k = [(JSV/m_o)^{n-1}(C\, d\Delta T/dt + J\Delta T)] / [(J(S - s) - C\Delta T]^n$$

where V is the volume of the sample, m_o the initial amount of sample in moles, S is the total DTA peak area, and s is the partial area up to the time t at which the peak height ΔT and the slope $d\Delta T/dt$ are measured. J is the heat transfer coefficient and C is the total heat capacity of the sample and the holder. Because of the problems of obtaining values of J and C, the assumptions are made [71] that $n = 1$ and that $C(d\Delta T/dt)$ and $C\Delta T$ are small compared with the terms to which they are added (or from which they are subtracted). It follows that:

$$k = \Delta T/(S - s)$$

For DSC the equivalent form would be:

$$k = (dH/dt)/(S - s)$$

where H is the enthalpy, so that (dH/dt) is the DSC signal. The values of k obtained are then used in a conventional Arrhenius plot. Shishkin [73] has discussed the similarities between the Borchardt and Daniels and the Kissinger approaches.

Flynn and Wall [74] suggested another method based on second derivatives. Equation (5.7) above may be modified to give:

$$T^2(d\alpha/dT) = (AT^2/\beta)\exp(-E_a/RT)(1 - \alpha)^n$$

When differentiated with respect to α, this gives:

$$d[T^2\, d\alpha/dT]/d\alpha = (E_a/R) + 2T + [n/(1 - \alpha)]\,[d\alpha/d(1/T)]$$

At low values of α, the last term on the right-hand side is negligible and even $2T \ll E_a/R$. By use of finite differences instead of the derivative:

$$\Delta(T^2\, d\alpha/dT)/\Delta\alpha = (E_a/R) + 2T_{ave}$$

where T_{ave} was the average temperature over the interval. The value of E_a was then calculated from the slope of a plot of $\Delta(T^2 d\alpha/dT)$ against $\Delta\alpha$ for the early stages of the reaction only.

5.5.8. Examples of integral methods

The use of equation (5.4) requires evaluation of the temperature integral. The variable $x = E_a/RT$ is introduced so that:

$$ {}_{0}\int^{T} \exp(-E_a/RT)\mathrm{d}T = (E_a/R)\,{}_{x}\int^{\infty}(e^{-x}/x^2)\mathrm{d}x = (E_a/R)\,p(x) $$

and equation (5.4) becomes:

$$ \mathrm{g}(\alpha) = (AE_a/R\beta)\,p(x) $$

Zsakó [29] has suggested sub-classification of integral methods on the basis of the means of evaluation of the temperature integral in equation (5.4). The three main approaches are the use of: (i) numerical values of $p(x)$; (ii) series approximations for $p(x)$; and (iii) approximations to obtain an expression which can be integrated.

Tables of values of the integral $p(x)$ have been provided [75,76]. Much attention has been directed towards finding suitable approximations for the above temperature integrals [29,58,77-79]. Gorbachev [80] and Sesták [49,81] have suggested that there is little value in trying to find more accurate approximations considering the experimental uncertainties in the original α, T data. Representative examples of the series suggested for approximating $p(x)$ are given in Table 5.3. (see also references [34,82,83]).

Flynn [34] has emphasized the importance of using accurate calculated data for the temperature integral in determining the magnitudes of E_a and A from non-isothermal measurements. He points out that modern computer methods make the use of approximations unnecessary.

Table 5.3.
Some approximations for the temperature integral, $p(x)$, with $x = E_a/RT$

$p(x) = e^{-x}/x^2 \approx 1 - (2!/x) + (3!/x^2) - (4!/x^3) + ... + (-1)^n(n+1)!/x^n + ...$	[29]
$p(x) = e^{-x}/x(x+1) \approx 1 - 1/(x+2) + 2/(x+2)(x+3) - 4/(x+2)(x+3)(x+4) + ...$	[84]
$p(x) \approx e^{-x}/(x - d)(x - 2)$ with $d = 16/(x^2 - 4x + 84)$	[85]
$p(x) \approx e^{-x}/\,x(x^2 + 4x)^{1/2}$	[86]
Doyle's approximation (for $x > 20$) is: $\log_{10}p(x) \approx -2.315 - 0.4567x$	[87]

5.5.9. Comparison of derivative and integral methods

The use of derivative methods avoids the need for approximations to the temperature integral (discussed above). Measurements are also not subject to cumulative errors and the often poorly-defined boundary conditions used for integration [74]. Numerical differentiation of integral measurements normally produces data which require smoothing before further analysis. Derivative methods may be more sensitive in determining the kinetic model [88], but the smoothing required may lead to distortion [84].

5.5.10. Non-linear regression methods

Vyazovkin and Lesnikovich [42] have emphasized that the majority of NIK methods involve linearization of the appropriate rate equation, usually through a logarithmic transformation which distorts the Gaussian distribution of errors. Thus non-linear methods are preferable [89]. Militky and Sesták [90] and Madarász *et al.* [91] have outlined routine procedures for non-linear regression analysis of equation (5.5) above by transforming the relationship:

$$\mathrm{d}\alpha/\mathrm{d}t = A\,T^{\mathrm{b}} \exp(-E_{\mathrm{a}}/RT)\,\alpha^{m}\,(1-\alpha)^{n}\,[-\ln(1-\alpha)]^{p}$$

to:

$$y_i = C_o + \sum_{j=1}^{5} C_j x_{ij}$$

with $y_i = \ln(\mathrm{d}\alpha/\mathrm{d}t)_i$; $x_{i,1} = \ln(T_i)$; $x_{i,2} = 1/T_i$; $x_{i,3} = \ln(\alpha_i)$; $x_{i,4} = \ln(1-\alpha_i)$; $x_{i,5} = \ln[-\ln(1-\alpha_i)]$; $C_o = A$; $C_1 = b$; $C_2 = E_{\mathrm{a}}$; $C_3 = n$; $C_4 = m$; $C_5 = p$.
The values of C_j are then optimized so that:

$$\sum_{i=1}^{n} [\alpha_{i,calc} - \alpha_{i,expt}]^2$$

is a minimum. Militky and Sesták [90] have discussed the statistical software available and the estimation of errors in the derived parameters, C_j.

Karachinsky *et al.* [92] have discussed some of the problems of non-linear regression. In determining the degree of coincidence between experimental and theoretical data, a situation of minimum deviation is sought. The minimum may, however, often be a "flat pit" because of a compensation relationship between E_{a} and A, and there is also the ever present possibility of finding a *local* rather than a *global* minimum. They therefore suggested a search strategy which takes into account the individual effects that each of the parameters has on the theoretical curve.

5.5.11. Kinetic analysis of complex reactions

Some attention has been given to the NIK analysis of complex reactions (i.e., rate processes incorporating contributions from reversible, concurrent and consecutive reaction steps). The occurrence of complex reactions is detected in isothermal studies by Arrhenius plots that are curved or give two linear regions [93]. The shapes of plots of α against reduced time also vary systematically with temperature. Experiments at different heating rates also show up complexities.

Elder [45] has modelled several multiple reaction schemes, including mutually independent concurrent first-order reactions, competitive first-order reactions, mutually independent n-th order reactions, and mutually independent Avrami-Erofeev models with $n = 2$ or 3. The criteria identified for recognizing the occurrence of multiple reactions were: (i) the apparent order of reaction, n, varies with the method of calculation, and (ii) the kinetic parameters, A and E_a, vary with the extent of reaction, α.

Vyazovkin and Lesnikovich [93] proposed that the type of complex process encountered in non-isothermal experiments could be identified by analysis of the shape of the curve of the dependence of the apparent E_a on α, found by isoconversional methods. Concurrent competitive reactions are characterized by an increasing dependence of the apparent value of E_a on α, but detailed shapes are dependent on the ratios of the contributing rates. A decreasing dependence of E_a on α, was found for intermediate reversible processes [93].

Reversible reactions. Many solid-gas reactions are reversible, e.g., dehydration of crystal hydrates, so that rate equations for such processes should include terms for the rate of the reverse reaction. If the rates of contributing forward and reverse reactions are comparable, the general set of kinetic models (Table 3.3.) will not be applicable. The decomposition step in a reversible reaction thus needs to be studied [94] under conditions as far removed from equilibrium as possible (e.g. low pressures or high flow rates of carrier gas) and sensitive tests are required for determining whether the kinetics vary with the prevailing conditions. Sinev [95] has calculated that, for the decomposition of calcium carbonate, the rate of the reverse reaction is comparable with that of the forward reaction even when small sample masses (10 mg) and high flow rates (200 cm^3 s^{-1}) of inert gas are used. Interpretation of observations becomes more difficult and the reliability of conclusions decreases if local inhomogeneities of kinetic behaviour develop within the reactant mass.

Consecutive reactions. Marcu and Segal [96] considered two consecutive reactions of the reaction order (RO) type. The α,T curves were particularly influenced by the E_a values (which determine the onset temperatures of the reactions). Differences of greater than 10 kJ mol^{-1} between E_a values of the two processes led to their total separation. Different values for the pre-exponential factor, A, could compensate for this effect. The value of the reaction order also influenced separation. Thus, at low α the kinetic parameters are approximately those of the first reaction step.

Concurrent reactions. Separation of the contributions to complex reactions is most easily achieved when the activation energies of the individual processes are considerably different. Reactions with low E_a values dominate the kinetics at low temperatures and slow heating rates, while those with high E_a values dominate at high temperatures and high heating rates. At the isokinetic temperature the rates of the participating reactions are equal. Contributions from individual reactions with similar E_a values are not separable by changing the heating rate and deconvolution has to be attempted by mathematical methods [97]. Many of the methods that have been proposed in the literature are restricted to RO models.

Reaction complicated by diffusion. Vyazovkin [98] has provided an analysis of consecutive reaction steps in which there is the formation of a surface layer followed by diffusion through that layer.

5.5.12. Kinetic analysis of reactions of complex materials

Burnham and Braun [99] have provided a valuable review of the approaches used in the kinetic analysis of decompositions of complex materials such as polymers, minerals, fossil fuels and biochemicals. Mathematical characterization of these reaction systems, recognized as being too complex to be characterized in any fundamental way, is termed *global kinetic analysis*. One of its main objectives is to predict the reactivities of such materials at temperatures different from those for which kinetic measurements were made (see Section 5.5.13.).

The decompositions of complex materials are treated as combinations of a series of parallel and/or consecutive pseudo-unimolecular reactions representing the rates of formation of the individual products. Detail is given [99,100] for pseudo-nth order and autocatalytic reaction types.

The parallel reaction model [99] assumes that the reaction complexity can be represented by a set of independent parallel processes each with its own values of A and E_a. Such a reaction system can lead to non-Arrhenius behaviour [99] so the following simplifications may be used. (i) All of the reactions are assumed to have the same pre-exponential factor and the differences in reactivity are represented by a distribution of activation energies. The distributions considered include discrete, Gaussian, Weibull and Gamma distributions. Such an approach has been particularly successful in characterizing the kinetics of coal pyrolysis [99]. (ii) E_a is assumed to have a single value and the reactions have a distibution of pre-exponential factors. (iii) E_a and ln A are assumed to show compensation behaviour. The applications of these approaches have been described in considerable detail [99-102].

5.5.13. Prediction of kinetic behaviour

A major practical aim of kinetic studies is to enable predictions to be made of kinetic behaviour under conditions other than those used for the original experimental measurements [103]. The reliability of predictions depends upon the values of the kinetic parameters, E_a, A and f(α) (or g(α)), not varying with T and also the precision with which these values are known [104].

Vyazovkin and Linert [104] have described some of the implications of attempting to predict kinetic behaviour when the kinetic model g(α) or f(α) has been incorrectly chosen, and when the reactions are complex. Flynn [105] has discussed reasons for the failure of kinetic predictions. These include extrapolation beyond temperatures at which phase changes (and accompanying changes of physical properties) occur.

5.6. CONCLUSIONS

5.6.1. Identification of rate equations

The first part of this Chapter deals with the methods by which α - time data, measured during isothermal experiments, are compared with the kinetic expressions listed in Table 3.3. This set of expressions can depict most of the characteristics of the experimental α - time relationships observed for decompositions (and other reactions) of solids, including the various acceleratory and deceleratory contributions. The geometric models of interface advance on which these rate equations are based are sufficiently versatile to apply to most rate processes that proceed to completion without discontinuities or irregularities. Interpretations of behaviour are usually based on these interface models, although other patterns of reactivity leading to similar kinetics cannot be excluded without complementary evidence. Identification of the geometric pattern of interface advance is only *part* of the fuller description required to provide a comprehensive reaction mechanism. Additional information is needed on the structural changes and bond redistribution steps involved (these are discussed in Chapter 6). This information is often very difficult to obtain. Interpretation may be complicated by factors such as melting, the presence of imperfections and impurities at low concentrations, mass and heat transfer processes, contributions from the reverse reaction in the presence of gaseous products, secondary reactions between primary gaseous products, etc. Sometimes useful information can be deduced from examination of the changes in surface texture of the reacting solid, and the sensitivity of the reaction to the prevailing conditions such as sample mass and disposition and the gaseous environment. Microscopic examinations have been particularly successful in quantitative studies of the formation and growth of product nuclei.

It is most unlikely that any of the expressions in Table 3.3. could describe the kinetics of processes containing several of the contributory factors listed above, so the possibility of more complicated expressions being needed for adequate analysis of a set of experimental results should not be ruled out. Mathematical description of results through the use of expressions with several adjustable parameters is, however, of little value if these parameters cannot be given a definite physical significance.

One major objective of kinetic studies is to contribute towards the development of the theory of solid state chemistry. Identification of the factors that determine reactivity requires a foundation of reliable kinetic measurements.

5.6.2. Comparison of methods of analysis of isothermal and non-isothermal data

Arguments about the relative values of non-isothermal and isothermal methods of kinetic analysis are generally unproductive [106]. The results of some kinetic studies using non-isothermal techniques are in good agreement with those obtained for the same system using isothermal methods. By contrast, other systems are reported as giving kinetic results which are very dependent on the technique used. Thus the complementary use of the two approaches can be valuable in revealing potential sources of difference. Both techniques can provide valuable insights into the processes occurring, provided that the experimenter is critically aware of the shortcomings and limitations of each approach. Vyazovkin and Lesnikovich [42] recommend that the Arrhenius parameters should be determined from isothermal measurements and the kinetic model from non-isothermal measurements

The reported lack of agreement amongst kinetic parameters calculated from the same set of experimental data using different methods of mathematical analysis [37,107] is disturbing. Some of the commercially available programs for kinetic analysis do not even specify the algorithms on which they are based, while other packages use kinetic expressions restricted to the reaction order (RO) type.

Vyazovkin and Lesnikovich [42] emphasize the need for careful statistical testing of the significance of the calculated parameters. Such tests may, at least, decrease the number of kinetic models which need to be considered. They specifically warn against the practice of forcing the model to be of the reaction order (RO) type where the value of n may not possess physical significance. Coincidence of the parameters calculated by alternative methods confirms only the equivalence of the methods of calculation and not the validity of the parameters obtained.

One of the arguments against the use of "discrimination" methods [42] is that the set of models from which the 'best model' is to be chosen is too limited. Hence one of the set must be identified as the 'best model' even if the set does not contain the true model. The formal kinetic models in the accepted set (e.g., Table 3.3.) are too simple to account for all the features of real processes. Modification of the models, however, results in an increased number of adjustable parameters.

Generalized descriptions include restriction to one class of model (e.g., reaction order or Avrami - Erofeev), i.e. using only one of the terms in equation (5.5), but placing no restriction on the value of the exponent. Also suggested are linear combinations of several formal models. The Avrami - Erofeev model, for example, is equivalent to linear combinations of some of the other formal models [43] and hence may serve as a generalized description. Generalized descriptions may also include approximating functions, such as polynomials, splines, etc., but the number of adjustable parameters becomes unwieldy and the parameters lose both their independence and their physical significance [108].

REFERENCES

1. P.D. Garn, J. Thermal Anal., 7 (1975) 475; 10 (1976) 99; 13 (1978) 581; Thermochim. Acta, 135 (1988) 71; 160 (1990) 135.
2. B. Delmon, Introduction a la Cinétique Hétérogène, Technip, Paris, 1969.
3. A.K. Galwey and M.E. Brown, Thermochim. Acta, 269/270 (1995) 1.
4. M.E. Brown and A.K. Galwey, Thermochim. Acta, 29 (1979) 129.
5. A.K. Galwey and P.W.M. Jacobs, Proc. R. Soc. (London), A254 (1960) 455.
6. R.E. Carter, J. Chem. Phys., 34 (1961) 2010; 35 (1961) 1137.
7. L.F. Jones, D. Dollimore and T. Nicklin, Thermochim. Acta, 13 (1975) 240.
8. A.K. Galwey, M.J. McGinn and M.E. Brown, Reactivity of Solids, (Eds J.S. Anderson, M.W. Roberts and F.S. Stone), Chapman and Hall, London, 1972, p.431.

9. L. Reich, Thermochim. Acta, 173 (1990) 253; 164 (1990) 1,7; 143 (1989) 311; 138 (1989) 177; Amer. Lab., 19(9) (1987) 23.
10. C. Rozycki and M. Maciewjewski, Thermochim. Acta, 96 (1985) 91.
11. W.H. Davis and W.A. Pryor, J. Chem. Educ., 53 (1976) 285.
12. H. Tanaka and N. Koga, Thermochim. Acta, 173 (1990) 53.
13. B.R. Wheeler and A.K. Galwey, J.Chem.Soc., Faraday Trans.I, 70 (1974) 661.
14. M.E. Brown and A.K. Galwey, Anal. Chem., 61 (1989) 1136.
15. M. Selvaratnam and P.D. Garn, J. Amer. Ceram. Soc., 59 (1976) 376.
16. M. Letort, J. Chim. Phys., 34 (1937) 206.
17. J.H. Flynn, J. Thermal Anal., 34 (1988) 367.
18. G. Várhegyi, Thermochim. Acta, 57 (1982) 247.
19. C. Popescu and E. Segal, J. Thermal Anal., 24 (1982) 309.
20. J.M. Criado and A. Ortega, Thermochim. Acta, 103 (1986) 317.
21. J. Rouquerol, Thermochim. Acta, 300 (1997) 247.
22. M. Reading, D. Dollimore, J. Rouquerol and F. Rouquerol, J. Thermal Anal., 29 (1984) 775.
23. A. Ortega, L.A. Pérez-Maqueda and J.M. Criado, Thermochim. Acta, 239 (1994) 171; J. Thermal Anal., 42 (1994) 551.
24. O. Toft Sørensen, Thermochim. Acta, 50 (1981) 163.
25. J.H. Flynn and B. Dickens, Thermochim. Acta, 15 (1976) 1.
26. E.V. Boldyreva, Thermochim. Acta, 110 (1987) 107.
27. M. Maciejewski, J. Thermal Anal., 38 (1992) 51; 33 (1988) 1269.
28. J. P. Elder, J. Thermal Anal., 30 (1985) 657; Analytical Calorimetry, (Eds P.S. Gill and J. F. Johnson), Vol. 5, Plenum, New York, 1984, p.269.
29. J. Zsakó, Thermal Analysis, (Ed. Z.D. Zivkovic), University of Beograd, Bor, Yugoslavia, 1984, p.167.
30. J.R. MacCallum, Thermochim. Acta, 53 (1982) 375.
31. J. Sesták, Thermochim. Acta, 83 (1985) 391.
32. E. Segal, Thermochim. Acta, 148 (1989) 127.
33. A. K. Galwey and M. E. Brown, Proc. R. Soc. (London), A450 (1995) 501.
34. J.H. Flynn, Thermochim. Acta, 300 (1997) 83.
35. J. Krís and J. Sesták, Thermochim. Acta, 110 (1987) 87.
36. W-L. Ng, Aust. J. Chem., 28 (1975) 1169.
37. J. Sesták and G. Berggren, Thermochim. Acta, 3 (1971) 1.
38. J. Málek and J.M. Criado, Thermochim. Acta, 175 (1991) 305.
39. R.K. Agrawal, Thermochim. Acta, 203 (1992) 93.
40. J.M. Criado, A. Ortega and F. Gotor, Thermochim. Acta, 157 (1990) 171.

41. J. Malék, Thermochim. Acta, 200 (1992) 257.
42. S.V. Vyazovkin and A.I. Lesnikovich, J. Thermal Anal., 35 (1989) 2169.
43. S.V. Vyazovkin and A.I. Lesnikovich, J. Thermal Anal., 36 (1990) 599.
44. P. Budrugeac and E. Segal, Thermochim. Acta, 260 (1995) 75.
45. J.P. Elder, J. Thermal Anal., 29 (1984) 1327; 34 (1988) 1467; 35 (1989) 1965; 36 (1990) 1077.
46. C.M. Wyandt and D.R. Flanagan, Thermochim. Acta, 197 (1992) 239.
47. J. Zimmerman, MS Thesis, Illinois State University, Normal, Illinois, USA, 1983.
48. P.D. Garn, Crit. Rev. Anal. Chem., 3 (1972) 65.
49. J. Sesták, J. Thermal Anal., 16 (1979) 503.
50. J Meindl, I.V. Arkhangelskii and N.A. Chernova, J. Thermal Anal., 20 (1981) 39.
51. V. Vyazovkin, A.I. Lesnikovich and V.I. Goryachko, Thermochim. Acta, 177 (1991) 259.
52. N.S. Felix and B.S. Girgis, J. Thermal Anal., 35 (1989) 743.
53. G. Várhegyi, Thermochim. Acta, 110 (1987) 95.
54. H. Friedman, J. Polym. Sci., 50 (1965) 183.
55. B. Carroll and E.P. Manche, Thermochim. Acta, 3 (1972) 449.
56. J.H. Flynn, J. Thermal Anal., 37 (1991) 293.
57. E.S. Freeman and B. Carroll, J. Phys. Chem., 62 (1958) 394; 73 (1969) 751.
58. J. Sesták, V. Satava and W.W. Wendlandt, Thermochim. Acta, 7 (1973) 333.
59. J.M. Criado, D. Dollimore and G. R. Heal, Thermochim. Acta, 54 (1982) 159.
60. A. Jerez, J. Thermal Anal., 26 (1983) 315.
61. A.A. Van Dooren and B.W. Muller, Thermochim. Acta, 65 (1983) 257, 269.
62. R.M. Fuoss, O. Sayler and H.S. Wilson, J. Polym. Sci., 2 (1964) 3147.
63. H. E. Kissinger, J. Res. Nat. Bur. Stand., 57 (1956) 217.
64. H. E. Kissinger, Anal. Chem., 29 (1957) 1702.
65. J.A. Augis and J.E. Bennett, J. Thermal Anal., 13 (1978) 283.
66. J. Llopiz, M.M. Romero, A. Jerez and Y. Laureiro, Thermochim. Acta, 256 (1995) 205.
67. T. Ozawa, Bull. Chem. Soc. Japan, 38 (1965) 1881.
68. T. Ozawa, J. Thermal Anal., 2 (1970) 301.
69. H. Tanaka, Thermochim. Acta, 267 (1995) 29.

70. H.J. Borchardt and F. Daniels, J. Amer. Chem. Soc., 79 (1957) 41.
71. R.L. Reed, L. Weber and B. S. Gottfried, Ind. Eng. Chem. Fundam., 4 (1965) 38.
72. E. Koch and B. Stikerieg, Thermochim. Acta, 17 (1976) 1.
73. Y. L. Shishkin, J. Thermal Anal., 30 (1985) 557.
74. J. H. Flynn and L. A. Wall, J. Res. Nat. Bur. Stand., 70A (1966) 487.
75. C.D. Doyle, J. Appl. Polym. Sci., 5 (1961) 285.
76. J. Zsakó, J. Phys. Chem., 72 (1968) 2406.
77. J. Blazejowski, Thermochim. Acta, 48 (1981) 125.
78. A. J. Kassman, Thermochim. Acta, 84 (1985) 89.
79. R. Quanin and Y. Su, J. Thermal Anal., 44 (1995) 1147.
80. V. M. Gorbachev, J. Thermal Anal., 25 (1982) 603.
81. J. Sesták, Thermochim. Acta, 3 (1971) 150.
82. J. Zsakó, J. Thermal Anal., 34 (1988) 1489.
83. E. Urbanovici and E. Segal, Thermochim. Acta, 168 (1990) 71.
84. J.H. Flynn, J. Thermal Anal., 37 (1991) 293.
85. C.D. Doyle, Nature (London), 207 (1965) 290.
86. J. Zsakó, J. Thermal Anal., 8 (1975) 593.
87. C.D. Doyle, J. Appl. Polym. Sci., 6 (1962) 639.
88. J.M. Criado, J. Málek and J. Sesták, Thermochim. Acta, 175 (1991) 299.
89. L. Endrenyi (Ed.), Kinetic Data Analysis: Design and Analysis of Enzyme and Pharmacokinetic Experiments, Plenum, New York, 1981.
90. J. Militky and J. Sesták, Thermochim. Acta, 203 (1992) 31.
91. J. Madarász, G. Pokol and S. Gál, J. Thermal Anal., 42 (1994) 559.
92. S.V. Karachinsky, O.Yu. Peshkova, V.V. Dragalov and A.L. Chimishkyan, J. Thermal Anal., 34 (1988) 761.
93. S.V. Vyazovkin and A.I. Lesnikovich, Thermochim. Acta, 165 (1990) 273.
94. D. Beruto and A.W. Searcy, J. Chem. Soc., Faraday Trans. I, 70 (1974) 2145.
95. M.Yu. Sinev, J. Thermal Anal., 34 (1988) 221.
96. V. Marcu and E. Segal, Thermochim. Acta, 35 (1980) 43.
97. J.M. Criado, M. González, A. Ortega and C. Real, J. Thermal Anal., 34 (1988) 1387.
98. S.V. Vyazovkin, Thermochim. Acta, 223 (1993) 201.
99. A.K. Burnham and R.L. Braun, Energy and Fuels, submitted.
100. R.L. Braun and A.K. Burnham, J. Energy and Fuels, 1 (1987) 153.
101. A.K. Burnham, R.L. Braun and T.T. Coburn, Energy and Fuels, 10 (1996) 49.

102. A.K. Burnham, B.J. Schmidt and R.L. Braun, Org. Geochem., 23 (1995) 931.
103. S.V. Vyazovkin and A.I. Lesnikovich, Thermochim. Acta, 182 (1991) 133.
104. S.V. Vyazovkin and W. Linert, Anal. Chim. Acta, 295 (1994) 101.
105. J.H. Flynn, J. Thermal Anal., 44 (1995) 499.
106. K.N. Ninan, J. Thermal Anal., 35 (1989) 1267.
107. M.E. Brown, D. Dollimore and A.K. Galwey, Reactions in the Solid State, Comprehensive Chemical Kinetics, (Eds C.H. Bamford and C.F.H. Tipper), Vol. 22, Elsevier, Amsterdam, 1980.
108. J.H. Taplin, J. Chem. Phys., 68 (1978) 3325.

Chapter 6

CHARACTERIZATION OF REACTANTS, DECOMPOSITION INTERMEDIATES AND PRODUCTS, AND THE FORMULATION OF MECHANISMS

6.1. INTRODUCTION

Heterogeneous reactions, of which decompositions of single crystalline solids are relatively simple examples, involve, by definition, more than one phase and chemical changes occur at interfaces between these participating phases. Such processes are expected to proceed through more steps, and hence have more complex mechanisms, than homogeneous reactions of comparable stoichiometry. Consequently more experimental evidence is required to identify the steps contributing to the overall chemical change. Information directly relating to some of these processes may not be experimentally accessible and it may be necessary to make inferences based on such evidence that can be obtained. Microscopic examination of the textural changes that accompany reaction (Section 6.3.) can provide evidence on the geometry of interface development. The chemical processes at the molecular level, occurring within zones of locally-enhanced reactivity are, however, much more difficult, or even impossible, to characterize. The total amounts of any intermediates present at any time, will be very small, perhaps a monomolecular layer (or less) at the reactant/product interface. Any attempt to expose or isolate intermediates by mechanical separation of the interface may result in its destruction, together with the loss of its immediate environment. Furthermore, the location of such species at the reactant/product interface and/or the properties of the residual product (which may be metallic, opaque, poorly crystallized, etc.) may preclude the use of spectroscopic and/or resonance techniques for the identification of intermediates.

Any technique, or combination of techniques, which can provide information on the physical properties and/or the structure of the bulk or the surface of the solid, including adsorbed layers at the surfaces and, particularly, the nature of any defects present, is potentially useful in characterizing the reactant. The same techniques may be used to characterize the solid residue present at any stage of the decomposition, including the final product. There are thus numerous experimental methods which could be considered. Most of these, including surface-area measurement, X-ray diffraction, optical and electron microscopy, and the diverse spectroscopic methods, are specialist fields with their own vast literature. In this chapter it is only feasible to outline some of the possibilities, to provide examples of techniques which have already been shown to be useful, and to refer to more detailed accounts [1-3].

6.2. CHARACTERIZATION OF SOLID REACTANTS, INTERMEDIATES AND PRODUCTS

6.2.1. Introduction

Each decomposition studied presents its own particular problems. It is important to confirm the identity and chemical composition of the reactant (Section 2.3.) and to note any pretreatments, such as crushing, dehydration, irradiation, ageing, etc., to which it may have been subjected. Once the overall stoichiometry of the decomposition has been determined (Chapter 2), and extensive kinetic analyses have been carried out (Chapter 5), formulation of detailed decomposition mechanisms generally requires investigation of some or all of the following properties of the solids involved in the decomposition, be they reactant, intermediates or products: (i) the crystal structures; (ii) the physical form of the sample, ranging from the dimensions of individual single crystals to particle size distributions, surface areas and porosities of powders; (iii) the identities and concentrations of any defects, including impurities, present; and (iv) those physical properties of the sample, such as electrical and thermal conductivities, magnetic susceptibility, etc., which directly relate to any mechanism formulated for the reaction under consideration.

6.2.2. Crystal structure

The determination of crystal structures by X-ray diffraction (and occasionally neutron diffraction) and the detailed examination of the surface regions using electron diffraction are all well-documented and widely used methods [1-4]. In decomposition studies, it is an advantage, if good single crystals are available, to be able to start with a knowledge of the crystal structure which may be of use in interpretation of the behaviour observed. X-ray diffraction data does not, however, provide information about local atypical features, such as defects, which play a major role in some decomposition mechanisms. In many studies large relatively perfect single crystals of reactant are not available and X-ray powder diffraction techniques may be used to characterize the sample. The powder diffraction patterns obtained may be used to provide information about the particle sizes and shapes from the widths of the lines. Strain in the crystallites, as a result of mechanical treatment such as grinding, may also influence the widths of the diffraction lines. Equipment has been developed to monitor the crystal structure continuously while the sample is being heated [5,6].

Reactant characterization also includes determination of any changes, such as loss of water of crystallization and/or phase transformations, prior to decomposition. Such processes may cause changes in composition (e.g. water loss) and/or surface textures, and in the numbers and distributions of imperfections. Some dehydrations yield an anhydrous salt, or lower hydrate, that is amorphous to X-rays, and a single crystal reactant is converted to an assemblage of fine crystallites. The consequences of annealing on defect concentrations and the decomposition of surface material of low stability may also require consideration.

X-ray diffraction measurements enable the presence or absence of epitactic and topotactic relationships between reactant and solid product to be established [7]. Electron diffraction may be useful when only small particles are available, and is especially valuable for the examination of selected small areas during electron microscopic examination of the sample. The detailed surface structure, including the presence of chemisorbed species, may also be revealed.

6.2.3. Spectroscopic information

A wide variety of spectroscopic techniques [1-4] are used to provide information about the identities and amounts of reactant, the solid product(s), or the solid phase(s) present at intermediate stages of the decomposition. The cooling of reactant samples after partial or complete decomposition may be accompanied by changes such as crystallographic transformations, solidification of any molten material and particle disintegration. Interpretations of observations must take account of these possibilities.

Information on the elemental composition and distribution at intermediate stages of the decomposition will only be useful if it indicates an abnormal accumulation or depletion of one of the known constituents, e.g. a local concentration of metal corresponding to the formation of decomposition nuclei in the form of metal particles. More useful are indications of differences in the electronic states of atoms, the coordination involved and the presence of impurities and defects, which can provide information on the decomposition mechanism. Usually the results of several complementary techniques are combined to give maximum information.

Mikhail and Robens [4] discuss the techniques available with reference to the Propst diagram, Figure 6.1. Only a few of the many possible combinations of probe and response, indicated in Figure 6.2, are in common use. Photons and electrons are the most important, as both probes and responses, for obtaining information about the electronic structures of the surface atoms of samples.

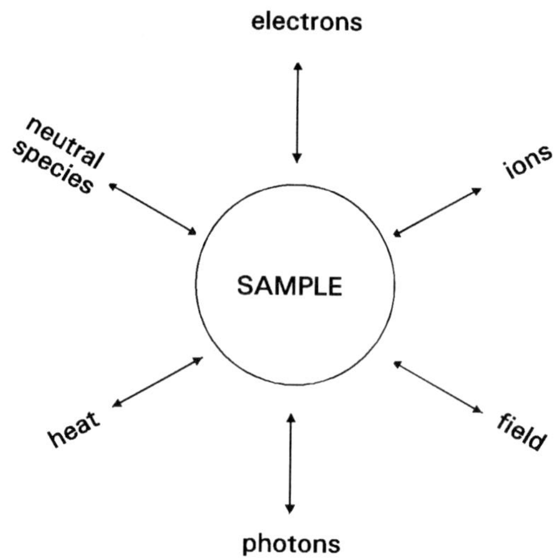

Figure 6.1.
The Propst diagram [4] summarises the probes that may be used (the inward arrows) on the sample (the circle) to produce the various responses (the outward arrows) which provide information about the sample.

The absorption or reflection by the solid sample of photons of various energies, from the ultraviolet to the infrared regions, are probably the most accessible and widely used techniques [8]. Absorption of infrared radiation by mulls or dispersions in alkali-halide discs are standard procedures. The sample holder may be capable of being heated so that peaks of interest can be monitored during the progress of decomposition. Hisatsune *et al.* [9] have successfully used infrared measurements to follow the decompositions of metal carboxylates (Chapter 16) incorporated in KBr discs. Spectra of powders may differ significantly from those of the same material in the form of larger crystals.

When electrons are used as probes, the alternative experimental configurations listed in Table 6.1. may be used. The various methods of reactant/product characterization described throughout this section contribute to, or confirm the reaction stoichiometry and may give information of value concerning the structural changes.

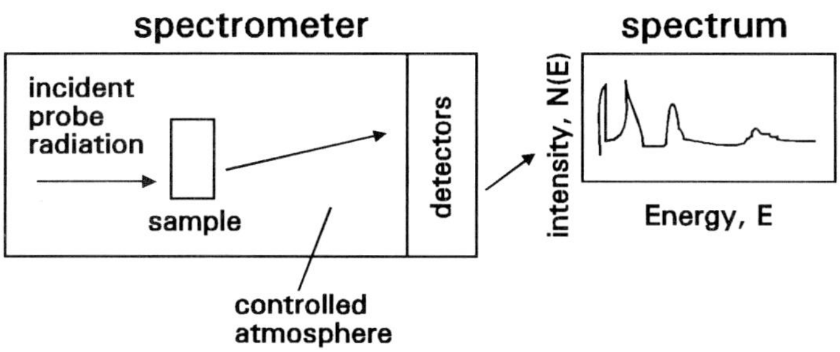

incident probe radiation		electron response
electron	→	Auger electron spectroscopy, AES
X-ray	→	X-ray photoelectron spectroscopy, XPS
ultraviolet	→	Ultraviolet electron spectroscopy, UPS
ions	→	Ion neutralization spectroscopy, INS

Figure 6.2.
Spectroscopic techniques [4].

Auger electron spectroscopy is used mainly for determining the elemental composition of the outer layers of a solid. Each element, when excited by the ionization of an inner electron level, emits Auger electrons with energies characteristic of that element and virtually independent of the chemical environment of the atom concerned.

Using an electric field as a probe can cause emission of electrons from a suitable sample. Measurement of the distribution of kinetic energy of the emitted electrons is called *field-emission spectroscopy* (FES).

Table 6.1.
Electron spectroscopies

Measurement	Technique	Abbreviation
Changes in energy of incident electrons after scattering by a solid	Energy loss spectroscopy	ELS
Energy spectrum of secondary electrons	Appearance potential spectroscopy	APS
Energy spectrum of electrons from inner electron ("core") levels	Auger electron spectroscopy	AES
Emission of photons	Soft X-ray appearance spectroscopy	SXAPS
Emission of X-rays characteristic of individual elements	Electron probe microanalysis	EPMA

Photons as probes may also cause the emission of electrons from the sample, and the kinetic energy distribution of these photoelectrons can be recorded and analysed. This technique is called *photoelectron spectroscopy* (PES) in general, but if the photons are in the ultraviolet range, UPS, or in the X-ray range, XPS. Photoelectron spectroscopy permits direct examination of electronic orbitals of atoms by providing information about the electrons in both the valence bands and the core levels of the constituent elements of solids. Under suitable conditions the electronic states of bulk and surface atoms can be distinguished.

West [10] gives examples of XPS spectra of sodium thiosulfate, $Na_2S_2O_3$, (see Figure 6.3.) where the central and terminal S atoms can be distinguished (sulfur 2p spectrum) and the Cr 3s and 3p spectra of KCr_3O_8, where the contributions from Cr(III) and Cr(VI) are clearly distinguished.

Van Attekum and Trooster [11] studied the X-ray induced decomposition of gold(III) dithiocarbamate complexes, using XPS. Unusually thin crystals of the reactant were prepared by vacuum sublimation or by evaporation of a solution.

Such samples give appreciably narrower spectral lines than are obtained from powders. Shifts of the peaks in the Au(4f) spectra showed that Au(III) is reduced to Au(I). The percentage of decomposition product formed at any time could be derived from the intensity ratios and was shown to be dependent upon both the intensity of the X-radiation and the temperature during irradiation.

Moroney *et al.* [12] used XPS, combined with mass-spectrometric analyses of the evolved gases, to study the thermal decomposition of β-NiO(OH). The O(1s) and Ni(2p) spectra of pressed powder samples were monitored during thermal decompositions between 423 and 773 K. No kinetic measurements were made, but the following mechanistic information was obtained. The O(1s) spectrum initially showed a single peak attributed to equivalence of oxygen species in O-H-O groups, stabilized by intercalated water in the reactant. At the lower temperatures (423 K), water was the main product evolved and, as the temperature was raised, the proportion of molecular oxygen increased. The deconvoluted O(1s) spectra were interpreted as contributions from the oxygen in water, from the O-H-O interaction, and from O^{2-}. The first two contributions decreased as the O^{2-} peak increased. The final solid decomposition product at 773 K contained a contribution from O^- which indicated the defective nature of the NiO formed. The Ni(2p) spectra supported these conclusions.

Copperthwaite and Lloyd [13] used XPS to study the radiolysis of sodium chlorate and of sodium perchlorate by monitoring the Cl(2p) and O(1s) spectra of powdered samples held in high vacuum at 263 ± 5 K. The decay in intensity of the Cl(2p) signal with time of exposure to X-rays was found to be a first-order process and rate constants for the decompositions of $NaClO_3$ and of $NaClO_4$ were derived. For both salts the rate constants increased linearly with the intensity of the X-radiation. The consecutive first-order processes were $ClO_4^- \rightarrow ClO_3^- \rightarrow Cl^-$ and there was evidence of trapping of molecular or atomic oxygen within the sample. Variable temperature studies between 113 and 293 K of $NaClO_3$ irradiated at a fixed X-ray intensity, showed an increase in decay rate of the Cl(2p) signal with decreasing temperature and hence a negative apparent activation energy. This was interpreted as the formation of a precursor to decomposition, having a probability of formation that increased with decreasing temperature.

Mössbauer spectroscopy [14-17] involves the use of γ-rays as probes. The resonant re-absorption by atomic nuclei of a highly monochromatic beam of γ-rays, whose energy can be varied by use of the Doppler effect on a moving source, is monitored as a function of energy. The spectrum obtained [10] (Figure 6.4.) is usually a series of poorly-resolved peaks from which information on oxidation

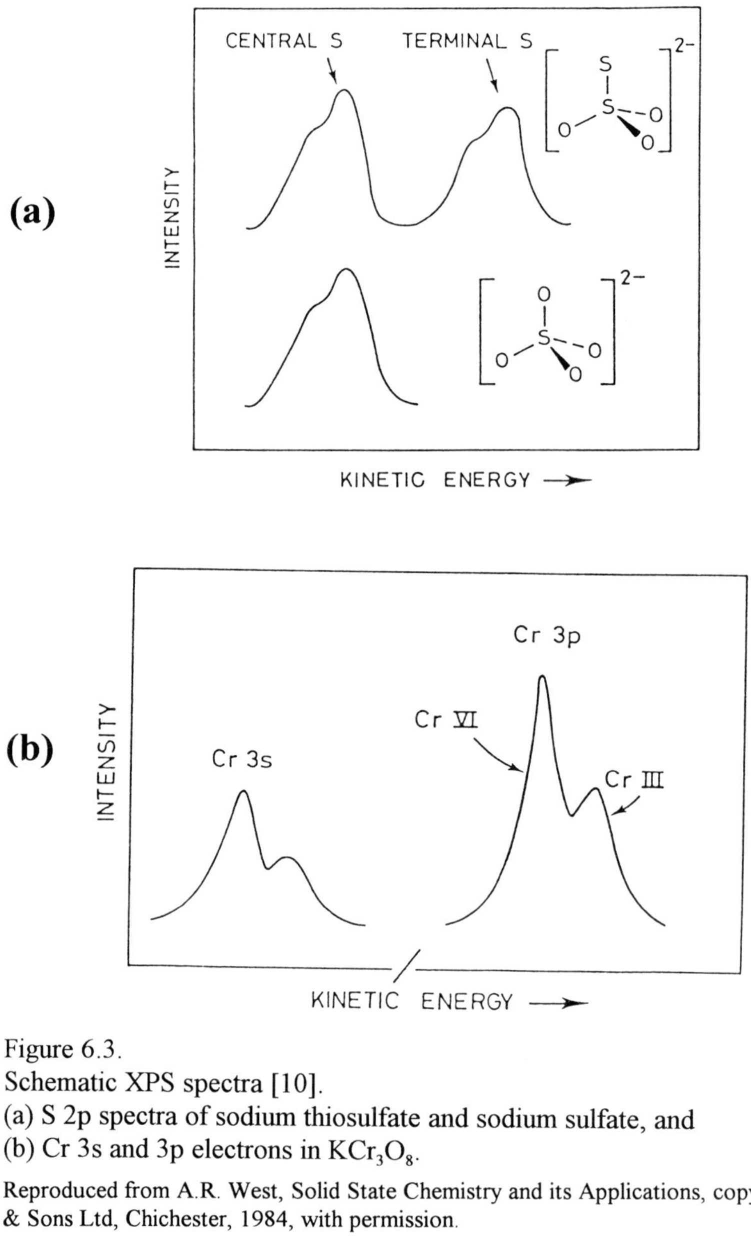

Figure 6.3.
Schematic XPS spectra [10].
(a) S 2p spectra of sodium thiosulfate and sodium sulfate, and
(b) Cr 3s and 3p electrons in KCr_3O_8.

states, coordination numbers and bond character may be deduced. The number of isotopes which emit suitable γ-rays is limited, and most studies have been concerned with compounds containing the isotopes ^{57}Fe or ^{119}Sn.

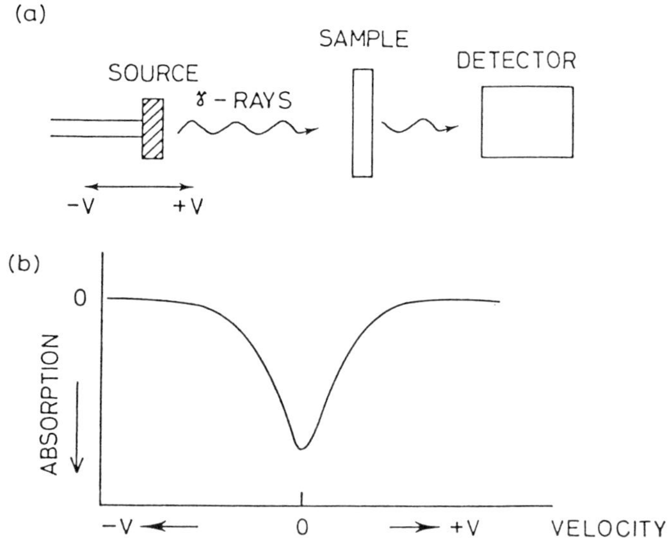

Figure 6.4
(a) The Mössbauer effect [10]. (b) Typical single line spectrum obtained when the source and the sample are identical.

Electron spin resonance spectroscopy (ESR) depends upon the presence of unpaired electrons in the sample [10] and the ESR spectra obtained for solid samples often consist of broad absorption peaks. These broad peaks may arise from spin-spin interactions between neighbouring unpaired electrons. To sharpen these peaks, the paramagnetic material may be diluted using a diamagnetic host structure, or spectra may be recorded at very low temperatures. The complexity of the ESR spectrum increases with the number of unpaired electrons per atom. Interpretation of ESR spectra [10] can provide information on the oxidation state, the electronic configuration and coordination number of the paramagnetic ion, as well as subtle details of the bonding of the ion with surrounding ions or ligands.

ENDOR (*electron nuclear double resonance*) is a combination of NMR and ESR [10] and is used to study crystal defects, radiation damage and doping.

Extended X-ray absorption fine structure (EXAFS) can provide information on bond distances and coordination, even for powdered samples of crystalline or amorphous materials [2]. The absorption edge itself is a characteristic of the element concerned, but the fine structure provides information on the short-range order around the absorbing atoms. Absorption of X-ray photons is followed by emission of photoelectrons which are diffracted to some extent by the environment of the absorbing atom. The phase of this photoelectron wave is dependent upon the wavelength (and hence the X-ray photon), the atomic number of the scattering atom and the distance between the absorbing and scattering atoms. The scattered waves also interfere with the emitted waves [2] (Figure 6.5.).

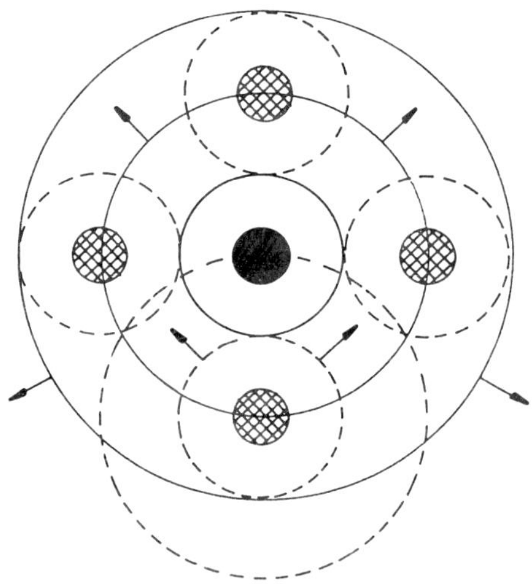

Figure 6.5.
The EXAFS process: one of the atoms in the sample (solid circle) absorbs X-rays and emits a photoelectron wave which is back-scattered by neighbouring atoms (hatched circles) [2].
Reproduced from C.N. Rao and J. Gopalakrishnan, New Directions in Solid State Chemistry, Cambridge University Press, Cambridge, 1986, with permission.

The EXAFS spectrum [10] (Figure 6.6.) shows the variation of absorption with energy in the region up to about 1 keV beyond the absorption edge. This typically ripple-like pattern is analysed mathematically to produce a radial distribution function [10] (Figure 6.7.) which provides information about the local structure. The technique is particularly useful for the study of amorphous solids such as glasses. EXAFS is more sensitive to structural disorder than XRD.

Synchrotron radiation is widely used for EXAFS. Rao and Gopalakrishnan [2] have reviewed the results obtained in studies of amorphous materials, catalysts, fast ionic conductors and intercalation compounds.

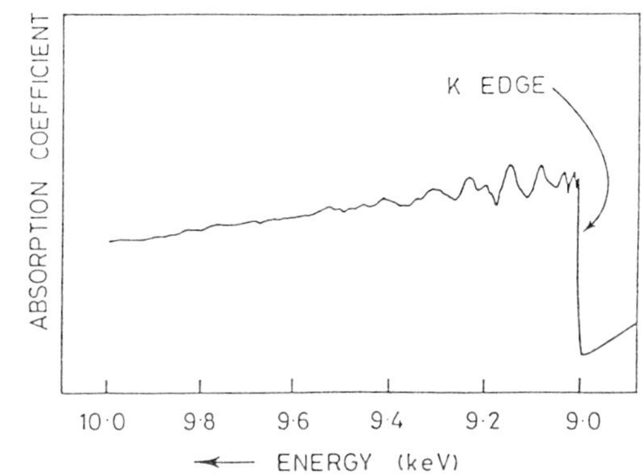

Figure 6.6.
EXAFS spectrum of copper metal [10].

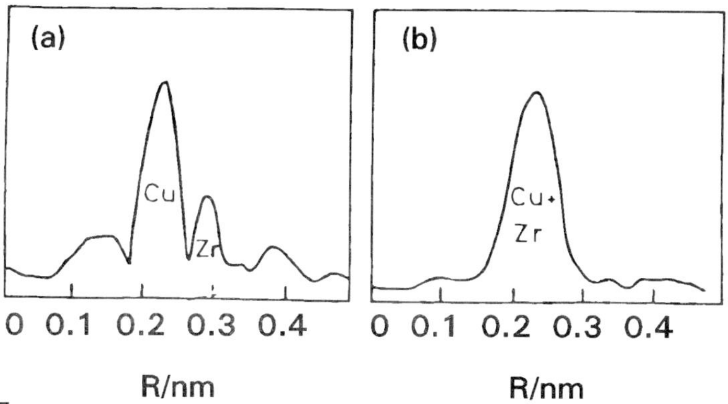

Figure 6.7.
Radial distributions derived from EXAFS spectra of an amorphous $Cu_{46}Zr_{54}$ alloy [10]. (a) Using the Zr *K* edge, the spectrum shows that each Zr atom is surrounded by an average of 4.6 Cu atoms at 0.274 nm and 5.1 Zr atoms at 0.314 nm. (b) Using the Cu *K* edge, the Cu - Cu distances are 0.247 nm.

Electron probe microanalysis [18] is used for quantitative analysis of small volumes (μm^3) of solid surfaces. The technique is most valuable for studying reactions between solids (A + B → AB or AB_2, A_2B, etc.) but can be useful in identifying decompositions which yield more than a single phase. EPMA can be used in two modes, either for quantitative analysis of a series of small volumes selected as being of interest, or for elemental mapping where a density distribution across a thin surface layer of the solid can be obtained for the element selected. The technique is particularly valuable for determining the elemental composition of atypical particles or textures where the total amount of an impurity present is very small. The technique is less accurate for rough surfaces, is inapplicable to the lightest elements and some combinations of elements are difficult to resolve.

Nuclear magnetic resonance (NMR) spectroscopy. NMR spectra can only be obtained for elements with non-zero nuclear spin. Early NMR studies on solids [1,2] gave broad and featureless spectra [19]. High resolution spectra [20,21] are, however, now obtainable using one or more of the line-narrowing techniques [22], including magic-angle spinning (MAS), dipolar decoupling and multiple-pulse averaging. Sensitivity may be increased by using cross-polarization (CP).

The discussion here will emphasize the problems involved in obtaining information about solids from NMR measurements. NMR has also been used to identify solvent-extracted decomposition intermediates [23], but allowance has to be made for possible solvent-intermediate interactions.

According to Maciel [22] "The availability of high-resolution NMR techniques for solids permits the use of NMR as a structure-determination bridge between the solid and liquid states of matter". Structural information is provided which is not available from X-ray diffraction. For decomposition studies, variable temperature facilities are required. Haw [20] has described the problems associated with such measurements, including rotor design and also temperature measurement and control.

Solid-state NMR of silicates using ^{29}Si has had notable success [24]. From the spectra it is possible to distinguish clearly between isolated (SiO_4^{4-}) tetrahedra and tetrahedra linked through oxygen at one, two, three or four corners [10] (Figure 6.8.). The fine detail of the chemical shifts depends on other features of the crystal structure. The detailed structures of aluminosilicates have been determined [2,24] through the effects on the ^{29}Si spectrum of the substitution of Si by Al.

X-ray topography. X-ray imaging techniques, such as X-ray topography [25], can provide useful information on the spatial relationship between dislocations present initially in the crystalline reactant and the sites at which thermal decomposition begins. The experimental arrangement [26], in either transmission or reflection modes, involves orienting the crystal in a finely-collimated X-ray beam to give Bragg diffraction for a particular set of lattice planes. The crystal size may be tens of mm across and 0.5 to 1.0 mm thick, but exposure times are long (measured in hours). For a transmission image, the crystal and film are scanned across the beam. For reflection measurements, longer wavelength radiation is used and only the surface layers are sampled.

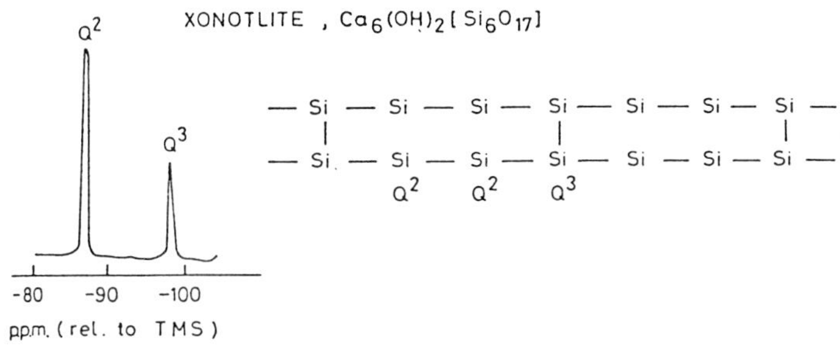

Figure 6.8.
The ^{29}Si NMR spectrum of xonotlite [10,24] and the structure of the silicate double chain anion involved. (Oxygen atoms are omitted). Q^n refers to the number of tetrahedra to which the silicon atom is joined by sharing oxygen at a corner.
Reproduced from A.R. West, Solid State Chemistry and its Applications, copyright John Wiley & Sons Ltd, Chichester, 1984, with permission.

The X-ray images of extended lattice imperfections within the crystal arise from variations in beam intensity caused by local distortions of the lattice. Dislocation densities of up to 10^3 mm^{-2} can be resolved in transmission studies or ten times this by the reflection technique. This resolution is lower than that obtainable by transmission electron microscopy, but the sample used may be thicker and does not have to be examined under high vacuum. X-ray beams also produce less radiation damage in the sample. When decomposition proceeds beyond $\alpha > 0.01$, distortion of the lattice is such that the X-ray image loses resolution and the exposures required become even longer. Reflection data obtained at several different diffraction angles may be required to characterize the imperfections present.

Synchrotron radiation [26] can provide the more intense X-ray beams required for X-ray topography and real-time examinations of samples, during thermal or radiolytic decomposition, are possible.

Sherwood *et al.* [25,26] have demonstrated the usefulness of X-ray topography in examining the microtextural displacements that occur during the initial stages of decomposition of single crystals of sodium chlorate. They showed that onset of decomposition is associated with centres of strain in the crystal, but, in contrast to several other studies on other compounds, these were not associated with the sites of dislocations. Between ambient temperature and 450 K the crystals are brittle, there is no dislocation multiplication and no decomposition occurs. Above 450 K the crystals and the strain centres become more plastic, and the strain centres expand forming dislocation loops associated with the [100] planes. Between 450 K and 520 K these loops develop into helices by vacancy flow along dislocation cores.

6.3. MICROSCOPIC EXAMINATION OF THE REACTANT SAMPLE BEFORE, DURING AND AFTER DECOMPOSITION

6.3.1. Introduction

Detailed examination of the textural changes occurring in a solid sample during reaction, by use of optical and/or electron microscopy, can provide a photographic record of important information that can be used to complement and to confirm kinetic interpretations. Decompositions generally produce residual solid phases which do not fill completely the reactant volume from which they were derived. Textures of the product phase can be characterized by examination of fully reacted material. If there are sufficient textural differences from the original reactant to enable the residual solid to be reliably recognized, the amounts and spatial distributions of product, in material reacted to various known values of α, can be used to confirm geometric conclusions based on kinetic measurements.

Microscopy is the most appropriate technique for studying the kinetics of nucleation. The shapes, sizes, textures and distributions of nuclei can be determined and the kinetics of nucleation can be distinguished from the kinetics of growth. Details of the intranuclear material, which is often porous with small crystallites separated by fine channels that provide routes for escape of product gas, may be discernable. Changes in particle-size, topochemical relationships and the possibility of melting of the solid reactant can also be recognized.

It may be necessary to section or lightly crush partially decomposed samples to reveal textural features of the reactant/product interface. This interface is generally assumed to be a thin laminar advancing zone, although recent studies of dehydrations (see Chapter 7) have shown that, for some hydrates, water loss occurs ahead of the advancing interface. In other hydrates the product recrystallization step, when recognizable under the microscope, appears to be distinctly separated from the water elimination steps. Such complications may apply to other decompositions. Extreme care and minimal delays are needed in the handling of reacted material between preparation and examination. Mechanical damage before and during examination and exposure to reactive gases (water, CO_2 and oxygen in the atmosphere) should be avoided. Finely-divided metal particles formed as decomposition products may be pyrophoric. During examination, heating and/or irradiation of samples by intense light or the electron beam may damage unstable materials, as may the vacuum conditions of the electron microscope.

Any reaction mechanism proposed must be consistent with all the evidence from microscopy.

6.3.2. Optical microscopy

Many observational configurations may be used in optical microscopy [1,27]. These include the use of transmitted or reflected light for examination of the sample in a bright or a dark field. The use of polarized illumination is essential in examining anisotropic and hence birefringent materials. Isotropic materials may become birefringent when stressed. Phase-contrast techniques may be used for revealing surface detail in opaque specimens. In reflection interference microscopy, differences in levels across surfaces are revealed by interference-fringe contours. Measurements from these contours can provide quantitative information on the depths and heights of surface features [28,29].

Optical microscopic examinations of the reactant at room temperature can provide information on the shapes and sizes of the crystallites and structural information from features of observed symmetry. The degree of perfection of the crystallites can be assessed and damage, major defects and inclusions may be identified. Surface defects, such as the points of emergence of dislocations, may be revealed by etching the sample surface with a suitable solvent. Cleaving of a crystal gives two closely related faces and one section may be etched to reveal the location of defects while the other section is partially decomposed and then re-examined [30,31] to enable the distributions of the different surface features (usually dislocations and nuclei) to be compared. Such techniques have been used to investigate the role of defects in the initiation of decomposition [32,33].

Mehta [34] compared the characteristics of chemical (0.2% acetic acid) and thermal (20 K min^{-1} in air) etch patterns on cleavage surfaces of calcite. Pits on corresponding surfaces matched only for the chemical etching of both. There was no matching of thermal pits on the two surfaces, or of thermal and chemical etch pits. Chemical etch pits were all of similar size and more regular in shape than thermal pits which were of assorted sizes. Mehta concluded that chemical etching occurs at emergent dislocations, but that thermal etch pits are associated with impurities or aggregations of defects. Thermal treatment may also cause movements, including aggregations, of dislocations.

Use of a controlled temperature heating-stage permits the sample to be observed up to and during decomposition. This technique can be used to detect the occurrence of phase transitions, sintering, sublimation or melting prior to, or concurrently with, decomposition as well as the geometry of interface development. Microscopy provided the evidence from which the rate laws of nucleation and growth (Chapter 3) were originally developed [35,36]. The conclusion that rates of interface advance are generally constant during growth of decomposition nuclei is based on sizes of nuclei, measured from micrographs taken at known time intervals during isothermal dehydrations [37]. Kinetic measurements from microscopy are usually in good agreement with measurements from other experimental techniques.

6.3.3. Electron microscopy

A major limitation of optical microscopy is the small depth-of-field at high magnifications. Use of electron microscopy, in either the transmission (TEM) or scanning (SEM) modes, provides a greater depth-of-field and permits higher magnifications than are possible by optical methods. These techniques, however, present their own practical problems. Samples have to be examined under high-vacuum conditions and while being irradiated by an energetic electron beam. Both these conditions may cause visible changes in potentially unstable materials, especially hydrates. Coating of the surface of the sample with a thin metal layer to improve the conductive properties of the sample is usually necessary in SEM. Some of these limitations can be overcome, in part, by examining plastic replicas of the surface of the sample. The textures of such replicas have been shown to be in good agreement with those of the original surface [38]. Internal features of nuclei have been revealed using Formvar (or Piloform) to penetrate and set to form a model of the crack structure generated in partially dehydrated alums [39]. Further detail is sometimes obtained by cleaving nuclei and replicating the surface revealed [40]. Evidence for the occurrence of localized melting, often a froth-like texture, can be revealed by fracture of the cooled products [41,42].

Temperature-controlled specimen stages are available and the cooling facilities may be used to counter the heating effect of the electron beam on the sample which may cause temperatures to rise as high as 670 K. Limited control of the atmosphere surrounding the specimen is also possible. TEM, although it requires thin samples (< 100 nm), also allows the recording of electron diffraction patterns for selected areas. Controlled atmosphere specimen stages enable decomposition studies to be made in the presence of gaseous species including possible reactants, but resolution is decreased.

Direct observations of the radiolysis of a wide range of compounds, especially azides and silver halides, in the electron beam have provided information on the mechanisms of radiolysis. These are often closely related to the mechanisms of thermal decomposition (see later Chapters on specific groups of compounds).

Cater and Buseck [43] compared the results reported for the thermal decomposition of dolomite in vacuum with observations of the beam-induced decomposition in the TEM. Only about 2% of the energy from the electron beam results directly in increased atomic movements, the greater proportion goes into electronic excitations. Radiolysis of the carbonate ion is a topotactic process, forming a metastable solid solution, $Ca_{0.5}Mg_{0.5}O$, which is initially amorphous before randomly-oriented crystallites of CaO and MgO are formed. Temperatures during radiolysis are low enough to prevent cation diffusion and consequent phase separation, which is apparent in the thermal decomposition, so decomposition in the electron beam may be a method of preparing solid products of high reactivity.

Kim *et al.* [44] reported results from a similar study of the decompositions of $Mg(OH)_2$ and $MgCO_3$ in the TEM. They were able to determine in detail the product/reactant orientation relationships, by careful preparation of specimens, using ion-beam milling where necessary. They detected no intermediate phase in either decomposition. The decomposition of $Mg(OH)_2$ produces MgO with a single orientation relationship to the substrate, while the decomposition of $MgCO_3$ yields MgO in two principal orientations. These topotactic relationships are explained in terms of the orientations of oxygen octahedra in the reactants and product.

Electron microscopy may be complemented by measurements of X-ray fluorescence and other spectroscopic methods. The development of high-resolution electron microscopy (HREM) has led to the identification of detail on the atomic scale and the precise structure of solids, rather than the averaged representations obtainable from X-ray diffraction.

6.3.4. Atomic force microscopy

The relatively new technique of atomic force microscopy (AFM) offers the possibility of high-resolution characterization of surface textures of solids. The absence of requirements for vacuum conditions or irradiation during examination decreases the chance of damage to the sample by the investigative process. Information on the surface topology, in sufficient detail to reveal features of atomic dimensions, is obtainable. In combination with other experimental observations, AFM may be capable of advancing understanding of problems that have not been resolvable by lower-resolution techniques. The earliest steps in nucleation, for example, may be detectable by the development of atomic-scale holes in reactant crystal surfaces (compare the electron micrographs of replicas of nuclei on alum surfaces in Figures 8 and 12 of reference [37] and in Figure 1.8.). Similarly, high-resolution AFM may be able to reveal the textures of the contact zones between reactant and products at interfaces that have been exposed by cleavage after reaction. A recent general survey of AFM [45] includes background information but is not directly applied to the decompositions of solids.

6.4. SURFACE AREA AND PARTICLE SIZE

The solid product of decomposition of a solid reactant often has a considerably increased specific surface area [46]. Increases in surface may result from the formation of cracks and internal pores, while sintering and recrystallization may decrease the area [47]. The atmosphere prevailing locally during reaction, e.g. self-generated atmospheres of the gaseous product, such as water-vapour in the neighbourhood of the solid reactant during dehydration, may influence the surface area of the solid. Information on the primary products together with any subsequent interactions is important in formulating the reaction mechanism.

Most surface area determinations are based on measurements of the low temperature adsorption of nitrogen or krypton on the solid and use of the BET theory. This procedure may not give reliable results because the products are chilled well below reaction temperature, possibly resulting in the sealing of internal pores. Volumes of gases adsorbed are sometimes small, as observed for dehydrated alums [37] and decomposed ammonium perchlorate [48], where the areas are consistent with product crystallites of linear dimensions between 1 and 3 μm. The results indicate, however, that little, if any, zeolitic material is formed [36]. The surface area of a solid may also be estimated from electron micrographs. Density measurements may be used to complement area measurements.

The total surface area of a decomposing solid must be distinguished from the effective area of the active reactant/product interface which, in most cases, is an internal structure. This active area is not usually accessible to measurement, but if it could be determined, this would enable the Arrhenius pre-exponential factor to be expressed in dimensions of $area^{-1}$, as in catalytic studies [49].

6.5. OTHER PHYSICAL PROPERTIES

6.5.1. Electrical conductivity

The considerable benefits in understanding the behaviour of solids that can result from establishing their transport properties are emphasized by Hamnett [50]. Information on transport properties can be used to gain insights into electronic structures but deductions must be interpreted in the context of other available physical measurements and structural information. This review [50] provides a useful introduction to the methods that are used to identify the charge-carrying species and their numbers in solid materials. Electrical measurements also provide information concerning the energy band levels, together with the acceptor or donor states. Further aspects of electrical characteristics of solids are discussed by West [51].

The use of electrical conductivity data in the formulation of mechanisms of solid state decompositions is not straightforward. More than a single species may be mobile within the solid and the behaviour of the same species will generally differ in regions of imperfection, including the surfaces, from that in the bulk crystal. The various influences of all the participants in the several chemical steps occurring may thus be difficult to distinguish and to characterize unambiguously.

Reviews of the application of electrical measurements in solid state decompositions have been given by Kabanov [52]. Electrical conductivity measurements, both a.c. and d.c. studies, have been used to characterize the species that participate in the thermal decomposition of ammonium perchlorate [53,54] (this reaction is discussed in Chapter 15). Other studies have been concerned with the mechanisms of oxide decompositions [55,56]. Torkar *et al.* [56] conclude from electrical conductivity evidence that the decompositions of alkali oxides are more complicated than exciton formation processes.

6.5.2. Magnetic properties

Solids may be classified, on the basis of the magnitudes of their magnetic susceptibilities (χ), as: *diamagnetic* (about -1×10^{-6}, slightly repelled by the field), *paramagnetic* (positive and below 10^{-2}), *ferromagnetic* (between 10^{-2} and 10^{6}) and *antiferromagnetic* (below 10^{-2}). The magnetic susceptibilities of both paramagnetic and ferromagnetic materials decrease with increasing temperature, while values for antiferromagnetic materials increase and for diamagnetic materials do not change. The background theory of this subject is given by West [57] and by Hatfield [58] who also describe experimental techniques and provide access to the literature. Measurements of magnetic susceptibility [8,59] are of value in determining oxidation states of transition metal ions, both in the reactant and the changes that occur as a result of reaction. For many paramagnetic solids, the magnetic moment may be used to estimate the number of unpaired electrons in constituent ions from which some structural inferences may be possible. Darrie *et al.* [60] showed that χ values could be used to measure extents of reaction in decompositions of selected chromates(VI). Roy *et al.* [61] used magnetic data, together with other observations to identify the products of isomerizations and decompositions of several nickel(II) coordination compounds. Changes in oxidation states have been measured for decompositions of iron, cobalt and chromium oxides and hydroxides [62] and on ferrite formation [63].

The main use of magnetic measurements in investigations of solid state decompositions has been through the technique of *thermomagnetometry* (TM) [17,64] in which the apparent weight gain or loss of the sample in a magnetic field is monitored as a function of time at constant temperature, or, more commonly, as a function of temperature while the sample is heated (or cooled) at a constant rate. Examples of the use of TM are the detection of the formation of intermediate spinel structures during the thermal decomposition of $FeCO_3$ [65] and of alloys as the products of the thermal decompositions of coprecipitated metal oxalates [66].

Kabanov and Skrobot have shown [67] that magnetic fields (200 to 500 oersteds) caused a slight diminution in the rate of $KMnO_4$ decomposition. Relatively few studies of this type have been made but these workers mention that magnetic fields increase the rate of barium azide decomposition, decrease the rate of decomposition of silver oxalate and do not change the rate of decomposition of silver azide.

6.6. REACTANT DOPING

Doping is the controlled introduction of specific additives into the reactant crystal as a solid solution. The additive, present at a low concentration, is selected to change the numbers of imperfections, usually lattice vacancies and electronic anomalies. The impurity must be compatible with the accommodating structure (similar ionic sizes and character). The introduction of altervalent ions is expected to result in either an increase or a decrease in the concentration of unoccupied sites and/or electronic defects. Through this the electronic conductivity of the solid, including mobilities of the constituents ions, will be changed. Studies of decompositions of doped reactants most usually compare the behaviour of the pure reactant with those of preparations that contain a range of concentrations of additives selected to both increase and to decrease the numbers present of the vacancy, the activity of which is being investigated.

Reports of the effects of doping are scattered through the literature and some representative examples are mentioned in Brown *et al.* [68]. The influences exerted on the courses of reactions in solids by additives present at low concentrations are diverse and complicated. This topic would benefit from a thorough survey. Impurities introduced by doping may enhance nucleation, result in lattice strain, promote reaction and/or modify the electronic properties and mobility of component ions in the reactant.

Non-isothermal kinetic studies [69] of the decomposition of samples of nickel oxalate dihydrate doped with Li^+ and Cr^{3+} showed no regular pattern of behaviour in the values of the Arrhenius parameters reported for the dehydration. There was evidence that lithium promoted the subsequent decomposition step, but no description of the role of the additive was given.

The incorporation of K^+ or Br^- (mole fractions 10^{-4} to 10^{-1}) both increased [70] the rate and decreased the activation energy for $NaBrO_3$ decomposition. The influence of the incorporated K^+ ions was ascribed to the presence of vacancies (through supporting evidence was not provided). Reaction was then identified as being promoted at the weakened bonds in the vicinity of these defects. The bromide ions facilitate eutectic formation thus enhancing the decomposition rate. The introduction of Ba^{2+} ions into $CsBrO_3$ is proposed [71] to generate free space and local strain. This facilitates melting in which decomposition proceeds more readily. The decompositions of strontium bromate doped with K^+, Mg^{2+}, Al^{3+} or KBr (10^{-4} to 10^{-1} M) were compared with the reactivity of the pure salt [72]. The incorporation of Al^{3+} or KBr increases the rate and decreases the activation energy of $Sr(BrO_3)_2$ breakdown. In contrast the presence of K^+ or of Mg^{2+} decreased the reactivity of the salt. It is concluded that the stabilities of the doped samples diminish with increase in surface charge densities of the added ions K^+, Mg^{2+} and Al^{3+}. Promotion by KBr is ascribed to eutectic formation in the presence of the Br^- ion.

The effect of doping with Ba^{2+} on the decomposition of cubic NH_4ClO_4 has been shown to be complicated [73]. As the concentration of added Ba^{2+} is increased, the reaction rate first decreases, then increases to a maximum and finally diminishes. It is concluded that barium acts as an electron trap, Ba^+. (This appears inherently improbable in the presence of such a strong oxidizing agent and an alternative reaction mechanism is discussed in Chapter 15.)

These recent reports of the effects of doping on decompositions contribute relatively little to the understanding of the mechanisms of the reactions investigated. The presence of the additive significantly complicates behaviour and unless sufficient evidence concerning the chemistry of all participating processes is obtained, the steps controlling reaction cannot be reliably characterized.

6.7. PRE-IRRADIATION STUDIES

Irradiation of some solids prior to, or even during, thermal decomposition may have a marked effect on the kinetics of decomposition [68]. Such effects are usually interpreted in terms of the imperfections generated by the treatment. Types of radiation used have ranged from visible light to ionizing radiation to neutrons. Solids whose stabilities have been shown to be particularly sensitive to irradiation include the azides (Chapter 11), the permanganates (Chapter 14) and some metal carboxylates (Chapter 16). Comparisons of kinetics of radiolysis with those of pyrolysis can provide useful mechanistic information. Examples of such comparisons are given in the chapters mentioned above.

6.8. INTERFACE CHEMISTRY

6.8.1. Fundamentals

Obtaining insights into interface chemistry is difficult. Many decomposition reactions occur within a zone of contact between reactant and product; these are usually regarded as having different crystal structures, see Figure 6.9. The thickness of this reaction zone may be of molecular dimensions and thus the chemical changes, reaction intermediates, textures and other factors may be difficult to characterize. Reaction precursors and participating intermediates may represent only a very small fraction of the total amount of material present and are consequently difficult to identify and to measure quantitatively. Responses to spectral and other probes may be weak, or unobtainable, when opaque or conducting products, such as metals, are present. Isolation of participants may be impossible because these species may be stabilised by the reaction-promoting properties of the interface. Even non-destructive methods of investigation may alter the unstable conditions at the interface (e.g. spectroscopy relies on molecular responses to radiation) and uncertainties in interpretation will always exist if the temperature during investigation deviates from that of the reaction. Intrusive methods, such as wet-chemical tests, introduce even greater uncertainties.

Reaction conditions at an interface may differ from those in homogeneous rate processes in (at least two) further respects that are important in formulating reaction mechanisms. (i) Very small total amounts of intermediates immobilised within the active zone may be disproportionately effective in promoting reaction. This "super-cage" effect, to use the homogeneous kinetics analogy, ensures repetitive collisions of a type that cannot arise in reactions of gases or liquids where there is diffusive separation of constituents after a collision. (ii) Interface processes can, in principle, involve high molecularity encounters, where three, or perhaps even four molecules may participate in an activating collision. This is a consequence of the restricted movements of effectively immobilized interfacial species and may be significant in interpreting the mechanisms of solid state rate processes.

The inaccessibility of reaction interfaces to investigation means that indirect methods must be used to explore the chemical reactions that occur within these active zones. The determination of the sequence of bond redistribution steps that results in the transformation of a crystalline reactant into a (usually different but often crystalline) product phase is the fundamental objective of mechanistic studies [74]. All intermediates and the factors that determine the rate and energetics of the transformation (reactivity) must be identified.

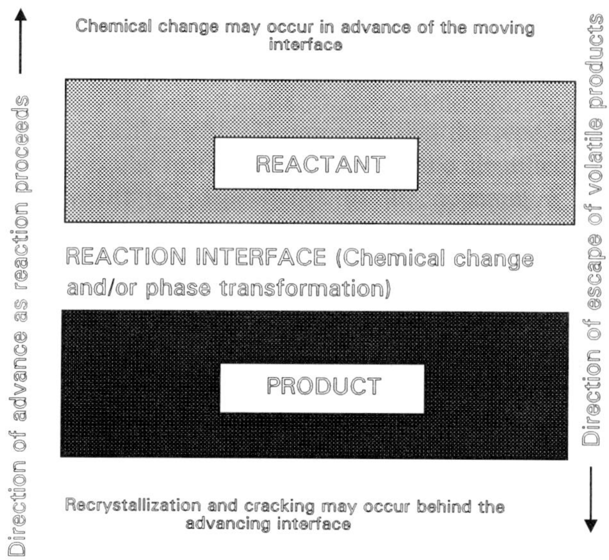

Figure 6.9.
Representation of a section across an advancing interface. Within the interface both chemical reactions and phase transformations may occur. In some decompositions, these processes may be spacially separated. The thickness of the interface may range from molecular dimensions (e.g. in topotactic reactions) to much larger distances when there is melt formation.

'Reaction mechanism' as currently used in solid state chemistry is an ambiguous term. Sometimes it is used to refer to the geometrical progress of the reactant-product interface, i.e. the fit of data to a particular rate equation (Table 3.3.), rather than the chemical steps taking place at the interface. Because the first interpretation is experimentally more accessible, many studies do not proceed further. Throughout this book the *reaction mechanism* refers to the sequence of chemical steps through which the reactant is transformed into products.

Some of the experimental approaches which may provide evidence concerning interface chemistry are as follows.

(i) *Stoichiometry.* Consideration of the nature and distribution of bonds in both the reactant and the product constituents identifies the electronic reorganisation steps that are required to convert the one into the other.

(ii) *Crystallography.* The phase transformation that accompanies reaction can provide information about the nature of the interface changes. Most significant is the detection of topotaxy [7,74] where the spatial disposition of one component of the reactant (usually the larger anion) apparently undergoes little displacement during passage of the interface. The product, therefore, is crystallographically orientated with respect to the reactant.

(iii) *Kinetics.* The magnitudes of activation energies are sometimes associated, by analogy with homogeneous reactions, with the energy of bond rupture and this evidence has been used to identify the rate controlling step. Similarly, the pre-exponential term can be used as a measure of the entropy of activation or the numbers of molecules per unit area participating in the chemical change. This approach assumes that the Arrhenius and transition state models can be applied to solids (see Chapter 4), but there is a problem in providing a theoretical explanation of the term "mole".

(iv) *Microscopy.* Microscopic examination of the reaction zone/interface may be useful in providing evidence of the occurrence of melting, product disintegration (cracks and pores), interface sizes, textual interrelationships between reactant and product, sintering, etc. Textures based on microscopic observations were used to characterize most of the nucleus structures described below. The possibility of obtaining mechanistic information from inspection of textures is a feature of the chemistry of solids which can be explored with imagination [75]. Sesták [76] has pointed out the careful interpretation needed when attempting to correlate the results of macroscopic (and averaged) kinetic measurements based on mass losses or evolved gas pressures with microscopic detail for a limited fraction of the sample, such as the surface, which may also be susceptible to damage during investigation.

(v) *Chemical methods.* As in homogeneous kinetics, the addition of an identified or inferred intermediate can be expected to promote reaction and the successful prediction of the influence of an additive may be used as evidence in support of a proposed mechanism. For example, the demonstration that the anhydrous salt acted as seed crystals for the dehydration of *d*-$LiKC_4H_4O_6.H_2O$ provided essential evidence in support of the reaction mechanism proposed [77]. Similarly, the role of copper acetate as a possible intermediate in the decomposition of copper(II) malonate was confirmed [41] by comparative studies of reaction rates in appropriate mixtures of these salts. (Such mixtures may be distinguished from reactant doping because the additive is not necessarily incorporated into the reactant as a solid solution.)

6.8.2. Classification of interface structures and their modes of action

Classification of solid state decompositions is not a simple problem, on account of the number of variables involved, particularly the unique crystal structure of each solid. It is thus debatable whether any set of criteria is likely to provide a suitable classification for more than limited purposes of comparison within relatively small groups of compounds. Some of the attempts at establishing a framework for comparison of the large number of decompositions that have been studied are described in Chapter 18. Recognition of the similarities, as well as the differences, of the structures and chemical characteristics of *interfaces* (or surfaces [78]) offers one potentially useful approach to classifying decomposition mechanisms. Different types of interfaces, distinguished largely through microscopic and kinetic evidence [75], are described below and possible reasons for the preferential location of chemical changes within such active zones are discussed.

(i) *Fluid-flux nuclei.* These nuclei are regarded as structures that retain temporarily a proportion of a volatile product at intranuclear surfaces where the adsorbed/condensed material promotes recrystallization of the residual solid. This is autocatalytic behaviour and structural reorganisation of chemically modified material at the advancing interface is regarded as the difficult step. The reaction enhancing function of the nucleus therefore results from the delay in the escape of a volatile product. Reaction proceeds more rapidly at the nucleus boundary than on pristine crystal faces where the chemical change cannot progress until a product crystal is generated by the nucleation step. Extensive surface reaction can be initiated on hitherto unchanged surfaces by exposure to water vapour, as observed for alum dehydration [79,80]. There is also evidence that restricted amounts of water are lost from the crystal faces of many hydrates [37,81,82] but reaction ceases in the absence of recrystallization of dehydrated product with the generation of the intranuclear product-retaining structure. Although not the decomposition of a solid, the reaction: $2KBr + C\ell_2 \rightarrow 2KC\ell + Br_2$ below 370 K is a nucleation and growth process in which nuclei become filled with liquid halogen product. This is cited [32] as an extreme example of a reaction of this type.

(ii) *Fusion nuclei.* Reaction may proceed in a local and temporary zone of melt resulting from the presence of either an intermediate that is liquid at reaction temperature or the formation of a eutectic melt between reactant and product. Some decompositions proceed more rapidly in a melt, where the stabilizing forces of intracrystalline attraction are relaxed and there is greater freedom for participating species to adopt configurations favourable to chemical reactions [75]. Examples are (a) copper(II) malonate, in which copper acetate is the molten intermediate [41], and (b) ammonium dichromate [42] where reactant dissociation yields CrO_3 which is molten close to the decomposition temperature.

(iii) *Functional nuclei.* These nuclei grow by the promoted decomposition of a reactant constituent adsorbed at the surface of a product particle. The mechanism is formally identical to models widely applied in heterogeneous catalysis. Examples include the decompositions of transition metal carboxylates, where the anion (oxalate, formate, etc.) bonds to the surface of particles of metallic product which promotes its breakdown. Examples include the decompositions of (a) nickel oxalate, (b) nickel formate, (c) nickel squarate [83], all of which yield nickel metal product which is a catalyst for anion breakdown (see Chapter 16).

(iv) *Flux-filigree nuclei.* Here the product recrystallization occurs in partially, or perhaps completely, reacted material and is characterized by relatively rapid onset of reaction on the surface, followed by advance of the interface into the bulk of the crystal. An example is the dehydration of $Li_2SO_4.H_2O$ [84].

(v) *Fixed interface.* Control of reaction rate by the desorption step only (no geometric factor) means that the boundaries of the original crystal are the effective interface and do not advance as reaction proceeds. This pattern of behaviour has been recognised [78] for the dehydration of uranyl nitrate hexahydrate.

(vi) *No reaction interface.* During decompositions that yield volatile products only, reaction (or sublimation) proceeds at the advancing boundaries of growing cavities as observed on heating NH_4HCO_3 [85]. In other systems there may be a physical change only as, for example, in lignite dehydration [86] where there is evaporation of water immobilized in a coherent matrix but there is no chemical reaction.

6.8.3. The rate limiting step

A central feature of the theory of homogeneous reaction kinetics is that the overall reaction rate is regarded as being controlled by the bond redistribution step in which the largest energy barrier (activation energy) must be overcome. The concept of a rate controlling step should also be generally applicable to solid state reactions, but theoretical developments have been hindered by the experimental difficulties attending the detailed investigation of the reaction mechanism. For most reactions at interfaces, the complexity of the rate processes themselves, together with the inaccessibility and small size of the reaction zone, make it difficult to determine the sequence of steps by which a crystalline reactant is transformed into a crystalline product. The number of variables that must be considered as possibly influencing the reactivity of heterogeneous processes is increased. However, there exists an important parallel between homogeneous and heterogeneous reactions in that both exhibit comparable temperature dependencies through the fit of rate constants to the Arrhenius equation (even if particular conditions apply to solids: see Chapter 4). The reported instances of experimental values of activation energies for solid decompositions being similar to the values representing the strength of the bond ruptured in reaction, could only be valid for mechanisms where bond rupture was considerably slower than any of the other steps in the the sequence making up the overall reaction.

The classification by Boldyrev *et al.* [87,88] in 1965 of thermal decompositions of inorganic solids according to the character of the electronic transformation steps in the reaction, has received less attention than it deserves. The main features of this classification, to which has been added the possibility of control through proton transfer, are summarized in Table 6.2.

6.8.4. Complex interfacial structures

Interface reactions involve, in general, two different types of chemical steps: (i) breakdown of a reactant constituent, such as rupture of a bond in an anion, $C\ell O_4^-$, MnO_4^-, etc., or between a ligand (H_2O, etc.) and its central atom, followed by (ii) recrystallization of this chemically changed material from the reactant structure to the more stable product structure. In some reactions this recrystallization may involve only relatively minor structural displacements and modifications to some of the lattice components (topotactic processes), whereas, in others, an extensive lattice reorganisation may be accompanied by strain induced cracking which may even result in disintegration of the particles.

These steps may occur concurrently, sometimes in close proximity across a thin interface, or be appreciably separated. However, many other reactions are known where the recrystallization step is absent, such as during the dehydrations of zeolites and silicates where the dispositions of the highly stable, extended anions are modified only slightly by the removal of the relatively weakly held intercalated water molecules. The rates of such reactions are controlled predominantly by diffusion which may include an influence of reactant structure, where water migration may be possible in specific directions only, and geometric parameters, see for example the dehydrations of mica, talc [68] and gypsum [89].

Table 6.2.
Scheme for the classification of solid state decompositions, based on proposals by Boldyrev *et al.* [87,88] and extended to include the possibility of control through proton transfer

Reactions in which cleavage of chemical bonds is not accompanied by the transfer of electrons.	Irreversible decompositions: alkali and alkaline earth formates and oxalates.	
	Reversible decompositions: complex salts: hydrates, ammines etc. Alkaline earth carbonates.	
Reactions in which cleavage of chemical bonds is accompanied by electron transfer (oxidation-reduction).	Electron transfer within anionic or cationic sub-lattice: permanganates and perchlorates.	
	Intermediate formation of radicals.	Radical formed in cationic sublattice: (none listed).
		Radical formed in anionic sublattice: silver oxalate, silver azide.
Reactions accompanied by proton transfer.	Proton transfer between cation and anion in ammonium salts.	
	Proton transfer between anions in hydroxides.	

Examples of intermediate behaviour, where the reacted zone and the recrystallised product do not coincide, are to be found in the literature [90]. From diffraction studies using synchrotron radiation, lattice spacings in the vicinity of the reaction interface during dehydration of $Li_2SO_4.H_2O$ have been determined. Spacings were found to change in *advance* of the crystallographic discontinuity with which the interface is usually identified (because the textural change can be observed microscopically). Water loss is detected from the reactant, which initially shows modification of the original structure, before transformation into the product phase, perhaps accompanied by completed water elimination. This demonstrates the existence of two distinct and complementary reaction zones that contribute to the overall reaction.

Other examples of reaction occurring in advance of the product recrystallization interface include the dehydrations of $NiSO_4.6H_2O$ [91] and of $Ca(OH)_2$ [92]. For dehydration reactions, a complete sequence of types of behaviour can be distinguished, based on the degree of separation of the zones of water loss and of product recrystallization. This pattern of reactivity may have wider application.

(i) *Reaction and recrystallization in close proximity.* Water losses from alum appear to be initiated across all crystal surfaces. This modification of boundary material is demonstrated by the retexturing (orange-peel structures) that develops on subsequent exposure to water vapour [37,81, 82]. The extent of initial surface reaction is limited by the absence of recrystallization, which is achieved at only a few specialized sites and, once established, dehydration continues at interfaces that advance unhindered within the crystal bulk. Nucleus structures, by retaining water, promote recrystallization, but the water-loss zone is thin so that water release and structure change are closely spaced.

(ii) *Reaction and recrystallization separated.* In many dehydrations, the onset of significant water loss may proceed within the crystal at some distance ahead of an advancing recrystallization surface that yields the anhydrous (or lower hydrate) product. In some solids, the reaction zone may be so thin that the kinetics of water loss may be represented by the rate equation based on the observed generation and growth of nuclei. Where the interface is thicker, however, or the reaction is reversible, these factors must be allowed for in the interpretation of rate measurements. The dehydration of $Li_2SO_4.H_2O$, for example, occurs by rapid and extensive nucleation followed by growth [84], but agreement of kinetic results amongst various studies, using different salt preparations, has been unsatisfactory [93]. Boldyrev *et al.* [90] have confirmed the development of a reaction zone of appreciable thickness (about 150 μm) in this salt. The dehydration zone may be even further ahead of the advancing recrystallization front during the dehydrations of $NiSO_4.6H_2O$ [91] and of $Ca(OH)_2$ [92]. The overall kinetic behaviour (based on water release) may thus not be quantitatively related to the nucleation and growth process which involves the recrystallization of partly (or completely) dehydrated product that has retained the reactant structure.

(iii) *No recrystallization*. When the reactant structure is very stable (as in zeolites, silicates, etc.), dehydration results in only minor reorganisation. Kinetic behaviour is dominated by the diffusive escape of water vapour [94].

(iv) *Nucleation*. Although it is often stated that nuclei appear at the sites of intersections of line dislocations with surfaces, the evidence in support of this is not entirely unequivocal. The numbers of growth nuclei seen on partly-reacted crystals are often comparable with the numbers of etch pits revealed, but these are not always in the same places [30]. The stresses associated with crystal cleavage, required to expose the matched surfaces used to compare the locations of etch pits (dislocations) and nuclei, are believed to displace line imperfections. It is also probable that initial surface reactions on hydrates (often selected as model reactants for work of this type) may, similarly, displace sites of imperfection.

6.8.5. Alternative mathematical description of solid state reactions

Korobov, in a review [95] that summarizes earlier work [96,97], emphasizes the chemical foundations of solid state reactions. He points out that the mathematical approach to this subject currently used is based on variables that have a geometric-probabilistic rather than a chemical significance. Accordingly he suggests that these phenomena would be more appropriately described by different mathematical concepts using the "language of planigons, random mosaics, and also the third variety of Dirichlet tesselations, Wigner-Seitz cells". While these may, at present, be unfamiliar to most chemists, the concepts are no more difficult than those currently employed. The advantage identified is that these methods lead to "deeper insight into the chemical essence of solid-phase reactions through representing the chemical individuality of a solid reagent in the mathematical kinetic modules".

6.9. MECHANOCHEMISTRY

Mechanochemical reactions, in which chemical changes are induced by energy supplied to the solid from external sources, are reviewed by Butyagin [98], who focusses attention on the generation of non-equilibrium states. Such energy may be supplied as ultrasound, in various types of crushing mills, during crystal disintegration when compressed between anvils, or from shock waves. These sources introduce into the reactant larger than equilibrium concentrations of activated states, including vibrational and/or electronic excitation, centres retaining electronic charges and various types of structural imperfections, all of which enhance reactivity. The review cited above discusses, in the context of the available theory of the topic, a number of representative heterogeneous reactions on SiO_2 and MgO together with aspects of the use of the method for synthesis.

The theory of electronic excitation, which occurs during plastic deformation and fragmentation of crystals, is discussed by Molotskii [99]. The mechanochemistry of inorganic solids has been reviewed by Boldyrev *et al.* [100] with particular reference to the work done in The Institute of Solid State Chemistry at Novosibirsk. The consequences of mechanical activation on the structures and properties of selected spinels have been examined and the rate of the reaction forming barium tungstate investigated. Prospects for the future development of this subject are assessed.

6.10. REACTANT MELTING

Reasons why reactions of solids may proceed more rapidly in a molten zone than within a crystalline reactant [75], include: (i) relaxation of the regular stabilising intracrystalline forces; (ii) establishment of a favourable configuration for chemical change may be possible due to mobility in a liquid but inhibited within a rigid crystal structure, and (iii) the influences of intermediates and impurities may be greater (or different) in a molten phase.

When an intermediate is generated above its melting point, the inevitable chemical similarities can be expected to result in reactants and products forming a eutectic. The rate of reaction is then controlled by the amount of liquid phase present and the concentrations of participating species. Melting may be local, temporary and influenced by the precipitation of any residual solid formed. Detection of comprehensive melt formation on heating a solid is straightforward by observation, or by thermal (e.g., DSC) techniques where the response is sharp and may be reversible. The detection of partial/temporary/local liquefaction is less straightforward. In some crystals, for example copper(II) malonate [41], a coherent layer of unreactive salt (a "skin") preserves the identity of individual particles, so that the product is pseudomorphic with the reactant. The occurrence of melting was detected by light crushing of samples of partially reacted salt followed by microscopic examination of surfaces exposed. The appearance of a froth-like texture is evidence of control by surface tension forces in a liquid. Features characteristic of solids, such as flat faces with sharp edges, parallel or other aligned surfaces were absent. It was concluded that the reaction intermediate detected (copper acetate) melts at the reaction temperature. Similar froth-like textures observed [42] during the decomposition of $(NH_4)_2Cr_2O_7$ were interpreted as melting of the CrO_3 intermediate. The presence of liquid halogen in nuclei developed during the reaction $2KBr + C\ell_2 \rightarrow 2KCl + Br_2$ [32] (not a decomposition) was confirmed by analytical measurements.

Bawn [101] has proposed a kinetic equation for reactions of solids involving melting. He considered the situation where reactant A is soluble in product B with solubility, s. When the extent of decomposition is α, the fraction of A in the liquid phase is αs, and in the solid phase is $(1 - \alpha - \alpha s)$. If the rate coefficients for reactions in the solid and in the liquid phases are k_s and k_l, respectively, then:

$$d\alpha/dt = k_s\,(1 - \alpha - \alpha s) + k_l\,\alpha s = k_s + K\,\alpha$$

where $K = k_l\,s - k_s s - k_s$, and:

$$\alpha = (k_s / K)\,[\exp(K\,t) - 1]$$

The expected ratio k_l / k_s can be estimated [101] as follows. Using:

$$k_s = A_s \exp(-E_s/RT) \qquad \text{and} \qquad k_l = A_l \exp(-E_l/RT)$$

and assuming that the difference $E_s - E_l$ is equal to the molar enthalpy of melting [101], then if $A_s \approx A_l$:

$$k_l / k_s \approx \exp(\Delta H_{melting}/RT)$$

For a typical value of $\Delta H_{melting} \approx 20$ kJ mol^{-1} and $T = 400$ K, $k_l / k_s \approx 410$.

As Carstensen [102] has pointed out, the alpha-time curves for the Bawn model are sigmoidal, being acceleratory up to the liquefaction point, where $\alpha = \alpha' = 1/(1 + s)$ at $t = t'$, and deceleratory beyond, according to:

$$(1 - \alpha)/(1 - \alpha') = \exp[-k_l(t - t')]$$

To test the Bawn model, the early stages of decomposition should give a linear plot of $\ln \alpha$ against t and the later stages a linear plot of $\ln(1 - \alpha)$ against t. The same holds for the Prout-Tompkins model (Table 3.3.), so the two models cannot readily be distinguished on kinetic grounds.

It would be experimentally very difficult to measure amounts of liquid present (including variations with α and temperature) in mixtures that have undergone partial melting. The kinetics of processes that demonstrably involve the participation of a liquid phase often resemble those for solid state decompositions, as illustrated by the following examples.

(i) *Thin molten layer.* If a thin layer of liquid phase is generated between reactant and product, kinetic behaviour must be identical with that already described above for interface advance. The crystalline zone of enhanced reactivity is replaced by a layer of liquid in which the chemical change occurs. The reaction geometries are identical but the reaction mechanisms are fundamentally different.

(ii) *Comprehensive melting.* Studies of dehydration in which there is complete fusion sometimes exhibit approximately zero-order kinetics [103]. Reactions in a solvent-free homogeneous phase are expected to exhibit a concentration dependence on rate similar to other gas or liquid phase reactions.

(iii) *Partial, temporary or local melting.* The kinetics of reactions of this type, where the rate depends on the amount of liquid present [101,102], which can vary with place and time throughout the reactant mass, require further investigation. Such reactions probably include the decompositions of copper(II) malonate [41], $(NH_4)_2Cr_2O_7$ [42] and many chlorates and perchlorates [104] (Chapter 13).

Thus a clear recognition of the occurrence of melting cannot be based on kinetic analyses. The phase in which the chemical transformation occurs should always be investigated and regarded as a central feature of any proposed mechanism.

6.11. ANALYSIS OF GASEOUS PRODUCTS

Most of the decompositions described in this book involve the formation of gaseous products and, hence, it is essential to determine the identities and amounts of gaseous products formed so that the stoichiometry of the decomposition is established (Section 2.8.3.). If it is possible to determine the profiles of evolution with time (or temperature) of individual gases, these data can be used to provide valuable kinetic information. Normalization and comparison of such profiles for the range of products formed immediately reveals any change of reaction mechanism, which may then be able to be associated with information on the solid phases present at that stage of the decomposition.

Any suitable method of repetitive gas analysis [105] can be used (see Section 2.8.3.). Like most of the other determinations described as necessary above, analysis of evolved gases has some inherent difficulties. The condensation of less-volatile products before reaching the detector and the possibility of secondary reactions between the primary gaseous products are the two major problems. In view of the suggestions that decompositions may involve initial sublimation of reactant (see Sections 2.4.5. to 2.4.7.), identification and measurement of the primary gaseous species formed, particularly in the very early stages of decomposition, are of great importance in formulating reaction mechanisms.

6.12. CONCLUSIONS

6.12.1. Experimental methods

Of the experimental techniques described, microscopic observations have had a significant impact on subject development, deriving from the earliest recognition of nucleation and growth processes in solids. Use of microscopy has increased with the wider availability of scanning electron microscopes. However, much more detailed textural (and other) information is required to characterize interface structures. It is possible that atomic force microscopy may be capable of providing detailed evidence of textures within and adjoining the reaction zone, perhaps in molecular detail. Whether such observations will provide the information required to formulate detailed interface mechanisms remains to be seen. A series of X-ray diffraction measurements across the reaction zone is a further most promising investigative approach, but requires a high-intensity radiation source such as a synchrotron [90] thus restricting its application.

The other experimental techniques surveyed generally have more specific applications. All possible investigative methods, both direct and indirect, have to be imaginatively applied in the search for evidence to characterize mechanisms of reactions in solids. Magnetic and electrical measurements can be used to identify transport properties, as well as aiding product recognition, including compositional and structural features and aspects of surface properties. The small concentrations of mobile electronic or ionic species that may participate in nucleation or interface processes are very much more difficult to identify. Similarly the consequences of reactant doping must be interpreted with care. Detailed knowledge of the changes that result from the accommodation of impurities are required (but are not always available) if their effects on the decomposition are to be determined. The methods mentioned above have yielded valuable and relevant complementary information for many individual reactions, but considerable skill and care are required to reach useful conclusions.

6.12.2. Formulation of decomposition mechanisms

Formulation of mechanisms for interface reactions is the greatest challenge of this subject. The evidence from which such reaction chemistry can be deduced is often indirect. Consequently, in the absence of theoretical models capable of indicating the nature and sequence of steps through which a solid reactant is transformed into a solid product, inferences may be incomplete or uncertain. From the extensive information available, it may be possible to classify different types of interface reactions and use the patterns of behaviour discerned to develop the theory.

In formulating a reaction mechanism, contributions from processes preceding and following the chemical reaction must be incorporated into the full description, as illustrated in Figure 6.10.

The reliability of mechanistic and other theoretical conclusions is increased through the use of data obtained using complementary experimental techniques in the formulation of reaction models. The above account also stresses the value of making comparisons between the behaviours observed for different, perhaps related, reactants. From this approach, the systematic classification of diverse decompositions might be expected to reveal patterns in reactivities that could be used to develop the theory. The absence of a systematic approach in the literature is a possible reason for the small number of reviews (or *vice versa*) and both may be associated with the lack of a coherent theoretical foundation to the subject. By drawing attention to the significant differences in interface forms and functions (Section 6.8.) we hope to encourage interest in this feature common to many solid state decompositions as a possible approach to classification. There is also value in considering the comparative chemistry of different solid reactants. Mechanisms form an important feature of the reactions that are described in the chapters that follow.

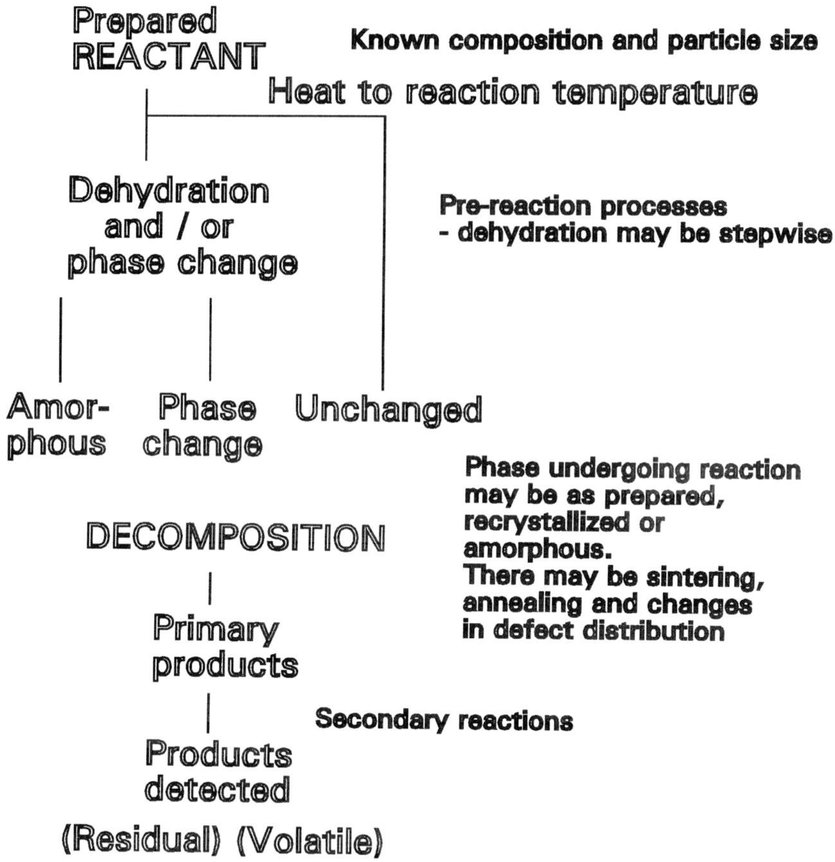

Figure 6.10.
Representations of the textural and chemical changes that may accompany a solid state decomposition [8]. Contributions from these processes must be considered in formulating a reaction mechanism.

REFERENCES

1. A.R. West, Solid State Chemistry and its Applications, Wiley, Chichester, 1984.
2. C.N.R. Rao and J. Gopalakrishnan, New Directions in Solid State Chemistry, Cambridge University Press, Cambridge, 1986.
3. A.K. Cheetham and P. Day (Eds), Solid-State Chemistry : Techniques, Clarendon, Oxford, 1987.
4. R.S. Mikhail and E. Robens, Microstructure and Thermal Analysis of Solid Surfaces, Wiley-Heyden, Chichester, 1983.
5. H.G. Wiedemann, Thermal Analysis, (Ed. H.G. Wiedemann), Birkhaüser, Basel, 1972, Vol.1, p.171.
6. W. Engel, N. Eisenreich, M. Alonso and V. Kolarik, J. Thermal Anal., 40 (1993) 1017.
7. H.R. Oswald, Thermal Analysis, (Ed. H.G. Wiedemann), Birkhaüser, Basel, 1980, Vol.1, p.1; V.V. Boldyrev, React. Solids, 8 (1990) 231.
8. N.J. Carr and A.K.Galwey, Thermochim. Acta, 79 (1984) 323.
9. I.C. Hisatsune and N. Haddock Suarez, Inorg. Chem., 3 (1964) 168; D.L. Bernitt, K.O. Hartman and I.C. Hisatsune, J. Chem. Phys., 42 (1965) 3553; K.O. Hartman and I.C. Hisatsune, J. Phys. Chem., 69 (1965) 583; 70 (1966) 1281; 71 (1967) 392; J. Chem. Phys., 44 (1966) 1913; F.E. Freeberg, K.O. Hartman, I.C. Hisatsune and J.M. Schempf, J. Phys. Chem., 71 (1967) 397; I.C. Hisatsune, R. Passerini, R. Pichai and V. Schettino, J. Phys. Chem., 73 (1969) 3690; I.C. Hisatsune, E.C. Beahm and R.J. Kempf, J. Phys. Chem., 74 (1970) 3444; I.C. Hisatsune, T. Adl, E.C. Beahm and R.J. Kempf, J. Phys. Chem., 74 (1970) 3225; I.C. Hisatsune and T. Adl, J. Phys. Chem., 74 (1970) 2875; I.C. Hisatsune and D.J. Linnehan, J. Phys. Chem., 74 (1970) 4091.
10. A.R. West, Solid State Chemistry and its Applications, Wiley, Chichester, 1984, Chap.3.
11. P.M.Th.M. van Attekum and J.M. Trooster, J. Chem. Soc., Dalton Trans., (1980) 201.
12. L.M. Moroney, R.St.C. Smart and M. Wyn Roberts, J. Chem. Soc., Faraday Trans. I, 79 (1983) 1769.
13. R.G. Copperthwaite and J. Lloyd, J. Chem. Soc., Dalton Trans., (1977) 1117; Nature (London), 271 (1978) 141; J. Chem. Soc., Faraday Trans. I, 74 (1978) 2252.

14. A.K. Nikumbh, M.M. Phadke, S.K. Date and P.P. Bakare, Thermochim. Acta, 239 (1994) 253.
15. K.J.D. MacKenzie and D.G. McGavin, Thermochim. Acta, 244 (1994) 205.
16. P.K. Gallagher, Application of Mössbauer Spectroscopy, (Ed. R.L. Cohen), Academic Press, New York, 1876, p.199; P.K. Gallagher, K.W. West and S.St.J. Warne, Thermochim. Acta,50 (1981) 41.
17. P.K. Gallagher, Thermal Characterization of Polymeric Materials, (Ed. E. Turi), 2nd Edn, Academic Press, New York, 1997, Chap.1.
18. C. Tang, P. Georgopoulos and J.B. Cohen, J. Amer. Ceram. Soc., 65 (1982) 625.
19. B.C. Gerstein, Anal. Chem., 55 (1983) 781A; 899A.
20. J.F. Haw, Anal. Chem., 60 (1988) 559A.
21. C.A. Fyfe and R.E. Wasylishen, Solid-State Chemistry : Techniques, (Eds A.K. Cheetham and P. Day), Clarendon, Oxford, 1987, Chap.6.
22. G.E. Maciel, Science, 226 (1984) 282.
23. A.K. Galwey, S.G. McKee, T.R.B. Mitchell, M.A. Mohamed, M.E. Brown and A.F. Bean, React. Solids, 6 (1988) 173,187.
24. E. Lippmaa, M. Magi, A. Samoson, G. Engelhardt and A.R. Grimmer, J. Amer. Chem. Soc., 102 (1980) 4889.
25. D.B. Sheen and J.N. Sherwood, Chem. Brit., 22 (1986) 535.
26. I.D. Begg, P.J. Halpenny, R.M. Hooper, R.S. Narang, K.J. Roberts and J.N. Sherwood, Proc. R. Soc. (London), A386 (1983) 431.
27. R. Haynes, Optical Microscopy of Materials, International Textbook Company, Glasgow, 1984.
28. N.H. Hartshorne and A. Stuart, Practical Optical Crystallography, Arnold, London, 2nd Edn, 1969; Crystals and the Polarizing Microscope, Arnold, London, 4th Edn, 1979.
29. S. Tolansky, Microstructures of Surfaces using Interferometry, Arnold, London, 1968; Multiple-beam Interference Microscopy of Metals, Academic Press, New York, 1970.
30. J.M. Thomas and G.D. Renshaw, J. Chem. Soc. A, (1967) 2058; (1969) 2749, 2753, 2756.
31. J.M. Thomas, Adv. Catal., 19 (1969) 293.
32. A.K. Galwey and L. Pöppl, Phil. Trans. R. Soc. (London), A311 (1984) 159.
33. A.K. Galwey, P.J. Herley and M.A. Mohamed, React. Solids, 6 (1988) 205.
34. B.J. Mehta, Indian J. Pure Appl. Phys., 19 (1981) 1206.

35. A. Wischin, Proc. R. Soc. (London), A172 (1939) 314.
36. P.W.M. Jacobs and F.C. Tompkins, Chemistry of the Solid State, (Ed. W.E. Garner), Butterworths, London, 1955, Chap.7.
37. A.K. Galwey, R. Spinicci and G.G.T. Guarini, Proc. R. Soc. (London), A378 (1981) 477.
38. A.V. Hodgson, Lab. Pract., 19 (1970) 897; 20 (1971) 797.
39. A.K. Galwey, R. Reed and G.G.T. Guarini, Nature (London), 283 (1980) 52.
40. A.K. Galwey and M.A. Mohamed, Thermochim. Acta, 121 (1987) 97.
41. N.J. Carr and A.K. Galwey, Proc. R. Soc. (London), A404 (1986) 101.
42. A.K. Galwey, L. Pöppl and S. Rajam, J. Chem. Soc., Faraday Trans. I, 79 (1983) 2143.
43. E.D. Cater and P.R. Buseck, Ultramicroscopy, 18 (1985) 241.
44. M.G. Kim, U. Dahmen and A.W. Searcy, J. Amer. Ceram. Soc., 70 (1987) 146.
45. S. Chiang (Ed.), Chem. Rev., 97 (1997) 1015 to 1230.
46. D. Dollimore and D. Tinsley, J. Chem. Soc. A, (1971) 3043; D. Dollimore and P. Spooner, Trans. Faraday Soc., 67 (1971) 2750; D. Dollimore, P. Spooner and A. Turner, Surf. Tech., 4 (1976) 121.
47. D. Dollimore and J. Pearce, J. Thermal Anal., 6 (1974) 321.
48. A.K. Galwey and P.W.M. Jacobs, Proc. R. Soc. (London), A254 (1960) 455.
49. A.K. Galwey, Adv. Catal., 26 (1977) 247.
50. A. Hamnett, Solid-State Chemistry: Techniques, (Eds A.K. Cheetham and P. Day), Clarendon, Oxford, 1987, Chap.8.
51. A.R. West, Solid-State Chemistry and its Applications, Wiley, Chichester, 1984, Chaps 13 to 15.
52. A.A. Kabanov, Russ. Chem. Rev., 40 (1971) 953; A.A. Kabanov and E.M. Zingel, Russ. Chem. Rev., 44 (1975) 538.
53. P.W.M. Jacobs, F.E. Lovatt and W.L.Ng, Canad. J. Chem., 50 (1972) 2031.
54. G.P. Owen and J.M. Thomas, J. Chem. Soc., Faraday Trans. I, 68 (1972) 2356.
55. P.W.M. Jacobs and F.C. Tompkins, J. Chem. Phys., 23 (1955) 1445.
56. K. Torkar and A. Isenberg, Monatsh. Chem., 98 (1967) 298; 97 (1966) 1339; 98 (1967) 245.
57. A.R. West, Solid state Chemistry and its Applications, Wiley, Chichester, 1984, Chap.15.

58. W.E. Hatfield, Solid-State Chemistry : Techniques, (Eds A.K. Cheetham and P.Day), Clarendon, Oxford, 1987, Chap. 4.
59. A.L. Cavalieri, E.F. Aglietti, A.N. Scian and J.M. Porto, React. Solids, 2 (1986) 315.
60. R.G. Darrie, W.P. Doyle and I. Kirkpatrick, J. Inorg. Nucl. Chem., 29 (1967) 979.
61. S. Roy, A. Ghosh and N.R. Chaudhuri, Transition Metal Chem., 12 (1987) 453.
62. M.A. Alario-Franco, J. Fenerty and J.W.S. Sing, Reactivity of Solids, (Eds J.S. Anderson, M.W. Roberts and F.S. Stone), Chapman and Hall, London, 1972, p.327.
63. D. Shamir and M. Steinberg, Prog. Vac. Microbalance Tech., 2 (1973) 19.
64. S.St.J. Warne and P.K. Gallagher, Thermochim. Acta, 110 (1987) 269.
65. P.K. Gallagher and S.St.J. Warne, Thermochim. Acta, 43 (1981) 253.
66. P.K. Gallagher, P.T. Chao, L. Zhong and J. Subramanian, J. Thermal Anal., 39 (1993) 975.
67. A.A. Kabanov and V.N. Skrobot, Russ. J. Phys. Chem., 48 (1974) 1508.
68. M.E. Brown, D. Dollimore and A.K. Galwey, Reactions in the Solid State, Comprehensive Chemical Kinetics, (Eds C.H. Bamford and C.F.H. Tipper), Vol. 22, Elsevier, Amsterdam, 1980.
69. W.J. Wong and B. Hall, Thermochim. Acta, 147 (1989) 251.
70. S.M.K. Nair and P.D. Jacob, Thermochim. Acta, 181 (1991) 269.
71. M.K. Sahoo and D. Bhatta, Thermochim. Acta, 197 (1992) 391.
72. S.M.K. Nair and T.S. Sahish, Thermochim. Acta, 250 (1995) 207.
73. T. Ganga Devi, M.P. Kannan and B. Hema, Thermochim. Acta, 285 (1996) 269.
74. J.R. Günter and H.R. Oswald, Bull. Inst. Chem. Res. Koyoto Univ., Japan, 53 (1975) 249.
75. A.K. Galwey, Thermochim. Acta, 96 (1985) 259; React. Solids, 8 (1990) 211; J. Thermal Anal., 41 (1994) 267; Pure Appl. Chem., 67 (1995) 1809.
76. J. Sesták, J. Themal Anal., 36 (1990) 1997.
77. A.K. Galwey, G.M. Laverty, N.A. Baranov and V.B. Okhotnikov, Phil. Trans. R. Soc. (London), A347 (1994) 139,157.
78. M.L. Franklin and T.B. Flanagan, J. Chem. Soc., Dalton Trans., (1972) 192; J. Phys. Chem., 75 (1971) 1272.
79. W.E. Garner, Chemistry of the Solid State, (Ed.W.E. Garner), Butterworths, London, 1955, Chap.8.

80. A.K. Galwey and G.G.T. Guarini, Proc. R. Soc. (London), A441 (1993) 313.
81. G.G.T. Guarini and L. Dei, J. Chem. Soc., Faraday Trans. I, 79 (1983) 1599.
82. M.E. Brown, A.K Galwey and G.G.T. Guarini, J. Thermal Anal., 49 (1997) 1135.
83. A.K. Galwey and M.E. Brown, J. Chem. Soc., Faraday Trans. I, 78 (1982) 411.
84. A.K. Galwey, N. Koga and H. Tanaka, J. Chem. Soc., Faraday Trans, 86 (1990) 531.
85. M.M. Pavluchenko, E.A. Prodan and T.M. Samoseiko, Kinet. Katal., 15 (1974) 796.
86. M.E. Brady, M.G. Burnett and A.K. Galwey, J. Chem. Soc., Faraday Trans., 86 (1990) 1573.
87. V.V. Boldyrev, Kinet. Katal., 6 (1965) 934.
88. V.V. Boldyrev and A.A. Medvinskii, Kinet. Katal., 4 (1963) 549.
89. V.B. Okhotnikov, S.E. Petrov, B.I. Yakobson and N.Z. Lyakhov, React. Solids, 2 (1987) 359.
90. V.V. Boldyrev, Y.A. Gaponov, N.Z. Lyakhov, A.A. Politov, B.P. Tolochko, T.P. Shakthtshneider and M.A. Sheromov, Nucl. Inst. Method. Phys. Res., A261 (1987) 192.
91. G.G.T. Guarini, J. Thermal Anal., 41 (1994) 287.
92. A.K. Galwey and G.M. Laverty, Thermochim. Acta, 228 (1993) 359.
93. M.E. Brown, A.K. Galwey and A. Li Wan Po, Thermochim. Acta, 203 (1992) 221; 220 (1993) 131.
94. V.B. Okhotnikov, I.B. Babicheva, A.V. Musicantov and T.N. Aleksandrova, React. Solids, 7 (1989) 273.
95. A. Korobov, Heterogeneous Chem. Rev., 3 (1996) 477.
96. A. Korobov, Thermochim. Acta, 224 (1993) 281; 243 (1994) 79; 254 (1995) 1; J. Math. Chem., 17 (1995) 323.
97. A. Korobov, J. Thermal Anal., 39 (1993) 1451; 44 (1995) 187; 46 (1996) 49.
98. P. Yu. Butyagin, Chemistry Rev., 14 (1989) 1; React. Solids, 1(1986) 345, 361.
99. M.I. Molotskii, Chemistry Rev., 13 (1989) 1.
100. V.V. Boldyrev, N.Z. Lyakhov, Yu. T. Pavlyukhin, E.V. Boldyreva, E. Yu. Ivanov and E.G. Avvakumov, Chemistry Rev., 14 (1990) 105; Russ. Chem. Rev., 40 (1971) 847.

101. C.E.H. Bawn, Chemistry of the Solid State, (Ed. W.E. Garner), Butterworths, London, 1955, Chap.10.
102. J.T. Carstensen, Drug Stability, Principles and Practice, 2nd Edn, Marcel Dekker, New York, 1995, Chap.9.
103. S.D. Bhattamisra, G.M. Laverty, N.A. Baranov, V.B. Okhotnikov and A.K. Galwey, Phil. Trans. R. Soc. (London), A341 (1992) 479; A.K. Galwey, G.M. Laverty, V.B. Okhotnikov and J. O'Neill, J. Thermal Anal., 38 (1992) 421.
104. F. Solymosi, Structure and Stability of Salts of Halogen Oxyacids in the Solid Phase, Wiley, London, 1977.
105. E. Kaiserberger, Thermochim. Acta, 295 (1997) 73.

PART B

THERMAL DECOMPOSITIONS OF SELECTED IONIC SOLIDS

Part B of this book (Chapters 7 to 17) is a selective survey from the formidable literature concerned with decompositions of ionic solids, in the context of the theoretical background provided in Part A (Chapters 1 to 6).

During the early development of this field, interest tended to focus on the decompositions of a limited number of compounds, including $CaCO_3$, BaN_6, $KMnO_4$, $NH_4C\ell O_4$, $CaC_2O_4.2H_2O$ and $Li_2SO_4.H_2O$. These 'model' compounds, have served as references as the field has widened. The compounds whose decompositions are surveyed in the following chapters have been selected from the much wider range of substances for which studies have now been reported, to illustrate as wide a range of different reactions as possible. Compounds have been grouped on the basis of a common anion, which is the component that most usually undergoes breakdown. Broadly the sequence of presentation is based on increasing complexity of the anion. Decompositions of ammonium salts have been considered separately, because behaviour of this cation is also important. Other exceptions to this scheme are the group of dehydrations (Chapter 7), where water-cation coordination is involved and this constituent is discussed separately, and in greater depth, than the decompositions of compounds containing other coordinated ligands (Chapter 17).

Much of the interest in the field of thermal decompositions has been in studies of relatively simple solid reactants to minimise the problems of interpretation of behaviour in contributing towards a set of fundamental principles for the subject. In spite of such an approach, chemical correlations are often not readily discernable. Reactants containing chemically similar components do not always give comparable reactions, whereas resemblances in kinetic behaviour are sometimes very clearly apparent on heating materials with very different chemical constituents. One of the major aims of decomposition studies, namely the prediction of thermal behaviour from chemical and other properties, thus still remains unfulfilled, as discussed in the concluding Chapter 18.

Chapter 7

THERMAL DEHYDRATION OF HYDRATED SALTS

7.1. INTRODUCTION

Water is an important constituent of many solids and can be accommodated in crystal lattices in several distinct bonding situations. Water is also strongly held when in the form of constituent hydroxyl ions. Hydroxides are discussed in Chapter 8. At the other extreme, intercalated molecular water is loosely retained between the aluminosilicate layers in many clays [1] and may be reversibly withdrawn and replaced. (Clays also contain structural hydroxyl groups). Between these markedly different bonding situations, there are innumerable crystalline hydrates in which water molecules are incorporated either as individual units, or in the form of ligands coordinated to a particular ion, usually the cation. Such water stabilizes the solid and in many compounds it may be removed partially or completely to yield a lower hydrate or anhydrous salt. The overall change may proceed in a single or in several discrete steps. It is with this group of substances that the present chapter is almost exclusively concerned. The dehydrations of ionic compounds, notably inorganic salts and metal carboxylates containing water of crystallization, have attracted the greatest interest. Coordinated water may be withdrawn from the host lattice without change in the nominal chemical composition other than a decrease in the degree of hydration. This removal may be accompanied by immediate, or by delayed, reorganization of the structure with consequent changes in the stability, bonding and reactivity of the residual solid.

Some of the earliest kinetic studies of solid state reactions were concerned with the dehydrations of ionic crystals and the progress made in determining the mechanisms of these reactions was an important stimulus towards the investigation of the reactions of a wider range of crystalline materials [2].

The decompositions of series of related solids, containing ligands other than water, e.g. ammoniates, alcoholates, etc., have been investigated and some show behaviour that is parallel to the reactions of hydrates. Some examples of such reactions are discussed in Chapter 17.

7.2. STRUCTURAL CONSIDERATIONS

The term *hydrate* will be restricted to those compounds having compositions in which it is possible to distinguish a stoichiometric number of water molecules. This water of crystallization, often retained during growth of the crystal from aqueous solution, may be incorporated into the lattice either through coordination with the cation or, less frequently, to fill voids in the structure. In copper sulfate pentahydrate, for example, four molecules of water are coordinated with each copper(II) cation and one occupies a space within the structure. Similarly, nickel sulfate heptahydrate incorporates $[Ni(OH_2)_6]^{2+}$ ions and an additional water molecule elsewhere. Water associated with a cation decreases the effective charge through coordination with the lone pair of electrons on the oxygen. This tendency towards hydration increases (i) the greater the ionic charge and (ii) the smaller the cation or the larger the anion. A highly hydrated cation may be effectively surrounded by a "shell" of water, but when fewer water molecules are incorporated as ligands, the molecules available may be shared by more than a single cation, or fill gaps in the anionic environment. Removal of some, or all, of this water from the salt changes the stability of the crystal and there may be reorganization of the lattice. The relationship between the the removal of water and the recrystallization step is of fundamental significance in determining the mechanisms of dehydrations. The loss of salt stability [3] may result in decomposition involving other constituents following or accompanying removal of water.

Water is an asymmetric molecule, possessing a large dipole moment, so that the electrostatic force fields, including contributions from the ligands, around a hydrated ion exert an important control on the disposition of ions within the crystal lattice. Infrared spectroscopy has been used [4] to identify water bonding with anions in various crystalline hydrates. Within the regular array there is a tendency for water to adopt a tetrahedral configuration, with the formation of the largest possible number of hydrogen bonds, and the preferred structure derives from the maximization of the stabilization energy arising from all contributory factors. This includes the electrostatic forces (attractive and repulsive) of the ions present and also the positions, shapes and polar properties of the water molecules. On removal of these water molecules, the changes in participating forces are often sufficiently great to cause the adoption of a new crystal structure.

The most suitable basis for the classification [5] of the structures of hydrates is by grouping together substances in which water molecules are in similar environments. One group of hydrates has water arranged in a three-dimensional lattice framework (ice-like hydrates). Another contains infinite arrays of aqua- complexes. Finally, there are structures which contain isolated groups of water molecules. Many solids are known in which water occurs in more than a single environment, and the variety of possible combinations decreases the value of this classification.

Alternatively, hydrates may be grouped on the basis of their cation:water ratio even though this does not derive from any single chemical property. Wyckoff [6] emphasizes the arbitrary nature of considering hydrates as a separate class of compounds, because these substances should, in principle, be discussed with reference to other comparable molecular groupings.

Early (1930 to 1940) kinetic studies of dehydrations contributed much to the development of the concept of the reaction interface as the important feature of nucleation and growth reactions [2]. Kinetic equations applicable to the decompositions of a wide range of crystalline substances were developed. Large, well-formed crystals of hydrates could be prepared relatively easily and studies of these were particularly rewarding. The interpretation of kinetic data was supplemented by microscopic evidence concerning the formation and development of product nuclei. Recent work on dehydrations has included more precise determinations of the crystal structures of reactants, products and their interrelationships, including interface textures, in the attempt to resolve unanswered questions.

Although many dehydrations occur by nucleation and growth processes, few generalizations have yet been made. The theoretical framework at present available does not allow prediction of kinetic behaviour for untested systems. More extensive systematic comparisons of related compounds are needed to establish behaviour patterns of wider applicability. General reviews of the subject have been given by Garner [2], by Lyakhov and Boldyrev [7], and less extensive but more recent surveys by Galwey and Laverty [8] and Galwey [9].

7.3. EXPERIMENTAL METHODS

Most of the techniques described in Chapters 2 to 6 have been applied to the study of dehydrations. The fractional reaction, α, has often been measured using thermogravimetry, preferably with continuous evacuation, or in a strongly flowing carrier gas, to ensure rapid removal of product water vapour. Measurements of the pressure of evolved water vapour in an initially evacuated constant volume apparatus have also been used [10], which enables the influence of the reverse (hydration) reaction to be determined. Microscopic observations [11-13] continue to be an important approach to the investigation of the formation and the growth of product nuclei. The numbers, sizes, shapes and distributions of nuclei have been used to establish the characteristics of nucleus development, including detection of an initial period of relatively slow growth which is found for development of the product phase in some systems [12].

Most dehydrations are strongly endothermic so that the progress of reaction can be followed using DSC or DTA (Chapter 2). Within large crystals of a poorly conducting medium, such as an ionic salt, there may be appreciable self-cooling. Kinetic behaviour in these reversible reactions is also sensitive to the prevailing water vapour pressure, $p(H_2O)$. These two factors, the availability of water at, and the temperature of the reaction interface, have remained a continuing source of uncertainty for experimentation and discussion throughout studies of the kinetics and mechanisms of dehydrations.

Non-isothermal measurements (Chapter 2) have yielded valuable information about reaction temperatures and the successive steps in the removal of water from crystalline hydrates, e.g. oxalates [14], sulfates [15-17]. DTA and DSC studies have sometimes provided additional information on the recrystallization of the dehydrated product [18]. The problems of relating kinetic parameters obtained by non-isothermal measurements to those from isothermal experiments are discussed in Chapter 5. The effects of heat transfer and diffusion of water vapour may be of even greater consequence in non-isothermal work. Rouquerol [19,20] has suggested that some of the above problems may be significantly decreased through the use of constant rate thermal analysis.

The problem of water removal and the influence of inhomogeneous distribution of water within the mass of reactant can be serious where the dehydration reaction is reversible. Misleading results may sometimes arise through the design of the sample container, because the reactant geometry and packing can influence the ease of removal of product gas. When kinetic behaviour of the salt is sensitive [21,22] to $p(H_2O)$, non-isothermal data must be interpreted with due allowance for the conditions prevailing at the reaction interface. Many recent studies of dehydrations have included careful control of the prevailing $p(H_2O)$ in the vicinity of the reactant. Often two or more features of the reaction being investigated are studied concurrently: e.g. structure by X-ray diffraction and thermal events by DSC. Crystallographic transformations may sometimes be temporarily delayed after water loss, and the existence of intermediate phases may also be recognised. Detailed comparisons of the crystallographic structures involved can be used to identify the movements of lattice constitutents which follow water elimination, including the detection of topotactic relationships. Concurrent measurements of enthalpy changes enable the progress of reaction to be determined.

Apparatus for such concurrent studies of more than a single reactant parameter has been reported by Barret, Watelle-Marion, Gérard and their co-workers [23-28]. The design of cells for X-ray diffraction studies during reaction of a sample heated in a controlled atmosphere has also been described [26-28].

In selecting material from the literature on dehydration reactions, emphasis has been given here to work in which two or more complementary experimental approaches have been used to obtain mechanistic information. Systems for which the experimental observations are limited to non-isothermal data, unsupported by other measurements, and where atmosphere control in the reaction vessel is not specified, have not been included.

Most of the reactions discussed in this chapter involve the removal of a stoichiometric proportion of the water of crystallization without chemical change, other than concurrent or consecutive recrystallization. At higher temperatures, there may be a reaction between the evolved water and other constituents of the solid, particularly where the product water is not rapidly removed. The possible occurrence of hydrolysis is not discussed, because complex reactions of this type are difficult to characterize in detail and little mechanistic information is available.

7.4. STOICHIOMETRY OF REACTION

The first step in any kinetic study is to identify all the products and intermediates of the reaction. Dehydration often involves several distinct steps which may be very dependent upon reaction conditions, e.g., copper sulfate pentahydrate may yield trihydrate and/or monohydrate [29,30]. Metastable intermediates may be formed, e.g. amorphous magnesium carbonate produced on dehydration undergoes [18] exothermic recrystallization at higher temperatures.

7.5. KINETICS OF DEHYDRATION

At least three major interrelated aspects of kinetic behaviour are important in investigating dehydration reactions. (i) Kinetic equations based upon nucleation and growth models have been developed and found application to a wide range of reactants [31]. (ii) Theoretical explanations of the magnitudes of calculated Arrhenius parameters have been proposed. (iii) The influence of water vapour pressure on reaction rates has been investigated in detail (the Smith-Topley effect). Topics (ii) and (iii) are expanded below.

Dunning [32] has reviewed earlier interpretations of nucleation and growth phenomena in crystalline hydrates and has developed a theory of product phase generation based on that applicable to the homogeneous formation of nuclei in a supersaturated vapour. The production of dehydration nuclei in a crystalline solid is treated theoretically as the diffusion and aggregation of vacant lattice sites, following the removal of water molecules, in the vicinity of the surface. In the solid these vacancies may be trapped and aggregate at lattice imperfections. Nuclei are formed on surfaces at points of emergent dislocations or other defective regions and, later, develop as growth nuclei [32].

The steps involved in the formation of a stable nucleus have also been considered by Niepce and Watelle-Marion [33]. (There are several points of similarity with the treatment given by Dunning [32]). Four stages in the production of a nucleus are identified. (i) "Free" water molecules are generated, which retain only weak bonding links with the hydrate structure. (ii) Elimination of this "free" water results in the production of vacancies. (iii) Growth of strain energy in the vicinity of aggregates of vacancies culminates in the separation of this matrix from the host lattice. (iv) Recrystallization of the dehydrated material yields the stable product phase.

Few studies have been specifically directed towards identifying the factors which control nucleation in dehydration reactions [7]. The possibility that (i) elimination of water from the reactant structure and (ii) recrystallization of the product to form the crystalline lower hydrate or anhydrous phase may be separable kinetic processes has been discussed [34].

7.6. ARRHENIUS PARAMETERS CALCULATED FOR INTERFACE REACTIONS

Experimental activation energies calculated for many reversible endothermic dehydrations are close to values of the enthalpy of dehydration. These reactions have been termed "normal" and the energy is identified as that required to rupture the water-cation bond. For such "normal" reactions, the rate of dehydration is expressed [2] by the Polanyi-Wigner [2,35] equation:

$$\text{Rate} = \mathrm{d}x_o/\mathrm{d}t = \nu C \exp(-E_\mathrm{a}/RT) \tag{7.1}$$

where x_o is the unit step for interface advance and C is calculated for the reactant-product interfacial area in a crystal of known geometry, composition and density. ν is the frequency [36] associated with the principal vibration mode responsible for initiating the decomposition. Equation (7.1) was originally developed by extension of the consideration of frequency factors for unimolecular reactions and was applied with apparent success in early studies of dehydrations. The value of ν for the dehydration of copper sulfate pentahydrate falls just within the range of vibrational frequencies which is considered acceptable [29]. This range was determined in a very detailed theoretical treatment which included assumptions about the interface structure and estimated strengths of chemical bonding between the participants in the desorption step. From this agreement between theory and experiment it was concluded that the dissociative adsorption mechanism applies to the dehydration of $CuSO_4.5H_2O$. Agreement with the Polanyi-Wigner equation and reasonable values of ν have also been found for the dehydrations of some alums [37,38], again with values of E_a that are close to the enthalpies of dehydration [2].

Not all dehydrations are satisfactorily described by equation (7.1) and the so-called "abnormal" rate processes are characterized by anomalously large values of both A and E_a. The frequency factor for the dehydration [39] of $CaCO_3.6H_2O$ is particularly large, $\nu = 10^{34}$ s^{-1} and $E_a = 170$ kJ mol^{-1}. Large values of A were also obtained for dehydrations of some alums [37,38,40] and $2KHC_2O_4.H_2O$ [11,41]. The magnitudes of E_a were not attributable to a simple reversible dissociation step. Various attempts have been made to explain these divergences of experimental results from theory. Kassel [42] interpreted the results for $CaCO_3.6H_2O$ by assuming the intermediate formation of lower hydrates through a sequence of reactions involving low values of E_a. Such an explanation would not be expected [2] to be applicable to potassium hydrogen oxalate hemihydrate because the degree of hydration is small and presumably the required intermediate sequence of hydrates does not exist. The application of the terms "normal" and "abnormal" to these reactions presupposes the general acceptability of the underlying theoretical model, an assumption that Galwey [43] regards as insufficiently demonstrated.

Acock *et al.* [38] reviewed the explanations given for the anomalously high A and E_a values for dehydration reactions which are as follows. (i) A high energy nucleation step is required to initiate the subsequent rapid decomposition of all reactant components comprising that individual crystallite (perhaps 10^{12} molecules). (ii) Several co-operative steps may be necessary to form the product. (iii) The thickness of a product layer which hinders the escape of water may be temperature dependent. Anous *et al.* [37] later discussed the same problem and preferred the first of these explanations. In a study of the dehydration of manganese formate dihydrate, Clarke and Thomas [44] concluded from the low entropy of activation that no exceptional configurational requirements were involved (see also [45]). The value of E_a was close to that expected if proton motion in advance of the interface were the rate-limiting process.

Shannon [46] ascribed the abnormally high values of A reported for some dehydrations to the formation of a mobile surface layer of molecules. Rate processes for which values of A exceeded expectation by more than $x10^{10}$ were excluded from consideration by this form of the activated complex theory.

Much of the discussion of the magnitudes of A and E_a in the literature is based on theoretical models developed in the field of homogeneous kinetics and these concepts are not necessarily always applicable to reactions involving solids. While the model from which equation (7.1) was developed may provide acceptable representations of the mechanisms of some rate processes, it is always possible that the chemical change may be more complicated. The reversible adsorption/desorption of water at or near the reaction interface may influence the apparent magnitude of the values of A and E_a [47]. Similarly, structural constraints and diffusion phenomena may exert some control over the temperature dependence of reaction rate.

7.7. THE SMITH-TOPLEY EFFECT

For many hydrated salts, the rate of dehydration is markedly dependent upon the pressure of water vapour in the vicinity of the reactant. A common form of the variation of rate with $p(H_2O)$ is shown in Figure 7.1. (based on observations for $MnC_2O_4.2H_2O$ [48,49] and for $CuSO_4.5H_2O$ [2,21]) and is referred to as the Smith-Topley effect (S-T). The occurrence of the maximum (C) on the rate-$p(H_2O)$ curve has been attributed [2] to systematic changes of the texture of the dehydrated product in the wake of the advancing interface. Important features of this are schematically represented in Figure 7.2. The relative rate of recrystallization may depend on both temperature and $p(H_2O)$. The initial product of dehydration is believed to be highly-divided and disorganized material which is unstable and perhaps retains structural features of the reactant lattice. The transformation of this material into the more stable crystalline product phase involves a nucleation and growth process which is accompanied by the opening of wider channels and cracks in the product layer through which product water can escape. Such reorganization of an initially amorphous and metastable product has been demonstrated and discussed by Dell and Wheeler [18].

The S-T effect (Figure 7.1.) is conveniently discussed with reference to Figure 7.2. Relatively rapid reaction occurs during dehydration in a continuously evacuated system, though there may be some impedance [50] of the escape of water vapour along the narrow channels within an initially amorphous product. X-ray studies have confirmed [51] that these residual solids may be amorphous, with little

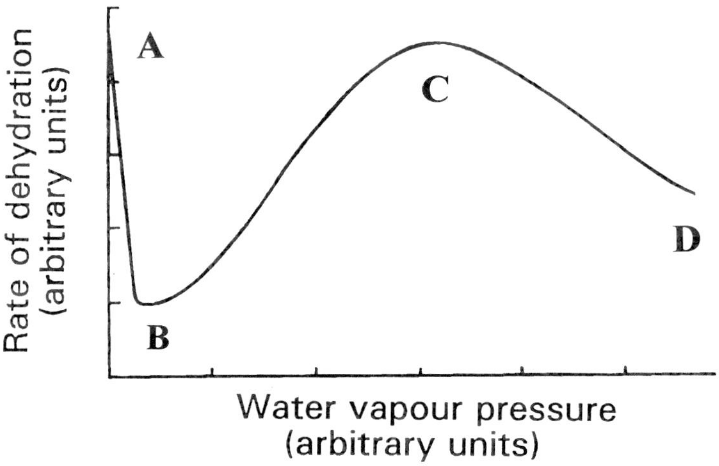

Figure 7.1.
The Smith-Topley (S-T) effect. The characteristic pattern of variation of the rate of some dehydrations with prevailing water vapour pressure [2,21,48,49].

recognizable long-range order. If water removal is not rapid, and a low value of $p(H_2O)$ is maintained, adsorption on the surfaces of the fine channels strongly inhibits [38,50] dehydration. This obstruction accounts for the characteristic marked decrease in rate (A to B in Figure 7.1.). At higher values of $p(H_2O)$, the available water catalyzes structural reorganization. The growth of product crystals (Figure 7.2.) and the consequent formation of wider intergranular spaces allow the more rapid escape of product (B to C in Figure 7.1.). When $p(H_2O)$ increases beyond C, the reverse reaction (rehydration) becomes significant, explaining the subsequent decrease of the rate (C to D) and beyond, reaching zero at equilibration. Whether or not there is the intermediate production of a lower hydrate may depend on the degree of impedance within the transition layer [52], because an increase in $p(H_2O)$ within the growing nucleus may be sufficient for the system to enter the zone of stability of a lower hydrate (e.g. formation of $CuSO_4.3H_2O$ during the dehydration of the pentahydrate).

This widely-accepted explanation of the S-T effect has been critically reappraised by Bertrand *et al.* [21] The rates of dehydration of five different salts, all exhibiting S-T behaviour, ($Li_2SO_4.H_2O$, $MgSO_4.4H_2O$, $CuSO_4.5H_2O$, $CuSO_4.3H_2O$ and

$Na_2B_4O_5(OH)_4.8H_2O$) within appropriate ranges of temperature and water-vapour pressure, have been compared. S-T behaviour was ascribed [21] to control by physical processes, such as heat transfer and/or gas diffusion influencing the rate of reaction. Such suggestions should be tested by the systematic introduction into kinetic experiments of physical changes such as variations in the mass, distribution and particle size of the sample, the presence of inert additives and the presence of gases (e.g., H_2, He) having high thermal conductivity.

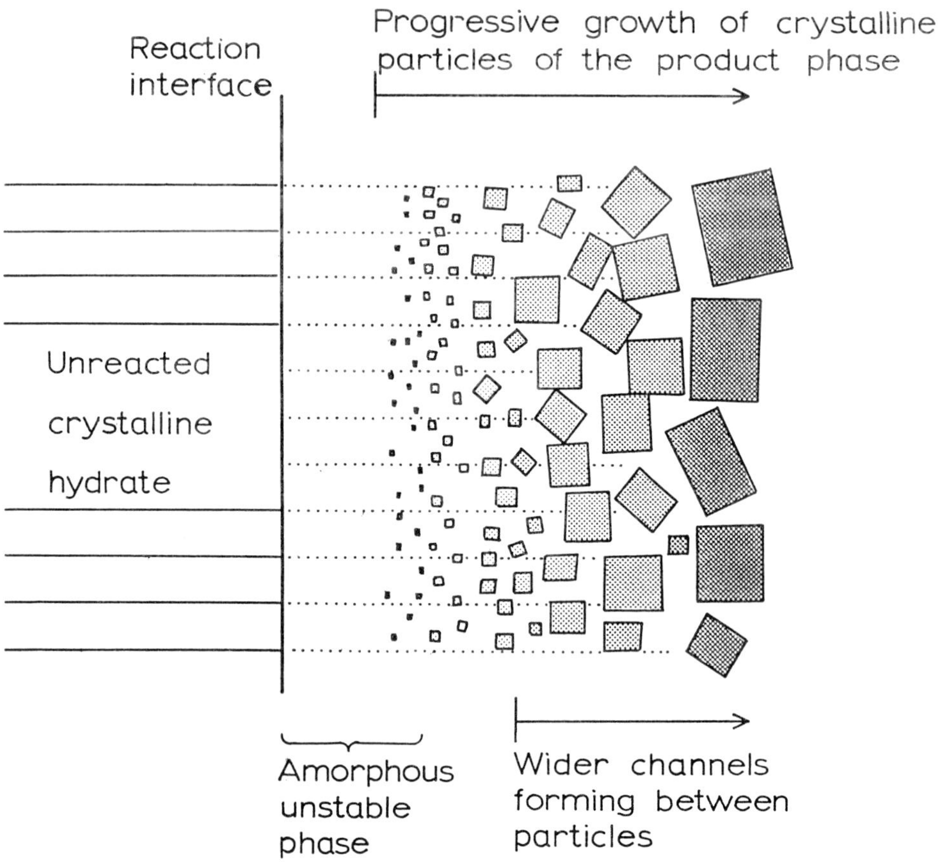

Figure 7.2.
Schematic representation of retexturing of the dehydration product in the wake of an advancing interface. The growth of crystals of the residual phase enables more rapid escape of the water vapour evolved.

The occurrence of S-T behaviour has been demonstrated for the following different types of systems: (i) hydrates which invariably yield a crystalline product, even at very low values of $p(H_2O)$; (ii) hydrates which yield an amorphous phase in a vacuum, but a crystalline product in appreciable pressures of water vapour; (iii) reactions for which there is evidence that the the advance of the interface is rate controlling; (iv) reactions regarded as being diffusion-controlled; (v) dehydrations which proceed to the anhydrous salt in a single stage, and finally (vi) reactions in which there is the intervention of a lower hydrate. Smith-Topley behaviour cannot, therefore, be identified specifically with any of these features of solid state reactions.

The S-T effect is thus of wide occurrence and the magnitudes of the rate changes measured for diverse systems are generally comparable. The effect cannot therefore be associated with specific chemical properties of the solids concerned and probably [21] results from changes in the ease of intra- and intergranular gas and heat transfer in microdomains at the reaction interface. The temperature and water vapour pressure in the immediate vicinity of the reaction zone may differ considerably from those elsewhere within the reactant mass. Such local variations cannot be measured with the experimental techniques currently available. Though self-cooling of large crystals has been considered, the effects of small inhomogeneities, particularly in the immediate vicinity of the interface, have not been studied in detail and may be significant in controlling the overall kinetic behaviour.

In addition to the two theoretical explanations of S-T rate - $p(H_2O)$ curves given above (Figure 7.1.), Lyakhov *et al.* [53], using adsorption data, have ascribed the S-T effect to variations in the structure of the adsorbed layer of water on the product solid, caused by systematic changes in polarization with increased coverage. Thomas *et al.* [54] have provided a kinetic analysis for reactions in which evolved H_2O catalyzes the water release step which accounts for the occurrence of a maximum in the rate - $p(H_2O)$ curves. A full mechanistic and generally applicable explanation for S-T behaviour, however, still has to be provided. It is also possible that different factors may control the properties of different hydrates.

7.8. HYDRATES OF METAL SULFATES

7.8.1. Copper sulfate pentahydrate and trihydrate

During the earliest (1930) studies of the mechanisms of decomposition of solids, Garner and Tanner [52] worked with copper sulfate pentahydrate. To simplify the interpretation of rate-time measurements, the surfaces of the prepared crystals were initially activated (i.e. rapidly nucleated) by abrasion with product solid. Because the rate of interface advance decreased as reaction proceeded, it was concluded that the measured value of E_a (43 kJ mol^{-1}) may not be characteristic of a single interface step, and the possible influence of reversible processes is mentioned. Smith and Topley [29,50] extended the range of kinetic measurements and first demonstrated a variation in rate with $p(H_2O)$.

Bright and Garner [12] (1934) measured the rate of formation and the rate of growth of nuclei on the (110) face of $CuSO_4.5H_2O$. Later investigations by Garner and Pike [55] extended measurements to nine different crystal faces of the same hydrate. The advance of the reaction interface was found to proceed at different rates in different lattice directions.

Colvin and Hume [56] showed that, on dehydration, copper sulfate pentahydrate yielded amorphous trihydrate which, in vacuum, reacted further to form the monohydrate, but in water vapour the crystalline trihydrate could be formed (see also Guenot *et al.* [57]). Cooper *et al.* [58] attributed an observed decrease in rate during dehydration of the trihydrate to impedance of the escape of water by the residual solid.

Heats of dissolution of copper sulfate samples, which had been dehydrated at various known values of $p(H_2O)$, were measured by Frost *et al.* [59] who concluded that the rate of crystallization of the amorphous product increased with the availability of water vapour during reaction. Smith-Topley behaviour, similar to that described for manganese oxalate dihydrate [48], was attributed [51] to the transition of an initially amorphous product to a crystalline phase accompanied by the opening of channels which facilitate the continued escape of water. Other aspects of the formation and growth, including shapes, of nuclei were discussed by Garner and Jennings [60] and by Garner [2].

Lallemant and Watelle-Marion [61] investigated the interrelationships between water release and structural transformations, using DTA and X-ray measurements on samples of $CuSO_4.5H_2O$ heated in controlled pressures of water vapour. When $p(H_2O) < 10$ Torr, the water elimination and recrystallization steps did not occur simultaneously. The DTA trace for the pentahydrate showed two consecutive endothermic peaks. During the first, $2H_2O$ was eliminated but there was an appreciable delay before lattice reorganization. When $p(H_2O) > 10$ Torr, the endotherms coincided, showing that water release is accompanied by immediate recrystallization. The ascending part of the rate-$p(H_2O)$ curves for the penta- and trihydrates, both of which exhibit Smith-Topley behaviour, is the domain within which the water elimination and structural transformation processes do not synchronize [62]. During the dehydration of the pentahydrate to yield the trihydrate, the period of existence of the metastable intermediate, which retains structural features of the reactant lattice, depends on $p(H_2O)$. The kinetic features of these rate processes were later discussed [21,30,62] with reference to the copper sulfate-water phase diagram. There is evidence [63] that there are localized fluctuations of temperature and water vapour pressure within microdomains of the reacting crystal.

$CuSO_4.5H_2O$ (vacuum) ⇌ $CuSO_4.H_2O$
↘↖ (in water vapour) ↗↙
$CuSO_4.3H_2O$

The role of the solid product ($CuSO_4.H_2O$) in catalyzing the dehydration of $CuSO_4.5H_2O$ was investigated by Lyakhov *et al.* [53]. The rate of reaction was increased by admixture with the product. It was also shown, however, that an identical effect occurred during experiments in which alternate layers of pentahydrate and monohydrate reactant were separated from direct contact by glass wool. The catalytic properties of the product therefore do not arise at points of solid-solid contact, but result from changes in composition of the gases present, because water is adsorbed by the lower hydrate. Lyakhov *et al.* [63] concluded that S-T behaviour can be explained by variations in the polarization of surface-held water with coverage and $p(H_2O)$. At low coverages, the initial decrease in reaction rate (Figure 7.1.) may be ascribed to surface blocking and a polarizing effect, between neighbouring molecules, which increases the energy barrier to water release. As water coverage is increased, however, water bonded to sulfate ions with the opposite orientation allows an increase in movement and thus of removal. At still higher coverages the adsorbed layer is stablized and the rate of reaction is decreased. (A similar model was proposed [63] for the dehydration of $MgSO_4.7H_2O$.)

Ammonium copper(II) sulfate hexahydrate [64] loses $4H_2O$ between 339 and 393 K with no influence of water vapour pressure on reaction rate below 172 N m^{-2}. The Avrami-Erofeev equation $n = 2$ fitted the data below 350 K with E_a = 170 kJ mol^{-1}. Above this temperature the contracting area equation applied and E_a = 82 kJ mol^{-1}.

7.8.2. Nickel sulfate heptahydrate and hexahydrate

The dehydration of nickel sulfate heptahydrate [13] resulted in nuclei being formed on the (110) face of crystals. These nuclei were fairly uniformly distributed half-elipses. The number of nuclei increased with time according to a square law and the initial relatively slow rate of growth increased up to a constant value. The catalytic effect resulting from the addition of monohydrate to the heptahydrate was again attributed [53] to changes in gas composition and not to reactant-product contacts.

Both geometric and energetic factors are important in explaining [65] the observed enhancement of reactivity in the vicinity of regions of lattice imperfection, especially dislocations. The role of dislocations (which can be identified by a wet-etching technique) was investigated [65] in the dehydration of $NiSO_4.6H_2O$. In vacuum (303 to 323 K) this is a nucleation and growth reaction which yields five molecules of product water. Patches of product formed on the (00ℓ) faces were not randomly distributed and the number of such growth nuclei did not increase during the early stages of reaction. Product formation commenced at both screw and edge dislocations and there were indications of preferred initiation at those dislocations which glide on (100) and (11ℓ) planes. Nucleation also occurred on cleavage steps and at inclusions. There was no multiplication of dislocations at low values of α. Surface impurities could either influence the ease of onset of reaction at a particular site or, in other situations, stabilize the disordered lattice regions and prevent the development of a nucleus. However, [65] "it is difficult to deduce whether or not the nucleation at the flat (00ℓ) faces is always centred at dislocations....".

The diameters of nuclei grew linearly during isothermal reactions (305 to 322 K and $E_a = 70$ kJ mol^{-1}), but below 313 K there was an initially slow rate which then accelerated to a constant value. The sigmoid α - time curves for powders and single crystals were fitted by the Avrami-Erofeev equation ($n = 2$) over a particularly large range, $0.001 < \alpha < 0.96$. The mean value of E_a (70 kJ mol^{-1}) was about the magnitude expected if proton motion was the rate-limiting step in interface advance. The overall kinetic behaviour was not entirely consistent with the rate of nucleation on the single crystal surface examined. Randomly distributed dehydration nuclei must have been formed at other locations within the reactant. It was not possible to compare the value of E_a for reaction at a dislocation with that for the similar process at an ideal lattice site. Etch pit formation in ethanol at surfaces of the same solid [66] showed that there was a significant decrease (by 95 kJ mol^{-1}) in E_a at an emergent dislocation, compared with the more perfect surface. This decrease was ascribed to the core energy rather than to the elastic strain energy associated with the lattice imperfection.

Striking evidence of phenomena that are clearly dependent on the formation of a superficial layer across the crystal surface is provided [67] by a microscopic examination of the changes that occur during the dehydration of cleaved single crystals of α-$NiSO_4.6H_2O$. DSC observations (4 K min^{-1}) between 400 and 415 K detected a series of small endotherms, identified from observations as bubble development with swelling of a relatively impermeable and elastic surface layer.

Such bubbles sometimes swelled and deflated several times due to water vapour accumulation and spasmodic escape. Photomicrographic evidence of bubble production is included in the article.

Guarini [67] confirmed that this dehydration is not only a nucleation and growth process but is more complex and is governed by the properties and behaviour of a modified outer layer of the crystal. The overall transformation involves three steps: (i) reaction, (ii) water migration, and (iii) product crystallization. The last is the slowest step, but the changes brought about by crystallization may result in diffusion, step (ii), being rate controlling [34].

Koga and Tanaka [68] used X-ray diffraction and infrared measurements to characterize the intermediate phases formed during the dehydration of α-$NiSO_4.6H_2O$ under different conditions. These reactions can be strongly influenced by the self-generated atmosphere of water vapour and its diffusive removal. The isothermal dehydration of small reactant masses (less than 15 mg) in flowing N_2 between 350 and 370 K yielded the dihydrate by a nucleation and growth process and data fitted the Avrami-Erofeev equation ($n = 2$). Values of E_a increased from 94 to 110 kJ mol^{-1} with decreasing particle sizes. During non-isothermal dehydrations in higher (self-generated) water vapour pressures, reaction proceeds to completion (the anhydrous salt) through three distinguishable steps involving (approximately) the tetrahydrate and the monohydrate. Reaction paths and mechanisms change with conditions so that the different surface compositions and textures influence the observed kinetic characteristics. The roles of the several phases that participate are discussed. This study illustrates the potential complexity of dehydrations and the need for extensive observations.

7.8.3. Magnesium sulfate hydrates

Lallemant and Watelle-Marion [69] distinguished two sequences in the dehydration of $MgSO_4.7H_2O$. When $p(H_2O) < 40$ Torr, five endothermic peaks were detected during progress of dehydration to the anhydrous salt. Each endothermic step was accompanied by recrystallization and the intermediates identified were the hexa-, tetra-, di- and monohydrates. Between $50 < p(H_2O) < 200$ Torr, reaction was similar, except that water loss from the tetrahydrate gave the compound $MgSO_4.2.5H_2O$ which reacted further to yield an alternative form of the dihydrate, then the monohydrate, before the ultimate production of the anhydrous salt.

The dehydration of $MgSO_4.7H_2O$ is [31] a nucleation and growth process and exhibits the S-T effect [70,71]. At low values of $p(H_2O)$ the tetrahydrate yields [30] the monohydrate direct. At intermediate pressures, during which the rate of reaction increases with $p(H_2O)$, the dihydrate was formed as an unstable intermediate. At higher pressures, the intermediate $MgSO_4.2.5H_2O$ became stable and the reaction rate decreased with a further rise of $p(H_2O)$. The domains of stability of the intermediates are dependent on both temperature and availability of water, and reaction proceeds to completion in two steps when the hydrate has been abruptly subjected to conditions that are sufficiently remote from those with which the intermediate phase may equilibrate [30].

The shapes of dehydration nuclei and rates of interface advance at different crystal surfaces of $MgSO_4.7H_2O$ are different [72]. Reaction rates at 323 K for several crystals of equal size, but varied habits, were compared [72] and it was shown that the rate of penetration of the reactant-product interface normal to the (111) face was considerably greater than that normal to the (110) face. This anisotropy was attributed [73] to the influence of the crystal lattice. The rate of dehydration measured [74] by microscopy from the advance of the interface agreed with that measured from mass-losses.

7.8.4. Calcium sulfate dihydrate and hemihydrate

The dehydration and rehydration reactions of calcium sulfate dihydrate (gypsum) are of considerable technological importance and have been the subject of many studies. On heating, $CaSO_4.2H_2O$ may yield the hemihydrate or the anhydrous salt and both the product formed and the kinetics of the reaction are markedly dependent upon the temperature and the water vapour pressure. At low temperatures (i.e. < 383 K) the process fits the Avrami-Erofeev equation ($n = 2$) [75]. The apparent activation energy for nucleation varies between 250 and 140 kJ mol^{-1} in 4.6 and 17.0 Torr water vapour pressure, respectively. Reactions yielding the anhydrous salt (< 10^{-5} Torr) and the hemihydrate ($p(H_2O) > 17$ Torr) proceeded by an interface mechanism, for which the values of E_a were 80 to 90 kJ mol^{-1}. At temperatures > 383 K the reaction was controlled by diffusion with E_a = 40 to 50 kJ mol^{-1}.

The α and β forms of calcium sulfate hemihydrate, prepared either from solution or by heating the dihydrate, respectively, differ slightly. The dehydration of β-calcium sulfate hemihydrate, 388 to 413 K, and $10^{-5} < p(H_2O) < 42$ Torr is diffusion controlled [76]. The activation energy increased with the prevailing water-vapour pressure from 26 to 200 kJ mol^{-1} between 10^{-5} and 42 Torr. This is believed to be due to the progressive blocking of pores preventing escape of water vapour. The reaction at temperatures above about 400 K exhibits Smith-Topley behaviour. The dehydration of the α form has been studied [77] between 341 and 490 K and in water vapour pressures of 10^{-5} to 360 Torr. Below 384 K the rate controlling factor varies with reaction conditions and may be diffusion or interface advance. Diffusion control was important in the higher temperature range and E_a increased with

$p(H_2O)$. The magnitudes of the kinetic parameters change at 415 K owing to the γ-$CaSO_4$ → β-$CaSO_4$ phase transformation.

The interrelationship between the reactions of hydrated calcium sulfates may be summarized schematically as follows [77] (see also [78]):

DIHYDRATE → $CaSO_4$(HEXAGONAL) → $CaSO_4$(ORTHORHOMBIC)

↘↖ ↑↓

HEMIHYDRATE
(α and β forms)

McAdie [79] showed that the Arrhenius parameters for the dehydration of $CaSO_4.2H_2O$ increased linearly with water vapour pressure. Molony *et al.* [80] studied the same reaction in vacuum and found approximate fits to both the contracting-volume and contracting-area equations.

The heats of formation and of reaction of the α and β forms of the hemihydrate are perceptibly different [79]. The surface structure of the hemihydrate has been investigated [80] by electron microscopy and adsorption measurements. During ageing the surface area of this material decreased, the fine pores disappeared and geometric features developed. The topography and surface area were not greatly influenced by dehydration-rehydration cycles, which suggests that the solid has zeolitic character. Calcium sulfate transforms from the hexagonal to the orthorhombic structure by two different mechanisms [81] over different temperature ranges.

A study of the dehydration of single crystals of $CaSO_4.2H_2O$ in vacuum using a quartz crystal microbalance [82] led to identification of a chemical step as controlling the rate of interface advance in the [001] direction for which E_a = 82 kJ mol^{-1} and A = $1.5x10^{14}$ s^{-1}. The rate of interface advance in the [010] crystallographic direction was controlled by H_2O diffusion through the product layer.

The helium pycnometric technique was applied [83] in a study of the dehydration of $CaSO_4.2H_2O$ in vacuum. For losses greater than 12.8% the water molecules simply vacate the micropores without changing the reactant structure and the phase change to anhydride is accompanied by only a slight volume change. Removal of water from the β-hemihydrate is 'zeolite like' and small amounts of tightly retained water block entrances to microspace in this structure. These observations are consistent with the conclusions discussed in the previous paragraph.

The rate of rehydration of the hemihydrate [84] in liquid water passes through a maximum, the position of which depends on the rate of reactant dissolution and of product precipitation. A general kinetic equation was derived which also described previously published data.

$Na_2Ca_5(SO_4)_6.3H_2O$ dehydrates [85] between 450 and 490 K by a process that fits the parabolic law (E_a = 75 kJ mol^{-1}) at low $p(H_2O)$. Arrhenius parameters were sensitive to $p(H_2O)$. At low $p(H_2O)$ the reaction is surface controlled but is bulk diffusion controlled at higher $p(H_2O)$.

7.8.5. Lithium sulfate monohydrate

The dehydration of $Li_2SO_4.H_2O$ ($\rightarrow Li_2SO_4(s) + H_2O(g)$) is a predominantly deceleratory process usually studied between 350 and 400 K. This reaction was proposed in 1990 as a model system for use in comparative kinetic investigations between different laboratories. The suggestion [86] stimulated a number of dehydration rate studies.

Huang and Gallagher [87] carried out a comprehensive study, using TG and DSC, of the influence of $p(H_2O)$ on the kinetics of dehydration of samples of powder, pressed pellets and single crystals (platelets and cubic). The rate of dehydration of pellets lay between those of crystals (slow) and powders (fast). The apparent activation energies varied with α, from 220 kJ mol^{-1} for powder at low α, to 61 kJ mol^{-1} for pellets at high α. Because the rate is influenced by self-generated water, the results for crystals are the most reproducible.

Gaponov *et al.* [88], using a quartz crystal microbalance, studied the role of diffusion of H_2O during the initial stages of dehydration. Masuda *et al.* [89] examined the behaviour of 100 to 150 mesh powder in vacuum, using TG (343 to 403 K). Results were well represented by the Avrami-Erofeev (n = 2) and contracting area equations. This conclusion was confirmed by Tanaka and Koga [90]. Microscopic studies of replicated surface textures [91], immediately after the onset of reaction (α = 0.005), showed that there had been rapid and dense initial nucleation. Crack development in advance of the recrystallized product zone showed the strong tendency towards generating a comprehensive layer of product covering all surfaces at low α. The subsequent inward advance of this boundary interface explained the fit of isothermal rate data to the contracting volume and contracting area equations. It is difficult to detect the initial acceleratory stage of reaction, probably completed by about α = 0.03, during reactant heating and obscured by the rapid progression towards the maximum reaction rate. Rate measurements during the final stages of dehydration are also subject to error, being detectably influenced by the onset of the reverse reaction, particularly at lower temperatures. Reaction rates for powder dehydrations were significantly greater than those for larger single crystal reactants.

Synchrotron radiation studies [92] have identified systematic changes in the interplanar spacing in the vicinity of the microscopically visible reaction interface, showing that the thickness of the reaction zone is about 150 μm. There is apparently significant loss of water from the zone ahead of the recrystallization plane identified as the advancing reaction interface.

From a series of complementary and comparative rate measurements [93] it was concluded that the contracting volume equation gave the most accurate and extended representation of the α-time data. Rate coefficients varied in replicate experiments by up to 10% for single crystals, but variations were much greater for powdered reactants, up to 50%, particularly at low temperatures. Significant variations in the magnitudes of apparent A and E_a values for dehydrations of single crystals or powders were obtained using the different experimental techniques of product pressure measurements, thermogravimetry or differential scanning calorimetry in the isothermal or rising temperature modes. The overall spread of calculated E_a values ranged between about 60 to 130 kJ mol^{-1}, and a compensation effect was identified.

Recently Simakova *et al.* [94] have reported a detailed study of the influence of a series of dehydration/rehydration cycles on the morphology of the solid product. Results show the stability of the "vacancy structure" produced by dehydration.

L'vov [95] has re-analyzed the published results for $Li_2SO_4.H_2O$ dehydration in terms of his dissociative evaporation theory (see Section 2.4.6.) and is able to explain the variations in the reported E_a values by taking into account the inhibiting effect of water vapour on the rate of dehydration and the contributions from the condensation of Li_2SO_4(g).

7.8.6. Other hydrated sulfates

Boldyrev and Schmidt [96] observed that the shape of nuclei formed during the dehydration of $ZnSO_4.7H_2O$ depended on the conditions. At low temperatures (280 K) and low $p(H_2O)$ (1 Torr), rectangular nuclei were formed with no cracking, whereas rounded nuclei with deep cracking were apparent at 308 K and 4 Torr, which indicates that the product phase had recrystallized. Wheeler and Frost [71] showed that there were two maxima on the plot of reaction rate against $p(H_2O)$. Guenot *et al.* [57] found some evidence for the intermediate formation of $ZnSO_4.4H_2O$ during dehydration.

Several lower hydrates were identified during the dehydration of $Al_2(SO_4)_3.16H_2O$. Barret and Thiard [15], using TG and DTA, supported by crystallographic studies, and with small reactant masses (50 to 80 mg) identified phases containing 16, 14, 12 to 11, 9 and 6 H_2O. Similarities between this sequence of reactions and the properties and structures of the corresponding hydrates of gallium [16] and chromium [17] sulfates have been discussed.

There have been many studies of the dehydration of iron(II) sulfate heptahydrate [97-100]. In an inert atmosphere, vacuum or nitrogen, the monohydrate is formed as an intermediate before water is lost to yield the anhydrous salt (473 to 573 K) [71,101]. In an oxidizing atmosphere, the changes are more complex and products include [97-101] $FeSO_4$, $FeOHSO_4$ and $Fe_2O(SO_4)_2$. Oxidation proceeds to α-Fe_2O_3 at higher temperatures (see Chapter 14.). Activation energies for the thermal decomposition of various hydrates of $FeSO_4$ (9 reactants, 23 E_a values) have been listed [100] and reactions presented.

Studies of the dehydrations of many other hydrates of simple sulfates are to be found in the literature.

7.9. ALUMS

The dehydrations of alums have been of particular interest in kinetic studies because large, relatively perfect crystals can be prepared, and the cubic lattice, common to different alums, should simplify the interpretation of kinetic data. (Some lattice modification does, however, accompany cation variation [102]). Despite these structural similarities, the extents of water loss, the kinetic parameters and the shapes of nuclei developed all show significant differences with changes in cation composition of the alum.

The dehydration of chrome alum in vacuum [58] has an induction period, followed by nucleation at a constant rate. Surface nuclei were circular, but because the rate of bulk penetration was less than that of surface advance, the growth nuclei were flattened hemispheres. The rate of initial growth of each nucleus was exponential until a diameter of about 0.1 mm was attained and remained constant thereafter. Two types of nuclei were recognized [40], but in only one of these had the product undergone reorganization or recrystallization with the appearance of surface cracking.

Kinetic studies [37,40] of the dehydration of chrome alum (260 to 270 K) yielded Arrhenius parameters of greater magnitude than predicted by the Polanyi-Wigner equation (E_a = 125 kJ mol^{-1}, whereas the enthalpy of dehydration is 42 kJ (mol $H_2O)^{-1}$). This was ascribed to systematic changes in the thickness of the transition layer and thus on impedance of the escape of water vapour. At higher temperatures (288 to 308 K) a lower value of E_a was found (96 kJ mol^{-1}) and the Arrhenius parameters were in approximate conformity with the requirements of the Polanyi-Wigner expression. No explanation for this change in magnitudes of A and E_a was provided.

Arrhenius parameters for the dehydrations of ammonium and potassium aluminium alums [38] were in agreement with the Polanyi-Wigner equation. Comparisons of the shapes of nuclei on different crystal surfaces indicated that reaction proceeds along (100) planes. The observed decrease of the rate in water vapour is attributed to the blocking of pores by adsorbed molecules. No intranuclear cracking was apparent and the product-reactant boundaries became irregular because of the influence of water on reorganization of the product phase. The appearance of these nuclei contrasted markedly with those in mixed potassium chromium/aluminium alums, where there is an approximately concentric structure.

Garner and Jennings [60] demonstrated that water vapour catalyzed nucleation of alum surfaces following evacuation. Defects generated at the surface diffuse and accumulate at dislocations to develop germ nuclei [32]. The initiation and development of reaction in potassium and chrome alums were contrasted. The nuclei in the potassium alum were bounded by planar faces and the Arrhenius parameters were "normal", while in chrome alum the nuclei were approximately hemispherical, A and E_a values were large and the nucleus boundaries were not oriented in a simple relationship to the lattice structure.

The dehydration of thin crystals of potassium alum [103] in dry air (323 to 343 K) showed different rates of reaction following nucleation of different surfaces. This anisotropy was attributed to the variation in density of packing of water molecules with crystallographic direction. At low pressure [104] the adsorption of water vapour was reversible, but at larger values of $p(H_2O)$ multilayers were formed and uptake of water was controlled by diffusion into the bulk of the crystal ($E_a < 2$ kJ mol^{-1}).

A subsequent investigation [105] of alum dehydrations generally confirmed the previously reported kinetic results. A constant rate of nucleation was observed for $KCr(SO_4)_2.12H_2O$, whereas the initiation of reaction in $KAl(SO_4)_2.12H_2O$ was more rapid. The distribution of nuclei on these cleaved surfaces was not random, but sites of onset of reaction could often be associated with areas of damage or dislocations. More perfect surfaces developed fewer nuclei. Onset of reaction within cracks increased the probability of impingement between close neighbouring reaction zones. Growth of larger nuclei proceeded at a constant rate of interface advance, that was equal for different nuclei on the same face. For small nuclei, instances of initial growth rates that were both smaller and larger than those subsequently attained, were observed. The reaction model proposed is illustrated in Figure 7.3.

Figure 7.3.
Schematic representation of surface textural changes and suggested reaction model proposed to explain observations for the dehydrations of $KAl(SO_4)_2.12H_2O$ and of $KCr(SO_4)_2.12H_2O$

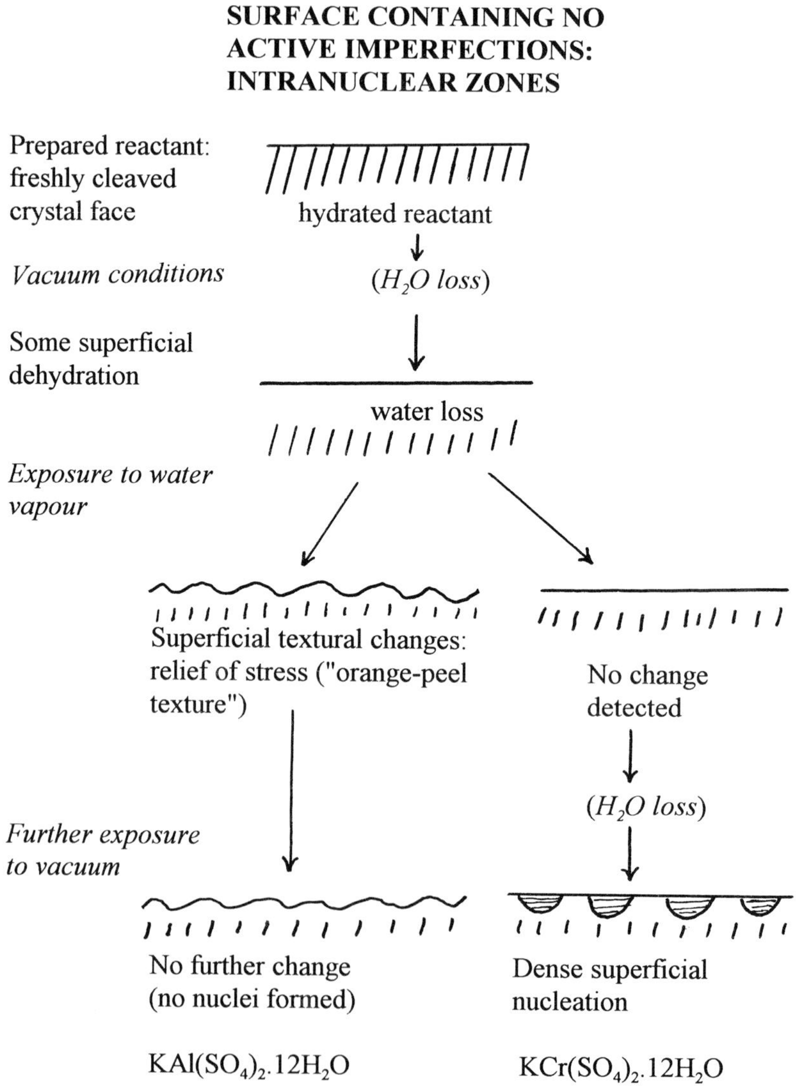

Figure 7.3. (continued)
Schematic representation of surface textural changes and suggested reaction model proposed to explain observations for the dehydrations of $KAl(SO_4)_2.12H_2O$ and of $KCr(SO_4)_2.12H_2O$

SURFACE IMPERFECTION AT WHICH A GROWTH NUCLEUS IS GENERATED

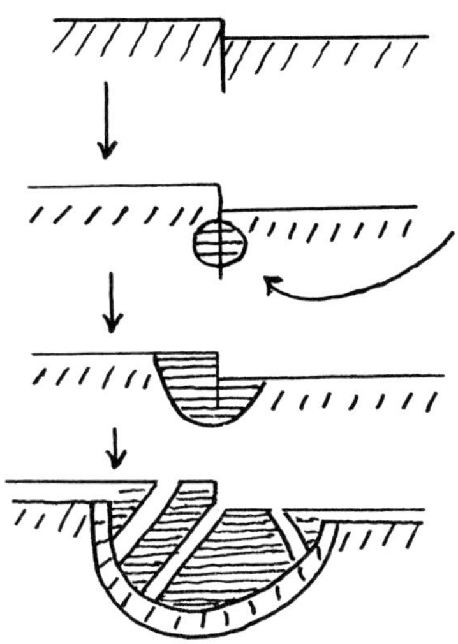

Prepared reactant: freshly cleaved crystal face

Vacuum conditions maintained throughout

Some superficial dehydration

Zone of stress sufficient to initiate recrystallization giving lower hydrate product

Growth nucleus established

Development of growth nucleus

Nucleus is a zone of water retention and recrystallized product (seed crystals)
Both possible promotors of water loss and the phase change

Both alums:
$KAl(SO_4)_2.12H_2O$ and $KCr(SO_4)_2.12H_2O$

NOTES: (i) There is a different pattern of change of intranuclear surface by the two salts, but subsequent textural development is enhanced by exposure to water vapour after previous evacuation. (ii) A specialized reaction zone is required to initiate nucleation in both salts. The shapes of nuclei developed are slightly different but both grow in three dimensions. (iii) Water is rapidly and easily reaccommodated in the Al^{3+} ion coordination shell but reaction of H_2O with Cr^{3+} after dehydration proceeds much more slowly [105]. (iv) The crystalline products of alum dehydrations, double or separated single sulfates, have not been characterized.

These kinetic measurements were complemented with microscopic examinations that provided new insights on the reaction mechanism. On exposure of partly dehydrated $KAl(SO_4)_2.12H_2O$ crystal faces to low pressures of water vapour there was superficial retexturing ('orange peel' texture) of internuclear, apparently unreacted, surface. Similar treatment of $KCr(SO_4)_2.12H_2O$ resulted in no obvious textural change but restoration of dehydration conditions yielded a remarkable rise in nucleation density [2]. It was concluded that exposure of crystals to dehydration conditions resulted in some loss of water of crystallization from *all* surfaces. Reaction could continue only in zones where the difficult step, identified as recrystallization of the dehydrated product, could occur. This was inferred, from the above observations, as being promoted by the presence of water vapour. 'Orange peel' texture development has since been shown [106] to occur on the surfaces of many different crystalline hydrates. Accepting the ability of water vapour to promote structural reorganization of dehydrated material, the nucleus can be regarded as a texture specifically generated to retain temporarily sufficient water vapour to promote the recrystallization that is the essential feature of interface advance. Concurrent generation of pores and channels permits the volatile product to escape. This accounts for the occurrence of reaction in regions below the outer surfaces of the solid reactant that are themselves inactive because the transition to product phase cannot occur in the absence of adsorbed water. Nucleation occurs within zones of structural distortion or damage where transition to product solid is possible, believed to be promoted by local water vapour retention. Nucleus growth is, therefore, an autocatalytic reaction dependent on sufficient availability of the volatile product at the interface.

A more detailed examination of the intranuclear channel structures and interface textures developed during these reactions demonstrated that no detectable cracking penetrated the reactant in advance of the reactant/product contact [107]. This is consistent with the above conclusions. No considerable strain is associated with the interface, so that reaction is not a consequence of water loss at surfaces exposed following crack propagation through stress relief in the vicinity of and associated with the phase transition.

A recent kinetic study [108] of the overall dehydration rates of $KAl(SO_4)_2.12H_2O$ and $KCr(SO_4)_2.12H_2O$ showed that measured α-time data were well described by the Avrami-Erofeev equation with $n = 2$. This was not consistent with expectation for the growth of three dimensional nuclei for which $\lambda = 3$ and $\beta = 0$ or 1, respectively. In accordance with the reaction models described above, there must be limited water losses from *all* surfaces together with an overall reaction controlled by product recrystallization that confers the apparent topotacticity on the overall process. Evidence was also obtained to indicate that the reactant temporarily adsorbed a proportion of the water released, consistent with the suggested participation of this product in the restructuring steps at the active interface boundaries of all growth nuclei. The rate of dehydration is consistent with diffusive loss by evaporation of intranuclear water. This model is supported by the observation that the rate of H_2O evolution from within chrome alum nuclei (about 1.8×10^{-7} mol min^{-1} mm^{-2} at 305 K) is close to that measured for the evaporation of unbound water from lignite (7.9×10^{-7} mol min^{-1} mm^{-2} at 293 K) [109]. Water in lignite is retained within a carbonaceous matrix from which it is volatilized without chemical change in vacuum.

7.10. HYDRATES OF OTHER INORGANIC SALTS

7.10.1. Carbonates

The exothermic effects associated with some dehydrations [18] may result from delayed recrystallization, or the oxidation of a product of dehydration. Dehydration of magnesium carbonate trihydrate, followed by heating in carbon dioxide, resulted in the exothermic transformation of an anhydrous pseudomorph of the reactant into magnesite, $MgCO_3$.

The thermal dehydration of $Na_2CO_3.H_2O$ between 336 and 400 K fits the Avrami-Erofeev equation with $n = 2$ ($E_a = 71.5$ kJ mol^{-1} and $A = 2.2 \times 10^7$ s^{-1} [110]). The apparent reduction in rate resulting from an increase of $p(H_2O)$ is ascribed to competition from the rehydration reaction. Electron micrographs confirm the nucleation and growth mechanism indicated by the kinetic behaviour, nucleation develops from circular defects that may be occluded solution.

7.10.2. Uranyl(VI) nitrate hydrates

A number of unusual kinetic characteristics have been described [111,112] for the dehydration of $UO_2(NO_2)_2.6H_2O$. The necessity for strict control of reaction conditions is stressed. The onset temperatures [113] of the four successive transformations through the tri-, di-, monohydrate and, finally, to the anhydrous salt are 332, 345, 414 and 449 K, respectively.

Isothermal dehydrations of single crystals (about 1 mm^3) of the hexahydrate [111] in vacuum between 213 and 243 K gave α-time curves for conversion to the trihydrate with no induction period and a constant rate of water loss during a large fraction of reaction. The rate of dehydration decreased linearly with the prevailing pressure of water (contrasting with the behaviour of many other hydrates) attributed to the occurrence of the reverse reaction. The activation energy for conversion of the hexahydrate to the trihydrate was 46 kJ mol^{-1}, under conditions believed to be free of inhibition caused by product water vapour or self-cooling of the reactant. Rate studies using crystals with specific faces covered with a resin impermeable to water yielded no evidence that the dehydration process was anisotropic. The rehydration reaction was diffusion controlled.

An unusually large diffusion coefficient for the movement of water within the structure of the hexahydrate, via vacancies at the water sites, was noted [111]. The slow step in dehydration was identified as water desorption from the surface at which a constant concentration was maintained when $\alpha < 0.65$. Subsequent recrystallization to the trihydrate lattice resulted in a marked decrease in reaction rate.

At higher temperatures, 313 to 343 K, and at higher water vapour pressures, dehydration of single crystals [112] proceeded by a contracting volume process with a larger activation energy (105 kJ mol^{-1}). As in the low pressure studies, the rate decreased in direct proportion to $p(H_2O)$. At the melting point of the hexahydrate (333 K) reaction ceased because the molten reactant was encapsulated within a sheath of the solid trihydrate. When later cooled below 333 K, water evolution recommenced, a somewhat unexpected occurrence. Above 355 K, reaction was similarly restarted due to conversion of trihydrate to the dihydrate. Unlike the behaviour of single crystals, the reaction of the powder differs in that the rate increases across the melting point. Kinetic control is by diffusion through the reactant bed. Apparently crystals larger than some minimum size are required before significant inhibition becomes apparent.

7.10.3. Halides

The dehydration-hydration of $BaCl_2.2H_2O$ is one of the few fully reversible processes. Water is lost from barium chloride dihydrate [114] in two distinct processes, through the intermediate formation of the monohydrate. There is also a hemihydrate. Kinetic data for both reactions were obtained by DSC, isothermally and non-isothermally using different rates of temperature increase. The early stages ($\alpha < 0.15$) of both processes fitted the power law with $n = 2$ and, thereafter, the data fitted the contracting area or the Avrami- Erofeev equation. Values of E_a for the isothermal reactions (75 kJ mol^{-1}) were somewhat less than those obtained (90 kJ mol^{-1}) by non-isothermal methods. This may be due to increased overlap of the processes contributing to the overall change during the temperature increase and an enhanced rate of nucleation at higher temperatures. Reaction on flat (010) faces of large $BaCl_2.2H_2O$ crystals proceeded through the initial rapid two-dimensional growth of large nuclei ($n = 2$). Thereafter the reaction interface advanced into the bulk of the crystal.

Isothermal (317 to 333 K) studies [115] were conducted in vacuum and 40 to 667 Torr $p(H_2O)$. Apparent E_a values varied with sample size and α within the range 92 to 150 kJ mol^{-1}. Enthalpies of reaction ranged from 116 to 137 kJ mol^{-1}. Kinetic characteristics were found to be sensitive to transient inhomogeneities in $p(H_2O)$ resulting from constraints on diffusion.

Dehydrations of crystalline hydrates do not invariably yield the anhydrous salt as a product through the elimination of water, as exemplified by the dehydration of $MgCl_2.2H_2O$ [116]:

$$MgCl_2.2H_2O \rightarrow MgO + 2HC\ell + H_2O$$

Reaction (623 to 703 K) is believed not to proceed through the intervention of $Mg(OH)_2$ and no significant amount of $Mg(OH)C\ell$ is formed. There was no evidence of melting. Yield-time data were well expressed by the contracting volume equation with $E_a = 110 \pm 5$ kJ mol^{-1} in dry N_2, or 75 kJ mol^{-1} in 10 Torr H_2O. Kinetic characteristics of the reaction in water vapour were closely similar to behaviour reported by Ball [117] for the reaction $MgC\ell_2 + \frac{1}{2}O_2 \rightarrow MgO + C\ell_2$. It is concluded [116] that both reactions proceed with common controls based on the increasing stability of the MgO product at the reaction interface.

TG studies in N_2 [118] showed that the dehydration of $CuCl_2.2H_2O$ went to completion in a single step, T_{max} at 363 K at 3 K min^{-1} with $E_a = 70$ to 75 kJ mol^{-1} which is less than the reaction enthalpy (107 kJ mol^{-1}). Water loss from $NiCl_2.6H_2O$ involved two overlapping rate processes (peak maxima at 321 and 343 K heating at 1 K min^{-1}) with loss of 4 H_2O, followed by dehydration of the dihydrate (T_{max} at 428 K with $E_a = 76$ to 83 kJ mol^{-1} and reaction enthalpy slightly larger, 91.5 kJ mol^{-1}).

TG and DSC have been used to study the dehydration of $K_2CuCl_4.2H_2O$ at about 330 K and behaviour was interpreted through consideration of microscopic observations [119]. The reaction interfaces proceed inwards from crystal surfaces accompanied by some random bulk nucleation. The kinetics are generally described by a contracting geometry expression, although the Avrami-Erofeev equation is applicable at low α. Diffusion of water vapour becomes an important reaction control at high α and in the larger reactant crystals. Apparent magnitudes of E_a decrease systematically with increasing α, most values were in the range 140 to 90 kJ mol^{-1}.

7.10.4. Phosphates

Removal of water (409 to 537 K) from $CaHPO_4.2H_2O$, yields the anhydrous salt and, at higher temperatures, the pyrophosphate [120]. Below 478 K the value of E_a for the diffusion-controlled reaction was 190 kJ mol^{-1} and the rate was independent of $p(H_2O)$. Between 477 and 537 K the value of E_a was much smaller and depended on $p(H_2O)$. The possible formation of an intermediate, the monohydrate is discussed.

The dehydration of $MnPO_4.1.3H_2O$ ($\rightarrow 0.5Mn_2P_2O_7 + 1.3H_2O + 0.25O_2$) proceeded [121] in three stages: - $0.3H_2O$; $Mn^{3+} \rightarrow Mn^{2+}$ with oxygen evolution, and - H_2O. The high temperature of the onset of reaction (543 K) suggests that reaction is dehydroxylation, rather than dehydration, of a reactant that can be expressed as $Mn(OH)(HPO_4).0.3H_2O$ with the water strongly retained within the structure. This reformulation of the reactant is supported by infrared spectra.

The layered phosphonates $M(O_3PCH_3).H_2O$(M=Mg,Zn,Co) undergo topotactic dehydrations [122]. Rehydration is shape selective to the incoming molecule, the Mg salt rehydrates rapidly, Zn is unreactive to H_2O and the behaviour of the Co salt is intermediate.

7.10.5. Borates

The structure of borax ($Na_2B_4O_5(OH)_4.8H_2O$) is composed of polyanions and chains surrounded by water molecules [123]. On heating to 350 K, approximately 3.2 H_2O is internally retained and evolved at about 375 K. At 406 K a further 1.5 H_2O is released. X-ray diffraction studies show that the salt heated to 413 K loses the lines characteristic of borax and those for tincalconite appear. Further water is lost above 472 K and crystalline $Na_2B_4O_7$ is formed at 848 K which melts at 996 K. Dehydration of borax at 319 K was described [124] by the contracting volume equation and the plot of rate against $p(H_2O)$ included a single maximum at low pressure. Reaction was suggested to proceed by two parallel paths: direct desorption and desorption catalyzed by water vapour. Later work [21] (302 to 311 K) showed the occurrence of S-T behaviour. Pradhananga and Matsuo [125] found that dehydration (- $8H_2O$) was a first-order process and $E_a = 84$ kJ mol^{-1}. This was ascribed to control by first-order nucleation of the finely divided reactant crystallites.

Structural reorganizations [126] during the thermal decompositions of hydrated borates in general, and of calcium borate (pandermite) and potassium magnesium borate in particular are complex, proceeding through step-by-step processes. The parent structure is gradually transformed into more stable forms by apparently the energetically most convenient route. Dehydration depends on the strength of links with anion rings or cations and the permeability of the structure to H_2O.

A survey of the dehydrations of 14 boron-containing minerals [127] revealed that only colemanite ($CaB_3O_4(OH)_3.H_2O$) decrepitates. This solid loses all the constituent water in a narrow temperature interval causing stress in the crystal. This contrasts with the gradual water loss in ulexite ($NaCaB_5O_6(OH)_6.5H_2O$) which does not decrepitate.

7.10.6. Sodium thiosulfate pentahydrate

The kinetics of dehydration [128] of $Na_2S_2O_3.5H_2O$ were difficult to interpret because the course of the reaction was markedly influenced by the perfection of the initial reactant surface and the reaction conditions. No reliable Arrhenius parameters could be obtained. The mechanism proposed to account for behaviour was the initial formation of a thin superficial layer of the anhydrous salt which later reorganized to form dihydrate. The first step in the reaction pentahydrate $\rightarrow$ dihydrate was satisfactorily represented by the contracting area ($0.08 < \alpha_1 < 0.80$) expression. The second reaction, giving the anhydrous salt, fitted the Avrami-Erofeev equation ($n = 2$) between $0.05 < \alpha_2 < 0.8$. The product layer offers no impedance to product water vapour escape and no evidence of diffusion control was obtained. The mechanistic discussions are supported by microscopic observations of the distributions and development of nuclei as reaction proceeds.

7.10.7. Sodium tungstate dihydrate

A non-isothermal DSC investigation [129] of the dehydration of $Na_2WO_4.2H_2O$ using slow heating rates (less than 1 K min^{-1}) below 373 K showed that data fitted the contracting area rate equation. This kinetic interpretation conflicted with the microscopic observation that there was random production and subsequent growth of nuclei. These apparent inconsistencies were resolved by a reaction model that identifies reaction as first generating a dehydrated surface layer that later recrystallizes with the formation of circular growth nuclei. Such nuclei facilitate product water escape and the interface advances towards crystallite centres, fitting the contracting area model.

7.10.8. Molybdic acid

Single crystals of $MoO_3.H_2O$ ("α-molybdic acid") were transformed [130] on heating at 433 K into perfect pseudomorphs composed of MoO_3 crystallites. This reactant to product transition is explained by a topotactic reaction in which the MoO_3 structure was formed by corner linking of isolated double chains after water elimination. This reaction model provided evidence of value in identifying the hitherto unknown $MoO_3.H_2O$ structure which was confirmed independently by X-ray diffraction. The value of topotactical relationships in understanding structural changes during solid state rate processes is emphasized [130].

7.11. HYDRATES OF METAL OXALATES

7.11.1. Introduction

The crystallographic transformations which follow water removal from Group IIA metal oxalate hydrates have been studied in some detail, but structural [131] changes have not always been considered in kinetic investigations of water removal from transition-metal oxalate hydrates.

DTA curves for the reactions of 25 different oxalates on heating in nitrogen or in oxygen have been reported [14], from which the onset temperatures of dehydration can be found. The measured changes in surface area during decomposition of various different oxalates [7,14] have shown that, although the area increased on dehydration, the maximum area did not occur until after the onset of anion breakdown. Shkarin *et al.* [132] concluded that the dehydrations of seven oxalates (Ni, Mn, Co, Fe, Mg, Ca and Th) proceeded by identical mechanisms, though the evidence on which this generalization is based, i.e. the common range of values of E_a reported (83 to 96 kJ mol^{-1}), is not consistent with other literature values reported for the same reactions.

7.11.2. Manganese(II) oxalate dihydrate

The reaction of this solid [48] was the first [133] example of Smith-Topley behaviour recognized and studies of this rate process have continued. Flanagan and Kim [133] showed that irradiation decreased the induction period to dehydration and the rate of water evolution rapidly reached a maximum value which was maintained between $0 < \alpha < 0.4$. Water evolution was more rapid than that found for unirradiated salt and the value of E_a was decreased. Irradiation damage to the crystal promoted nucleation and there was rapid initial establishment of a constant area of reaction interface (the contracting volume equation approximates to zero-order kinetics at low values of α). There was also evidence [134] that preirradation aided recrystallization during vacuum dehydration.

The dehydration of $MnC_2O_4.2H_2O$ in air and in nitrogen (373 to 473 K) was described [135] by the contracting volume equation with $E_a = 70$ kJ mol^{-1}. Similar values were obtained from non-isothermal measurements. Recrystallization of the product to give crystalline anhydrous salt may occur more readily at higher reaction temperatures. The value of E_a found (90 kJ mol^{-1}, between 320 and 350 K) was less than previously reported [133] values. Some discussion and photomicrographs [136] of surface textures of salt dehydrated in air and in vacuum were given in studies more directly concerned with the influence of conditions of dehydration on subsequent reactions of the anhydrous salt.

7.11.3. Nickel oxalate dihydrate

The isothermal dehydration (358 to 397 K) of $NiC_2O_4.2H_2O$ in vacuum [22], with rapid removal of the water vapour released and small masses of reactant (2.0 to 0.2 mg), gave an activation energy, from extrapolation to zero mass, of 130 kJ mol^{-1}. This value was significantly larger than values reported for the same dehydration in higher temperature intervals. (Le Van [137], however, reported a value of 121 kJ mol^{-1}). Flanagan *et al.* [22] emphasized the severity of the inhibition resulting from the presence of water vapour and its effect on the magnitudes of the apparent Arrhenius parameters when the dehydration is reversible. The careful attention directed here towards ensuring that the measured reaction rates referred to the water elimination step only, makes this investigation especially significant. Attention is drawn to possible uncertainties in kinetic measurements for processes where the volatile product can participate further.

7.11.4. Hydrates of Group IIA metal oxalates

The dehydration of $MgC_2O_4.2H_2O$ between 425 and 500 K in flowing dry nitrogen is a predominantly deceleratory reaction [138]. Reaction starts at the outer edges of the crystals and moves inwards by two-dimensional advance, confirmed by optical microscopic observations. α-time data fit the contracting area equation with E_a = 111 kJ mol^{-1} and $A = 3.4 \times 10^9$ s^{-1}. Precise kinetic analysis of the brief initial acceleratory stage ($\alpha < 0.1$) showed that a nucleation process could be distinguished, expressed by the exponential law, for which E_a = 430 kJ mol^{-1} and $A = 3.7 \times 10^{49}$ s^{-1}. The sample masses (12 mg) were larger than those used by Flanagan *et al.* [22] and the rate of this reaction in N_2 could be very sensitive to water vapour pressure.

The stability domains and equilibrium relations between $CaC_2O_4.3H_2O$ and $CaC_2O_4.H_2O$ and the α and β forms of the anhydrous salt, have been established [24]. Two molecules of water are lost from the trihydrate to yield the monohydrate in a single endothermic process. Dehydration of the monohydrate involves two endothermic steps, of which the second is the greater. Reaction may proceed by a nucleation and growth process which is more rapid than the subsequent volatilization of the water released, or there may be a sudden disintegration of water bonding within the solid during recrystallization of the salt so that removal of water is slightly delayed. Dehydration of $CaC_2O_4.H_2O$ is a topotactic reaction in which appreciable quantities of water can be retained by the residue. The thermal reactions of $CaC_2O_4.H_2O$ are often used as a model system in testing thermal analysis equipment.

Two hydrates of strontium oxalate are known [139], the monohydrate and the polyhydrate, $SrC_2O_4.(2+n)H_2O$. Their behaviour is comparable with that described above for calcium oxalate monohydrate.

Despite some crystallographic similarities with the acid salt (see below), the water molecules in barium oxalate dihydrate are accommodated differently in the lattice [140]. The extent of water loss from $BaC_2O_4.2H_2O$ varied with $p(H_2O)$. The products were the anhydrous salt below 4 Torr, the hemihydrate between 4 and 100 Torr, and the monohydrate above 100 Torr. The rate of water evolution at 353 K exhibited Smith-Topley behaviour and the changes in rate occur at values of $p(H_2O)$ (5 and 100 Torr) which are close to those characteristic of the stability range of the hemihydrate.

An infinite number of salt compositions, $BaC_2O_4.H_2C_2O_4.(2-n)H_2O$, exist [141] between $0 < n < 2$. The water content depends on both the pressure and the temperature. The H_2O molecules occupy well-defined crystallographic sites within the unit cell and may be removed without fundamental reorganization of the structure. The partially dehydrated salt is not zeolitic and the stability of the solid results from the strong linkages between barium and oxygen of the anion. Water molecules are situated in molecular channels interposed between alternate layers of cations and anions and hydrogen bonding is probably not significant. During removal of water, the dimensions of the unit cell change gradually from values characteristic of the dihydrate to those of the anhydrous salt. A single crystal of the dihydrate yields a single crystal of the anhydrous salt. There was no evidence of nucleation and growth of a product phase. Water release is believed to occur after diffusion through the reactant solid which undergoes comparatively minor structural modification. The upper temperature limit of hydrate stability (about 370 K) is attributed to the $\alpha \rightarrow \beta$ transition of $BaC_2O_4.H_2C_2O_4$. At 430 K, oxalic acid is sublimed leaving a residue of anhydrous barium oxalate.

Isothermal and non-isothermal kinetic experiments on the dehydration of $(NH_4)_2C_2O_4.H_2O$ were compared [142] using multiple sets of data. Isothermal experiments identified the contracting area expression as giving the best fit for the predominantly deceleratory reaction and $E_a = 73$ kJ mol^{-1}. Different expressions were identified (A1.5, A3 or D3) using rising temperature methods.

7.12. HYDRATES OF METAL FORMATES

7.12.1. Copper(II) formate hydrates

Copper(II) formate forms tetra- and dihydrates, each of which loses water to yield different crystallographic modifications of the anhydrous salt. Some thermodynamic [143] and structural [144] properties of these (and other related) compounds have been given. The dihydrate was not formed [144] by the partial removal of water from the tetrahydrate.

The dehydration of copper(II) formate tetrahydrate is a topotactic process [145]. The anhydrous crystalline product of dehydration retains two-dimensional features of the reactant structure but the stacking mode is changed. This structure is different from that obtained by the direct preparation of anhydrous $Cu(HCOO)_2$.

Fichte and Flanagan [146] studied the kinetics of dehydration of single crystals of $Cu(HCOO)_2.4H_2O$ (228 to 298 K) in vacuum by gravimetric and microscopic measurements. Water is accommodated between extended polymer-like planar layers containing both cations and anions. Crystallographic evidence suggests the accommodation of the water in two alternative coordination environments, but all the water is evolved in a single kinetic process. Details of the structure of the anhydrous product were not available, but it is probably also of a layer-type because it reverts to the original structure on rehydration.

During dehydration of $Cu(HCOO)_2.4H_2O$ (228 to 298 K) in vacuum, the reaction interface penetrates the crystal at a constant rate in two directions, parallel to the extended {001} planes which constitute the layer structure. Nucleation occurs most rapidly on the (111) faces. The self-cooling of single crystals was signficant above 268 K. The rate of reaction decreased with an increase in $p(H_2O)$. There was no S-T effect and this is understandable considering the ease of structural reorganization following water removal or repacement. The measured value of E_a (47 kJ mol^{-1}) was unexpectedly small and this was the first well-documented example of a solid state dehydration which has a value of E_a less than the dissociation enthalpy (52 kJ mol^{-1}). (Other low values in the literature have been identified to be the consequences of self-heating). The evolved water remains temporarily between lattice layers, thus decreasing the enthalpy of reaction to a value close to that for the vaporization of water (44 kJ mol^{-1}). Because the rate of advance of the interface was constant, it was concluded that reactions at the reactant-product contact are rate-limiting and escape of water by diffusion across the product layer is relatively rapid.

7.12.2. Manganese(II) formate dihydrate

The dehydration of $Mn(HCOO)_2.2H_2O$ is an anisotropic process [147] (as in the copper salt above) in which the reaction interface advances into the crystal as a contracting parallelogram. The value of E_a for reaction (293 to 348 K) was 72 kJ mol^{-1}, which is close to the enthalpy of dissociation. This value of E_a is consistent with the requirements of the P-W equation. The rate decreased with increasing $p(H_2O)$ and no S-T effect was detected. Similar results were reported by Clarke and Thomas [148]. The anhydrous salt rehydrates, readily and completely, to the dihydrate by a diffusion-controlled process for which $E_a = 50$ kJ mol^{-1}.

7.12.3. Yttrium formate dihydrate

A gravimetric study of the thermal dehydration of $Y(HCOO)_3.2H_2O$ was undertaken [149] in various water vapour pressures between 5×10^{-4} to 8 Torr (387 to 407 K). Following an initial short acceleratory process, the reaction predominantly fitted the contracting volume equation. The reaction rate increased with $p(H_2O)$, reached a maximum value and thereafter decreased to a constant value. This is a pattern of behaviour similar to the Smith-Topley effect. The results are explained on the basis of the crystallinity of the dehydrated residual product.

7.13. HYDRATES OF METAL SALTS OF OTHER ORGANIC ACIDS

7.13.1. Lithium potassium tartrate hydrates

Complementary measurements have been made of the dehydration kinetics of three distinct lithium potassium tartrate hydrates : those containing the *d* and *dℓ* anions were monohydrates and the *meso* salt was a dihydrate. These reactants were of interest because they contained chemically identical components, but had different crystal structures and reactivities [150].

d **- $LiKC_4H_4O_6.H_2O$:** Dehydration proceeded to completion through two distinct rate processes [151]. The first was deceleratory and identified as the diffusive loss of H_2O from the outer layers of a crystal structure that is at first ordered, but later becomes more disordered as the water site vacancy concentration increases. Kinetic measurements (363 to 423 K) for single crystals were irreproducible, due to individual differences in imperfection structures, but dehydration rates of powders showed less variation. E_a for both single crystals and powder was 153 ± 4 kJ mol^{-1} between 400 and 460 K. This reaction resulted in loss of about 4% of the constituent water from single crystals and much more, up to 50%, from powdered reactant. It was estimated that dehydration extended to a depth of about 10 μm below all crystal faces and, in large single crystals, a discontinuity was detected microscopically at about 1 μm below the solid surfaces.

The final stages of the above reaction overlapped with the onset of the nucleation and growth process that continued to complete the dehydration. Growth of three dimensional nuclei was confirmed microscopically. This second rate process was well described by the Avrami-Erofeev equation with $n = 2$ and E_a for crystals was 175 ± 30 kJ mol^{-1} (with a considerable scatter of data) below 460 K and a more reproducible reaction rate, with $E_a = 153 \pm 10$ kJ mol^{-1}, for powder. Above about 450 K there were some indications of intracrystalline melting of single crystals and the value of E_a increased markedly to 350 ± 50 kJ mol^{-1} (again with significant scatter of data).

After completion of water losses from all faces, local crystallization at a small number of surface sites resulted in the establishment of nuclei of solid product. In separate experiments it was shown that the significant induction periods to establishment of nuclei in large single crystal reactants were eliminated by the presence of anhydrous reaction product. This additive is regarded as providing seed crystals for the rapid initiation of the active interface which advances inwards, in this particular reaction system, by a predominantly deceleratory rate process. Nucleus growth is explained by a reaction model in which (effectively seeded) product recrystallization results in the detachment of superficial anhydrous material as a product particle, thereby, exposing a new surface. Water loss from this area then becomes possible by the diffusive mechanism of the first reaction to a stage at which it, in turn, is seeded, is recrystallized and detaches. The cycle repeats, resulting in interface advance that is probably intermittent and spasmodic.

This reaction mechanism is supported by the observed quantitative agreement between the rate of the first, diffusion-controlled reaction and the rate of interface advance, together with the similarity of E_a values. Diffusive loss of water from within the disorganized solid is therefore identified as the controlling step in both rate processes. The extent of this is limited during the first deceleratory process but subsequently seeded detachment of the recrystallized product enables reaction to be maintained during continued nucleus growth.

The chemistry of this reaction differs from that discussed for alum dehydration above for two main reasons. At the temperature of this reaction, water is more volatile and is not adsorbed within the interface in sufficient quantities to promote the difficult recrystallization step. Secondly, the present reactant releases water close to its melting point and dynamic reorganization of the surface layer may reduce local strains compared with those that may be developed at lower temperatures within alum dehydration interfaces of residual sulfates.

***dℓ*- $LiKC_4H_4O_6.H_2O$:** Dehydration ($-H_2O$) (350 to 460 K) was accompanied by melting to yield a glassy solid, amorphous to X-rays. [152]. An initial rapid release of some water (about 6%) was followed (18 to 80%) by a zero-order rate process before the final deceleratory (first-order) approach to completion. Dehydration rates for the deceleratory reaction of the powdered reactant were about eight times more rapid than that of the crystal. The value of E_a for both crystals and powder was 330 ± 50 kJ mol^{-1}. The constant rate of this reaction proceeding with at least some melting is discussed with reference to two possible models. In one the reactant is regarded as developing two phases consisting of unreacted hydrate embedded within, and in equilibrium with, a phase that contains only a small proportion of water. In the alternative model, rate control is ascribed to a constant rate of water loss across a peripheral barrier layer. The system is of interest as a semi-homogeneous reaction proceeding with melting but in the absence of a solvent. The system may also have relevance to the theory of drying.
***meso*-$LiKC_4H_4O_6.2H_2O$:** This reactant also melted during dehydration [153]. Reaction of crystals yielded 1.2 H_2O by a zero-order process with $E_a = 230 \pm 10$ kJ mol^{-1} between 350 and 380 K. The remainder of this reaction and the complete dehydration ($- 2H_2O$) of powdered samples were deceleratory. The reaction of this salt was significantly more rapid that of the *dℓ*-isomer.

All the kinetic data obtained in this series of comparative and complementary studies are given Table 2 of reference [151]. More detailed mechanistic discussions are given in the individual papers cited.

7.13.2. Hydrates of barium and lead styphnates

Barium styphnate trihydrate [154] loses water readily and the anhydrous product decomposes at a lower temperature than that obtained from the monohydrate. Dehydration of the monohydrate (389 to 439 K) proceeds through rapid nucleation of the surface and rapid progress of reaction through intergranular material during the initial acceleratory process. Thereafter a constant rate of reaction is attributed to growth of nuclei proceeding predominantly along one preferred crystal axis. The activation energy (54 kJ mol^{-1}) was ascribed to breaking of hydrogen bonds between water molecules and styphnate ions at a non-coherent interface that did not promote the reaction.

The approximately constant rate of dehydration of lead styphnate monohydrate [155] is (like that of the barium salt above) attributed to the advance of the reaction zone into the sub-grain boundaries and crystallite penetration by formation of product in a single crystallographic direction. The value of E_a for dehydration of the samples as prepared (91 kJ mol^{-1}) was appreciably greater than that of irradiated salt

(68 kJ mol^{-1}). After several successive dehydration-rehydration cycles, the water removal process obeyed first-order kinetics. The rate was increased by a factor of about 7 and the value of E_a was decreased to 61 kJ mol^{-1}.

7.14. COMPLEX MATERIALS

7.14.1. Vermiculite

The dehydration of vermiculite crystals (267 to 290 K) was studied [156] at low pressure using a quartz crystal microbalance. A detailed theoretical discussion of reactions controlled by diffusion processes, within a layer-type reactant structure that does not recrystallize during reaction, is provided. Water removal is treated as gas diffusion within the semi-infinite crystal medium. A quasi-linear expression applied during the early stages of reaction of large crystals ($E_a = 27$ kJ mol^{-1}) and the later decrease of the rate of mass loss was proportional to $t^{½}$, attributable to control by H_2O diffusion. The rate of diffusive dehydration of reactant samples composed of small crystals was satisfactorily expressed by the first-order equation.

7.14.2. Lignite

Lignite, also known as brown coal, is a natural sedimentary deposit composed of finely divided carbonaceous material within which water fills the labyrinthine interparticulate channels. Kinetic studies of water evolution [157], which involves no chemical change, showed that the rate fitted the contracting volume equation, suitably modified to allow for the prevailing water vapour pressure. These observations, supported by examinations of sections of blocks of partially dehydrated lignite, are consistent with the evaporation of water at an interface that systematically advances from particle boundaries to their centres. Water removal resulted in a colour change from dark to light brown, observed in microscopic examinations. There was no evidence of diffusion control. The apparent value of E_a between 273 and 313 K was 35 ± 5 kJ mol^{-1} and the dehydration rates of different lignite samples were the same. It is interesting that this physical process fits a kinetic expression more usually regarded as applicable to the reactions of solids. The dehydration rate is controlled by water volatilization at an advancing interface. The number of water molecules evaporating per unit area is somewhat greater than that observed in the intranuclear reaction during chrome alum dehydration [157], which suggests that water vapour pressures within the alum [105] nucleus approach the saturation value.

7.15. PERHYDRATES

7.15.1. Introduction

A number of compounds which are frequently, but inaccurately, referred to as persalts, are more correctly described as perhydrates, because a peroxy anion structure is absent, but hydrogen peroxide of crystallization is present in the structure, e.g., sodium percarbonate is $Na_2CO_3.1.5H_2O_2$. Crystalline materials containing both (hydrogen bonded) H_2O and H_2O_2 are also known. Interest in the stabilities and mechanisms of decomposition of these salts has resulted from the technical use of specific materials as sources of oxygen, for example in washing powder. While such compounds could alternatively be considered in the chapter on oxides, they are included here as phases from which oxygen is released on heating. Such breakdown of H_2O_2 may occur in the solid phase, and is believed to proceed through a free radical mechanism involving the intervention of (amongst other intermediates) HO_2. Because the homogeneous decomposition of hydrogen peroxide in water is rapid at these temperatures, any small amount of liquid phase present in inclusions, capillary condensation, etc., would provide an alternative route to oxygen evolution. Water is a product of the reaction also, so that the autocatalytic homogeneous mechanism of decomposition, resulting from acceleration through the accumulation of this product, is not readily distinguished, on kinetic evidence alone, from the operation of a solid phase nucleation and growth process.

7.15.2. Group IA carbonate perhydrates and peroxocarbonates

In a thermogravimetric investigation of the products of reaction between $NaHCO_3$ and H_2O_2, Firsova *et al.* [158] concluded that $Na_2CO_3.1.5H_2O_2$ decomposed exothermically, 383 to 413 K, followed by the endothermic desorption of water. Firsova and Filatov [159] reported a gravimetric study of the isothermal decomposition of sodium and potassium peroxocarbonates ($M_2C_2O_6$) within the temperature interval 373 to 413 K. These compounds are prepared by the reaction of carbon dioxide with hydrogen peroxide in the presence of alkali, e.g.:

$$Na_2O_2.8H_2O + 2CO_2 \rightarrow Na_2C_2O_6 + 8H_2O$$

and the irreversible decomposition is expressed as:

$$M_2C_2O_6 \rightarrow M_2CO_3 + CO_2 + \tfrac{1}{2}O_2$$

Following a short period of acceleration, decomposition fitted the contracting volume equation between $0.1 < \alpha < 0.7$ with values of $E_a = 88$ and 96 kJ mol^{-1} for the sodium and potassium salts, respectively.

Galwey and Hood [160] showed that $Na_2CO_3.1.5H_2O_2$ decomposed in vacuum (360 to 410 K) to produce $Na_2CO_3 + 1.5H_2O + 0.75O_2$. α-time curves were sigmoidal and the kinetics could be described by the Avrami-Erofeev equation with $n = 2$ or 3. The activation energy was 112 ± 8 kJ mol^{-1}. The reaction rate between 313 and 343 K was significantly increased by the presence of small amounts of liquid water. This deceleratory reaction was fitted by the first-order equation ($E_a = 80 \pm 10$ kJ mol^{-1}) and it was concluded that breakdown of hydrogen peroxide proceeded in the liquid water, possibly with trace amounts of impurity transition-metal ions acting as catalysts.

Tsentsiper *et al.* [161] found the decomposition of $K_2CO_3.3.05H_2O_2.0.16H_2O$ to be more rapid in vacuum ($E_a = 136$ kJ mol^{-1}, 333 to 348 K) than in nitrogen. The rate is largely controlled by crystallite size, and the distribution of decomposition centres is related to the dislocation structure. Oxygen release proceeds through the intervention of free radicals (OH and HO_2) as shown by EPR measurements. The pyrolyses [162] of $Rb_2CO_3.3H_2O_2$ and the corresponding caesium salt were autocatalytic, 328 to 358 K, yielding water, oxygen and the carbonate.

7.15.3. Other perhydrates

Systematic investigations of the structures and stabilities of a wide variety of other solids which contain hydrogen peroxide have been reported by Bogdanov and his co-workers: compounds of group III elements [163], of sodium and potassium phosphates and arsenates [164], and of rubidium tungstates [165].

Comparative studies [166] of the decompositions of alkali (Na,K,Rb and NH_4) oxalate perhydrates ($M_2C_2O_4.H_2O_2$) showed that behaviour was closely similar to the reactivities of the related hydrates (where these exist). Several reactions showed sigmoid α-time curves and E_a values for the perhydrates (reactions commenced at 350 to 415 K) were 64 to 208 kJ mol^{-1}, whereas those of the hydrates (reacting between 315 and 355 K) were 98 to 250 kJ mol^{-1}.

7.16. CONCLUSIONS

7.16.1. Introduction

Dehydrations were amongst the earliest solid state reactions investigated in detail and interest continues. The indices of Chemical Abstracts list approximately 200 entries per year concerned with reactions of this type. Searches of this extensive literature identify no generally accepted criteria for the classification of dehydration reactions of crystalline solids. This absence of a classification framework may account for the small number of reviews of the subject [2,7-9,167]. Several individual hydrates and small groups of related reactants have been the subject of repeated and intensive research, however no general pattern of behaviour or theoretical explanation of reactivities has yet been accepted. As a contribution towards developing a theoretical structure for the subject, the scheme shown in Figure 7.4. is proposed. The scheme is based on the reported diversity of dehydration models, classified by distinctive features of the steps contributing to water expulsion. Some of the dehydroxylations described in Chapter 8 are included for completeness.

The identification of similarities and of differences of behaviour of reactant hydrates may provide insights into not only the mechanisms of dehydration, but also a wider range of crystolysis reactions. Inconsistencies of behaviour may be recognized and directions for future research identified. The progress achieved towards a general classification of solid state decompositions is discussed in Chapter 18.

7.16.2. Reversibility of dehydrations

Many (although not all) dehydrations are reversible and it is essential that rate measurements, from which mechanisms are to be formulated, should account quantitatively for any contribution from the reverse process. Identification or removal of such contributions requires carefully designed experimental techniques and precise observations [22]. The variation of kinetic results with reaction conditions has been discussed for the dehydration of $Li_2SO_4.H_2O$ [93,95] and also the approach to equilibrium conditions for the dehydrations of alums [108].

If reversibility is ignored, the significance of reported Arrhenius parameters is diminished because of a possible compensation effect dependent on reaction conditions [93,168]. Assessment of the reliability of measured Arrhenius parameters, particularly when using non-isothermal methods, requires demonstration that they are not dependent upon experimental variables, other than temperature, such as sample mass, $p(H_2O)$, etc. Evidence that a reaction is (or is not) reversible is a necessary observation in discussions of reaction mechanisms.

7.16.3. Factors controlling dehydrations

Factors which may control the reactivities of crystalline hydrates and hence determine the mechanisms of elimination of water from the solids are summarized below. Many have been mentioned in describing specific reactions above.

Reactant structure. The reactivity of water within crystalline hydrates varies widely. Types of bonding include: hydrogen bonds, coordination of H_2O ligands and ionic bonding of hydroxyl ions. Constituent water may be essential in stabilizing the packing of other crystal components, and it may be present in several different environments, particularly in higher hydrates. Release of water may proceed through several steps occurring at different temperatures.
Product structure. Dehydration may result in relatively minor changes in structure such as decreases in some or all unit cell dimensions in topotactic processes. Alternatively, reaction may result in fundamental reorganization through recrystallization, or conversion to an amorphous or zeolitic material.
Dehydration. Dehydration mechanisms encompass a wide variety of routes, summarized in Figure 7.4. The following features require comment.
For reaction types 1(a)-(d), water loss commences at the original crystal surfaces and proceeds inward towards the centres of particles.
1(a). *Topotactic reactions.* Dehydrations continue to completion without recrystallization. Extended, relatively stable structures in the reactant are retained, while less-strongly bonded (intercalated) water is released. General aspects of the preservation of structures during topotactic reactions have been discussed by Oswald [169]. Water loss may (1a(iii)) or may not (1a(ii)) be controlled by diffusion. Examples include the dehydrations of some metal carboxylates, layer structure minerals, etc.
2. *Homogeneous reaction.* The occurrence of water loss throughout the bulk of the crystal has been discussed [172]. This may be followed by recrystallization by a separate rate process (as in 1(c) and 1(d))[171].
3(a)-(c). *Nucleation and growth.* In contast with 1(c) and (d), dehydration and recrystallization occur together at, or very close to, a single advancing interface and the two steps appear to be interdependent. Interface advance may be promoted by water retention [105], seed crystals [151], or strain [92].
4. *Other reactions.* Water elimination may be accompanied, or rapidly followed, by chemical interaction of H_2O with other reactant constituents (hydrolysis) [116]. Few such reactions have been studied in detail.
5. *Melting.* Many hydrates melt, form a eutectic, or dissolve in constituent water on heating under conditions not favouring removal of water [152,153]. In some systems water may be retained temporarily under a superficial barrier layer [67].

Some dehydration mechanisms may be more complicated than described above, for example stepwise dehydration may occur in a sequence of concurrent/ consecutive rate processes. The structures and textures of the residual solids may vary with reaction conditions (availability of water vapour, rate of heating, etc.).

Figure 7.4.
Mechanisms of dehydrations of crystalline hydrates
(proposed basis for classification of these reactions)

1. Water loss from original reactant surfaces without immediate recrystallization

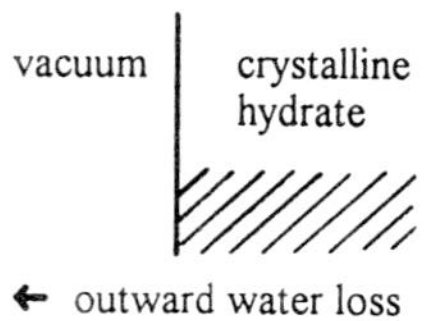

(a) **Topotactic reaction**
structural features of reactant retained as reaction proceeds to completion [169]

→ (i) water loss from original crystal surfaces: no interface developed (uranyl(VI) nitrate hexahydrate [111])
(ii) interface advance at constant rate (copper formate [146])
(iii) diminishing dehydration rate: diffusion control (vermiculite [156])

(b) **Loss of structural order**
amorphous or zeolitic product
($CaSO_4.2H_2O$ [79,80])

→ (i) product not recrystallized (metal salts of mellitic acid [170])

(c) **Progressive modification of crystal spacings**
($CuSO_4.5H_2O$ [59,93],
$Li_2SO_4.H_2O$ [92,93])

(d) **Dehydrated phase formed**
($NiSO_4.6H_2O$ [67])

→ **Later product recrystallization**
Proceeds at an advancing interface and is not directly associated with the dehydration step (water elimination and product recrystallization are independent rate processes)

2. Water loss throughout reactant: homogeneous reaction

(a) **Modified reactant structure**
($Ca(OH)_2$ [171,172])

(b) **Retained water**
(colemanite, with explosive release at 648 K [173])

Figure 7.4.(continued)
Mechanisms of dehydrations of crystalline hydrates
(proposed basis for classification of these reactions)

3. Nucleation* and growth reactions

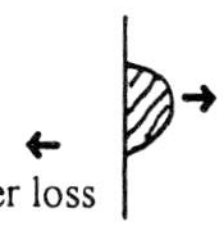

(a) **Intranuclear retained water**
(promotes recrystallization, alums [105])

(b) **Seed crystals of product**
(promotes recrystallization, *d*-$LiKC_4H_4O_6.H_2O$ [151])

(c) **Interface strain**
promotes crack propagation, ($Li_2SO_4.H_2O$ [92])

→ Water elimination and product recrystallization **linked** together in the vicinity of a single advancing reaction interface
(The mechanistic interrelationship between the movement of H_2O from its stable structural position and the recrystallization of the product phase may (in different reactions) be closely connected or quite independent and, probably, all shades of intermediate behaviour are possible)

4. Other reactions

Product reaction
On heating, water may react with other crystal constituents, e.g. hydrolysis ($MgCl_2.2H_2O$ [116])

Anhydrous product may or may not influence the rate of water loss

5. Reactant melting

(a) **Comprehensive melting**
($LiKC_4H_4O_6.H_2O$ [152,153])

(b) **Melting of outer product layer**
($NiSO_4.6H_2O$ [67])

* Nucleation may occur within a surface layer that has undergone water loss. Dehydration occurs at an advancing recrystallization interface

These residual products are the reactants in many studies of the further decomposition processes described in later Chapters.

Reactant surfaces. In many nucleation and growth dehydrations (Type 3 in Figure 7.4.) the dominant reaction occurs at an advancing interface. There is, however, evidence that for a wide range of hydrates [106] limited loss of water occurs early in reaction from all crystal surfaces [34,105]. The energy barrier for recrystallization is identified as inhibiting continued water loss at the crystal outer faces, so that water release continues preferentially (and unexpectedly) at the subsurface peripheries of advancing nuclei. At these peripheries, retained water [105] or seed crystals [151] promote the recrystallization that enables the escape of product water vapour required to maintain the advance of the interface. Nucleation occurs only infrequently within an outer zone that can be retextured on exposure to water vapour ('orange peel' texture [105]). Measurement of energy changes during the earliest stages of dehydration might provide information on the steps involved in nucleation that culminate in a growth nucleus.

Dehydration-anation reactions of coordination compounds. House [174] has proposed a general mechanism for the solid state dehydrations and anations of coordination compounds. The generation of point defects, analogous to Schottky or Frenkel defects (Chapter 1), is considered as the transition state. Reactions with low E_a values may occur by an S_N1 dissociative mechanism and reactivity is influenced by the availability of free space in the crystal structure. Corbella and Ribas [175] found support for such mechanisms [174,176].

7.16.4. Dehydration mechanisms

Theories of solid state dehydrations are currently undergoing reappraisal. Recent evidence shows that some mechanisms are more complicated than was previously recognized. The simple reaction model, assumed in applications of the Polanyi-Wigner equation, is no longer so widely accepted. For many reactions of interest there is insufficient information available concerning the rate controlling step to warrant attempts to apply transition state theory. X-ray diffraction, particularly with the high power and short exposures from synchrotron sources [93], can provide invaluable information on the structures of reactants, intermediates and products, together with any structural relationships (topotaxy). Microscopy can reveal the distribution, shapes and textures of reactant/product interfaces. Effects of reactant pretreatments such as crushing, irradiation, the presence of additives (including product) etc., can provide additional mechanistic information. Comparisons of dehydrations with analogous rate processes involving ligands other than water is an area of considerable potential interest.

The use of isotopes has found surprisingly few applications in the determination of mechanisms of dehydrations. Below the temperature of appreciable water evolution, D_2O/H_2O exchange could, in principal be used to measure, in partially decomposed salt, the quantities of water held at surface sites of different reactivities, i.e. water physically adsorbed, chemisorbed, within amorphous phases and at the reaction interface. Similarly, kinetic isotope effects in dehydration reactions have not been extensively investigated. If E_a corresponds to the enthalpy of dissociation, the substitution of a deuterium bond for a hydrogen bond could result in a change of the magnitude of E_a.

More quantitative measurements of the systematic variations of dehydration rates with $p(H_2O)$, referred to as Smith-Topley behaviour, could lead to support for one or more of the several theoretical explanations that have been proposed [2,21,49,54,63] based on recrystallization of solid product, local self-cooling and/or diffusion (effects expected to occur in all dehydration reactions) and adsorption of the volatile product. Dehydrations may also involve the intervention of a zeolitic residue and/or an amorphous phase, the formation and recrystallization of one or more lower hydrates as intermediates, and diffusive escape of water through various channels of the barrier layer of product may be slow.

In a limited number of solids, water can be released internally into a structure that is maintained and also prevents escape. Under such conditions a temperature may be achieved at which there is a violent outburst of water (explosive dehydration [173]).

Studies of dehydrations continue to contribute much to the theory of solid state reactions.

REFERENCES

1. R.E. Grim, Clay Mineralogy, 2nd Edn, McGraw Hill, New York, 1968.
2. W.E. Garner, Chemistry of the Solid State, (Ed. W.E. Garner), Butterworths, London, 1955, Chap.8.
3. D.M. Adams, Inorganic Solids, Wiley, London, 1974, Chap.6.
4. A.V. Karyakin and G.A. Muradova, Zh. Fiz. Khim., 42 (1968) 2735.
5. A.F. Wells, Structural Inorganic Chemistry, 3rd Edn, Clarendon, Oxford, 1962, p.575 ff; 4th Edn, 1975, p.548ff.
6. R.W.G. Wyckoff, Crystal Structures, 2nd Edn, Vol. 3, Interscience, New York, 1965, p.529.
7. N.Z. Lyakhov and V.V. Boldyrev, Russ. Chem. Rev., 41 (1972) 919.
8. A.K. Galwey and G.M. Laverty, J. Chim. Phys., 87 (1990) 1207.

9. A.K. Galwey, J. Thermal Anal., 38 (1992) 99.
10. M.G. Burnett, A.K. Galwey and C. Lawther, J.Chem. Soc., Faraday Trans., 92 (1996) 4301.
11. J. Hume and J. Colvin, Proc. R. Soc. (London), A125 (1929) 635.
12. N.F.H. Bright and W.E. Garner, J. Chem. Soc., (1934) 1872.
13. W.E. Garner and W.R. Southon, J. Chem. Soc., (1935) 1705.
14. D. Dollimore and D.L. Griffiths, J. Thermal Anal., 2 (1970) 229.
15. P. Barret and R. Thiard, C. R. Acad. Sci. Paris, 260 (1965) 2823.
16. A. Roux and G. Watelle-Marion, C. R. Acad. Sci. Paris, C, 264 (1967) 1466.
17. G. Watelle-Marion and R. Thiard, C. R. Acad. Sci. Paris, C, 261 (1965) 4105.
18. R.M. Dell and V.J. Wheeler, Reactivity of Solids, (Ed. G.M. Schwab), Elsevier, Amsterdam, 1965, p.395.
19. J. Rouquerol, J. Thermal Anal., 5 (1973) 203.
20. J. Rouquerol, Bull. Soc. Chim. Fr., (1964) 31; J. Thermal Anal., 2 (1970) 123; Thermochim. Acta, 144 (1989) 209.
21. G. Bertrand, M. Lallemant and G. Watelle-Marion, J. Inorg. Nucl. Chem., 36 (1974) 1303 and 40 (1978) 819 (with A. Mokhlisse).
22. T.B. Flanagan, J.W. Simons and P.M. Fichte, Chem. Commun., (1971) 370.
23. R. Hocart, N. Gérard and G. Watelle-Marion, C. R. Acad. Sci. Paris, 258 (1964) 3709.
24. N. Gérard, G. Watelle-Marion and A. Thrierr-Sorel, Bull. Soc. Chim. Fr., (1968) 4367.
25. N. Gérard, G. Watelle-Marion and A. Thrierr-Sorel, Bull. Soc. Chim. Fr., (1967) 1788.
26. P. Barret, N. Gérard and G. Watelle-Marion, Bull. Soc. Chim. Fr., (1968) 3172.
27. N. Gérard, M. Lallemant and G. Watelle-Marion, C. R. Acad. Sci. Paris, C, 267 (1968) 1679.
28. N. Gérard, Bull. Soc. Chim. Fr., (1970) 103.
29. B. Topley, Proc. R. Soc. (London), A136 (1932) 413.
30. G. Watelle-Marion, M. Lallemant and G. Bertrand, Reactivity of Solids, (Eds J.S. Anderson, M.W. Roberts and F.S. Stone), Chapman and Hall, London, 1972, p.772.
31. P.W.M. Jacobs and F.C. Tompkins, Chemistry of the Solid State, (Ed. W.E. Garner), Butterworths, London, 1955, Chap. 7.
32. W.J. Dunning, Mater. Sci. Res., 4 (1969) 132.

33. J-C. Niepce and G. Watelle-Marion, C. R. Acad. Sci. Paris, C, 276 (1973) 627.
34. M.E. Brown, A.K. Galwey and G.G.T. Guarini, J. Thermal Anal., 49 (1997) 1135.
35. M. Polanyi and E. Wigner, Z. Phys. Chem., 139 (1928) 439.
36. R.S. Bradley, Phil. Mag., 7/12 (1931) 290.
37. M.M.T. Anous, R.S. Bradley and J. Colvin, J. Chem. Soc., (1951) 3348.
38. G.P. Acock, W.E. Garner, J. Milsted and H.J. Willavoys, Proc. R. Soc. (London), A189 (1947) 508.
39. B. Topley and J. Hume, Proc. R. Soc. (London), A120 (1928) 211.
40. J.A. Cooper and W.E. Garner, Proc. R. Soc.(London), A174 (1940) 487.
41. R.S. Bradley, J. Colvin and J. Hume, Proc. R. Soc. (London), A137 (1932) 531.
42. L.S. Kassel, J. Amer. Chem. Soc., 51 (1929) 1136.
43. A.K. Galwey, Thermochim. Acta, 242 (1994) 259.
44. T.A. Clarke and J.M. Thomas, J. Chem. Soc. A, (1969) 2227, 2230.
45. R.C. Eckhardt and T.B. Flanagan, Trans. Faraday Soc., 60 (1964) 1289.
46. R.D. Shannon, Trans.Faraday Soc., 60 (1964) 1902.
47. A.K. Galwey, Adv. Catal., 26 (1977) 247.
48. B. Topley and M.L. Smith, J. Chem. Soc., (1935) 321.
49. M. Volmer and G. Seydel, Z. Phys. Chem., 179 (1937) 153.
50. M.L. Smith and B. Topley, Proc. R. Soc. (London), A134 (1932) 224.
51. G.B. Frost and R.A. Campbell, Canad. J. Chem., 31 (1953) 107.
52. W.E. Garner and M.G. Tanner, J. Chem. Soc., (1930) 47.
53. N.Z. Lyakhov, È.M. Frid and V.V. Boldyrev, Kinet. Katal., 15 (1974) 803.
54. G. Thomas, J-J. Gardet, J-J. Gruffat, B. Guilhot and M. Soustelle, J. Chim. Phys., 69 (1972) 375.
55. W.E. Garner and H.V. Pike, J. Chem. Soc., (1937) 1565.
56. J. Colvin and J. Hume, Trans. Faraday Soc., 34 (1938) 969.
57. J. Guenot and J-M. Manoli, Bull. Soc. Chim. Fr., (1969) 2663; J. Guenot, J-M. Manoli and J-M. Bregeault, Bull. Soc. Chim. Fr., (1969) 2666.
58. M.M. Cooper, J. Colvin and J. Hume, Trans. Faraday Soc., 29 (1933) 576.
59. G.B. Frost, K.A. Moon and E.H. Tompkins, Canad. J. Chem., 29 (1951) 604.
60. W.E. Garner and T.J. Jennings, Proc. R. Soc. (London), A224 (1954) 460.
61. M. Lallemant and G. Watelle-Marion, C. R. Acad. Sci. Paris, C, 267 (1968) 1775.

62. M. Lallemant, N. Gérard, J. Gaffodio and G. Watelle-Marion, C. R. Acad. Sci. Paris, C, 272 (1971) 1737.
63. N.Z. Lyakhov, V.V. Boldyrev and V.P. Isupov, Kinet. Katal., 15 (1974) 1224.
64. M.C. Ball, Thermochim. Acta, 257 (1995) 139.
65. J.M. Thomas and G.D. Renshaw, J. Chem. Soc. A, (1969) 2749, 2753, 2756.
66. J.M. Thomas, E.L. Evans and T.A. Clarke, J. Chem. Soc. A, (1971) 2338.
67. G.G.T. Guarini and M. Rustici, React. Solids, 2 (1987) 381; J. Thermal Anal., 41 (1994) 287.
68. N. Koga and H. Tanaka, J. Phys. Chem., 98 (1994) 10521.
69. M. Lallemant and G. Watelle-Marion, C. R. Acad. Sci. Paris, C, 264 (1967) 2031; 265 (1967) 627; 272 (1971) 642.
70. R.W. Ford and G.B. Frost, Canad. J. Chem., 34 (1956) 591.
71. R.C. Wheeler and G.B. Frost, Canad. J. Chem., 33 (1955) 546.
72. V.V. Boldyrev, Bull. Soc. Chim. Fr., (1969) 1054.
73. V.V. Boldyrev, A.V. Safiulina, A.V. Boldyreva and B.I. Treskova, Zh. Fiz. Khim., 34 (1960) 2184.
74. Yu.P. Savintsev and V.V. Boldyrev, Kinet. Katal., 10 (1969) 538.
75. M.C. Ball and L.S. Norwood, J. Chem. Soc. A, (1969) 1633; (1970) 1476.
76. M.C. Ball and R.G. Urie, J. Chem. Soc. A, (1970) 528.
77. M.C. Ball and L.S. Norwood, J. Chem. Soc. A, (1969) 1633; (1970) 1476; Reactivity of Solids, (Eds J.S. Anderson, M.W. Roberts and F.S. Stone), Chapman and Hall, London, 1972, p.717.
78. M.J. Ridge and H. Surkevicius, Austr. J. Chem., 17 (1964) 1197.
79. H.G. McAdie, Canad. J. Chem., 42 (1964) 792.
80. B. Molony, J. Beretka and M.J. Ridge, Austr. J. Chem., 24 (1971) 449.
81. M. Goto, B. Molony and M.J. Ridge, J. Chem. Soc. A, (1966) 9.
82. V.B.Okhotnikov, S.E. Petrov, B.I. Yakobson and N.Z. Lyakhov, React. Solids, 2 (1987) 359.
83. J.J. Beaudoin and R.F. Feldman, J. Chem. Soc., Faraday Trans. I, 79 (1983) 2071.
84. J.H. Taplin, Nature (London), 205 (1965) 864; J. App. Chem. Biotechnol., 23 (1973) 349.
85. M.C. Ball, Thermochim.Acta, 24 (1978) 190.
86. M.E. Brown, R.M. Flynn and J.H. Flynn, Thermochim. Acta, 256 (1995) 477.
87. J. Huang and P.K. Gallagher, Thermochim. Acta, 192 (1991) 35.

88. Y.A. Gaponov, B.I. Kidyarov, N.A. Kirdyashkina, N.Z. Lyakhov and V.B. Okhotnikov, J. Thermal Anal., 33 (1988) 547; React. Kinet. Catal. Lett., 23 (1983) 125; 33 (1988) 547; 36 (1988) 417; 39 (1989) 345.
89. Y. Masuda, H. Takeuchi and A. Yahata, Thermochim. Acta, 228 (1993) 191.
90. N. Koga and H. Tanaka, J. Phys. Chem., 93 (1989) 7793; Thermochim. Acta, 185 (1991) 135; 224 (1993) 141; J. Thermal Anal., 40 (1993) 1173; H. Tanaka and N. Koga, J. Thermal Anal., 36 (1990) 2601.
91. A.K. Galwey, N. Koga and H. Tanaka, J. Chem. Soc., Faraday Trans. I, 86 (1990) 531.
92. V.V. Boldyrev, Y.A. Gaponov, N.Z. Lyakhov, A.A. Politov, B.P. Tolochko, T.P. Shakhtshneider and M.A. Sheromov, Nucl. Inst. Methods Phys. Res., A261 (1987) 192.
93. M.E. Brown, A.K. Galwey and A. Li Wan Po, Thermochim. Acta, 203 (1992) 221; 220 (1993) 131.
94. N.A. Simakova, N.Z. Lyakhov and N.A. Rudina, Thermochim. Acta, 256 (1995) 381.
95. B. L'vov, Thermochim. Acta, 315 (1998) 145.
96. V.V. Boldyrev and I.V. Schmidt, Kinet. Katal., 1 (1960) 537.
97. K. Skeff Neto and V.K. Garg, J. Inorg. Nucl. Chem., 37 (1975) 2287.
98. A. Bristoti, J.I. Kunrath, P.J. Viccaro and L. Bergter, J. Inorg. Nucl. Chem., 37 (1975) 1149.
99. A.H. Kamel, Z. Sawires, H. Khalifa, S.A. Saleh and A.M. Abdallah, J. Appl. Chem. Biotechnol., 22 (1972) 591, 599.
100. T.P. Prasad, J. Thermal Anal., 35 (1989) 2539.
101. P.K. Gallagher, D.W. Johnson and F. Schrey, J. Amer. Ceram. Soc., 53 (1970) 666.
102. J.R.I. Hepburn and R.F. Phillips, J. Chem. Soc., (1952) 2569, 2578.
103. V.V. Boldyrev and Yu.A. Zakharov, Zh. Fiz. Khim., 34 (1960) 446.
104. A. Bielanski and F.C. Tompkins, Trans. Faraday Soc., 46 (1950) 1072.
105. A.K. Galwey, R. Spinicci and G.G.T. Guarini, Proc. R. Soc. (London), A378 (1981) 477.
106. G.G.T. Guarini and L. Dei, J. Chem. Soc., Faraday Trans. I, 79 (1983) 1599.
107. A.K. Galwey and M.A. Mohamed, Thermochim. Acta, 121 (1987) 97.
108. A.K. Galwey and G.G.T. Guarini, Proc. R. Soc. (London), A441 (1993) 313.

109. M.E. Brady, M.G. Burnett and A.K. Galwey, J. Chem. Soc., Faraday Trans. I, 86 (1990) 1573.
110. M.C. Ball, C.M. Snelling and A.N. Strachan, J. Chem. Soc., Faraday Trans. I, 81 (1985) 1761.
111. M.L. Franklin and T.B. Flanagan, J. Phys. Chem., 75 (1971) 1272; J. Chem. Soc., Dalton Trans., (1972) 192.
112. P.T. Moseley and C.J. Seabrook, J. Chem. Soc., Dalton Trans., (1973) 1115.
113. W.H. Smith, J. Inorg. Nucl. Chem., 27 (1965) 1761.
114. G.G.T. Guarini and R. Spinicci, J. Thermal Anal., 4 (1972) 435.
115. J.A. Lumpkin and D.D. Perlmutter, Thermochim. Acta, 249 (1995) 335; 202 (1992) 151.
116. A.K.Galwey and G.M. Laverty, Thermochim. Acta, 138 (1989) 115.
117. M.C. Ball, Thermochim. Acta, 21 (1977) 349.
118. M.A. Mohamed and S.A. Halawy, J. Thermal Anal., 41 (1994) 147.
119. H. Tanaka and N.Koga, J. Phys. Chem., 92 (1988) 7023; Thermochim. Acta, 163 (1990) 295.
120. M.C. Ball and M.J. Casson, J. Chem. Soc., Dalton Trans., (1973) 34.
121. M.A.G. Aranda and S. Bruque, Inorg. Chem., 29 (1990) 1334.
122. G. Cao and T.E. Mallouk, Inorg. Chem., 30 (1991) 1434.
123. I. Waclawska, J. Thermal Anal., 43 (1995) 261.
124. G. Thomas and M. Soustelle, J. Chim. Phys., 69 (1972) 1154.
125. T.N. Pradhananga and S. Matsuo, J. Phys. Chem., 89 (1985) 72.
126. L. Stoch and I. Waclawska, Thermochim. Acta, 215 (1993) 255, 265, 273.
127. M.S. Celik and F. Suner, Thermochim. Acta, 245 (1995) 167.
128. G.G.T. Guarini and S. Piccini, J. Chem. Soc., Faraday Trans. I, 84 (1988) 331.
129. G.G.T. Guarini and L. Dei, Thermochim. Acta, 250 (1995) 85.
130. H.R. Oswald, J.R. Günter and E. Dübler, J. Solid State Chem., 13 (1975) 330.
131. R. Deyrieux, C. Berro and A. Péneloux, Bull. Soc. Chim. Fr., (1973) 25.
132. A.V. Shkarin, N.D. Topor and G.M. Zhabrova, Zh. Fiz. Khim., 42 (1968) 2832.
133. T.B. Flanagan and C.H. Kim, J. Phys. Chem., 66 (1962) 926.
134. T.B. Flanagan and M.K. Goldstein, J. Phys. Chem., 68 (1964) 663.
135. D. Dollimore, J. Dollimore and J. Little, J. Chem. Soc. A, (1969) 2946.

136. A.K. Galwey, D.M. Jamieson, M.E. Brown and M.J. McGinn, Reaction Kinetics in Hetergeneous Chemical Systems, (Ed. P. Barret), Elsevier, Amsterdam, 1975, p.520.
137. M. Le Van, C. R. Acad. Sci. Paris, C, 268 (1969) 684.
138. Y.Masuda, K. Iwata, R. Ito and Y. Ito, J. Phys. Chem., 91 (1987) 6543.
139. N. Gérard and G. Watelle-Marion, Bull. Soc. Chim. Fr., (1969) 58.
140. J-C. Mutin, A. Thrierr-Sorel and G. Watelle-Marion, C. R. Acad. Sci. Paris, C, 268 (1969) 483.
141. J-C. Mutin, G. Watelle-Marion, Y. Dusausoy and J. Protas, Bull. Soc. Chim. Fr., (1972) 4498.
142. J.E. House and R.P. Ralston, Thermochim. Acta, 214 (1993) 255.
143. N.I. Ostannii, L.A. Zharkova and B.V. Erofeev, Zh. Fiz. Khim., 44 (1970) 1427.
144. R.L. Martin and H. Waterman, J. Chem. Soc., (1959) 1359.
145. J.R. Gűnter, J. Solid State Chem., 35 (1980) 43.
146. P.M. Fichte and T.B. Flanagan, Trans. Faraday Soc., 67 (1971) 1467.
147. R.C. Eckhardt and T.B. Flanagan, Trans. Faraday Soc., 60 (1964) 1289.
148. T.A. Clarke and J.M. Thomas, J. Chem. Soc. A, (1969) 2227, 2230.
149. Y. Masuda and Y.Ito, J. Thermal Anal., 38 (1992) 1793.
150. N.A. Baranov, V.B.Okhotnikov, L.I. Rynskaya, A.R. Semenov, A.K. Galwey and G.M. Laverty, Solid State Ionics, 43 (1990) 37.
151. A.K. Galwey, G.M. Laverty, N.A. Baranov and V.B. Okhotnikov, Phil. Trans. R. Soc. (London), A347 (1994) 139, 157.
152. S.D. Bhattamisra, G.M. Laverty, N.A. Baranov, V.B. Okhotnikov and A.K. Galwey, Phil. Trans. R. Soc. (London), A341 (1992) 479.
153. A.K. Galwey, G.M. Laverty, V.B. Okhotnikov and J. O'Neill, J. Thermal Anal., 38 (1992) 421.
154. F.C. Tompkins and D.A. Young, Trans. Faraday Soc., 52 (1956) 1245; J. Chem. Soc., (1957) 4281.
155. T.B. Flanagan, Trans. Faraday Soc., 55 (1959) 114; Canad. J. Chem., 44 (1966) 2941.
156. V.B.Okhotnikov, I.P. Babicheva, A.V. Musicantov and T.N. Aleksandrova, React. Solids, 7 (1989) 273.
157. M.E. Brady, M.G. Burnett and A.K. Galwey, J. Chem. Soc., Faraday Trans. I, 86 (1990) 1573; Fuel, 73 (1994) 1343.
158. T.P. Firsova, V.I. Sokol, V.M. Bakulina and N.N. Stasevich, Izv. Akad. Nauk, SSSR, Ser. Khim., 9 (1968) 1941.
159. T.P. Firsova and E. Ya. Filatov, Zh. Fiz. Khim., 42 (1968) 1812.

160. A.K. Galwey and W.J. Hood, J. Phys. Chem., 83 (1979) 1810; J. Chem. Soc., Faraday Trans. I, 78 (1982) 2815.
161. B.V. Erofeev, A.B. Tsentsiper, G.G. Dedova and I.I. Gorelkin, Vestsi Akad. Navuk BSSR, Ser. Khim. Navuk., 5 (1974) 14; 6 (1974) 5.
162. B.S. Dzyatkevich and A.B. Tsentsiper, Izv. Akad. Nauk. SSSR, Ser. Khim., 7 (1968) 1623; 2 (1967) 235.
163. G.A. Bogdanov, T.L. Garkushenko and D.P. Balyasova, Zh. Fiz. Khim., 47 (1973) 1395.
164. G.A. Bogdanov, Z.N. Dymova and T.P. Sidorova, Zh. Fiz. Khim., 48 (1974) 249.
165. L.I. Kozlova, N.A. Korotchenko and G.A. Bogdanov, Zh. Fiz. Khim., 49 (1975) 1267.
166. J.M. Adams, V. Ramdas, G.G.T. Guarini and C.J. Adams, J. Chem. Soc., Dalton Trans., (1980) 269.
167. M.E. Brown, D. Dollimore and A.K. Galwey, Reactions in the Solid State, Comprehensive Chemical Kinetics, (Eds C.H. Bamford and C.F.H. Tipper), Vol.22, Elsevier, Amsterdam, 1980.
168. A.K. Galwey and M.E. Brown, Thermochim. Acta, 300 (1997) 107.
169. H.R. Oswald, Proc. 6th Int. Conf. on Thermal Analysis, (Ed. W. Hemminger), Birkhaüser, Basel, 1 (1981) 1.
170. A.K. Galwey, J. Chem. Soc. A, (1965) 5433; (1966) 87.
171. A.K. Galwey and G.M. Laverty, Thermochim. Acta, 228 (1993) 359.
172. N.H. Brett, K.J.D. MacKenzie and J.H. Sharp, Q. Rev. Chem. Soc., 24 (1970) 185.
173. L. Stoch, J.Thermal Anal., 37 (1991) 1415.
174. J.E. House, Thermochim. Acta, 38 (1980) 59.
175. M. Corbella and J. Ribas, Inorg. Chem., 26 (1987) 2589.
176. H.E. Le May and M.W. Babich, Thermochim. Acta, 48 (1981) 147.

Chapter 8

THERMAL DEHYDRATION OF HYDROXIDES

8.1. INTRODUCTION

At first sight, the dehydrations of hydroxides are attractive potential reactant systems for fundamental mechanistic studies. Many hydroxides and their expected products (usually oxides) possess relatively simple, well-characterized structures. Reaction should involve proton transfer, but is not necessarily accompanied by a valence change of any cation present. Although studies of the dehydrations of hydroxides have provided much information concerning the rates of water elimination and the accompanying crystallographic changes, problems have arisen in determining detailed chemical mechanisms. The reversibility of the dehydration step, as in so many solid state reactions, introduces uncertainty into the reliability of kinetic measurements, because of the unknown availability of water retained within and between reactant and product crystals. If reversible proton transfer is possible within the reactant structure, the formation of water molecules may be transitory. This possibility has led to the proposal that dehydrations of hydroxides may proceed by homogeneous mechanisms [1] (discussed below). Unfortunately kinetic data are incapable of distinguishing such behaviour from interface reactions and the mechanisms remain uncertain.

Another complication in studies of the dehydrations of hydroxides is the number of different reactant and oxide phases that may exist, making recognition of the significant phase transformations difficult. For example, there are several different crystalline forms of aluminium hydroxide and of aluminium oxide, and similar problems are found with iron and some other transition metal hydroxides. Examples of the diversity of behaviour reported include a topotactic process for dehydration of magnesium hydroxide, accompanied by crystal disintegration due to the large volume change, which contrasts with the readily reversible dehydration of calcium hydroxide. The rate of water elimination from extended anion hydroxysilicates (e.g., clays), based on strong -O-Si-O- linkages, is controlled by diffusive escape of product water along linear or planar channels in the crystal.

8.2. GROUP IIA METAL HYDROXIDES

8.2.1. Magnesium hydroxide

The dehydration of $Mg(OH)_2$ is particularly suitable for experimental investigation because of the existence of only two well-crystallized phases, $Mg(OH)_2$ and MgO. The reaction mechanism has been studied in greater detail than has been found practicable for decomposition of most other hydroxides through a combination of complementary experimental approaches (kinetics, crystallography, microscopy and others).

Gordon and Kingery [2], from a kinetic and crystallographic study, concluded that the decomposition of brucite ($Mg(OH)_2$) is a nucleation and growth process, in which the product phase (MgO) grows in coherent contact with the reactant. The strain caused by the disparity of molar volumes between these phases results in extensive cracking of the crystallites. The dehydration of small crystals of reactant fitted the first-order rate equation ($E_a = 125$ kJ mol^{-1}), but the disintegration of larger particles resulted in more complex behaviour. Physical factors such as cracking must be expected to influence kinetic characteristics, so that this reaction may not be strictly represented by either an interface model or a simple kinetic equation.

Brett *et al.* [1], in the context of a more general review of dehydroxylation, discussed the chemical steps at the reaction interface which culminate in water elimination. The reaction involves the reorganization of the hexagonal close-packed lattice of anions in $Mg(OH)_2$ to a cubic close-packed structure in MgO, with the preservation of a high degree of topotaxy. The orientation relationship between the two phases is an important factor in formulating a reaction mechanism. The initially-produced MgO has a strongly deformed structure in which the (111) dimension is identical with the (0001) spacing in $Mg(OH)_2$.

Alternative reaction models were proposed [1]. The *homogeneous* mechanism suggests that throughout the crystal, hydroxyl groups in adjacent layers interact to yield water, which is subsequently lost by diffusion through the crystal. The *inhomogeneous* model postulates the development of 'donor' and 'acceptor' regions, referred to the migration of Mg^{2+}, so that the acceptor regions become MgO, while within the donor regions water is produced. The inhomogeneous model is consistent in many respects with the nucleation and growth process described by Gordon and Kingery [2], referred to above, but interface advance is influenced by cracking from induced strain.

Freund *et al.* [3,4] have discussed the chemical changes which occur in the reaction zone. The transfer [3] of a proton from OH^- to a neighbour by a tunnelling process is only possible if there is a vacant acceptor level of equivalent energy available; a condition which is not fulfilled when both OH^- groups are crystallographically equivalent. Coupling with opposite phase O-H bending leads to a broadening of levels and overlap which permits tunnelling. This process is enhanced at the surface and by the presence of an electric field [5]. The advance of the reaction interface [4] into the crystal bulk does not proceed at constant rate, but is intermittent. Kinetic data for small samples indicate a variable rate of product formation, which is ascribed to the generation of product in a state of increasing strain. The product also acts as a diffusion barrier. When a critical thickness has been reached, there is spontaneous reorganization of the dehydrated material, a new surface of the reactant is exposed, and the constraints on decomposition are temporarily relaxed.

The kinetics of dehydroxylation of $Mg(OH)_2$ are sensitive to the prevailing pressure of water vapour, a factor which lead Sharp [6] to conclude that the value of E_a should only be determined under conditions of controlled atmosphere. From consideration of the ranges of values of E_a which have been reported, 67 to 400 and 80 to125 kJ mol^{-1} from dynamic and from isothermal methods, respectively, it is concluded that more sophisticated analyses of the programmed temperature data are required. The most reliable value of E_a is identified [6] as 84 kJ mol^{-1}, comparable with, or slightly greater than, the reaction enthalpy.

Stable ions, including silicate, sulfate, aluminate and phosphate ions, hinder the dehydration of $Mg(OH)_2$. The influence depends on the particular ion present, its concentration and the crystallographic plane in which it is located. Both the induction and growth processes of the anisotropic water-elimination reaction are affected.

The apparent equilibrium pressures measured [7] for the reaction:

$$Mg(OH)_2 \rightleftarrows MgO + H_2O$$

were only x 10^{-4} the expected values when measured by the Knudsen effusion method. It was shown that the heat of solution of MgO in HCℓ increased with area, excess heats of formation of MgO with areas above 300 m^2g^{-1} were 73 kJ mol^{-1}. The MgO activities for these excess heats were sufficient to account for the low $p(H_2O)$ equilibrium values.

^{35}Mg NMR studies [8] of the decomposition of $Mg(OH)_2$ suggest that MgO particles grow coherently as small domains within the reactant brucite lattice. Later these coalesce to form crystallites that can be detected by X-ray diffraction.

8.2.2. Calcium hydroxide

The structure of $Ca(OH)_2$ is similar to that of brucite, although hydrogen-bonding is absent. Again reaction involves no intermediate crystalline phase. Brett [9], from X-ray studies on single crystals, identified some retention of orientation during dehydroxylation in vacuum (above 500 K at 10^{-6} Torr). The absence of topotaxy for the reaction in air, at the significantly higher temperature of 720 K, is attributed to the high nucleation rate. Decomposition commenced at crystal edges and defects and advanced inwards towards the centre. This is consistent with the requirements of the homogeneous model mentioned above for $Mg(OH)_2$. Brett *et al.* [1] cite the following as possible reasons for the reduced tendency towards topotaxy during the dehydration of $Ca(OH)_2$ (as compared with $Mg(OH)_2$): the lower reaction temperature, the disruptive effect of the larger Ca^{2+}ion, and the rapid nucleation of $Ca(OH)_2$ in air (which enhances the probability of lattice mismatch). Mikhail *et al.* [10] also conclude that the dehydroxylation of $Ca(OH)_2$ is an interface process, acceleratory for $0.1 < \alpha < 0.25$, though in the later stages, α about 0.8, the rate of growth diminishes. Because the value of E_a (62 kJ mol^{-1}) is signficantly lower than the enthalpy of reaction (105 kJ mol^{-1}), the overall chemical change must occur by more than a single step.

Changes in the $Ca(OH)_2$ structure that occur in the temperature interval below the onset of dehydration have been characterised and discussed by Chaix-Pluchery *et al.* [11]. Some 50 K below the temperature of onset of dehydroxylation there is a sudden sharpening of the (001) diffraction reflection. The formation of H_2O following proton transfer results in the later aggregation, at a critical temperature, of the residual material into domains. The coalescence of domains is accompanied by the appearance of diffusion pathways for water escape. These pre-reactional transformations for $Ca(OH)_2$ were represented [12] as a pressure-temperature phase diagram for the system: $Ca(OH)_2$-CaO-H_2O. The second-order transitions within the crystal are identified as necessary precursor changes to the onset of water elimination. The possibility that there may be progressive transformation from the stable crystal to a pre-reactional state in other solids is also discussed.

Galwey and Laverty [13] concluded, from complementary kinetic investigations, that dehydration of $Ca(OH)_2$ was satisfactorily represented by the first-order equation. The overall rate of H_2O release was sensitive to the locally prevailing water vapour pressure and its distribution within the reactant mass. Kinetic characteristics were strongly influenced by the rates of intracrystalline and intercrystalline diffusive escape of product water which strongly depended on reactant compaction/dispersal within the heated zone. Measured values of A and E_a for $Ca(OH)_2$ dehydration could not be identified with a single reaction step because the overall change is complex, involving proton transfer, intracrystalline diffusion and possibly a contribution from the reverse process. The data obtained exhibited a compensation effect for which the isokinetic temperature (683 K) was close to the value at which the onset of water release became significant.

Water loss is identified [13] as being diffusion controlled across a dehydration layer of significant thickness that advances into the reactant particles (see Figure 8.1.). The $Ca(OH)_2$ structure is maintained during the greater part of reaction [14]. The rate of this reaction is more deceleratory than the requirements of the contracting volume equation and data satisfactorily fit the first-order equation. The anhydrous residue later recrystallises to CaO but the principal water elimination zone and the phase transformation interface are separated and may advance independently.

8.2.3. Strontium and barium hydroxides

$Sr(OH)_2.8H_2O$ loses [15] water above 480 K to yield the hexahydrate and monohydrate. Above 800 K the hydroxide reacts in three stages to form SrO at 970 K. The equilibria between the hydrates of the oxides and the hydroxides of barium and strontium have been studied in detail by Tamaru and Shiomi [16].

8.2.4. Beryllium hydroxide

Particular interest [17] has been directed towards the pore structure of BeO formed from the decomposition of β-$Be(OH)_2$. The channels present in the residue are of molecular dimensions, though some increase in width is apparent after reaction in water vapour. On heating above 770 K the narrow pores were eliminated, a process which was facilitated by water vapour.

8.2.5. General

The sintering [15] of calcium oxide produced by decomposition of $Ca(OH)_2$ was rapid only after formation of the true oxide from the initially-generated pseudohydroxide. The unexpectedly low rate of sintering of SrO and BaO at 1270 K (oxides again produced from the hydroxides) was ascribed to the influences of CaO and SrO impurities, respectively.

8.3. ALUMINIUM HYDROXIDES

Several authors [1, 18-20] have surveyed the chemical and crystallographic changes which occur during the dehydroxylation of the three forms of aluminium hydroxide, ($Al(OH)_3$: gibbsite (γ), bayerite (α) and norstrandite) and the two forms of aluminium oxyhydroxide, (AlOOH: diaspore (α) and boehmite (γ)). These phases

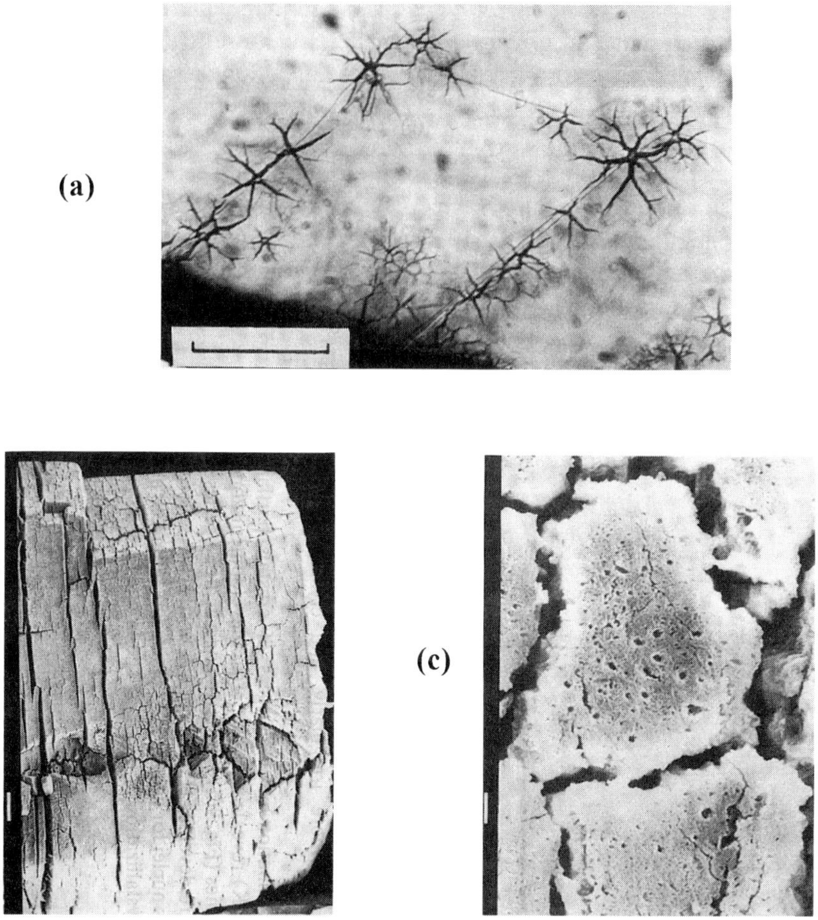

Figure 8.1.
Micrographs of the surfaces of single crystals of $Ca(OH)_2$ [13].
(a) Optical micrograph of crack features developed in the [001] face of a partly reacted (α = 0.2 at 673 K) crystal (scale bar = 70 μm). (b) and (c) Scanning electron micrographs of a completely dehydrated crystal (α = 1.00 at 668 K) (scale bars, 100 μm and 1.0 μm, respectively). The largest cracks are in the [001] direction, but particles are extensively cracked and penetrated by pores.

are alternative lattice forms which approximate to double layers of close-packed anions, between which the cations occupy octahedral sites. There may also be significant hydrogen-bonding. The existence of several polymorphs of the product of dehydration, alumina, Al_2O_3 (α (corundum), γ, δ, κ, θ and χ) caused initial difficulties in recognizing the series of changes which occurred during loss of water and subsequent ignition. These apparent inconsistencies have now been largely resolved and a wide measure of agreement has been reached on the qualitative lattice reorganizations, through the use of well-defined preparations of reactants and reliable X-ray diffraction data for all the phases concerned. Less information is available concerning the kinetics of these changes, which may be markedly influenced by the prevailing pressure of water vapour in the vicinity of the reactant, and by the particle sizes. The qualitative changes which occur on heating the several aluminium hydroxides, oxyhydroxides and oxides are summarized in Figure 8.1. [1].

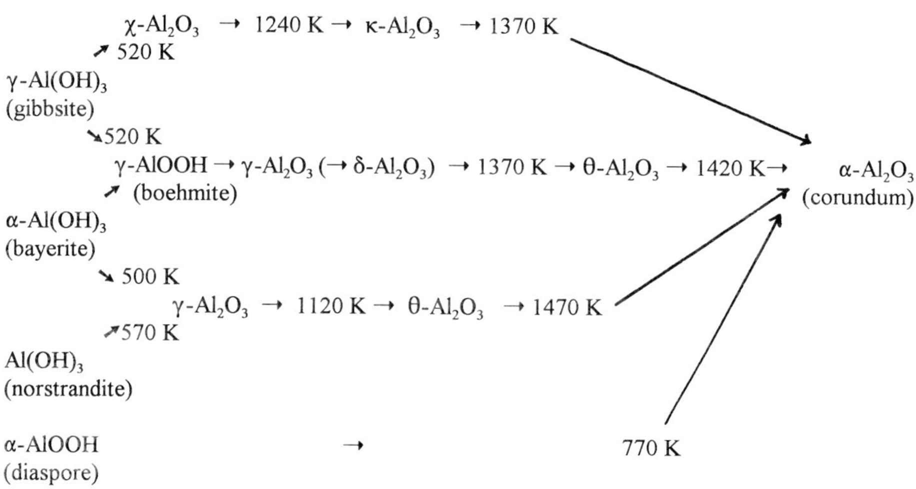

Figure 8.2.
Dehydration of aluminium hydroxides

Reaction temperatures, which depend on pressures of gases present, including water, vary from one literature report to another and some points of difference are apparent from comparison of the different sources.

Unlike $Mg(OH)_2$, the rate of dehydration of $Al(OH)_3$ (gibbsite) is slightly decreased [5] by an electric field, although it has been suggested that the dehydrations of both solids proceed through a proton-transfer mechanism. The difference is ascribed to greater affinity of the cation for a proton in $Al(OH)_3$ at low temperature, and lack of mobility of the oxide ion at the higher temperature. Drobot and Khazanov [21] report values of E_a for the conversion of gibbsite to Al_2O_3 and to boehmite as 75 and 167 kJ mol^{-1}, respectively. Rouquerol *et al.* [22] conclude that the conversion of gibbsite into boehmite requires hydrothermal conditions, which may prevail locally within the bulk of reactant particles, so that γ-AlOOH is formed towards the centre of the crystals of γ-$Al(OH)_3$. The disintegration of this assemblage, through cracking following dehydration, decreases the effectiveness of this barrier to water escape and the decreased availability of water results in the production of alumina. These alternative rate processes are mentioned in the above scheme. Clearly the balance between the reactions will be sensitive to particle size and ease of removal of water vapour.

While the qualitative changes which occur during the dehydrations of aluminum hydroxides have been characterized, little quantitative information concerning the kinetics and mechanisms of these processes has been obtained. The surface textures and reactivities of the alumina products have been investigated, but little is known about the detailed changes which occur on heating the precursor phases.

8.4. ZINC AND CADMIUM HYDROXIDES

Giovanoli *et al.* [23] conclude that there is a structural relationship between γ-$Zn(OH)_2.0.5H_2O$ and β-$Zn(OH)_2$, into which it is converted topotactically on dehydration. These hydroxides and γ-$Zn(OH)_2$ all decompose in an electron beam to yield highly-oriented ZnO. The crystal structure of the original hydroxide appears to exert a strong influence on the rate of oxide sintering at about 1070 K.

Low and Kamel [24] identified four steps in the dehydration of $CdO.1.65H_2O$ at 570 K: water loss destroys the reactant texture, cadmium oxide is formed, which then sinters and finally yields oxygen. The sintering process below 670 K is inhibited by Al^{3+} and enhanced by Li^+, attributed to control through vacancy concentration. Niepce *et al.* [25] affirm that the isothermal reaction of $Cd(OH)_2$, between 434 and 469 K, is deceleratory throughout, disputing the earlier suggestion [26] that α - time curves are sigmoid. The product oxide is developed in a topotactic relation to the hydroxide. The (111) planes of the former are parallel to (001) of the reactant and the behaviour resembles the dehydration of $Mg(OH)_2$. The most unusual feature of this reaction [25] was that a proportion of the crystallites did not decompose. While water evolution from many of the particles proceeded to completion in a matter of hours, other crystals failed to undergo any reaction at all. This is an unusual observation, though such behaviour could easily be overlooked.

8.5. TIN(IV) HYDROXIDES

Using electron diffraction and proton magnetic resonance measurements to supplement thermogravimetric data, Giesekke *et al.* [27] identified the following steps in the dissociation of tin(IV) hydroxides:

$$SnO_3H_2 \rightarrow 520\ K \rightarrow Sn_2O_5H_2 \rightarrow 610\ K \rightarrow Sn_4O_9H_3 \rightarrow 770\ K \rightarrow Sn_8O_{16}H_2 \rightarrow 870\ K \rightarrow SnO_2$$

The existence of crystalline phases was demonstrated by sharp diffraction patterns. Some overlap of the reactions mentioned was detected together with partial reduction in the intermediates. The first compound ($Sn_2O_3(OH)_2 . H_2O$) contains clusters of 3 to 4 protons, but after water loss the hydroxyl groups were more widely separated and the reaction rate decreased.

Ramamurthy and Secco [28] identified the following steps in the decompositions of some divalent metal (Ca, Cd, Co, Cu, Mg, Mn, Ni, Sr and Zn) hydroxystannates:

$$M(Sn(OH)_6) \rightarrow 570\ \text{to}\ 640\ K \rightarrow MSnO_3 + 3H_2O$$

$$MSnO_3 \begin{cases} \nearrow & \tfrac{1}{2}M_2SnO_4 + \tfrac{1}{2}SnO_2 \\ \searrow & MO + SnO_2 \end{cases}$$

8.6. TRANSITION METAL HYDROXIDES

8.6.1. Chromium hydroxide

Alario-Franco *et al.* [29] studied the hydrothermal (470 to 520 K) reactions of certain chromium hydroxide gels that yield orthorhombic CrOOH, which then undergoes topotactic oxidation in oxygen, 620 to 770 K, to form CrO_2 and Cr_2O_3 concurrently. Giovanoli [30] drew attention to the many phases which may participate in these reactions and emphasized the considerable sensitivity of behaviour to the prevailing, perhaps localized, conditions especially the availability of volatile reactants. Agreement has not been reached [29] concerning the steps involved in the dehydration of chromium hydroxide and the differences in behaviour reported still have to be resolved. Pribylov *et al.* [31] proposed the following reaction steps:

$$Cr_2O_3.4H_2O \rightarrow 376\ K \rightarrow Cr_2O_3.2H_2O \rightarrow 453\ K \rightarrow Cr_2O_3.H_2O$$

The monohydrate loses water gradually, over a considerable temperature range, to yield amorphous oxide that crystallizes at about 800 K. Giovanoli and Stadelmann [32] dehydrated $Cr(OH)_3.3H_2O$ to yield amorphous hydroxide and reported that the crystallization to α-Cr_2O_3 was temperature dependent, with nucleation influenced by the gases present. Cr_2O_3 may undergo some superficial oxidation, possibly to Cr(IV). In water vapour, $Cr(OH)_3.3H_2O$ develops short range order below 570 K, due to the production of trigonal CrOOH of extremely small crystallite dimensions.

Many aspects of the reactivity and crystallography of chromium hydroxides are incompletely understood and further work is undoubtedly required to resolve the many inconsistencies apparent in the literature.

8.6.2. Manganese hydroxides

Sato *et al.* [33] described the endothermic dehydration of γ-MnOOH in nitrogen as yielding $MnO_{1.63}$ at 548 K and further reaction to α- Mn_2O_3 at 913 K. In air at 493 K, this compound undergoes successive water loss and oxidation to form β-MnO_2 which decomposes to α-Mn_2O_3 at 823 K.

8.6.3. Iron hydroxides

There is a very considerable literature relating to the dehydrations of the several known iron hydroxides, which range from amorphous gels [19,34] to the various crystalline forms [1] (α, β, γ and δ) of FeOOH. Emphasis has been on the identification of the changes of crystal structure, including topotactic relationships, and studies have not yet included many kinetic measurements. The oxyhydroxides, FeOOH, are generally converted to α-Fe_2O_3, hematite, on heating at about 670 K in air, and there is a tendency [1] to preserve the existing framework of oxygen during such phase transformations, resulting in topotactic relationships between reactant and product. Studies in this field also consider the existence of two stable valence states of iron: $Fe(OH)_2$ is converted to FeO (wüstite) on heating below 470 K in an inert atmosphere, but in the presence of oxygen reacts [1] further:

$$FeO \rightarrow Fe_3O_4\text{(magnetite)} \rightarrow \gamma\text{-}Fe_2O_3\text{(maghemite)} \rightarrow 670\text{ K} \rightarrow \alpha\text{-}Fe_2O_3\text{ (hematite)}$$

Okamoto *et al.* [34] studied the crystallization of iron(III) hydroxide gel, containing 15 to 22% water, precipitated from aqueous solution, using X-ray diffraction and Mössbauer spectra. The structure of this material is identified as a hexagonal close-packed lattice of oxygen in which the iron(III) ions are distributed in an almost random manner amongst the available octahedral voids. The kinetics of recrystallization to α-FeOOH were measured from Mössbauer spectra and shown to fit the first-order equation. This change is identified as a dissolution-recrystallization process in which the dissolution step is rate limiting. For a review of earlier literature concerned with the thermal analysis of iron(III) hydroxide gels, see Mackenzie and Berggren [19].

During the dehydration of goethite (α-FeOOH) to hematite [35], the hexagonal arrangement of oxygen ions is preserved. The activation energy for the reaction, which is deceleratory throughout, is 88 kJ mol^{-1} (499 to 512 K) and the migration of protons is an important step in the elimination of water. Detailed structures within domain boundaries of synthetic goethite, α-FeOOH, have been examined [36] by high resolution electron microscopy and areas of intergrowth have been characterized. This technique may be extended and be applicable to examination of the early stages of dehydration to give consideration to the role of topotactic structures in product development.

Although it has been suggested [35] that β-FeOOH is converted to hematite directly, possibly [1] in the presence of a low pressure of water vapour, Mackenzie and Berggren [19] recommend further investigation of the system. A detailed discussion of the structure and imperfections present in the colloid system of β-FeOOH and β-Fe_2O_3 has been given.

Giovanoli and Brütsch [37] emphasize the necessity for supplementing thermoanalytical kinetic data for the dehydration of γ-FeOOH (→ γ-Fe_2O_3) with diffraction and electron microscopy studies. The deceleratory rate process can be described by either the first- order, or various diffusion-controlled expressions and thus a specific reaction model does not result from this kinetic analysis. Values of E_a varied (108 to 140 kJ mol^{-1}) with reaction conditions and the preparation used. It was concluded from electron microscopy, however, that reaction proceeded by random nucleation to yield oriented, but disordered crystals of γ-Fe_2O_3 with linear dimensions of about 7 nm. The strain developed during transformation resulted in structure collapse. At about 670 K in air, γ-Fe_2O_3 is converted to hematite [1]. The dehydration of δ-FeOOH involved [35] the intervention of δ-Fe_2O_3 which is then transformed to hematite. On heating between 388 and 418 K, γ-FeOOH lost bound water from the crystal surfaces by a first-order reaction with E_a = 100 kJ mol^{-1} [38]. At about 500 K, dehydroxylation to γ-Fe_2O_3 satisfactorily fitted the two-dimensional phase boundary controlled rate equation with E_a = 150 kJ mol^{-1}.

When there is effective removal of product water vapour, the decomposition of iron(II) hydroxide (at about 423 K) may be represented [39] by:

$$Fe(OH)_2 \rightarrow FeO + H_2O$$

Data fit the contracting area equation when α is less than 0.6 and $E_a = 94$ kJ mol^{-1}. The rate-limiting process is identified as the escape of water from the crystal. In the presence of water vapour, the reaction yields magnetite, ascribed to the occurrence of the second step:

$$3FeO + H_2O \rightarrow Fe_3O_4 + H_2$$

Inouye *et al.* [40] investigated the influence of copper(II) on the conversion of colloidal iron(II) hydroxide to magnetite in the absence of oxygen, 303 to 323 K, and concluded that the Cu^{2+} facilitates the structural transformation required to form the product phase.

8.6.4. Cobalt hydroxides

Cobalt(II) hydroxide has the brucite structure (as do $Fe(OH)_2$ and $Ni(OH)_2$) and from nonisothermal measurements [39], 423 to 443 K, E_a for the dehydration reaction was 75 kJ mol^{-1}. In air, two reactions occur:

$$Co(OH)_2 \rightarrow 443\text{ K} \rightarrow Co_2O_3.H_2O \rightarrow 528\text{ K} \rightarrow Co_3O_4$$

From a photomicrographic study, Figlarz and Vincent [41] concluded that the dehydration of $Co(OH)_2$ is initiated at randomly-distributed sites on the surfaces of the approximately hexagonal crystals of reactant. Following the topotactic advance of the reaction interface, alignment of the (111) face of CoO with (0001) of $Co(OH)_2$ preserves the external form of the reactant particles to yield pseudomorphic particles of product. Appreciable quantitites of water are retained in the residual material.

Avramov [42] has made an isothermal kinetic study (533 to 583 K) of the dehydration of CoOOH in air to form Co_3O_4. The deceleratory rate process was fitted by the contracting area equation, values of E_a being 145 and 79 kJ mol^{-1}, below and above 553 K, respectively. It was concluded that, in the lower temperature range, the rate-controlling process involves electron transfer, while above 553 K dehydration becomes rate determining and the value of E_a is comparable with that for the dehydration of $Co(OH)_2$. For the same reaction, O'Brien and Patel [43] also distinguished two separate rate processes, but found significantly different characteristics for the lower temperature reaction. Below 563 K the reaction was identified as a nucleation and growth process, fitting the Avrami-Erofeev equation ($n = 2$) with $E_a = 214$ kJ mol^{-1}. Above 563 K the contracting area expression was found satisfactory with a lower value of E_a (101 kJ mol^{-1}).

8.6.5. Nickel hydroxide

The reaction of $Ni(OH)_2$ resembled [39] that of $Fe(OH)_2$ in that the contracting area equation fitted the data and the value of E_a was 95 kJ mol^{-1}. The rate was appreciably decreased by water vapour. The textural changes that accompany water removal have been studied [41] by electron microscopy which identified rapid initial nucleation at crystallite edges to form a continuous interface. The dehydration is topotactic to yield particles of product which are pseudomorphs of the reactant. These textural changes are consistent with the earlier conclusions based on kinetic evidence.

The crystallinity of nickel hydroxide depends [44] on the method of preparation. The highly crystalline β-$Ni(OH)_2$ contains hydroxyl ions in the brucite lattice, whereas hydrogen bonding occurs in the poorly crystallized α-$Ni(OH)_2$. Kuznetsov [45] has decomposed the so-called higher oxides of nickel ($Ni_xO_y(H_2O)_z$) and concludes that the excess oxygen is present as OH groups. Such excess oxygen and water are simultaneously lost on heating between 400 and 720 K to yield NiO. Progressive heating has also been shown [46] to increase the crystallite size of nickel oxide produced from alkaline oxidation of $Ni(NO_3)_2$ with bromine. Interpretation of the X-ray diffraction data indicates the intermediate production of phases of the type Ni_2O_3 and Ni_3O_4, in which the vacant positions are progressively filled by Ni^{2+} to yield finally NiO.

8.6.6. Copper hydroxide

The texture and porosity of copper oxides are controlled by the water vapour pressure prevailing during the dehydroxylation of the hydroxide [47].

8.6.7. Hydrated rhodium oxide

Hydrated rhodium oxide ($Rh_2O_3.5H_2O$) decomposes [48] in vacuum or in air to yield $Rh_2O_3.H_2O$, while in water vapour it seems that the dihydrate is formed. Each of these compounds reacts further to produce the anhydrous oxide, which may be decomposed to the metal at higher temperatures.

8.7. GENERAL CONSIDERATIONS

Although some hydroxides are precipitated from solution in a finely-divided, disordered and perhaps even amorphous state, compounds containing the OH^- ion may also be prepared in a well-crystallized state, and studies of dehydroxylation reactions have often been concerned with such materials. Dehydration of compounds possessing the brucite lattice ($M(OH)_2$ where M = Mg, Ca, Fe, Co or Ni) are characterized as topotactic processes, through which reaction proceeds with minimum lattice disturbance, though concurrent cracking may be a consequence of water loss. The crystallography and mechanisms of these processes have been discussed by Brett *et al.* [1]. Several features of water elimination are understood, but care must be exercised in the interpretation of kinetic data, because the Arrhenius parameters and the reaction rates may be significantly influenced by the availability of water at the site of dehydration, presumably arising through some reversibility of the desorption or release step [13].

The role of proton transfer in dehydroxylation has been extensively discussed [11,12]. For many substances it has not yet been agreed which of the steps: water elimination ($2OH^- \rightarrow O^{2-} + H_2O$), or subsequent removal of this volatile product, is to be identified as rate limiting. Indeed, both are probably operative under different conditions. Interface advance and/or diffusion processes are thus alternative, and possibly complementary [13], controlling steps. Other possibilities involving crystallographic shear, although theoretically available, have not been explored in detail.

Although traces of hydrogen have been identified in the products of dehydroxylation of $Mg(OH)_2$, $Al(OH)_3$ and kaolinite, MacKenzie [49] detected this product only when the reactant contained impurities. The evolution of H_2 may also arise [1] in dehydroxylations where cation oxidation is possible:

$$2Fe^{2+} + 2OH^- \rightarrow 2Fe^{3+} + 2O^{2-} + H_2$$

Peroxide formation may intervene [50] during dehydroxylation of hydroxides of the following metals: Ba, Al, Cd, Zn and Pb. This conclusion was reached from the observation that during the reactions of these salts with alkali halides elemental halogen was liberated. These reactions need further investigation.

8.8. CATALYST PRODUCTION

There is an extensive literature devoted to the preparation of oxide and mixed-oxide catalysts. These compounds are often derived from hydroxides, though alternative processes are available, e.g., calcination of metals, oxalates etc. Some of the studies referred to above were undertaken to add to the understanding of the textural changes which accompany water removal during the preparation of catalytically-active phases of technological importance. Heating of mixed hydroxides, prepared by concurrent precipitation, yields spinels and other specific phases, including solid solutions, at lower temperatures than is possible with less-intimate mixtures. The thermal reactions of mixed hydroxides [51] differ considerably from those of the pure individual compounds.

8.9. DEHYDRATION OF LAYER SILICATES

8.9.1. General

Layer silicates, including clays, which occur in geological deposits are of considerable importance as source materials for many technological uses. Studies of the thermal dehydration and subsequent recrystallization of such materials have been undertaken to determine the mechanisms of changes which occur during firing. Such knowledge may be then exploited in the design of individual processes, the use of cheaper source minerals and the development of new products. Progress in determining the mechanisms of the water elimination reactions of layer silicates has been greater than for dehydration of silicates containing other anionic structures (fibres, three-dimensional lattices, zeolites) which will not be discussed here. The layer silicates include minerals which are widely-distributed constituents of rocks, sometimes as large crystals, such as muscovite (mica), chlorite, etc., where the extended planar anionic structures contain constituent hydroxyl groups. In contrast, clays (kaolinite, talc etc.) usually consist of very small crystallites. Again hydroxyl groups are constituents of the aluminosilicate condensed structures and between successive layers there may be molecular water. It is often difficult to identify the point in clay dehydration where the removal of (hydrogen bonded) molecular water is complete and structural dehydroxylation commences. A comprehensive survey of the chemistry of layer silicates is beyond the scope of the present review. An authoritative account of the chemistry of clays has been given by Grim [52]. Brett *et al.* [1] have critically discussed the mechanisms of dehydration of layer silicates.

Discussion of experimental studies of the dehydrations (and other) reactions of naturally-occurring minerals must include consideration of the following possibilities.
(i) Natural minerals are usually solid solutions; isomorphous replacement of both cationic and anionic constituents may be random or ordered with appreciable compositional inhomogeneity in any one crystal.
(ii) The compositions of successive units in the stacked sequence of anionic layers in ordered materials can be different. There may be a regular array of layers of alternate composition (e.g. the chlorite lattice, where $Mg(OH)_2$ is interposed between magnesium hydroxysilicate), or there may be less-regular stacking of varied groups (many natural clay assemblages contain irregularly sequenced layers of different structures). These materials deviate significantly from the ideal crystal.
(iii) Natural sediments contain impurities as separate phases, or as inclusions.

Crystals in the sediment sometimes have undergone mechanical damage and/or chemical modification, which may be restricted or more extensive. There are frequently wide variations of particle size in any one deposit.
(iv) The characterization of any clay sample is often difficult due to the presence of finely-divided material, local inhomogeneities and impurities. The identification of reaction products is often subject to even greater uncertainties.

Thus the well-defined crystalline reactant materials required for kinetic studies are not often readily available. Variations in composition and lattice order (of all types) may influence reactivity and result in irreproducibility amongst different samples. Because classification of the behaviour of silicates has almost invariably been based on anion structure, the significance of the cation array may have been underestimated.

The dehydration of hydrous layer silicates is often deceleratory throughout because water release is diffusion controlled and may be reversible. Arrhenius parameters are often sensitive to prevailing conditions, notably the availability of water vapour.

A comparison of the reaction rates for several representative phases, for which sufficient kinetic data were available in the literature, was made using values of the rates of water evolution/unit area of surface. It was assumed that water loss occurred at crystallite edges during the early stages of reaction and that dehydration may be an interface process [53]. Arrhenius parameters obtained (A in molecules m^{-2} s^{-1} and E_a in kJ mol^{-1}) are shown to exhibit a marked compensation-type relation in Figure 8.2. which refers to the dehydrations of illite [54] (x), kaolinite [54-58](+), montmorillonite [54] (Δ) and muscovite [59] (□). This marked interrelationship of kinetic parameters for the different systems is ascribed to control by a common, or similar, parameter. These data have been considered together with observations for catalytic reactions on clays [60] and have been interpreted as indicating an approximately common temperature of onset of both dehydroxylation and catalysis involving hydroxyl groups in different structures. Note the very wide range of variations of both A and E_a in the dehydration of kaolinite in Figure 8.2. The compensation line can be expressed as:

$$\log (A/\text{molecules m}^{-2}\text{ s}^{-1}) = 18.5 + 0.067\,(E_a/\text{ kJ mol}^{-1})$$

and the isokinetic temperature is 935 K, somewhat above the temperature interval used in many kinetic studies.

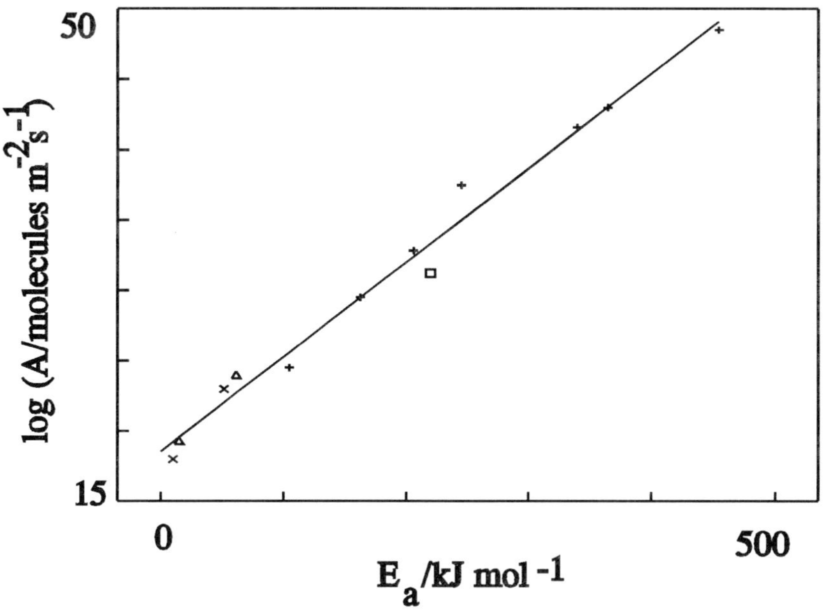

Figure 8.3.
Compensation behaviour identified for the dehydrations of a number of natural minerals (illite (x), montmorillonite (Δ), muscovite (□) and kaolinite (+)) estimated from literature data [53-60].

8.9.2. Kaolinite

The dehydration of kaolinite has been the subject of several kinetic studies and Brett *et al.* [1] summarize the salient features of the mechanisms proposed for the sequence of reactions by which kaolinite is converted to mullite (920 to 1370 K). The first step, water loss, is most satisfactorily described by a two-dimensional diffusion equation. Brindley *et al.* [57] proposed this model from isothermal kinetic measurements (670 to 810 K) and reported a marked increase of E_a in a maintained pressure of water vapour. Anthony and Garn [58] concluded that random nucleation is rate limiting at low pressures of water vapour and that this accounts for reports of first-order kinetic behaviour. Increase in the rate of nucleation, as the $p(H_2O)$ is increased, is ascribed to a proton transfer mechanism, and acceleration of the growth process may result from contributions due to the onset of the reverse reaction.

Mikhail and Guindy [54] found that water vapour present during the isothermal (370 to 460 K) dehydrations of montmorillonite and of illite increased the reaction rate relative to that in vacuum, increased the value of E_a and the total yield of water. Water lost in this temperature range is presumably intercalated between anionic layers, and higher temperatures would be required to remove structural hydroxyl groups.

8.9.3. Muscovite

From isothermal (818 to 847 K) kinetic studies [59] of the dehydration of muscovite in vacuum, it was concluded that the reaction rate was controlled by two-dimensional diffusion in the product with $E_a = 226$ kJ mol^{-1} at pressures below 0.1 Torr. Many phases have been reported [1] as occurring in the products of high temperature breakdown of mica.

8.9.4. Talc

Daw *et al.* [61] used microscopy and X-ray diffraction to investigate the dehydroxylation and formation of enstatite during the heating of talc. Dehydration proceeds inhomogeneously with the formation of bubbles, possibly initiated at dislocations. The orientation of the first-formed enstatite was more random than that subsequently attained, in which the product was preferentially aligned in three possible directions with respect to the reactant, controlled by oxygen lattice preservation and cation distribution.

The isothermal dehydroxylation of talc, 1100 to 1160 K, in flowing argon [62] fitted the first-order equation and $E_a = 422$ kJ mol^{-1}. This result is at variance with other reported observations and here the rate-limiting step is identified as Mg-OH bond scission followed by migration of the magnesium.

More recently, MacKenzie and Meinhold [63] confirmed the occurrence of dehydroxylation of talc at about 1150 K. The reaction is identified as being essentially simple, in which the elimination of structural water is accompanied by reorganization of the residual oxides to form orthoenstatite with the separation of SiO_2.

8.10. CONCLUSIONS

In contrast to the decompositions of many solids (including the hydrates discussed in the previous chapter), the dehydrations of hydroxides show some common patterns of behaviour in two broad groups: the dehydroxylations of (i) simple hydroxides ($Mg(OH)_2$, $Ca(OH)_2$, etc.) and (ii) extended silicates (clays, minerals, etc.). The range of reactants for which detailed kinetic studies have been reported does not include all types of hydroxides, so further different reactions may yet be recognized. Topotactic behaviour is often mentioned in discussions of reactions of both types. Comparison of the behaviour of the two groups identified does, however, provide information on the influences of structure on reactivities.

Experimental studies of dehydrations are not straightforward, because it is often difficult to distinguish the individual contributions from several participating rate processes. The initial step in the dehydration of many simple hydroxides is often identified as proton transfer [11-13] and this is probably reversible within a reactant zone that conserves structure. Water product within the unrecrystallized reactant (homogeneous reaction: type 2 in Figure 7.3.) may be released only after (reversible) diffusion to surfaces, followed by (reversible) desorption. Deeper understanding of these processes is not generally obtainable from kinetic studies of the reversible overall reactions, but rather from the identification of pre-reactional changes in crystals, as discussed for $Ca(OH)_2$ [11,12].

The dehydroxylations of many crystalline minerals with layer structures proceed without significant modification of stable extended bonded components, while intercalated water and/or hydroxyl groups are lost. These are reactions of type 1(a) in Figure 7.3. Again the reversibility of many such processes diminishes the value of kinetic observations (Figure 8.2.) in identifying reaction mechanisms. Water may be present in different bonding environments, so that distinguishing the rate processes contributing to clay dehydrations may be difficult [52]. For example in the layer silicates, such as vermiculite (Section 7.14.1. and Chapter 7, reference [156]), the kinetics of water evolution at low values of α may be different from that observed later in the reaction. This pattern of differing reactivity is applicable to other hydroxides, as shown by Giovanoli [64] for various manganese oxyhydroxides. Adsorbed water was lost between 300 and 400 K and interlayer water below 525 K. Dehydroxylation occurred up to 800 K followed at higher temperatures by decomposition of the manganese oxides.

The decompositions of many hydroxy salts of Cu, Pb, Cd and Zn have been studied [65].

Closely comparable experimental measurements on the dehydrations of $Ca(OH)_2$ and of $CaSO_3.\frac{1}{2}H_2O$ [66] have shown that the rates of dehydration and the values of A and E_a are almost identical. This could indicate similarities in mechanisms in spite of apparent chemical differences. 600 K is an unexpectedly high temperature for expulsion of molecular water of crystallization. One possibility is that water in the heated sulfite is present as hydroxyl groups, or that water release proceeds through these intermediates.

REFERENCES

1. N.H. Brett, K.J.D. MacKenzie and J.H. Sharp, Q. Rev. Chem. Soc., 24 (1970) 185.
2. R.S. Gordon and W.D. Kingery, J. Amer. Ceram. Soc., 49 (1966) 654; 50 (1967) 8.
3. F. Freund and H. Naegerl, J. Thermal Anal., 2 (1970) 387.
4. F. Freund, R. Martens and N. Scheikh-Ol-Eslami, J. Thermal Anal., 8 (1975) 525.
5. K.J.D. MacKenzie, J. Thermal Anal., 5 (1973) 19.
6. J.H. Sharp, Trans. Brit. Ceram. Soc., 72 (1973) 21.
7. D. Beruto, P.F. Rossi and A.W. Searcy, J. Phys. Chem., 89 (1985) 1695.
8. K.J.D. MacKenzie and R.H. Meinhold, Thermochim. Acta, 230 (1993) 339.
9. N.H. Brett, Mineral. Mag., 37 (1969) 244.
10. R.Sh. Mikhail, S. Brunauer and L.E. Copeland, J. Colloid Interf. Sci., 21 (1966) 394.
11. O. Chaix-Pluchery, J. Bouillot, D. Ciosmak, J-C. Niepce and F. Freund, J. Solid State Chem., 50 (1983) 247; 67 (1987) 225.
12. O. Chaix-Pluchery and J-C. Niepce, React. Solids, 5 (1988) 69.
13. A.K. Galwey and G.M. Laverty, Thermochim. Acta, 228 (1993) 359.
14. D. Beruto, L. Barco, A.W. Searcy and G. Spinolo, J. Am. Ceram. Soc., 63 (1980) 439.
15. R. Dinescu and M. Preda, J. Thermal Anal., 5 (1973) 465.
16. S. Tamaru and K. Shiomi, Z. Phys. Chem., A171 (1935) 221,229.
17. R.F. Horlock and P.J. Anderson, Trans. Faraday Soc., 63 (1967) 717,2316.
18. J. Beretka and M.J. Ridge, J. Chem. Soc. A, (1967) 2106.
19. R.C. Mackenzie and G. Berggren, Differential Thermal Analysis, (Ed. R.C. Mackenzie), Academic Press, New York, Vol 1., 1969, Chap.9.
20. R.M. Dell, I. Chem. E., Symp. Ser., No. 27 (1968) 3.
21. N.M. Drobot and E.I. Khazanov, see Chem. Abs., 78 (1973) 18381v.
22. J. Rouquerol, F. Rouquerol and M. Ganteaume, J. Catal., 36 (1975) 99.
23. R. Giovanoli, H.R. Oswald and W. Feitknecht, J. Microscopie, 4 (1965) 711.
24. M.J.D. Low and A.M. Kamel, J. Phys. Chem., 69 (1965) 450.
25. J-C. Niepce, G. Watelle-Marion and C. Clinard, C.R. Acad. Sci., Paris, C, 274 (1972) 96; 269 (1969) 632, 683; 270 (1970) 298.
26. M.M. Pavlyuchenko, M.Y. Novoselova and E.A. Prodan, Vestsi Akad. Navuk Belaruss SSR, Ser. Khim. Navuk, 1 (1969) 6.

27. W. Giesekke, H.S. Gutowsky, P. Kirkov and H.A. Laitinen, Inorg. Chem., 6 (1967) 1294.
28. P. Ramamurthy and E.A. Secco, Canad. J. Chem., 49 (1971) 2813.
29. M.A. Alario-Franco, J. Fenerty and K.S.W. Sing., Reactivity of Solids, (Eds J.S. Anderson, M.W. Roberts and F.S.Stone), Chapman and Hall, London, 1972, p.327.
30. R. Giovanoli, contributed discussion to reference 29.
31. K.P. Pribylov, L.V. Kovshova and E.G. Yarkova, Zh. Neorg. Khim., 17 (1972) 895.
32. R. Giovanoli and W. Stadelmann, Thermochim. Acta, 7 (1973) 41.
33. M. Sato, K. Matsuki, M. Sugawara and T. Endo, Nippon Kagaku Kaishi, 94 (1973) l655.
34. S. Okamoto, H. Sekizawa and S.I. Okamoto, Reactivity of Solids, (Eds J.S. Anderson, M.W. Roberts and F.S. Stone), Chapman and Hall, London, 1972, p.341.
35. G.V. Loseva and N.V. Murashko, Izv. Akad. Nauk SSSR, Neorg. Mater., 8 (1972) 485.
36. R.M. Cornell, S. Mann and A.J. Skarnulis, J. Chem. Soc., Faraday Trans.I, 79 (1983) 2679.
37. R. Giovanoli and R. Brütsch, Thermochim. Acta, 13 (1975) 15.
38. N. Koga, S. Okada, T. Nakamura and H. Tanaka, Thermochim. Acta, 267 (1995) 195.
39. I.F. Hazell and R.J. Irving, J. Chem. Soc. A, (1966) 669.
40. K. Inouye, K. Murata and T. Ishikawa, Z. Anorg. Allg. Chem., 386 (1971) 340.
41. M. Figlarz and F. Vincent, C.R. Acad. Sci. Paris, C, 266 (1968) 376; M. Figlarz, Ann. Chim., 9 (1974) 367; F. Fievet and M. Figlarz, J. Catal., 39 (1975) 350.
42. L.K. Avramov, Thermochim. Acta, 10 (1974) 409.
43. P. O'Brien and U. Patel, J. Chem. Soc., Dalton Trans., (1982) 1407.
44. M. Sakashita and N. Sato, Bull. Chem. Soc. Japan, 46 (1973) 1983.
45. A.N. Kuznetsov, Zh. Neorg. Khim., 13 (1968) 2027.
46. C. Aravindakshan, M. Misra and B.K. Banerjee, Technology, 8 (1971) 132.
47. J.C. Volta, V. Perrichon, A. Renouprez, P. Turlier and Y. Trambouze, J. Chim. Phys., 69 (1972) 196.
48. E. Morán, M. Alario-Franco, J. Soria and M. Gayoso, Reactivity of Solids, (Eds J. Wood, O. Lindqvist, C. Helgesson and N-G. Vannerberg), Plenum, New York, 1977, p.393.

49. K.J.D. MacKenzie, J. Inorg. Nucl. Chem., 32 (1970) 3731.
50. D.I. Ozherel'ev and M.E. Krasnyanskii, Zh. Neorg. Khim., 13 (1968) 1530.
51. R.I. Razouk, R. Sh. Mikhail and B.S. Girgis, J. Colloid Interf. Sci., 24 (1967) 470.
52. R.E. Grim, Clay Mineralogy, 2nd Edn, McGraw-Hill, New York, 1968.
53. H.B. Johnson and F. Kessler, J. Am. Ceram. Soc., 52 (1969) 199.
54. R.S. Mikhail and N.M. Guindy, J. App. Chem. Biotechnol., 21 (1971) 113.
55. G.W. Brindley and M. Nakahira, J. Am. Ceram. Soc., 40 (1957) 346.
56. F. Toussaint, J.J. Fripiat and M.C. Gastuche, J. Phys. Chem., 67 (1963) 26.
57. G.W. Brindley, J.H. Sharp, J.H. Patterson and B. N. Narahari, Am. Mineral., 52 (1967) 201.
58. G.D. Anthony and P.D. Garn, J. Am. Ceram. Soc., 57 (1974) 132.
59. H. Kodama and J.E. Brydon, Trans. Faraday Soc., 64 (1968) 3112.
60. A.K. Galwey, Adv. Catal., 26 (1977) 247.
61. J.D. Daw, P.S. Nicholson and J.D. Embury, J.Amer. Ceram. Soc., 55 (1972) 149.
62. J.R. Ward, Thermochim. Acta, 13 (1975) 7.
63. K.J.D. MacKenzie and R.H. Meinhold, Thermochim. Acta, 244 (1994) 195.
64. R. Giovanoli, Thermochim. Acta, 234 (1994) 303.
65. P. Ramamurthy and E.A. Secco, Canad. J. Chem., 47 (1969) 2181,3915; 48 (1970) 2617,3510; 51 (1973) 3882.
66. C. Anderson and A.K. Galwey, Canad. J. Chem., 70 (1992) 2468.

Chapter 9

THE THERMAL DISSOCIATION OF OXIDES

9.1. INTRODUCTION

This chapter is concerned with the thermal decompositions of oxides and peroxides. There are obviously very important connections with the reactions of hydroxides (Chapter 8) and so-called peroxysalts, which contain hydrogen peroxide of crystallization (included in Chapter 7 on hydrates). Hydrated oxides vary from compounds accurately represented by the stoichiometric formula $M(OH)_m$, to phases which contain discrete H_2O molecules. The chemistry of oxides should also be considered in the context of the other binary compounds (e.g. hydrides, nitrides, carbides, sulphides etc.) dealt with in Chapter 10.

9.2. CHEMISTRY OF OXIDES

9.2.1. Overview

The mechanisms of dissociation of solid oxides may include one or more of the following steps (some or all of which may be at least partially reversible):

O^{2-} (lattice) $\rightarrow$ O^{2-} (surface or interface)

This step may be accompanied by, or followed by, the transfer of electrons to acceptor sites, ultimately leading to cation reduction:

O^{2-} (surface) $\rightarrow$ 2e + O (surface)

2O (surface) $\rightarrow$ O_2 (adsorbed) $\rightarrow$ O_2 (gas) (or 2O (gas))

The reverse of these reactions is chemisorption and oxidation. Such chemical processes (in either direction) may also involve the transient formation of such intermediates [1] as O^-, O^{2-}, etc., perhaps localized at specific sites. Alternatively, the reaction may proceed by the elimination of molecular oxygen (also reversible), represented by:

$$2MO_x \rightarrow 2MO_{(x-1)} + O_2$$

Oxides can be classified into two groups [2]: (i) those which dissociate to yield molecular oxygen only, and (ii) those that yield a mixture containing both atomic and molecular oxygen (Table 9.1.). Other oxides melt unchanged at temperatures below the onset of dissociation, and sublimation may also occur.

Table 9.1
Occurrence of atomic oxygen in products of oxide decompositions [2-5]

Reactants which yield both atomic oxygen and molecular oxygen: $Co_3O_4 \rightarrow CoO$; $CuO \rightarrow Cu_2O$; α-$MnO_x \rightarrow \beta$-MnO_x; α-$PbO_x \rightarrow \beta$-PbO_x; $Pb_3O_4 \rightarrow PbO$; $U_3O_8 \rightarrow U_8O_{21}$; $UO_{2+x} \rightarrow UO_2$: detected by the darkening of a platinum crucible. Atomic oxygen, detected by blue colouration in MoO_3, was evolved during the decompositions of CuO, HgO, PtO_2 and ZnO; also oxides of Fe, Mn and Co.	Reactants for which there was no evidence of release of atomic oxygen: $Ag_2O \rightarrow Ag$; $CrO_2 \rightarrow Cr_2O_3$; β-$MnO_2 \rightarrow \alpha$-$Mn_2O_3 \rightarrow \beta$-$Mn_3O_4$; α- and β-$PbO_2 \rightarrow \alpha$-$PbO_x \rightarrow Pb_3O_4$; $UO_3 \rightarrow UO_{2.9} \rightarrow U_3O_8 \leftarrow \epsilon$-$UO_3$ and the following: AgO; Al_2O_3; Cu_2O; PbO; MgO; SiO_2 also oxides of Cr and Cd.

Removal of oxygen from the reactant oxide may be accommodated, to some extent, by a deviation from stoichiometry, because many oxides are capable of existing over an appreciable range of compositions. Continuation of reaction beyond limits sustainable by stoichiometric variation, however, necessitates a chemical change to a lower oxide, perhaps through a reaction involving an interface. The overall kinetics of these complex processes may be controlled initially by lattice-diffusion steps and subsequently by a topochemical process. The significance of the diffusion contribution varies widely amongst different oxide reactants. Complications arise where there is an effective overlap or interaction of two controlling processes, so that mechanistic interpretations of kinetic data must allow for contributions from both sources. Because such reactions may also be reversible, the influences of reaction conditions, notably the prevailing partial pressure of oxygen, $p(O_2)$, must be investigated. These possible complications have not prevented the profitable study of the solid-phase dissociations of many metallic oxides. The difficulties may be offset against the advantages which result from working with very fully-characterized binary compounds, solely comprised of M^{x+} and O^{2-} ions. The crystallographic structures, polymorphs and physical properties of numerous oxides are well known, and these must be regarded as being amongst the simplest and best defined reactant and product phases available. X-ray studies enable investigation of the interrelationships between reactant and product phases, including the detection of topotaxy.

In kinetic studies of the dissociations of many oxides, the cracking and disintegration of the product phase is expected [2] to be relatively unimportant, because the volume changes which arise in these transformations are often small. Furthermore, sintering may not influence kinetic behaviour where measurements are made in vacuum at moderate temperatures.

The thermodynamic properties of many oxides, including ranges of stability, phase diagrams, and changes in both composition and lattice structure over a wide range of conditions of oxygen partial pressure and temperature, have been particularly extensively documented [6,7]. This interest arises from the importance of these substances in many technological fields such as metallurgy, refining, corrosion, ceramics etc. Many oxides are refractory, however, and there are considerable difficulties in carrying out kinetic studies at high temperatures.

A wealth of information concerning the identities, mobilities, concentrations and properties of defects in many oxides is available through measurements of mass transfer made in association with metal oxidation studies [8,9]. The influences of crystal structure [10], temperature and oxygen partial pressure on cation and anion migration have also been investigated. Information on the reactivities of oxides, their polymorphism and the properties of the imperfections present, is often useful in the formulation of the mechanisms of oxide dissociation.

As for many solid state reactions, the properties of any particular oxide preparation may be influenced by its method of synthesis [11]. Oxides are often products of thermal treatment. Such heating may influence the surface area, impurity content (e.g. strongly-retained traces of water from hydroxide dehydration, oxidized species retained from the decomposition of a nitrate, carbonate etc.) and concentrations and distribution of defects (e.g. vacancies and non-stoichiometry arising during oxidation of a metal). Thus the preparative method exerts significant control over the numbers and distributions of imperfections which in turn, often influence reactivity in further solid phase reactions.

In addition to the many investigations of the bulk properties of crystalline oxides, there has also been considerable interest in the surface chemistry of these compounds. Mechanistic studies in this field often include discussion of heterogeneous catalytic-type processes and intermediates, similar to or identical with, those postulated above as occurring during oxide dissociation. Some reference is made below to relevant aspects of the surface chemistry of oxides.

9.2.2. Surface chemistry of oxides : catalysis

A measure of the reactivity of the surfaces of oxides can be obtained from the rate of exchange of the constituent oxide ions with ^{18}O enriched gaseous oxygen. From comparative studies of 38 inorganic oxides, Winter [12] concluded that most of these rate processes involve a dissociative atomic mechanism, restricted to the surface layer of the active solid, and that the slow step is desorption of the molecular oxygen. The Arrhenius parameters determined for these reactants exhibited compensation behaviour, and calculated values of E_a were related to crystallographic structures and parameters. The desorption of product (molecular) oxygen was similarly identified [13] as the rate-determining step in the decomposition of N_2O ($\rightarrow N_2 + \frac{1}{2}O_2$) on many oxides, and kinetic characteristics exhibit parallel behaviour with oxygen exchange reactions on the same solids. The values of E_a found were also shown to be related to the crystallographic parameters of the active phase. The reactions of NO on oxides are again generally similar to those of N_2O. Boreskov *et al.* [14] concluded that the bond energy of oxygen at the surfaces of oxide catalysts largely controls the activity of such solids in the promotion of oxidation reactions. Isotopic exchange studies can be useful in determining the mechanisms of oxide-catalyzed rate processes. Klier [15] has discussed the relationship between the catalytic activities and oxygen binding energies, in the bulk and at the surfaces, of a number of transition-metal oxides.

Several workers have observed a relationship between the the enthalpy of formation of an oxide and its activity in oxidation reactions. Kiyoura [16] has reported the occurrence of such a relationship in the (closely similar) exchange reactions of O_2 and of CO_2 on metal oxides of the fourth period. Moro-Oka *et al.* [17] correlated the catalytic properties of several oxides in the oxidation of various C_2 to C_4 hydrocarbons, with the energy released on formation of the solid concerned.

The literature concerned with hydrocarbon oxidation has been critically reviewed by Germain [1], who was specifically concerned with patterns of activity and selectivity of reactions on oxide catalysts. Although the relative activites of a particular solid for the reactions of several different oxidizable substrates may be comparable, the properties of each solid may also show significant variations for different preparations, perhaps arising from differences in oxidation states or defect density.

9.2.3. Reduction of nickel oxide

The reduction of nickel oxide with hydrogen is a well-established [18-20] example of a nucleation and growth process which occurs at a temperature well below that of nickel oxide dissociation. The chemical change is not readily initiated on the surface of NiO between 573 and 673 K. Nucleation is always partially rate-limiting, though the onset of reaction can be significantly accelerated by the addition of more electropositive metals (Cu, Pt, Pd). Once particles of metallic nickel product have appeared, however, reaction proceeds more readily at the advancing interface and α-time data exhibit the sigmoid curves characteristic of nucleation and growth. While the kinetics of reduction of nickel oxide have been considered in some detail, less information is available concerning the intermediates and interactions occurring at the oxide/metal interface. Presumably the role of the metallic phase is in the dissociation of gaseous hydrogen, and the adsorbed H-atoms thus generated may reduce oxide ions after transfer to the reactant across the reaction interface.

This reaction has been studied in considerable detail and is mentioned here as providing a link between the nucleation and growth mechanisms characteristic of solid state decompositions and surface processes of the type discussed in heterogeneous catalysis.

9.2.4. Comment

Two factors have been identified here as being of significance in consideration of the dissociations of solid oxides. (i) The correlations which have been found between the kinetic characteristics of heterogeneous catalytic changes and the surface, or even bulk properties, of the active oxide. These trends are (at best) semi-quantitative, but the existence of such patterns of behaviour is unusual in the field of heterogeneous catalysis [21]. (ii) The intermediates which have been postulated to occur on oxide surfaces during exchange, oxidation and, possibly, reduction or other catalytic processes, may also be participants in the dissociation of these solids. The ideas generated from heterogeneous catalysis may thus be useful in formulating mechanisms for the solid state decompositions of oxides (and possibly other solid reactants).

9.3. DISSOCIATION OF OXIDES

9.3.1. Introduction

The dissociations (reversible) and decompositions (irreversible) of oxides have not been the subject of many reviews, though the properties of these phases have been discussed with reference to allied topics, including the oxidation of metals, catalysis, etc. Dollimore [22] has reviewed the applications of the techniques of thermal analysis to the dissociation of single oxide systems.

The review by Malinin and Tolmachev [2] distinguishes two types of reaction mechanisms, one involving radicals ($\rightarrow O + O_2$), while the other is molecular (yielding O_2 only). Few kinetic studies are available for reactions of the first group (which are considered to include, for example, $Pb_3O_4 \rightarrow 3PbO + O$ and $2CuO \rightarrow Cu_2O + O$) and there are indications that, for these reactions, the apparent values of E_a may be smaller than the respective enthalpies of reaction. The dissociations of those oxides which yield molecular oxygen (e.g. $PbO_2 \rightarrow PbO_x$, $Ag_2O \rightarrow Ag$ and $CrO_2 \rightarrow Cr_2O_3$) are irreversible, consistent with the inability of the oxygen molecule to be (re)accommodated in the oxide structure. While this classification of reactions of oxides is convenient, it would clearly benefit from extension of the experimental evidence on which it is founded. The reactivities of oxides are influenced [2] by metal-metal interactions in unsaturated oxides, and O-O interactions between closely-spaced neighbours in higher oxides, most pronounced in peroxides. Malinin and Tolmachev [2] conclude that the mechanisms of decompositions of metal oxides are more complicated than the simple dissociation reaction. The more important of the proposed schemes for thermal reactions of oxides are:
(i) elimination of molecular oxygen;
(ii) dissociation to yield atomic oxygen;
(iii) volatility of both elements (sublimation);
(iv) equilibration with gaseous oxygen, and
(v) melting.
There is also the possibility of the concurrent and/or consecutive operation of more than one of these processes.

The overall rate resulting from the occurrence of more than a single process at the phase boundary may control the values of A and E_a so that they are not necessarily directly proportional to the active area of the solid, or equal to the enthalpy of formation of the reactant, respectively. Moreover, when oxygen elimination is accompanied by defect generation, resulting in a change in composition of a non-stoichiometric phase, the enthalpy of reaction cannot in general be expected to be identical with the value of this parameter measured for the oxide phase, which (presumably) refers to the stoichiometric compound.

The review of the decomposition of oxides, given below, focusses attention on those studies which have yielded information on the mechanisms of reactions proceeding in the solid state.

9.3.2. Group IA and IIA (alkali and alkaline earth metal) oxides

The alkali and alkaline earth metals form several series of compounds with oxygen in which the electropositive character of the cation stabilizes the different possible anions: O^{2-}, oxide; O_2^{2-}, peroxide; O_2^-, superoxide and O_3^-, ozonide. Surveys of the preparations, compositions, structures and properties (but not, however, specifically concerned with solid state behaviour) have been given for the peroxycompounds and the ozonides [23]. While some studies report the occurrence of breakdown of certain of these compounds without melting, few kinetic, microscopic or mechanistic studies have been described. The indications are that this is a field of considerable potential interest, but progress is slow due to the unusually high reactivity of the solids concerned and the difficulties which complicate attempts to prepare pure samples. Higher oxides tend to be stabilized by H_2O_2 and/or H_2O. Many preparations contain appreciable amounts of the reactants used in their preparation and perhaps also other related compounds (such as oxides and hydroxides). The accurate characterization of such materials poses very considerable analytical problems. Several studies have compared the stabilities of the higher oxides alone and their compounds with H_2O_2 and/or H_2O.

Two studies [24,25] identify the rate-limiting step in the decomposition of lithium peroxide (Li_2O_2), 540 to 600 K, as bond rupture in the O_2^{2-} ion, for which the measured value of E_a is 220 kJ mol^{-1}. The energy requirement for reaction is partially diminished by the decrease in the interionic distance (Li^+ to O^{2-}) in the product (Li_2O). This conclusion was supported by a later isotopic study [26]. Zero-order kinetic behaviour was reported [25] for Li_2O_2 decomposition and it was suggested [24] that solid-solution formation (Li_2O_2/Li_2O) occurred when α was less than 0.5. Li_2O sublimes below its melting point.

Sodium superoxide is decomposed [27] gradually on heating (400 to 520 K) to yield $Na_2O_{3.60}$ which reacts vigorously at about 520 K to give Na_2O_2. Below its melting point, this evolves oxygen and forms Na_2O. Rates of these reactions were appreciably diminished by the presence of CO_2. Variations in the stabilities of different preparations of sodium superoxide have been ascribed to differences of defect structure: onset of reaction in the most labile reactants occurred above 410 K (E_a = 117 kJ mol^{-1}) and, in the most stable, above 450 K (E_a = 155 kJ mol^{-1}).

$Na_2O_2.8H_2O$ fused and underwent hydrolysis [26] giving H_2O_2. There was no evidence that the sesquioxide was formed [27] during the decomposition of Na_2O_2, a conclusion also applicable [28] to breakdown of RbO_2:

$$2RbO_2 \begin{cases} \nearrow \ 553 \text{ to } 633\text{ K} \rightarrow Rb_2O_2\ (+O_2) \\ \searrow \ 573 \text{ to } 633\text{ K} \rightarrow Rb_2O(+1.5O_2) \end{cases}$$

(This reaction may be accompanied by melting). Sodium oxide, in the solid state, behaves [29] as a powerful oxidizing agent towards transition metals, converting them to alkali- rich highly oxidized ternary oxides. (The oxidizing tendency is diminished in solution in sodium metal, because the latter is a reducing medium.) Ozonides (K, Rb and Cs) underwent [30] hydrolysis more rapidly than decomposition (323 K) to yield metal hydroxide.

Hydrated magnesium peroxide loses water to yield $Mg(OH)_2$ and MgO_2. The peroxide is decomposed [31] above 640 K ($\rightarrow$ $MgO + \frac{1}{2}O_2$) or hydrolyzed to form $Mg(OH)_2$. On heating $CaO_2.2H_2O_2$, hydrogen peroxide is evolved [32] to yield $Ca(OH)_2$, $Ca(O_2)_2$, amorphous to X-rays, and some CaO_2. The onset of reaction over the crystal surface is uneven. Reaction is initiated at points of rupture of hydrogen bonds and propagates in the direction of chains formed by such linkages. $Ca(O_2)_2$ reacts with water to yield $Ca(OH)_2$, H_2O_2 and O_2. Decomposition of calcium peroxide is accompanied [26] by the rupture of the peroxide linkage. Yields of superoxide [32] which result from breakdown of salts of the type $MO_2.2H_2O_2$ increase in the sequence (M =) Ca, Sr, Ba due to diminuition of the polarization of the cation-to-anion bond. Retention of the peroxide linkage was found [26] by isotopic studies of the decomposition of $BaO_2.2H_2O_2$ at 373 K, whereas rupture occurred on reaction of BaO_2 (possibly by $O_2^{2-} \rightarrow O^{2-} + O$). Barium peroxide is a compound of considerable historical significance. It is formed by the (reversible) oxidation of BaO in air (slow at 770 K and dissociation occurs above 870 K). This behaviour provided an early method for the separation of oxygen from air and, later, the commercial production of H_2O_2 used the reaction:

$$BaO_2 + H_2SO_4 \rightarrow H_2O_2 + BaSO_4$$

Both industrial processes have long been superseded. There has been a recent renewal of interest in the solid state chemistry of barium oxides as reactants in the preparation of high-temperature superconductors.

Isothermal TG studies [33] of the thermal decompositions in N_2 of BaO_2 (763 to 883 K) and SrO_2 (653 to 803 K) showed overall deceleratory kinetics described by the Ginstling-Brounshtein diffusion model (low α) and the first-order equation at higher α values. E_a values were 185 ± 5 kJ mol^{-1} for BaO_2 and 119 ± 2 kJ mol^{-1} for SrO_2. Non-isothermal kinetic analyses gave similar E_a values for both decompositions (165 ± 5 kJ mol^{-1}). It is suggested [33] that the rate of removal of oxygen from the peroxide by diffusion could be drastically altered by the formation of a crystalline layer of oxide on the reactant surface.

9.3.3. Group IB and IIB oxides

Silver oxide: An early (1905) study by Lewis [34] of the kinetics of decomposition of Ag_2O was a notable contribution. The dissociation in oxygen (760 Torr, 593 to 623 K) showed a long induction period followed by a sigmoid α-time curve which fitted the Prout-Tompkins equation with E_a = 133 kJ mol^{-1}. Benton and Drake [35] studied the kinetics of the reversible dissociation using a sample of finely-divided active metal. The rate of reaction at 433 K fitted the expression:

$$d[O_2]/dt = k(1 - p/p_e)$$

and kinetic characteristics were independent of oxide content within wide limits (0.3 to10% oxygen). The difference between values of E_a determined for adsorption and for desorption (92 and 150 kJ mol^{-1}, respectively) corresponded to the enthalpy of formation of Ag_2O.

The induction period to decomposition of Ag_2O in vacuum was shorter than in oxygen, indicating the inhibition of growth of small metal nuclei by the gaseous product. While the work of Benton and Drake [35] confirms that reaction occurs at the Ag/Ag_2O interface, interpretation of kinetic observations must always include due allowance for any contribution from the reverse process. The influence of subdivision was investigated by Garner and Reeves [36] who obtained a higher value of E_a (192 kJ mol^{-1}) for the vacuum decomposition, 573 to 613 K, of silver oxide pretreated by annealing in oxygen at 478 to 600 K for several days. The acceleratory stages of these sigmoid α-time curves fitted the cube law. Several other authors support this kinetic analysis (n = 3 and 2), but apparent values of E_a show significant variation. Allen [37] identified three stages in the decomposition of Ag_2O in vacuum: the first two culminate in nucleation (studied below 570 K), while the third refers to interface advance (above 570 K) with E_a = 150 kJ mol^{-1}. Allen emphasized the value of structural and compositional measurements in supporting mechanistic conclusions. Under reversible conditions [37], 356 to 407 K, the dissociation also exhibited sigmoid α-time plots and E_a = 125 kJ mol^{-1}.

Herley and Prout [38] attributed differences in kinetic behaviour of different preparations of Ag_2O to the reactivity of the oxide with CO_2. Ag_2CO_3 in the reactant could be removed by preheating in vacuum for 3 h at 553 K. Decomposition curves (603 to 653 K) were then reproducible and sigmoidal.

Dubinin *et al.* [39] confirmed the autocatalytic participation of product silver nuclei in the decomposition of Ag_2O. Deposition of a thin layer of Ag or Ni on the reactant increased the rate of breakdown at 603 K, but mechanically-mixed metal provided inadequate contact. Lagier *et al.* [40] also confirmed the catalytic role of the product metal. The above discussion concludes that an interface reaction is rate limiting. Atomic oxygen was not identified [2,4,41] in the products.

Nakamori *et al.* [42] studied the isothermal decomposition of Ag_2O in vacuum (683 to 703 K), in ethene (409 to 418 K), in hydrogen (338 to 348 K) and in CO (265 to 283 K). All α-time curves were sigmoid and a linear free energy relationship was found between values of E_a (403, 161, 103 and 68 kJ mol^{-1}, respectively) and the enthalpy changes. The substantial influences of reactant gases on reaction temperatures unambiguously confirm the interactions of these additives with adsorbed intermediates in the oxide dissociation.

Apart from the work by Spath *et al.* [43], the powerful methods of microscopy have been little exploited in studies of the decomposition of Ag_2O.

Copper oxides: The dissociations of CuO and of Cu_2O resemble those of the silver analogues in being reversible and undergoing reduction (e.g. in H_2) more readily than decomposition. The relatively high value of E_a (205 kJ mol^{-1}) reported [44] for the reaction:

$$2CuO \rightarrow Cu_2O + O_2$$

is consistent with the presence of atomic oxygen in the gaseous products (in contrast with the reactions of AgO and Ag_2O, Table 9.1.). The values of E_a for the (H_2) reduction of both oxides of copper are significantly lower [44] (56 kJ mol^{-1}).

Zinc, cadmium and mercury oxides: The decomposition of zinc oxide was studied [45] thermogravimetrically, using sintered material and on (0001) and (1010) crystal faces. Evaporation results in the development of characteristic structures for the different crystallographic surfaces, including a remarkable anisotropy on faces vertical to the c-axis. This reaction yields [4] atomic oxygen. The dissociation of CdO is reversible [46].

Taylor [47] studied the decomposition of HgO (670 K) and concluded that the deceleratory reaction was well represented by the contracting volume expression. The value of E_a for fine powder crystallites (162 kJ mol^{-1}) was almost identical with the enthalpy of dissociation of the oxide, attributed to the breakdown of parallel zig-zag chains of cations and anions. The larger values of E_a (200 kJ mol^{-1}) characteristic of larger crystals may result from the need for a nucleation process to generate reaction fronts that advance normally to the constituent chain-like structure in well-crystallized material. Derbinskii *et al.* [48] also identified a deceleratory process which was inhibited by both mercury and oxygen, with mercury exerting the greater effect. The value of E_a was again 160 kJ mol^{-1}, decreased with particle size, 670 to 770 K, and was lower below 650 K. Results of a microscopic examination during the progress of reaction indicated that the interface advanced in diverse directions in differently-shaped crystals.

9.3.4. Group IVB oxides

Tin Oxides: In a study of the decomposition of tin(II) oxide (700 to1000 K):

$$2SnO \rightarrow SnO_2 + Sn$$

Lawson [49] identified the intermediate Sn_3O_4 (which may be $Sn_2^{2+} Sn^{4+}O_4$) from crystallographic data. On further heating, Sn_3O_4 dissociates to $2SnO_2$ + Sn. In a mass-spectrometric study of the decomposition of SnO_2, Colin *et al.* [50] measured the enthalpy of dissociation of SnO and the polymeric oxides formed:

$$SnO_2 \rightarrow 1/n\ (SnO)_n + \tfrac{1}{2}O_2 \qquad (n = 1,2)$$

$$Sn + SnO_2 \rightarrow 1/n\ (SnO)_n \qquad (n = 1, 2, 3, 4)$$

Lead oxides: Diverse changes of composition and of lattice structure occur during the several successive steps involved in the decompositions of the various lead oxides. Problems of characterization of the intermediates involved have slowed down progress in this topic, because a knowledge of the nature and properties of each intermediate phase is essential for the mechanistic interpretation of kinetic data. The literature includes many references to distorted lattices and non-specific formulae PbO_x to describe phases of indeterminate and variable composition. Some publications have been concerned with the identities of the solid phases, while other papers report on the gaseous products (O_2 or O_2 + O) and kinetic measurements.

Nishimura *et al.* [51] summarize the changes on heating (DTA, 5 K min^{-1}) as follows:

$$\beta\text{-}PbO_2 \rightarrow 633\text{ K} \rightarrow \alpha\text{-}PbO_x \rightarrow 723\text{ K} \rightarrow \beta\text{-}PbO_x \rightarrow 773\text{ K} \rightarrow Pb_3O_4$$

$$\rightarrow 858\text{ K} \rightarrow PbO(\text{red}) \rightarrow 893\text{ K} \rightarrow PbO(\text{yellow})$$

with three further phase transitions between 923 and 973 K. Gillibrand and Halliwell [52] cite several earlier reports on the decompositions of both forms (α and β) of PbO_2 and, from their own work, conclude that behaviour is dependent on the method of reactant preparation. The dissociations of four selected samples of PbO_2 in vacuum, in nitrogen and in oxygen were studied by DTA, supplemented by X-ray diffraction, surface-area measurements and optical microscopy. Intermediates identified included $Pb_{12}O_{19}$, $Pb_{12}O_{17}$ and Pb_3O_4 before the final product, PbO. The possibility of alternative paths, omitting a specific intermediate, is also mentioned. It is concluded that the mechanisms of transformation of each oxide depend on the nature and distribution of defects in the reactant phase, and that such defects control the nucleation of the next crystalline phase in the sequence of changes. This dependence on the imperfection structure prevents prediction of the behaviour of untested material.

The reactions of some lead oxides yield both atomic and molecular oxygen, while others yield molecular oxygen only (see Table 9.1.). Reactions of the former group are usually characterized by a high [53] value of E_a (e.g. 350 kJ mol^{-1} for Pb_3O_4), whereas those which yield O_2 only, often have E_a values of 125 to 200 kJ mol^{-1}.

9.3.5. Transition-metal oxides

The dissociations of transition-metal oxides have often been studied as later processes following the dehydration/dehydroxylation of hydroxides. The existence of polymorphic varieties of the oxide systems has inhibited rapid development of this field. Descriptions of behaviour tend to be predominantly qualitative, devoted to the recognition of the phases involved, the sequences of changes which occur and the crystallographic relationships (if any) between reactants and products in each transformation.

Chromium oxides: The behaviour of chromium(VI) oxide at temperatures below the melting point (470 K, with decomposition) has been described. Sublimation at

445 K yields [54] polymeric volatile species, $(CrO_3)_n$ where n = 3, 4 and 5, the average enthalpy of vaporization being 170 kJ mol^{-1}. Magnetic and X-ray examinations [55] of the vacuum pyrolysis of CrO_3 indicated the intermediate production of Cr_3O_8. Cr_2O_5 was formed only when oxygen was available. From these results it is suggested that processes involving Cr^{5+} cannot occur in the absence of oxygen. At higher pressures [56], a number of oxide phases of limited ranges of composition were identified: $CrO_{2.44}$ (α-oxide), $CrO_{2.65}$ (β-oxide), Cr_6O_{15}, Cr_5O_{12} and CrO_2.

The decomposition of chromium(IV) oxide [57] ($CrO_2 \rightarrow Cr_2O_3$, 723 to 823 K) commenced on the outer surfaces of reactant crystallites and progressed into the bulk. From X-ray evidence, it was concluded that a topotactic relationship was preserved between the reactant (CrO_2, rutile) and product (Cr_2O_3, corundum) lattice structures: both contain hexagonal close-packed layers of oxygen ions, although the cation distributions are different. The overall reaction can be represented as withdrawal of one oxygen atom in four. This is achieved by the removal of alternate lattice rows, involving only slight structural disturbance. Such a mechanism is consistent with the observed low temperature of reaction, low E_a (202 kJ mol^{-1}) and the shrinkage which occurs (15%) along the *c*-axis of CrO_2. CrO_2 is a product [58] of the decomposition of CrOOH in oxygen, and the amount produced depends on the availability of oxygen.

Manganese oxides: Malati, reviewing [59] some solid phase properties of manganese dioxide, summarizes the various transformations of MnO_2, which include the following:

$$
\begin{array}{ccccccc}
\delta\text{-}MnO_2 & \rightarrow & \gamma\text{-}MnO_2 & \rightarrow & \beta\text{-}MnO_2 & \leftarrow & \eta\text{- and } \epsilon\text{- } MnO_2 \\
\downarrow & & & & \downarrow & & \\
\alpha\text{-}MnO_2 & & \rightarrow & & \alpha\text{-}MnO_{3/2} & & \\
\downarrow & & & & \downarrow & & \\
\gamma\text{-}MnO_{3/2} & & \rightarrow & & (\alpha, \beta)\ MnO_{4/3} & & (\alpha \text{ and } \beta \text{ phases in equilibrium})
\end{array}
$$

Yanakiev and Zakharieva [60] noted considerable variations in reported values of E_a for the dissociation of MnO_2 and Mn_2O_3. This could arise through the influence of oxygen pressure on reaction temperature, studied by Tinsley and Sharp [61]:

$$\gamma\text{-}MnO_2 \rightarrow 720 \text{ to } 870\text{ K} \rightarrow Mn_2O_3 \rightarrow 1020 \text{ to } 1370\text{ K} \rightarrow Mn_3O_4$$

Giovanoli *et al.* [62] studied the reactions of hydrated manganese(III) manganate(IV) ($Mn_7O_{13}.5H_2O$) and manganese(II) manganate(IV) ($Mn_7O_{12}.6H_2O$). These crystallize in the form of platelets and heating resulted in two-dimensional ordered phases, which undergo topotactic processes to yield very finely-divided γ-MnO_2 and β-MnO_2, but the topotatic relation was not maintained in the subsequent transformation to α-Mn_2O_3.

More recently, Giovanoli has reported [63] comparative X-ray diffraction, electron microscopic and thermogravimetric studies of a number of synthetic samples of manganese dioxide. Three types of water present in the solid can be distinguished: (i) adsorbed water that is desorbed between about 300 and 400 K; (ii) interlayer water in phyllomanganates that is molecular but more strongly retained and is lost at temperatures up to 423 to 523 K; and (iii) hydroxyl water that may be retained to higher temperatures (up to 773 K). β, γ and ϵ-MnO_2 decompose to α-Mn_2O_3 above 773 K and sintering continues to 1273 K.

Iron oxides: The equilibrium dissociation pressure of oxygen over iron(III) oxide has been measured at temperatures up to 1300 K. Values are lower [64] when Fe_3O_4 is present in the Fe_2O_3 (900 to 1200 K). Braun and Gallagher [65] prepared β-Fe_2O_3 by dehydration of tetragonal β-FeOOH in vacuum at 440 K. This modification was more stable than the γ-form and transformed to α-Fe_2O_3 on heating in air at 670 K.

The decomposition [66] of FeO, ($4FeO \rightarrow Fe_3O_4 + Fe$), involves the initial enrichment of O or Fe in lattice planes to form a pseudoperiodic structure parallel to the (100) direction at about 600 K, and subsequently magnetite is precipitated.

Cobalt oxides: Dissociation of Co_3O_4 yields [2] some atomic oxygen as product. The activation energy for the dissociation in argon is reported [67] as 293 kJ mol^{-1} (600 to 1200 K). Ichimura and Komatsu [68] reported a value of 366 kJ mol^{-1} and identified a contracting interface process (970 to 1070 K) for dissociation of the pure oxide, or in the presence of Al_2O_3 or In_2O_3 additives.

A volumetric study [69] of the decomposition of Co_3O_4 between 1123 and 1200 K and 2.66 to 20.73 kPa of oxygen showed that reactions were deceleratory and a mixed rate control model was proposed. The values of E_a for reaction and diffusion were 160 and 10.7 kJ mol^{-1}, respectively.

Co_3O_4 also reacts [67] with carbon to yield CO_2 only, which implies a ready interaction between any CO generated and the oxide. Decomposition of Co_3O_4 was accelerated even when there was no direct contact between oxide and carbon. Incorporation of Li_2O or Na_2O in the reactant [68] resulted in the appearance of an induction period to decomposition during the delayed development of CoO nuclei.

Nishina *et al.* [70] showed that reactions of cobalt oxide were independent of the Al_2O_3 additive below 1120 K, but at higher temperatures the production of cobalt aluminate became significant.

Other transition-metal oxides: On heating V_2O_5, oxygen is initially evolved [71], and there is subsequent sublimation of the oxygen-deficient solid. The average enthalpy of dissociation is given as 330 kJ mol^{-1} and it is suggested that several polymeric species (e.g. V_4O_{10}, V_6O_{14}, V_6O_{12}, V_4O_8 and V_2O_4) are present in the vapour evolved by the vacuum distillation of this compound. MoO_3 and WO_3 sublime [72] as trimers, $(MoO_3)_3$, while the lower oxides disproportionate.

The dissociation of IrO_2 (1070 to 1370 K) has been described [73] as the equilibrium:

$$IrO_{2-x} \rightleftarrows Ir + (1 - x/2)O_2$$

and x varies with temperature. Thermodynamic data on the volatility of RuO_3, RuO_4 and other oxides of the platinum group of metals have been reported [74].

In a more general comparative study of oxide dissociations, including mass spectrometric analysis of the gases released, Kazenas *et al.* [75] conclude that the oxides of Fe, Ni, Cu, Zn and Pb usually give oxygen and lower oxide (+ metal), the oxides of Re are rearranged, and Mo and W oxides yield volatile polymeric species.

Uranium oxides: Interpretation of observations for the dissociations of uranium oxides (and hydroxides) are complicated by the large number of phases and solid solutions mentioned in the literature. Karkahnavala and Phadnis [76] conclude that the thermal stabilities of the following phases increase with symmetry. The same sequence is shown by the enthalpies, rates and values of E_a for these reactions, decomposition temperatures are shown in brackets:

$$\alpha\text{-}UO_3\ (673\ K) < \epsilon\text{-}UO_3\ (733\ K) < \text{amorphous } UO_3\ (773\ K) < \beta\text{-}UO_3\ (803\ K)$$

$$< UO_{2.90}\ (853\ K) < \delta\ (863\ K) < \gamma\ (903\ K)$$

Similar behaviour was found for the hydrates, α- and β-$UO_3.H_2O$. Trace impurities, defects and preparative methods all influence the thermal reactivity. (Comparable behaviour was mentioned above for the lead oxides).

A thermogravimetric kinetic study [77] of the decomposition of α-UO_3 identified two consecutive processes during oxygen loss:

$$\alpha\text{-}UO_3 \rightarrow UO_{2.9} \rightarrow UO_{2.67}$$

for which reported values of E_a were 134 and 450 kJ mol^{-1}, respectively. The value of E_a for the first step in:

$$\gamma\text{-}UO_3 \rightarrow U_3O_{8+x} \rightarrow U_3O_8$$

was 178 kJ mol^{-1}. Bessonov [78], from DTA, X-ray and conductivity measurements, concluded that amorphous UO_3 is converted to α-UO_3 at 640 K and above 810 K forms a series of solid solutions between α-UO_3 and a lower oxide. The kinetics of decomposition of $UO_3 \rightarrow U_3O_8$ depend on the pretreatment, most notably the washing procedure, of the ammonium uranate from which it was formed [79]. The apparent values of E_a for reaction between 873 and 973 K were high, 355 to 492 kJ mol^{-1}, and it was concluded that reaction was first order and not diffusion controlled.

Rundle [80] stated that:

$$UO_2(OH)_2 \rightarrow 563\text{ K} \rightarrow \alpha\text{-}UO_3.H_2O \rightarrow 620\text{ K} \rightarrow UO_3 \rightarrow 720\text{ K} \rightarrow UO_{2.8} \rightarrow 970\text{ K} \rightarrow U_3O_8$$

Table 9.1. includes reference to some of the reactions of uranium oxides.

The complexity of the reported behaviour, together with various inconsistencies apparent between proposed reaction schemes, shows that complete understanding of the kinetics and mechanisms of reactions of oxides of uranium is far from being established and agreed.

9.4. REFRACTORY OXIDES: EXTENDED IMPERFECTIONS - CRYSTALLOGRAPHIC SHEAR AND BLOCK STRUCTURES

Representations of lattice imperfections as individual and randomly-distributed entities within crystalline solids have provided satisfactory theoretical models for the explanation of a number of phenomena. This approach [81], however, has not so far been shown to be capable of accounting for observed reactivities, structures and thermodynamic properties of a number of the more highly defective and nonstoichiometric phases. Evidence has accumulated which indicates that properties of nonstoichiometric solids are determined by structures and mechanisms which depend on the interactions of imperfections which are distributed in a regular manner. Two related lattice features which may be relevant in explaining the observed decomposition (and reduction) reactions of certain oxides are described below. These are extended imperfections and ordered defect structures. There is a considerable literature on this topic, including several reviews [81-84].

The compositional changes in a crystal, which are a consequence of a departure from stoichiometry, may be accommodated by the generation of extended imperfections in the form of planes of atoms having different coordination (e.g. possessing the structure of a lower oxide) which are accommodated between regions of structure characteristic of the original compound (i.e. the higher oxide). Such behaviour, *crystallographic shear*, has been investigated in titanium oxide by Hyde *et al.* [82,83] and is discussed further below.

Increasing incidence of crystallographic shear, itself an extended and complex defect, is accompanied by a degree of interaction which may lead to an ordering process. Such reorganization may culminate, in favourable systems, in the development of regular structures wherein constant-sized blocks of one lattice type adhere to regions with an alternative structure, which themselves correspond to a different gross stoichiometry. This is regarded as coherent intergrowth (Wadsley defects) and the texture is sometimes referred to as a *block structure*. Work in this and related fields has been the subject of several reviews [83,84] concerned with the detailed textures of oxides (and some fluorides) of various metals, notably including those of Nb, W and Ti.

The cooperative movement of large numbers of atoms represents an alternative, and in some ways more precise [83], mechanism of reaction in addition to the well-established interface advance and diffusion-controlled processes which are considered throughout this book. Examination of the possible participation of crystallographic shear in the reactions of solids has been largely restricted to refractory oxides, but comparable or related behaviour could, in principle, operate in a variety of other solid state rate processes.

9.5. HIGH TEMPERATURE VOLATILIZATION OF OXIDES

L'vov [85], from published measurements of the rates of volatilization of many oxides, has proposed a mechanism of sublimation based on the Hertz-Knudsen-Langmuir model of vaporization (see Section 2.4.5.). This approach is similar in some respects to that developed by Searcy [86].

The experimental studies surveyed refer to the volatilization at high temperatures (mainly above 1200 K) of relatively stable compounds (18 oxides, including Li_2O, MgO, CaO, Al_2O_3, SnO_2, PbO, Cr_2O_3, MnO, ZnO, together with the alkali metal hydroxides and the metals Ni, Cu and Ag). Some of these melt before dissociation. Two modes of atomization of the oxides were distinguished, equimolar and isobaric. In the equimolar mode, the equilibrium partial pressures of both gaseous products conform to expectation from the reaction stoichiometry. For the reaction:

$$M_aX_b(\text{s or l}) \rightarrow aM(g) + bX(g)$$

in vacuum, the fluxes of both products are expressed by:

$$J_M / J_X = a/b$$

When, however, dissociative vaporization occurs in a foreign gas, the partial pressure (p'_i) of one of the gaseous components (O_2, OH, H_2O or halogen are cited as examples) may greatly exceed the equilibrium partial pressure resulting from decomposition (p_i). If the magnitude of p'_i remains unchanged with an increase in temperature, this reaction can be described as isobaric. The consequences of such a model have been explored by L'vov, who showed that the magnitude of E_a for the isobaric mode is (1 + b/a) times that for the equimolar vaporization mode. The influences of the presence of either O_2 or H_2 on these processes were also investigated. Although many assumptions are made, the apparent agreement between theory and experiment [85] is promising.

9.6. CONCLUSIONS

The structures of metal oxides and mixed oxides are often relatively simple, so that many features of reaction, such as the properties and dispositions of extended imperfections (Section 9.4.), can be characterized more easily than for more complex solids. The ability of these compounds to deviate from stoichiometry does, however, increase interpretational difficulties. Topotactic behaviour, arising from structures based on simple ions, is important in formulating mechanisms [87]. The surface chemistry and interface reactions of oxides are also of importance in heterogeneous catalysis and metal oxidations.

The decompositions of metal oxides may be classified (Table 9.1.) according to whether or not atomic oxygen is formed as a product. Atomic oxygen may be formed during the decompositions of other groups of solids, but this may arise through secondary reactions as, for example, in the breakdown of ClO_2 during the decomposition of NH_4ClO_4.

REFERENCES

1. J.E. Germain, Intra-Sci. Chem. Rept., 6 (1972) 101.
2. G.V. Malinin and Yu.M. Tolmachev, Russ. Chem. Rev., 44 (1975) 392.
3. G.V. Malinin and Yu.M. Tolmachev, Zh. Neorg. Khim., 14 (1969) 307; Zh. Fiz. Khim., 43 (1969) 2012.
4. Y. Harano, Nippon Kagaku Zasshi, 82 (1961) 19,152.
5. K. Mitani and Y. Asakura, Bull. Chem. Soc. Japan, 38 (1965) 901.
6. L. Brewer, Chem. Rev., 52 (1953) 1.
7. R.H. Lamoreaux, D.L. Hildenbrand and L. Brewer, J. Phys. Chem. Ref. Data, 13 (1984) 151; 16 (1987) 419.
8. See for example: O. Kubaschewski and B.E. Hopkins, Oxidation of Metals and Alloys, Butterworths, London, 1967.
9. P. Barret, Cinétique Hétérogène, Gauthier-Villars, Paris, 1973.
10. F.S. Stone and R.J.D. Tilley, Reactivity of Solids, (Eds J.S. Anderson, M.W. Roberts and F.S. Stone), Chapman and Hall, London, 1972, p.262.
11. V.V. Boldyrev, M. Bulens and B. Delmon, The Control of the Reactivity of Solids, Elsevier, Amsterdam, 1979.
12. E.R.S. Winter, J. Chem. Soc. A, (1968) 2889; (1969) 1832.
13. E.R.S. Winter, J. Catal., 15 (1969) 144; 19 (1970) 32; 22 (1971) 158.
14. G.K. Boreskov, V.V. Popovskii and V.A. Sazonov, Int. Congr. Catal., 4th, Moscow, 1968, p. 439; G.K. Boreskov, Adv. Catal., 15 (1964) 285; G.K. Boreskov and L.A. Kasatkina, Russ. Chem. Rev., 37 (1968) 613.
15. K. Klier, J. Catal., 8 (1967) 14.
16. T. Kiyoura, Bull. Soc. Chim. Japan, 39 (1966) 2135.
17. Y. Moro-Oka, Y. Morikawa and A. Ozaki, J. Catal., 7 (1967) 23; 5 (1966) 116.
18. D. Delmon and A. Roman, J. Chem. Soc., Faraday Trans. I, 69 (1973) 941 and references therein.
19. R. Frety, L. Tournayan and H. Charcosset, Ann. Chim., 9 (1974) 341.
20. B. Delmon, Introduction à la Cinétique Hétérogène, Editions Technip, Paris, l969.
21. A.K. Galwey, Adv. Catal., 26 (1976) 247.
22. D. Dollimore, Select. Ann. Rev. Anal. Sci., 2 (1972) 1.
23. S.A. Tokareva, Russ. Chem. Rev., 40 (1971) 165.
24. A.B. Tsentsiper and Z.I. Kuznetsova, Izv. Akad. Nauk SSSR, Ser. Khim., 10 (1965) 1902.
25. M.M. Pavlyuchenko and T.I. Popova, see Chem. Abs., 64 (1966) 13440c.

26. V.Z. Kuprii and V.A. Lunenok-Burmakina, Zh. Fiz. Khim., 45 (1971) 179.
27. T.V. Rode and G.A. Gol'der, Dokl. Akad. Nauk SSSR, 110 (1956) 1001; (Chem. Abs., 50 (1956) 12621d.)
28. D.L. Kraus and A.W. Petrocelli, J. Phys. Chem., 66 (1962) 1225.
29. M.G. Barker and D.J. Wood, Reactivity of Solids, (Eds J.S. Anderson, M.W. Roberts and F.S. Stone), Chapman and Hall, London, 1972, p.623.
30. A.B. Tsentsiper and M.S. Dobrolyubkova, Izv. Akad. Nauk SSSR, Ser. Khim., 7 (1972) 1471.
31. I.I. Vol'nov and E.I. Latysheva, Izv. Akad. Nauk SSSR, Ser. Khim., (1970) 13.
32. A.B. Tsentsiper and R.P. Vasil'eva, Izv. Akad. Nauk SSSR, Ser. Khim. (1968) 1437, 1610; (1967) 2563.
33. M.J. Tribelhorn and M.E. Brown, Thermochim. Acta, 255 (1995) 143.
34. G.N. Lewis, Z. Phys. Chem., 52 (1905) 310.
35. A.F. Benton and L.C. Drake, J. Amer. Chem. Soc., 56 (1934) 255.
36. W.E. Garner and L.W. Reeves, Trans. Faraday Soc., 50 (1954) 254.
37. J.A. Allen, Aust. J. Chem., 13 (1960) 431; 14 (1961) 20.
38. P.J. Herley and E.G. Prout, J. Amer. Chem. Soc., 82 (1960) 1540.
39. M.M. Dubinin, O. Kadlets and V. Ponets, Kinet. Katal., 8 (1967) 292.
40. J.C. Lagier, P. Bussiere and M. Prettre, Bull. Soc. Chim. Fr., (1969) 4289.
41. I.I. Vol'nov, Russ. Chem. Rev., 41 (1972) 314.
42. I. Nakamori, H. Nakamura, T. Hayano and S. Kagawa, Bull. Chem. Soc. Japan, 47 (1974) 1827.
43. H.T. Spath, H.G. Winkler and K. Torkar, Reactivity of Solids, (Eds J.S. Anderson, M.W. Roberts and F.S. Stone), Chapman and Hall, London, 1972, p.745.
44. E.P. Tatievskaya and G.I. Chufarov, Bull. Acad. Sci. URSS, Sci. Tech., (1946) 1005.
45. W. Hirschwald and F. Stolze, Z. Phys.Chem. (Frankfurt-am-Main), 77 (1972) 21; 42 (1964) 96.
46. O. Glemser and U. Stoecker, Berichte, 67 (1963) 505.
47. D. Taylor, J. Chem. Soc., (1962) 1047.
48. I.A. Derbinskii, M.M. Pavlyuchenko and E.A. Prodan, see Chem. Abs., 75 (1971) 133431m; 76 (1972) 18322k, 7057b and earlier papers.
49. F. Lawson, Nature (London), 215 (1967) 955.
50. R. Colin, J. Drowart and V. Verhaegen, see Chem. Abs., 63 (1965) 4954c.
51. N. Nishimura, T. Higaashiyama, S. Yamamoto and S. Hasegawa, Rept. Res. Lab. Surf. Sci., Okayama Univ., 3 (1971) 223.

52. M.I. Gillibrand and B. Halliwell, J. Inorg. Nucl. Chem., 34 (1972) 1143.
53. B.Ya. Brach, G.V. Malinin and Y.M. Tolmachev, Vestn. Lenningrad Univ., Fiz. Khim, 46 (1972) 126; see also: Zh. Neorg. Khim., 13 (1968) 1746; 14 (1969) 2899.
54. J.D. McDonald and J.L. Morgrave, J. Inorg. Nucl. Chem., 30 (1968) 665.
55. J.S. Tiwari, S.C. Sinha and P.K. Ghosh, Technology, 6 (1969) 209.
56. K.A. Wilhelmi, Acta Chem. Scand., 22 (1968) 2565.
57. R.D. Shannon. J. Amer. Ceram. Soc., 50 (1967) 56.
58. M.A. Alario-Franco and K.S.W. Sing, J. Thermal Anal., 4 (1972) 47.
59. M.A. Malati, Chem. Ind., (1971) 446.
60. N. Yanakiev and F. Zakharieva, Priroda (Sofia), 22 (1973) 46.
61. D.M. Tinsley and J.H. Sharp, J. Thermal Anal., 3 (1971) 43.
62. R. Giovanoli, E. Stähli and W. Feitknecht, Helv. Chim. Acta., 53 (1970) 453.
63. R. Giovanoli, Thermochim. Acta, 234 (1994) 303.
64. D.M. Chizhikov, Yu.V. Tsvetkov and E.K. Kazenas, Zh. Fiz. Khim., 46 (1972) 1577.
65. H. Braun and K.J. Gallagher, Nature (London), 240 (1972) 13.
66. J. Manenc, T. Herai and J. Bernard, Reactivity of Solids, (Ed. G.M. Schwab), Elsevier, Amsterdam, 1965, p.432.
67. A.K. Ashin and S.T. Rostovtseva, Izv. Vys. Ucheb. Zaved., Chem. Met., 10 (1967) 5.
68. H. Ichimura and W. Komatsu, Kogyo Kagaku Zasshi, 74 (1971) 147.
69. A. Malecki and B. Prochowska-Klisch, J. Thermal Anal., 41 (1994) 1109.
70. T. Nishina, M. Yonemura, T. Sekine and Y. Kotera, Zairyo, 21 (1972) 544.
71. V.P. Elyumin, Yu. A. Pavlov, V.P. Polyankov and P.A. Nesmerenko, Izv. Vyssh. Ucheb. Zaved., Chem. Met., 13 (1970) 5; 9 (1972) 5.
72. P.E. Blackburn, M. Hoch and H.L. Johnston, J. Phys. Chem., 62 (1958) 769.
73. E.H.P. Cordfunke and G. Meyer, Rec. Trav. Chim., 81 (1962) 670.
74. H. Schaefer, A. Tebben and W. Gerhardt, Z. Anorg. Allgem. Chem., 321 (1963) 41.
75. E.K. Kazenas, D. Chizhikov and Yu.V. Tsvetkov, see Chem. Abs., 79 (1973) 23676v.
76. M.D. Karkhanavala and A.B. Phadnis, India, A.E.C., Bhabha At. Res. Cent. Rept., (1972).
77. S.R. Dharwadkar and M.D. Karkhanavala, J. Indian Chem. Soc., 43 (1966) 363.

78. A.F. Bessonov, Zh. Neorg. Khim., 16 (1971) 3.
79. N.H. Rofail and B.S. Girgis, J. Thermal Anal., 39 (1993) 868.
80. D.L Rundle, Aust. A.E.C., AAEC/TM (Rept.) 1971, AAEC/TM 600.
81. J.S. Anderson, Reactivity of Solids, (Eds J.S. Anderson, M.W. Roberts and F.S. Stone), Chapman and Hall, London, 1972, p.1.
82. R.R. Merritt, B.G. Hyde, L.A. Bursill and D.K. Philp, Phil. Trans. R. Soc. (London), A274 (1973) 627.
83. B.G. Hyde, Reactivity of Solids, (Eds J.S. Anderson, M.W. Roberts and F.S. Stone), Chapman and Hall, London, 1972, p.23.
84. J.S. Anderson and R.J.D. Tilley, Surface and Defect Properties of Solids, (Eds M.W. Roberts and J.M. Thomas), Specialist Periodical Reports, The Chemical Society, London, 3 (1974) 1.
85. B.V. L'vov, Spectrochim. Acta B, 52 (1997) 1.
86. A.W. Searcy, Chemical and Mechanical Behaviour of Inorganic Materials, Wiley, New York, 1970, Chaps 4 and 6.
87. H.R. Oswald and J.R. Günter, Reactivity of Solids, (Eds P. Barret and L.-C. Dufour), Elsevier, Amsterdam, 1985, p.101; Bull. Inst. Chem. Res. Kyoto Univ., 53 (1975) 249; Pure Appl. Chem., 61 (1989) 1323.

Chapter 10

DECOMPOSITION OF OTHER BINARY COMPOUNDS

10.1. INTRODUCTION

The relatively small number of compounds within this classification for which kinetic studies have been reported (hydrides, carbides, nitrides and sulfides) exhibit an unusually wide range of properties. Behaviour can be classified into three very different groups.

(i) Intersitial carbides and nitrides are often extremely refractory and kinetic studies of their thermal reactions are experimentally inaccessible.

(ii) Interface advance reactions that are similar in character to many other decompositions.

(iii) Surface reactions in which the original boundary faces of the reactant remain immobile but are the site of chemical changes.

Reaction rates may be determined by the ease of intracrystalline transport to the surface, or by the chemical change on the surface. These surface reactions often resemble behaviour described in modelling heterogeneous catalytic processes and are usually reversible so that decomposition rates are sensitive to any gases present.

Behaviour of reactants of the present group is similar to that of the oxides (Chapter 9), which are in the same class. Little information is available for other binary compounds, fluorides, chlorides, etc., which usually melt rather than decompose.

10.2. METAL HYDRIDES

10.2.1. Overview

Although many metal hydrides decompose at temperatures well below their melting points [1], there have been comparatively few studies of the kinetics and mechanisms of these hydrogen evolution reactions. Much of the interest in solid hydrides has been concerned with their thermodynamic properties, as hydrogen sources in fuel cells, or as reducing agents, or for technological applications in nuclear processes.

On heating, many hydrides dissociate reversibly into the metal and H_2 gas. The rate of gas evolution is a function of both temperature and $p(H_2)$ but will proceed to completion if the volatile product is removed continuously [1], which is experimentally difficult in many systems. The combination of hydrogen atoms at the metal surface to yield H_2 may be slow [2] and is comparable with many heterogeneous catalytic reactions. While much is known about the mobility of H within many metallic hydride phases, the gas evolution step is influenced by additional rate controlling factors. Depending on surface conditions, the surface-to-volume ratio and the impurities present, the rate of H_2 release may be determined by either the rate at which hydrogen arrives at the solid-gas interface (diffusion control), or by the rate of desorption.

Palczewska [3] has discussed the role and activities of hydride phases in metal-catalyzed reactions. Attention is drawn to the influences of crystallite sizes and of impurities bonded to active sites on the stability of the hydride phases.

10.2.2. Nickel and palladium hydrides

The surface properties of nickel and palladium hydrides have been extensively investigated because of the interest in the mechanisms of heterogeneous reactions involving dissociated hydrogen on these metals [3] and, recently, the controversy over "cold fusion". Hydrogen at the surfaces of nickel and palladium exhibits considerable activity in hydrogenation, isomerization and cracking of hydrocarbons. Many properties of the phases in which hydrogen is incorporated in palladium have been thoroughly studied (e.g. see [4]). Molecular hydrogen is evolved slowly from $PdH_{0.99}$ at about 393 K. The surface desorption step is inhibited by small amounts of contaminants, thus permitting the preparation and investigation of hydrogen-rich phases [2]. In contrast, nickel hydride (H/Ni = 0.04 to 0.2) exerts a very considerable dissociation pressure [2,5] (> 3000 bar) at 298 K. This decomposition is first-order [6,7]. The rate is decreased when copper is deposited on regions of surface distortion [8] and decomposition is believed to proceed by hydrogen diffusion from non-defective regions as well as dislocations.

10.2.3. Chromium hydride

Chromium hydride evolves hydrogen by the irreversible reaction [8]:

$$2CrH_n \rightarrow 2Cr + nH_2$$

which, following a short induction period, proceeds at a constant rate between 273 and 304 K with $E_a = 9$ kJ mol^{-1}. The defective hexagonal structure is retained from $CrH_{<1.0}$ to $CrH_{0.1}$, after which there is lattice reorganization.

10.2.4. Other metal hydrides

Studies of the photodecompositions of alkaline earth hydrides (MgH_2, CaH_2, SrH_2 and BaH_2) $\rightarrow$ Metal + H_2 [9] showed that all exhibited generally similar behaviour. An initial maximum rate of release of hydrogen decreased to a constant value that approximated to a second-order dependency on light intensity. These observations are consistent with a reaction involving electronic carriers and/or energy transfer processes.

α-time curves for the isothermal decomposition of beryllium hydride [10] (BeH_2) between 478 and 523 K were sigmoid. The acceleratory process ($0.1 < \alpha < 0.35$) was fitted by the power law (n approximately 6), attributed to random nucleation followed by a constant rate of growth. The contracting volume equation applied for α between 0.35 and 0.8 and $E_a = 108$ kJ mol^{-1}. An increase in rate resulting from pre-irradiation of the reactant with X-rays was ascribed to an increase in the number of nuclei. Grinding also resulted in an increase of rate. The α-time curves for the decomposition of zinc hydride [11] were also sigmoid (333 to 373 K) and E_a was about 115 kJ mol^{-1}.

The decomposition of titanium hydride in vacuum between 523 and 773 K was slower than the rate predicted by diffusion calculations and the controlling step was identified [12] as the surface combination of hydrogen atoms. The rate of reaction was sensitive to traces of gaseous H_2, but not to O_2. The inhibiting effect exerted by the presence of helium was ascribed to opposition to the diffusive dispersal of product from the vicinity of the desorption interface. The rates of decomposition of the hydrides of four related metals [13] (TiH_2, ZrH_2, NbH and TaH) studied between 343 and 973 K pass through a temperature maximum. This was explained by the occurence of two consecutive reactions: first-order decomposition of the hydride, followed by second-order combination of the hydrogen atoms before desorption.

The decomposition of thorium hydride is controlled by kinetic rather than diffusion parameters [14]. The value of E_a is 58 kJ mol^{-1} for the decomposition of ThH_3 between 483 and 663 K and is 137 kJ mol^{-1} for the decomposition of $ThH_{1.5}$ in the interval 873 to 1073 K. Reaction orders were reported to be 0.34 and 0.25, respectively. The evolution of hydrogen from $PbH_{0.2}$ proceeds [15] at about 463 K in steps during an increasing temperature programme and the apparent magnitude of E_a increases from 6.7 to 15 kJ mol^{-1} as the hydrogen content decreases.

Studies of the rare earth hydrides have been largely concerned with the stabilities of the phases formed, and the thermodynamic properties of the solid solutions [16-18]. The decompositions of hydrides of metals with pronounced catalytic properties (Ni, Pd) are often limited by surface steps involving hydrogen atom reactions, which are sensitive to poisoning and resemble rate processes occurring at surfaces of heterogeneous catalysts. In contrast, the decomposition of hydrides of other metals (Be, Zn) apparently proceed by nucleation and growth mechanisms.

CuH, prepared by reduction of Cu^{2+} ions with hypophosphorous acid, consists of agglomerates of needle-like particles 10 nm in size [19]. It is unstable, i.e. the Cu-H bond is easily broken at 300 K and in an inert atmosphere decomposes to Cu + H_2. α-time plots were sigmoid, E_a = 8 to 36 kJ mol^{-1} and reaction at 373 K was almost instantaneous.

Aluminium hydride decomposes ($2AlH_3 \rightarrow 2Al + 3H_2$) between 403 and 423 K by an irreversible process exhibiting a sigmoid α-time curve [20,21]. During the induction period data fitted the cubic law ($0.001 < \alpha < 0.04$) and $E_a = 104 \pm 11$ kJ mol^{-1}. The Avrami-Erofeev equation satisfactorily expressed the rate of the main reaction with $n = 2$ during the acceleratory period ($0.06 < \alpha < 0.6$) and $n = 3$ during the decay process ($0.6 < \alpha < 0.97$) with $E_a = 157 \pm 13$ kJ mol^{-1} for both stages. Electron microscopic observations identified large numbers of subsurface colloidal particles which grow slowly during the initial stages of reaction to produce elongated acicular nuclei. These grow into ellipsoidal nuclei during the induction period, but details of the interface structure could not be resolved for the diffuse interfacial region. However, the generation of growth nuclei resulted in the appearance of sharp interfaces, the kinetic behaviour was characteristic of a nucleation and growth process and E_a increased. The reaction is accelerated by ultraviolet and γ-irradiation, due to the generation of additional decomposition sites, discussed in detail in the references cited above. Later work [22] showed that while the product from thermal and photolytic decomposition was in the form of needle-like particles, solid decomposed in an electron beam gave a residue having the appearance of metallic colloidal clusters.

10.2.5. Group IA metal aluminium hydrides

The decomposition of $LiA\ell H_4$ involves an initial rate process which may arise from impurities [23] (either $>A\ell\text{-}OH + H\text{-}A\ell< \rightarrow >A\ell\text{-}O\text{-}A\ell< + H_2$ or an organometallic compound). The possibility of melting was mentioned. When the contribution from the initial process had been subtracted, the subsequent isothermal reaction between 400 and 432 K gave sigmoid α-time curves that were fitted by the modified Prout-Tompkins equation and E_a was about 100 kJ mol^{-1}. The third stage in the decomposition was represented as $LiA\ell H_2 \rightarrow LiH + A\ell + ½H_2$. This was a first-order reaction with E_a = 195 kJ mol^{-1}.

Other studies [24,25] (between 398 and 423 K) gave sigmoid α-time curves in which five stages were distinguished. The initial rapid evolution of about 5% of the gaseous products was a deceleratory process. This was followed by an induction period after which the rate accelerated according to the cubic law for α from 0.09 to 0.6. After the point of inflection, the rate became deceleratory and data fitted the first-order equation ($0.6 < \alpha < 0.9$). This was ascribed to reactant particle disintegration leaving blocks without nuclei. The values of E_a for the acceleratory and decay stages were 94 to 99 and 81 to 87 kJ mol^{-1}, respectively, during the reaction $LiAlH_4 \rightarrow LiAlH_2 + H_2$. The final rate process was a constant slow rate of gas evolution ($LiAlH_2 \rightarrow$ products + H_2). It was concluded that the chemical steps occurring during decomposition of both unirradiated and γ-irradiated $LiAlH_4$ were the same, but prior irradiation increased the rate constants. In kinetic studies of the photolytic process it was concluded that two excited sites are required to produce a decomposition centre.

From non-isothermal (DSC) studies it has been shown [26] that $NaAlH_4$ decomposes in three steps :

$$3NaAlH_4 \rightarrow Na_3AlH_6 + 2Al + 3H_2 \qquad (502\ K)$$
$$Na_3AlH_6 \rightarrow 3NaH + Al + 1½H_2 \qquad (558\ K)$$

followed by NaH decomposition at 608 K. Below the melting point, $NaAlH_4$ reacted at 441 K with a sigmoid α-time curve which was satisfactorily represented by the Avrami-Erofeev equation with n about 2 during the initial stages and $n = 1$ (first-order behaviour) when α was greater than 0.45 [27]. γ-irradiation accelerated subsequent decomposition. Simultaneous ultraviolet irradiation and thermal decomposition gave complex plots of product yield against time. Electron microscopic examinations of $NaAlH_4$ [28] caused beam induced decomposition with H_2 evolution and growth of small (about 10μm) crystallites of metallic sodium.

10.3. METAL CARBIDES

10.3.1. Overview

Few detailed studies of the kinetics and mechanisms of carbide decompositions have been reported. The absence of a volatile product increases the experimental difficulties.

10.3.2. Cobalt, nickel and iron carbides

Hofer *et al.* [29] used a magnetic method to measure the isothermal kinetics of the decompositions of cobalt and nickel carbides. The reaction of Co_2C was zero-order ($0.20 < \alpha < 0.75$) with $E_a = 226$ kJ mol^{-1} in the range 573 to 623 K, and became deceleratory thereafter. The behaviour of Ni_3C differed in that there was a short induction period, but there was again a period of constant rate ($0.3 < \alpha < 0.9$) with $E_a = 295$ kJ mol^{-1} between 593 and 628 K and the final period was deceleratory. There was no evidence for the intervention of a lower carbide. The mechanisms of these reactions were not discussed.

According to Escoubes and Eyraud [30] the decomposition of nickel carbide in vacuum is deceleratory throughout to yield a residue consisting of particles of metal separated by carbon. Between 533 and 583 K the contracting volume ($0.10 < \alpha < 0.65$) and first-order equations ($0.10 < \alpha < 0.85$) applied over the ranges indicated, with $E_a = 206$ kJ mol^{-1}. In 200 Torr argon, α-time plots were sigmoid in the range 603 and 623 K and were fitted by the Prout-Tompkins equation ($E_a = 166$ kJ mol^{-1}). The induction period diminished as the gas pressure present decreased. The available kinetic data are compared in Figure 10.1.

In agreement with the above study, Dorémieux [32] found that the decomposition of finely-divided Ni_3C between 543 and 613 K was a first-order process, but, in contrast, no significant influence of nitrogen was detected. The reaction with hydrogen to yield methane proceeded more rapidly than the decomposition. The reaction of carbon atoms at the metal surface to yield amorphous carbon was identified as the slow stage, although when hydrogen is present this is unnecessary and the diffusion of carbon to the surface, followed by hydrogenation rather than combination, is rate limiting. When the carbon content is sufficiently decreased, nickel is formed and there is subsequent growth of the metal particles.

A related reaction, the decomposition of acetylene on small particles of nickel to yield deposits of amorphous carbon is of interest. This reaction [33] occurs at 870 K, which is above the stability range of Ni_3C. Acetylene is dissociatively adsorbed on restricted regions of small (< 1 μm) particles of nickel. Carbon dissolves in the metal, travels down a thermal gradient and is precipitated at the cooler end in the form of a cylindrical fibre. The exothermicity of these processes ensures melting of the participating metal particle. The relevance of these observations for a heterogeneous process is that small atypical fragments of catalyst are responsible for a disproportionately large fraction of the overall change (analogous in some respects to surface defects). The active surface of the solid cannot be represented as an immobile matrix, but may include local, perhaps mobile, fragments of considerably enhanced reactivity.

Several phases of iron carbide are known which differ in structure and/or carbon contents. The formation and physical disposition of these phases in steel and other alloys is of considerable technological importance and many systems have been considered in detail [34,35], although because melting occurs the subsequent decomposition has received less attention.

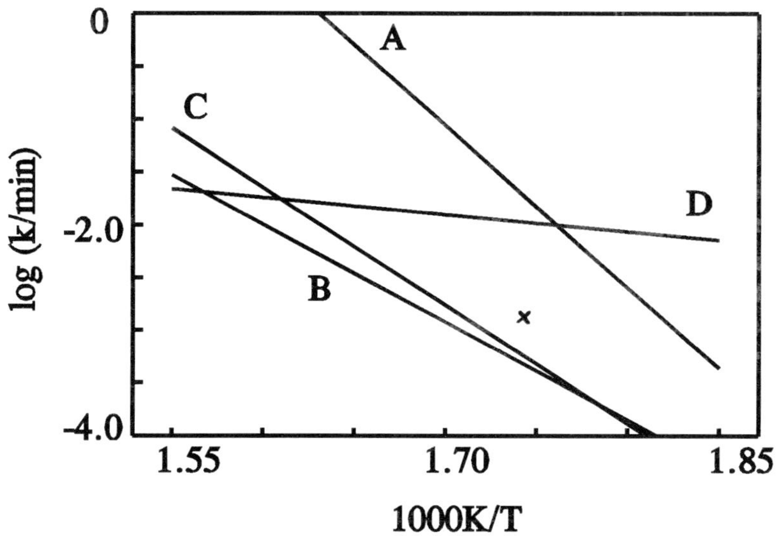

Figure 10.1.
Composite Arrhenius plot for the decompositions of nickel and cobalt carbides, using approximate zero-order rate constants for the interval $0.2 < \alpha < 0.7$. Lines A and B are for decomposition of Ni_3C in vacuum and in nitrogen, repectively (Escoubes and Eyraud [30]), Line C is for the decompositions of Ni_3C and Co_2C (data for both reactions are close to a single line, Hofer *et al.* [29]) and Line D is for the reaction of Ni_3C in hydrogen which yields methane (Galwey [31]). The single point (x) is estimated from Dorémieux [32].

10.3.3. Other carbides

The interstitial carbides are infusible, unreactive substances. Some temperatures of the onset of decomposition have been estimated using an argon plasma jet. Values reported [36] for WC, TaC and NbC are 4000, 4400 and 7000 K, respectively.

The decomposition of caesium graphite proceeds [37] to completion in several distinguishable steps, each of which results in a decrease in the Cs/C ratio. The rate of Cs evolution in each step remains constant, the controlling process being identified as metal desorption. The surface concentration is maintained constant by diffusion from within the crystal. The values of E_a (154 to 117 kJ mol^{-1}) decrease with decreasing caesium content of the reactants, but the kinetic data offer little insight into the mechanism of the solid state processes.

McCowan [38] used electron microscopic and X-ray measurements to study the thermal decomposition of silver acetylide. Although detailed α-time relationships could not be established, the value of E_a was estimated to be 170 kJ mol^{-1} in the interval 388 to 408 K. The rate-limiting step was identified as the production of an electron and an acetylide radical that react further to yield amorphous carbon. Decomposition is catalyzed by the product, probably metallic silver, and explosion was ascribed to the accumulation of catalyst rather than heat.

10.4. METAL NITRIDES

Many metal nitrides are interstitial [39] in character and, like many carbides borides, etc., tend to be hard, unreactive and have high melting points. Of the few decompositions of nitrides which have attracted interest, the iron compounds are important because such phases may participate in the synthesis of ammonia catalyzed by iron. The decomposition of iron nitride between 623 and 773 K is second-order [40] in concentration of interstitial nitrogen with E_a = 175 kJ mol^{-1}. Because the rate of diffusion of nitrogen to the surface was at least 10^4 times as fast as the rate limiting step for nitrogen release, a surface bimolecular interaction ($2N \rightarrow N_2$) was identified as rate controlling and this interpretation was supported by theoretical calculations. Logan *et al.* [41] reported E_a for decomposition of the nitride as 160 kJ mol^{-1}, the same value as that found for ammonia decomposition, but significantly greater than that for the hydrogenation of the nitride to yield ammonia (58 kJ mol^{-1}), which also involves transport of nitrogen from the bulk to the surface, but proceeds at a lower temperature. Again the nitrogen desorption step was identified as rate limiting in the decomposition of the nitride. A probable mechanism of decomposition of Fe_4N is diffusion of nitrogen atoms to the surface, followed by bimolecular combination and evolution of N_2.

The reaction of copper metal with ammonia [42] between 540 and 570 K gave copper nitride, Cu_3N. A reactant containing a high proportion of Cu_3N (92%) was more stable than that containing more copper metal (70% Cu_3N). Decomposition occurred at 700 K and above in nitrogen. Reaction ($\rightarrow NH_3$) commenced at about 380 K in H_2 and above 650 K in NH_3. Nickel metal reacted with ammonia to give Ni_3N at 623 K and the main decomposition occurred at 723 K. Ni_3N was stable only below about 430 K in H_2, but to 680 K in ammonia. Mechanisms of these reactions were not discussed but references to incorporation of nitrogen into metal lattices suggests that atomic diffusion processes control rates.

10.5. METAL SULFIDES

10.5.1. Overview

Decompositions of transition-metal sulfides, notably those of Fe, Ni, Cu and Co, have been of technological importance in ore refining. Some of the published work is concerned with naturally-occurring minerals, while other studies used synthetic preparations. Reactions often proceed by a contracting interface mechanism and the rates are decreased when gaseous product is present, or its escape is opposed by an inert gas. On heating in air, several metal sulfides form sulfates or oxysulfates as intermediates in a sequence of reactions which finally yield metal oxides [43].

10.5.2. Iron sulfide

There have been several studies [44-49] of the decomposition of FeS_2:

$$FeS_2 \text{ (pyrites)} \rightarrow FeS_{1+x} \text{ (pyrrhotite)} + (1\text{-}x)/2\ S_2(g)$$

Decomposition in vacuum between 724 and 749 K was shown [44] to fit the contracting volume equation (E_a = 311 kJ mol^{-1}) and the reaction rate decreased in the presence of an inert gas. Instantaneous nucleation was followed by rapid surface growth.

Coates and Bright [45] found that the value of E_a for reaction in argon between 873 and 926 K was 283 kJ mol^{-1}. The rate of advance of the interface into a cylindrical pellet of compressed reactant was constant, but attempts to investigate the early acceleratory process were unsuccessful. Samal [46], working in the temperature range 759 to 851 K, found a much smaller activation energy (112 kJ mol^{-1}). The contracting volume equation was applied and the area of the solid increased during reaction. Zhukovskii *et al.* [47] reported more complicated behaviour in which the value of E_a decreased with extent of reaction (E_a = 262, 196 and 141 kJ mol^{-1} when α = 0.2, 0.6 and 0.75, respectively). Rates were dependent on particle size and the progress of reaction was hindered by the presence of argon which opposed the ease of escape of the product sulfur.

From a kinetic thermogravimetric study [48], supported by microscopic observations, it was concluded that the decomposition of pyrites occurs in two stages starting above about 800 K. The first mass loss occurred during the formation of a surrounding layer of pyrrhotite as the reaction interface advanced inwards. Results gave an excellent fit to the contracting volume equation ($E_a = 285 \pm 5$ kJ mol^{-1} and $A = 10^{13}$ to $2\text{x}10^{14}$ s^{-1}). The second reaction was diffusion controlled during decomposition to lower members of the pyrrhotite series with $E_a = 190$ kJ mol^{-1} and $A = 10^5$ to 10^6 s^{-1}. A similar isothermal study by Jovanovic [49] of the thermal decompostion of FeS_2 between 873 and 1123 K in an inert atmosphere led to similar conclusions. Although Arrhenius-type plots are presented, activation energy values were not reported.

10.5.3. Cobalt sulfide

The kinetics of CoS_2 decomposition were similar [44] to the behaviour of FeS_2 in that the contracting volume equation applied, although the value of E_a was lower (241 kJ mol^{-1} in the interval 747 to 776 K). The decomposition of Co_4S_3 at low pressures (1393 to 1678 K) was complicated. The reaction was zero-order when α was less than 0.1, there was a parabolic dependence ($0.1 < \alpha < 0.35$) and a further linear dependence for α greater than 0.9, with values of $E_a = 75$, 162 and 201 kJ mol^{-1}, respectively, for the three stages.

10.5.4. Nickel sulfide

The decomposition of nickel sulfide ($NiS_2 \rightarrow NiS + \frac{1}{2}S_2$) in the range 668 to 700 K was deceleratory and data fitted the contracting volume expression, rather than a diffusion-controlled model [44] ($E_a = 243$ kJ mol^{-1}). Similar behaviour was reported by Pavlyuchenko and Samal [46,50], ($E_a = 233$ kJ mol^{-1}, 673 to 723 K). The decomposition of $Ni_{0.95}S$ (683 to 803 K) was controlled by the lattice diffusion of both constituent ions.

10.5.5. Copper sulfide

The kinetics of decomposition of CuS [50] in the interval 583 to 611 K were fitted by the contracting volume equation with $E_a = 250$ kJ mol^{-1}. Shah and Khalafalla [51] studied the decomposition in nitrogen (618 to 673 K, $E_a = 100$ kJ mol^{-1}) and concluded that the rate depended on the surface area rather than interface advance.

10.5.6. Zinc sulfide

Both α and β forms of ZnS sublime dissociatively [52] ($\rightarrow Zn(g) + \frac{1}{2}S_2(g)$, with a negligible concentration of ZnS(g)). Between 1010 and 1445 K the Gibbs energies of the α and β phases were $\Delta G°/\text{J mol}^{-1} = 376\,700 - 191.9T/\text{K}$ and $374\,200 - 190.4T/\text{K}$, respectively. The transition α-ZnS $\rightarrow$ β-ZnS was almost athermal at the transition temperature, 1293 K. Milling was found to decrease this transition temperature by more than 200 K.

Decompositions of the compounds $[M(SCH_3)_2]_n$ ($\rightarrow n(CH_3)_2S + nMS$ where M is Zn or Cd) fitted the first-order or Prout-Tompkins equations with E_a = 143 to 191 kJ mol^{-1} [53,54]. Kinetic behaviour depended upon the surface area of the reactant.

10.5.7. Other sulfides and some related compounds

The decomposition of MoS_2 occurs above 1070 K in vacuum to yield an epitaxial layer of metal on the disulfide, but oxides are formed if oxygen is present [55]. In a reducing atmosphere, the product is Mo_2S_3 [56] and, at higher temperatures, sulfur is volatilized from a melt and metal separates.

The phase relations, decomposition temperatures and/or enthalpies of formation of a wide range of solids e.g. $OsSe_2$, $RuSe_2$, $OsAs_2$ and $PtAs_2$ [57], $PdTe_2$ [58] etc., have been investigated using measurements of the dependence of dissociation pressure on temperature and sometimes including kinetic evidence. For example it is suggested [59] that in the sublimation of Zn_3P_2 ($\rightarrow$ Zn, P_4 and P_2 between 620 and 820 K) the activation step must involve considerable rearrangement of the intermolecular forces. Also, in the sublimation of GeSe, it is believed [60] that multistep desorption processes are involved. Facetting of the surfaces [61] of InSb and GaAs has been observed well below the melting points of these solids.

10.5.8. Evaporation rates

L'vov and Novichikhin [62] have explored the possibilities of treating decomposition as dissociative sublimation (see Section 2.4.5. for detail). Some of the reactants considered were binary compounds. Magnitudes of E_a calculated for the following reactions:

$BN(s) \rightarrow B(g) + \frac{1}{2}N_2(g)$ (1800 K)

$Be_3N_2(s) \rightarrow 3Be(g) + 1.4N(g) + 0.3N_2(g)$ (1800 K)

$Mg_3N_2(s) \rightarrow 3Mg(g) + N(g) + \frac{1}{2}N_2(g)$ (1200 K)

$AlN(s) \rightarrow Al(g) + 0.6N(g) + 0.2N_2(g)$ (1700 K)

$GaN(s) \rightarrow Ga(g) + 0.4N(g) + 0.3N_2(g)$ (1300 K)

$CdS(s) \rightarrow Cd(g) + 0.3S(g) + 0.35S_2(g)$ (1000 K)

$CdSe(s) \rightarrow Cd(g) + 0.7Se(g) + 0.15Se_2(g)$ (1000 K) agree reasonably well with experimental values at the temperatures indicated, except where evaporation was opposed by a condensed product. For some reactions allowance had to be included for changes in compositions of the volatile products.

10.6. CONCLUSIONS

The systems considered in the above survey exclude salts which melt and little information is available concerning the refractory interstitial compounds. Particular interest is evident in the literature in those phases which are probably intermediates in heterogeneous catalytic reactions (e.g. PdH, Ni_3C, Fe_4N, etc.). Reactivities within the groups of solids considered sometimes vary widely, e.g. nickel hydride is much less stable than palladium hydride; tungsten and niobium carbides are extremely refractory while Ni_3C decomposes at about 600 K. Kinetic characteristics and reactivities are not correlated in any obvious way with the chemical constituents of the solids. Although the shapes of α-time relations have been reported, and patterns of kinetic behaviour are often agreed for specific rate processes (agreements of E_a values are less satisfactory), the mechanistic significance of such information is apparently less easily interpreted here than in some other groups of solid state reactions. Microscopic evidence does not always materially supplement kinetic evidence and for some systems, e.g. the decomposition of Ni_3C, this approach has been unsuccessful. In surface-limited, reversible and diffusion-controlled reactions the important rate controlling function may vary with the particle size of the reactant and the prevailing reaction conditions.

While it is often possible to demonstrate that a surface process is rate limiting, identification of the step concerned is not always so readily achieved (as in heterogeneous catalysis which involve comparable mechanistic steps). Reaction rates are determined by reactant areas and are slow compared with the rate of diffusive transport of material to the appropriate boundaries. Surface limited reactions are also sensitive to the ease of removal of volatile products, which may be hampered by the presence of an inert gas. Readsorption may influence the effective concentrations of participating surface intermediates. As in catalytic heterogeneous reactions, the sequence of changes which precede product evolution may involve several interlinked steps, and the parameters which determine the overall progress of reaction are not always readily identified.

While several reactions exhibit sigmoid α-time relationships, including an induction period, such kinetic behaviour does not necessarily imply the occurrence of a nucleation and growth mechanism, but could be a consequence of initial impurity removal, or the necessity for the generation of an appropriate intermediate.

Microscopic evidence, enabling such distinctions to be made, is not available for many of the systems of interest, e.g. BeH_2 [10] and Ni_3C [29,30]. It is also dersirable that microscopic evidence should be used in support of mechanistic proposals based on advancing interface reaction models, for example in the decompositions of FeS_2 [44-47] and CoS_2 [44].

The control of reaction rates by a bulk diffusion process is not usually demonstrable by microscopic observations, but support may be obtained from measurements of diffusion coefficients of appropriate species within the structure concerned. This approach has been invaluable in formulating the mechanisms of oxidation of metals, where rates of reaction have been correlated with rates of transportation of ions across barrier layers of product. Sometimes the paths by which such movements occur correspond to regions of high diffusivity, involving imperfect zones within the barrier layer, compared with normal rates of intracrystalline diffusion across more perfect regions of material [63]. Diffusion measurements have been made for ions in nickel sulfide and it was concluded that the decomposition of NiS is diffusion controlled [50].

Comparative studies [62] of calculated and measured volatilization rates, on heating solids in vacuum, is an approach with potential value for determining reaction mechanisms.

REFERENCES

1. J.P. Blackledge, Metal Hydrides, (Eds W.M. Mueller, J.P. Blackledge and G.C. Libowitz), Academic Press, New York, 1968, Chap.5.
2. W.A. Oates and T.B. Flanagan, Canad. J. Chem., 53 (1975) 694.
3. W. Palczewska, Adv. Catal., 24 (1975) 245.
4. F.A. Lewis, The Palladium - Hydrogen System, Academic Press, London, 1967, Chap.5.
5. B. Baranowski and K. Bochenska, Rocz. Chim., 38 (1964) 1419.
6. A. Janko, Bull. Acad. Polon. Sci., Ser. Sci. Chim., 10 (1962) 617.
7. J. Pielaszek, Bull. Acad. Polon. Sci., Ser. Sci. Chim., 20 (1972) 611.
8. R.J. Roy and T.R.P. Gibb, J. Inorg. Nucl. Chem., 29 (1967) 341.
9. D. Dougherty and P.J. Herley, J. Less-Common Metals, 73 (1980) 97.
10. G.R. Schneider and W.G. Nigh, Combust. Flame, 15 (1970) 223.
11. R. Bluke, V. Breicis, B.Belicka and L.Liepina, Chem. Abs., 74 (1971) 130907q.
12. C.W. Schoenfelder and J.W. Swisher, J. Vac. Sci. Tech., 10 (1973) 862.

13. G.G. Zyryanov, B.M. Mugutnov and L.A. Shvartsman, Dokl. Akad. Nauk SSSR, 208 (1973) 888.
14. T.E. Os'kina and K.B. Zaborenko, Chem. Abs., 79 (1973) 97481z.
15. M.W. Roberts and N.J. Young, Trans. Faraday Soc., 66 (1970) 2636.
16. K.I. Hardcastle and J.C. Warf, Inorg. Chem., 5 (1996) 1719,1726, 1728,1736.
17. C.E. Messer and G.W.H. Hung, J. Phys. Chem., 72 (1968) 3958.
18. C.E. Messer and M.K. Park, J. Less-Common Metals, 26 (1972) 235.
19. N.P. Fitzsimons, W. Jones and P.J. Herley, J. Chem. Soc., Faraday Trans., 91 (1995) 713.
20. P.J. Herley, O. Christofferson and R. Irwin, J. Phys. Chem., 85 (1981) 1874, 1882, 1887.
21. P.J. Herley, O. Christofferson and J.A. Todd, J. Solid State Chem., 35 (1980) 391.
22. P.J. Herley and W. Jones, J. Mater. Sci. Lett., 1 (1982) 163.
23. M. McCarty, J.N. Maycock and V.R. Pai Verneker, J. Phys. Chem., 72 (1968) 4009.
24. P.J. Herley and D.A. Schaeffer, J. Phys. Chem., 82 (1978) 155.
25. P.J. Herley and D.H. Spencer, J. Phys. Chem., 83 (1979) 1701.
26. J.A. Dilts and E.G. Ashby, Inorg. Chem., 11 (1972) 1230.
27. D. Dougherty and P.J. Herley, J. Phys. Chem., 86 (1982) 4161.
28. W. Jones, T.G. Sparrow, B.G. Williams and P.J. Herley, Mater. Lett., 2 (1984) 377.
29. L.J.E. Hofer, E.M. Cohn and W.C. Peebles, J. Phys. Coll. Chem., 53 (1949) 661; 54 (1950) 1161.
30. M. Escoubes and C. Eyraud, Bull. Soc. Chim. Fr., (1966) 1369, 1374.
31. A.K. Galwey, J. Catal., 1 (1962) 227.
32. J.L. Dorémieux, Reactivity of Solids, (Eds J.S. Anderson, M.W. Roberts and F.S. Stone), Chapman and Hall, London, 1972, p.446.
33. R.T.K. Baker, M.A. Barber, P.S. Harris, F.S. Feates and R.J. Waite, J. Catal., 26 (1972) 51; 30 (1973) 86.
34. J.W. Christian, Transformations in Metals and Alloys, Pergamon, Oxford, 1965.
35. A.T. Davenport and R.W.K. Honeycombe, Proc. R. Soc.(London), A322 (1971) 191.
36. S.A. Panfilov and Yu.V. Tsvetkov, Chem. Abs., 68 (1968) 71813g.
37. F.H. Salzano and S. Aronson, J. Chem. Phys., 42 (1965) 1323.
38. J.D. McCowan, Trans. Faraday Soc., 59 (1963) 1860.

39. D.R. Glasson and S.A.A. Jayaweera, J. Appl. Chem., 18 (1968) 65.
40. C. Goodeve and K.H. Jack, Discuss. Faraday Soc., 4 (1948) 82.
41. S.A. Logan, R.L. Moss and C. Kemball, Trans. Faraday Soc., 54 (1958) 922.
42. A. Baiker and M. Maciejewski, J. Chem. Soc., Faraday Trans. I, 80 (1984) 2331.
43. J.G. Dunn, Thermochim. Acta, 300 (1997) 127.
44. G. Pannetier and L. Davignon, Bull. Soc. Chim. Fr., (1961) 2131; (1964) 1513.
45. A.W. Coats and N.F.H. Bright, Canad. J. Chem., 44 (1966) 1191.
46. G.I. Samal, Chem. Abs., 64 (1966) 15048d.
47. V.M. Zhukovskii, I.A. Montilo and A.A. Babadzhan, Chem. Abs., 66 (1967) 108758e.
48. I.C. Hoare, H.J. Hurst, W.I. Stuart and T.J. White, J. Chem. Soc., Faraday Trans. I, 84 (1988) 3071.
49. Dj. Jovanovic, J. Thermal Anal., 35 (1989) 1483.
50. M.M. Pavlyuchenko and G.I. Samal, Chem. Abs., 64 (1966) 13463a; 67 (1967) 26508e.
51. I.D. Shah and S.E. Khalafalla, Met. Trans., 2 (1971) 605.
52. P.J. Gardner and P. Pang, J. Chem. Soc., Faraday Trans. I, 84 (1988) 1879.
53. K. Imamura and M. Senna, J. Chem. Soc., Faraday Trans.I, 78 (1982) 1131.
54. K. Osakada and T. Yamamoto, Inorg. Chem., 30 (1991) 2328.
55. G. Poiblaud and M. Gillet, J. Microsc., 5 (1966) 543.
56. A.V. Gorokh, L.I. Klokotina and K.N. Rispel, Dokl. Akad. Nauk SSSR, 158 (1964) 1183.
57. J.J. Murray and R.D. Heyding, Canad. J. Chem., 45 (1967) 2675.
58. N.A. Subbotina, N.N. Zheligovskaya and A.S. Pashinkin, Zh. Neorg. Khim., 11 (1966) 2393.
59. R.C. Schoonmaker, A.R. Venkitaraman and P.K. Lee, J. Phys. Chem., 71 (1967) 2676.
60. E.A. Irene and H. Wiedemeier, Z. Anorg. Allg. Chem., 411 (1975) 182.
61. G.J. Russell, H.K. Ip and D.Haneman, J. Appl. Phys., 37 (1966) 3328.
62. B.V. L'vov and A.V. Novichikhin, Thermochim. Acta, 290 (1997) 241 and references therein.
63. M.J. Graham, D. Caplan and M. Cohen, J. Electrochem. Soc., 119 (1972) 1256.

Chapter 11

DECOMPOSITION OF AZIDES

11.1. INTRODUCTION

The inorganic azides constitute a particularly well-defined group of relatively simple ionic reactants whose decompositions have been extensively investigated both for their theoretical and practical importance. Large, relatively perfect, anhydrous crystals of many of these salts can be prepared [1] and these are ideal for fundamental studies of solid state reaction mechanisms. The stoichiometry of decomposition is normally simple because anion breakdown results in the formation of gaseous nitrogen and residual metal (sometimes together with some metal nitride). Prior to the 1970s, many important articles on azide decompositions contributed significantly to the theory of nucleation and growth processes, with particular reference to the role of identified crystal imperfections in the generation of new reaction sites and in interface advance. Revival of interest in these reactants is apparent in the recent literature, probably on account of the use of azide decompositions as gas generators in inflatable safety cushions ("air bags") in automobiles [2]. Difficulties encountered in studying the azides as reactants include their sensitivity to light, to hydrolysis by traces of water, and their unpredictable instability.

The optical properties of many metal azides permit the determination of the band structures of the solids. From these, the energy requirements for the generation of various types of defects can be estimated, and quantitative consideration can be given to the roles of these identified defects in formulating decomposition mechanisms.

Results obtained from thermal studies can often be compared with observations obtained in parallel investigations of the radiolysis of azides [3]. Additional information is also often available from measurements of the conductivities and photoconductivities of the solids, including the influences of various impurities in the crystals. Preirradiation generally increases the rate of subsequent thermal decomposition of azides and such studies can provide information on decomposition imperfections are recurrent features of many proposed decomposition mechanisms which are thus usually developed in greater detail than for most other solids.

The physical and chemical properties of the inorganic azides have been extensively reviewed [4-11]. Richter [12] has discussed the chemical classification of azides as (i) stable ionic azides, (ii) heavy-metal azides and (iii) unstable covalent azides. This classification is based on the percentage ionic character of the metal-azide bond, tabulated as formal ionicities in [12]. For example, the Na-N_3 and Ba-N_3 bonds are 70% ionic, but Pb-N_3 is only 34% and H-N_3 is 22%. Bertrand *et al.* [13] have reviewed the photochemical and thermal behaviour of organometallic azides. Richter has also given [12] an excellent review of the methods of preparation of HN_3 and other azides. He criticizes early workers for inadequate purification and characterization of their starting materials and their neglect of allowance for the possible formation of hydrates (e.g., barium azide may be present as $Ba(N_3)_2.1.5H_2O$ below 284 K, forms a monohydrate between 284 and 325.5 K and is anhydrous above 325.5 K).

The crystal structures of the azides have been reviewed by Choi [14] and tables of lattice dimensions are provided. He points out that the crystal structures of the metal azide hydrates are quite different from those of the anhydrous forms. He also notes a correlation between structure and impact sensitivity to explosion, which increases with the asymmetry of the azide ion, in the sequence:

$$KN_3 < Ba(N_3)_2 < \alpha\text{-}Pb(N_3)_2 < Cu(N_3)_2$$

with silver azide being an exception. The electronic structures of the azide ion and of the metal azides are discussed by Gora *et al.* [15] and they have warned of the care needed in attempting to correlate apparent activation energies from thermal kinetic studies with particular electronic states. Possible correlations between molecular vibrations or lattice dynamics and azide stability have been discussed by Iqbal *et al.* [16].

The relatively stable azides of the strongly electropositive metals undergo controlled (i.e. non-explosive or "slow") decomposition. For this reason, sodium and barium azides are probably the most completely studied salts in the azide group [12]. Their decompositions are characterized by relatively large apparent activation energies and well-defined induction periods to onset of reaction [10]. Mechanistic aspects of azide decompositions have been reviewed by Tompkins [17]. The radiation stabilities of the Group IA metal azides do not correspond to the sequence of thermal stabilities [10]. The catalytic decomposition of HN_3 has also been investigated [12].

The behaviour of the heavy-metal azides on heating is less predictable, some being extremely unstable and hence of value as primary explosives. Nevertheless it has been stated [18] that "All solid inorganic azides can be thermally decomposed at controllable and measurable rates". Azides which detonate under well-defined conditions have become model systems for the development of theories of "fast" reactions in solids [9,10].

Bowden and Yoffe [9] distinguished five stages in the response of solid primary explosives to suitable stimulation: (i) initiation of reaction in a definite region; (ii) growth of reaction to a burning region; (iii) acceleration of burning and transition to a low-velocity detonation; (iv) propagation of the low-velocity detonation; and (v) propagation of the high-velocity detonation. Heat, shock, friction and radiation could all cause localized heating ("hot spots") which lead to explosion. Walker [10] comments that the mechanisms of "slow" and of "fast" decompositions have often been assumed to be identical, but there is evidence, for some organic explosives, that different initiating stimuli lead to different reaction products. A detailed discussion of fast decompositions of inorganic azides has been given by Chaudhri and Field [19]. The stabilities of azides and the initiation and propagation of reactions in these salts have been discussed by Alster *et al.* [20].

11.2. GROUP IIA (ALKALINE EARTH) METAL AZIDES

11.2.1. Barium azide

The sigmoid α-time curves observed for the decomposition of barium azide are often characteristic of nucleation and growth processes and such a mechanism was here confirmed by microscopic observations [21]. This early study by Wischin (1939) has been described [11] as "one of the most important experimental papers on solid decompositions". Kinetic observations reported include the following:
(i) during isothermal reaction the number of nuclei increased with the third power of time (E_a = 310 kJ mol^{-1}); (ii) the increase in radius of each growing nuclei was directly proportional to time (E_a = 98 kJ mol^{-1}); and (iii) the rate of nitrogen evolution obeyed the power law with n = 6 to 8 and E_a = 695 kJ mol^{-1}.

Mott [22] suggested that the nucleation step required either the bimolecular combination of two interstitial barium ions, or the bimolecular trapping of two conduction electrons. The rate-determining step proposed for growth was the transfer of an electron from the azide valence band to the metal, subsequently attracting an interstitial barium ion into the developing nucleus. The positive holes (N_3) generated diffused to the surface and reacted to form N_2. Objections to this mechanism resulted from inconsistencies between the measured conductivity of the solid and the calculated concentrations and mobilities of the postulated participants required to sustain the observed rate of product evolution.

Both the photolytic and thermal decompositions of barium azide were studied by Thomas and Tompkins [23]. When allowance was made for an initial period of slow growth of nuclei, the reaction obeyed the power law $\alpha = k'(t - t_0)^6$ and the value of E_a (602 kJ mol^{-1}) was just below that reported by Wischin [21]. (Both values were calculated incorrectly and are unreasonably large!). The mechanism proposed was the combination of mobile F-centres to form a double F-centre as nucleus. Anion decomposition was identified as proceeding at the salt/metal interface from which the gaseous product N_2 escapes. The initial slow growth of nuclei was (probably incorrectly) attributed to the high value of E_a for growth.

Torkar, Spath *et al.* [24-26] confirmed the main features of the above behaviour but also noted a few points of difference. They found a fourth power dependence ($n = 4$) of α upon time rather than $n = 6$, as reported previously [23]. Nucleation according to an exponential law occurred on (100) and (001) planes, generating randomly distributed growth particles of the product. The rates of reactions of powdered samples were directly proportional to the surface area of the salt. A further difference from other reports was that yields of up to 73% of the nitride Ba_3N_2 (possibly arising through intermediate formation of Ba_2N_2) were found, whereas other workers mention only barium metal as the product. Salot and Warf [27] have described the preparation of barium pernitride from thermal decomposition of barium azide.

The steps which culminate in the generation of a metallic nucleus during the decomposition of BaN_6 have not been completely established [17]. Reaction probably occurs at or near the sites of surface imperfection, such as dislocations, and the probable participation of impurity Fe^{3+} ions, acting as electron traps, has been discussed [17,28]. Once established, the nucleus grows through an interfacial reaction [17], for which the activation energy is required to transfer an electron from the valence band of the azide ion to the Fermi level of the metal. Subsequent reaction of the azide radical with another thermally excited anion yields nitrogen gas. Interaction between barium ions and the charged surface of the metallic nucleus results in incorporation of Ba atoms and growth of the nucleus, by a mechanism showing a formal similarity to electrode processes.

11.2.2. Strontium azide

α-time curves for the decomposition of strontium azide [29] are also sigmoidal with the initial stage fitted by the power law with $n = 3$. The contracting volume equation described the decay period and $E_a = 97$ kJ mol^{-1}. Reaction was suggested to involve the interaction of positive holes and excited azide ions. The decreased stability resulting from γ-preirradiation was attributed to the introduction of a high concentration of vacancies. When decomposition in vacuum was interrupted and the sample was cooled to room temperature, decomposition continued immediately, as normal, on rapid reheating of the sample to reaction temperature. If, after interruption, the partially decomposed reactant was exposed to water vapour, further decomposition resembled that of a fresh sample, thus indicating the significant role that metal nuclei, which react readily with water, play in the decompositions of both the irradiated and unirradiated salts. Studies of the decompositions of barium and strontium azides (463 to 508 K) while being "co-irradiated" with high intensity UV [30] led to the conclusion that thermal and photolytic decomposition occurred concurrently during all stages of the decomposition.

11.2.3. Calcium azide

The kinetics of decomposition of calcium azide are irreproducible, but with careful control of conditions several contributory causes can be eliminated and the changes which result from ageing can be investigated [31]. Following an induction period, the acceleratory reaction can be described [31-33] by the power law with $n = 3$. The reaction below 370 K is believed [31] to occur through the growth of surface nuclei ($E_a = 75$ kJ mol^{-1}), but above this temperature, annealing could influence the magnitudes of measured Arrhenius parameters. Again decomposition is accelerated by preirradiation [33].

11.3. GROUP IA (ALKALI) METAL AZIDES

11.3.1. Lithium azide

Although the rates of decomposition of different preparations of lithium azide differed markedly [34], reproducible behaviour was observed for salt which had been crushed and pelleted. Such pretreatment was believed to produce a uniform concentration of defects within the reactant assemblages. The sigmoid α-time curves fitted the Avrami-Erofeev equation with $n = 3$ between $0.02 < \alpha < 0.58$ and the contracting volume expression across the wider interval $0.05 < \alpha < 0.95$, the value of E_a was 119 kJ mol^{-1}. Reaction involved the three-dimensional growth of a constant number of nuclei and it was suggested that acceleration of rate following preirradiation resulted from an increase in the number of such nuclei.

11.3.2. Sodium azide

Walker *et al.* [35,36] showed that the rate of decomposition of sodium azide was proportional to the surface area of the reactant crystallites. Reaction commenced at the edges of thin platelets and proceeded inwards parallel to the azide ions (i.e. in the [111] direction) as a contracting two-dimensional envelope composed of several distinct zones of different colours. The overall shape of the crystal is retained. Jacobs and Tariq Kureishy [37] showed that when argon (500 Torr) was present sublimation of the metallic (Na) product was suppressed and reproducible kinetic behaviour was observed. The α-time curve fitted the Avrami-Erofeev equation ($n = 2$) with $E_a = 159$ kJ mol^{-1} for the range 550 to 660 K. A mechanism involving the two-dimensional growth of linear filaments of sodium within the intergranular material was proposed. Later (at about 66% reaction) loss of liquid metal from the disintegrating crystal resulted in a change of the rate law. It is suggested that Fe^{+3}, incorporated in the reactant as an intentional impurity, provides electron traps which are the potential nucleus-forming sites. (Fe^{3+} is also believed to participate in the nucleation process during decompositions of BaN_6 and PbN_6). In general [38], cation additives, for which the radius is greater than that of the host cation, produce strain in the structure which is identified as the cause of the observed decrease in the induction period to decomposition.

Torkar, Hertzog *et al.* [39,40] observed an initial autocatalytic stage ($d\alpha/dt = k\alpha$) during nucleation on the (hk0) planes of the hexagonal reactant platelets which was accompanied by the removal of lattice defects. During the later deceleratory process, reaction proceeded by a diffusion-controlled mechanism along the preferred (001) planes. Sodium ions arriving at the salt-metal interface lowered the Fermi level of the metal, thus aiding electron transfer. From dielectric measurements on samples of sodium azide which had been heated to the point of onset of decomposition, Ellis and Hall [41] found evidence of the occurrence of three distinct processes during the very early stages of reaction ($\alpha < 0.02$). The 'dipole' structures identified are discussed with reference to the defects present in the solid. Preirradiation of NaN_3 with UV results in the development of centres of colloidal Na metal [42].

11.3.3. Potassium azide

The decomposition of potassium azide (530 to 623 K) is complicated by the volatility of the metallic product [43]. Potassium vapour catalyzes decomposition of the salt by the provision of electron traps, other than anion vacancies, with a decrease in E_a. The roles of charge carriers and ionic diffusion in the decomposition mechanism have been discussed [44].

11.3.4. Rubidium and caesium azides

Caesium azide melts with little decomposition (<1%) at 598 K. Slow decomposition was observed [45] when semiconducting oxides such as NiO were present in large amounts and was autocatalysed by caesium oxides formed in the reaction. The photochemical decompositions of both salts have been investigated [46].

11.4. OTHER AZIDES

11.4.1. Silver azide

The induction and acceleratory periods in the decomposition of silver azide are short, and reaction is predominantly deceleratory due to rapid initial covering of crystallite surfaces with metallic product [4,47]. Gray and Waddington [5] found that the reaction between 515 and 555 K could be described by the contracting volume model with E_a = 150 kJ mol^{-1}. AgN_3 is listed [7] as melting at 523 K. Bartlett *et al.* [48] also confirmed the applicability of the contracting volume model for reaction in the range 463 to 523 K with E_a = 130 kJ mol^{-1}.

McLaren and Rogers [49] identified the first step in AgN_3 decomposition as the production of free electrons and azide radicals (positive holes) and they discussed the mechanism in the context of the available spectroscopic and photochemical data. The rate of reaction is controlled by electron transfer and is increased [50] by the addition of semiconducting oxides, those with high work functions being the most effective. Incorporation of organic dyes [51] also accelerated decomposition, being greatest for the longest chain compounds, probably due to the increased delocalization of the *p* electrons. The catalytic effects result from promotion of the ease of electron transfer in the reactant.

Decomposition of silver azide in vacuum [52], induced by an electron beam, yields ultrafine particles of silver, predominantly less than 100 nm in diameter, and in the form of well-defined hexagonal or trigonal crystals and polygonized spheres. This technique of nanoparticle synthesis is recommended as having advantages over other methods.

11.4.2. Copper(I) azide

The initial acceleratory process in the decomposition of CuN_3 studied between 443 and 468 K could be described [53] by the power law ($n = 3$ and E_a = 110 kJ mol^{-1}) attributed to the three-dimensional growth of nuclei. The rate-limiting step is identified as the transfer of an electron from the azide band to the Fermi level of the metallic nuclei.

11.4.3. Lead azide

Many studies of the thermal stability of this reactant have been completed, due to its technical use as a primary explosive. Four polymorphic forms exist [12] of which the orthorhombic α-form is the most stable. The other forms can be prepared under specific conditions [12]. At high temperatures (above 613 K) the reaction accelerates to detonation. This autocatalytic process has been attributed [54] to an increased concentration of defect sites, rather than a mechanism controlled primarily by temperature increases due to self-heating.

α-time curves for the decomposition of α-lead azide, like those of the silver salt, are predominantly deceleratory. Griffiths and Groocock [55] studied this reaction using "well aged" α-lead azide samples of unusually homogeneous crystallite size-distribution. On heating there was an initial rapid evolution of gas (E_a = 63 kJ mol^{-1}) attributed to the decomposition of basic lead carbonate on crystallite surfaces. The acceleratory period of the main reaction was described by the power law (n = 5) with E_a about 1100 kJ mol^{-1} and, subsequently, the contracting volume equation. Nucleus formation, believed to continue during the acceleratory stage of reaction, involves the interaction of positive holes to yield nitrogen. Growth is attributed to the reaction of positive holes with metal at the reaction interface. The electron transfer step, therefore, controls salt decomposition. Some evidence for the existence of a reactant-product interface was obtained by microscopic examination.

Young [56] described a somewhat different pattern of kinetic behaviour for decomposition of the same salt. Again, there was an initial reaction of a surface impurity (about 2%) by a first-order process (E_a = 121 kJ mol^{-1}). The subsequent acceleratory reaction, perhaps delayed by the initial process, was fitted by the power law (n = 3). This is attributed to the growth of a constant number of nuclei present at time, t_0, and, because the number of such nuclei is slightly temperature dependent, the value of E_a for interface advance is 209 kJ mol^{-1}. The mechanism of reaction was not developed further.

Hutchinson *et al.* [57] have examined the effects of particle size on the decomposition of lead azide powder. Finer powders (radius = 8 μm) decomposed more than twice as rapidly as coarser powders (radius = 24 μm). A mathematical model was developed which incorporated the independently measured particle-size distribution and gave an excellent description of the experimental sigmoid α-time curves.

The influences of a variety of additives, either incorporated in the crystal or in contact with the surfaces, on the decomposition of PbN_6 have been studied. Hutchinson and Stein [58] showed that incorporation of Fe^{2+} accelerated the reaction more, but decreased the induction period less, than the presence of the $(FeN_3)^{2+}$ ion.

Savel'ev *et al.* [59] identified the rate-limiting step as the bimolecular combination of positive holes from the observed influences on reaction rates of added Cu^{2+} (which occupies interstitial positions and aids migration of positive holes) and Ag^{+} (which replaces Pb^{2+} at lattice sites). Coating the PbN_6 crystals with silver accelerated the rate of decomposition. Investigations have been made of the influences of low molecular mass additions [60] and larger organic dye molecules [61] on the rate of reaction.

The possible formation of intermediates during decomposition of lead azide has been discussed, but results were inconclusive [62]. Shechkov [63] suggested that up to 5% of the product nitrogen may be temporarily retained in the crystal and this may be expected to influence kinetic behaviour. The photolysis of lead azide has also been studied [64].

11.4.4. Thallium(I) azide

Following a short induction period, the decomposition of thallium(I) azide proceeded [35,65] at a constant rate (E_a = 150 kJ mol^{-1} between 507 and 529 K) and the zero-order rate constants were directly proportional to surface area. These kinetic observations were interpreted through reference to electron micrographs. Reaction occurred on preferred crystal planes, with the interface advancing predominantly in the (110) crystallographic direction and holes of uniform depth and constant cross-section were observed.

11.4.5. Mercury(I) azide

Deb and Yoffe [66] compared the photochemical decomposition of mercury(I) azide with that of triphenylmethyl azide. The first step in the decomposition is suggested to be the fission of the longer N-N bond of the azide group. Results were compared with data for reactions of inorganic azides and it was concluded that there is no clear distinction between the energy requirements of the two classes of azides, covalent and ionic. Deb [67] has determined the electron energy levels of several azides and correlated the band structures with observed stabilities.

11.5. METAL CYANAMIDES AND FULMINATES

The structures of the azide ion (N_3^-), the cyanamide ion (CN_2^{2-}) and the fulminate ion (CNO^-) are closely related and the decompositions of some metal cyanamides and fulminates are therefore considered here.

11.5.1. Cadmium cyanamide

$CdCN_2$ decomposed [68] in vacuum within the interval 865 and 989 K to yield metal, cyanogen and nitrogen (E_a = 159 kJ mol^{-1}) for which the α-time curves were fitted by the contracting volume equation. The reaction in oxygen, which formed metal oxide, nitrogen and carbon dioxide (E_a = 82 kJ mol^{-1}) followed the same kinetic model and both reactions are believed to proceed by an interface advance mechanism.

11.5.2. Thallium(I) cyanamide

The decomposition proceeded [69] to completion in four kinetically distinct stages studied between 733 and 1233 K and behaviour was further complicated by hydrolysis if constituent water in the crystal had not been completely removed. Only the first decomposition step, corresponding to 14.4% evolved nitrogen, proceeds below the melting point (793 K) of the dried solid. This step followed a first-order process and E_a was 129 kJ mol^{-1}.

11.5.3. Mercury(II) fulminate

The thermal decomposition of $Hg(CNO)_2$ was studied by Bartlett *et al.* [70] who were particularly concerned with the effects of ageing, preirradiation and crushing on reaction kinetics. The acceleratory period for the fresh reactant followed the exponential relation, whereas for the aged salt the acceleratory period was described by the cubic law ($n = 3$ and E_a = 113 kJ mol^{-1}). There was an initial small (< 0.4%) deceleratory evolution of gas.

11.5.4. Silver fulminate

Singh and Palkar [71] studied the decomposition of silver fulminate across the interval 463 to 483 K. The initial deceleratory reaction (E_a = 27 kJ mol^{-1}) was much more significant than the corresponding process in the mercury salt and overlapped with the onset of the subsequent acceleratory process, which fitted the power law, with $n = 2$ and E_a = 119 kJ mol^{-1}. The suggested mechanism is an initial electron-transfer step from the anion to the product metal during the two-dimensional advance of the interface.

11.5.5. Other fulminates

Iqbal *et al.* [72,73] have reviewed the physical properties and decompositions of selected inorganic fulminates. The sodium, potassium and thallium salts are predominantly ionic, whereas those of Hg and, possibly, Ag are covalent. Electronic transitions are believed [72] to play an important role in the effects of heat and light on the ionic salts, but to be less significant in controlling the reactivity of the covalent compounds. In a comparative study [73] of the slow thermal and photochemical decompositions of Hg, Ag, Na and Tℓ fulminates, it was concluded that these reactions are complex. The rate-limiting step is probably bond rupture in the covalent salts (Hg and Ag) whereas electron transfer probably controls the decomposition of the ionic (Na and Tℓ) solids.

11.6. CONCLUSIONS

The factors governing the slow thermal decompositions of inorganic azides have been discussed by Fox and Hutchinson [18]. They draw attention to the interest shown in early studies for fitting of kinetic results to rate equations based on nucleation and growth models. Support of kinetic interpretations by microscopic observations (e.g., [21]), contributed significantly towards establishment of the role of the active, advancing interface in solid state reactions. The kinetic characteristics of some of the metal azides are summarized in Table 11.1.

Table 11.1.
Selected kinetic data for the thermal decompositions of metal azides (based on Table 1 of Fox and Hutchinson [18])

Metal Azide	Value of n in the power law $\alpha = (kt)^n$	E_a/kJ mol^{-1}	Reference
NaN_3	1	193	[74]
KN_3	1	205	[43]
CaN_6	3	80 (< 370 K) 147 (> 370 K)	[75]
BaN_6	6	268	[26]
SrN_6	3	84	[75,76]
CuN_3	3	110	[53]
AgN_3	3	189 (< 460 K) 134 (> 460 K)	[48]
TlN_3	1	95	[77]
α-PbN_6	1	123	[78]

Dissimilarities in the behaviour of the reactants in Table 11.1. were attributed to variations in patterns of interface advance. Later developments lead to the recognition that quantitative kinetic analysis also required consideration of factors such as the sizes and distributions of reactant particles and their defect and impurity contents. The possible influences of inadequate purification and characterization of the materials used in early studies, mentioned by Richter [12], are also very important.

In addition to the general contributions made to the development of the kinetics of solid state reactions by studies on azides, Fox and Hutchinson state [18] that "the success of Mott [22] in describing the kinetics of barium azide decomposition in terms of the band structure of the material was crucial to the development of the science of solid state decomposition reactions." However, the generalizations by Yoffe *et al.* [9,79] that the activation energy for azide decomposition is related to the energy required to excite an electron from the azide ion to the conduction band of the solid, and that this energy decreases as the ionization potential of the metal increases, were criticized [18] as being oversimplification, because differences in the crystal structures of the azides are ignored. Fox and Hutchinson [18] comment further that "correlations between activation energies for decomposition and the energy-band structure of the solid are fraught with difficulties. There is disparity in the literature over the values of the activation energies and this is compounded with the problem of identifying the relevant solid state energies...".

Recent work by L'vov [80,81] proposes a mechanism of dissociative evaporation for use in interpreting the kinetics of decomposition of the azides of Na, K, Pb, Ag, Tℓ, Ba, Sr and Ca. All primary products, including metals, are assumed to be formed in the gaseous state, with nitrogen present as N atoms as well as N_2 and N_3 molecules. This gasification mechanism assumes that there will be subsequent condensation of non-volatile components following collisions with each other or with any surface. L'vov points out that generation of intense UV radiation is a feature of azide decompositions [82]. The UV spectrum with peaks at 198, 215, 230 and 240 nm has been identified [83] as the Vegard-Kaplan band system originating from a ground state transition in excited nitrogen molecules, and is observed when electric discharges are passed through molecular dinitrogen gas. If it is assumed that the UV emission results from recombination of ground state nitrogen atoms, the emission should increase with the square of the number of N atoms produced by azide decomposition. The intensity of the UV emission during decomposition of alkaline earth azides is reported [80] to be about one third to one quarter of the emission from other azide decompositions. L'vov also notes that the correlation between the observed activation energies for azide decomposition and the energy of the first exciton level of the azide ion leading to the formation of neutral azide radicals [9] applies to only about half of the azides studied [18].

Studies of the decompositions of metal azides have provided many of the fundamental concepts and much of the background to solid state mechanisms that are discussed in a broader context in Chapter 18.

REFERENCES

1. W.L. Garrett, Energetic Materials 1: Physics and Chemistry of the Inorganic Azides, (Eds H.D. Fair and R.F. Walker), Plenum, New York, 1977, Chap.2.
2. R.T. Barbour, Pyrotechnics in Industry, McGraw-Hill, New York, 1981, p.50.
3. W.L. Garrett, P.L. Marinkas, F.J. Owens and D.A. Wiegand, Energetic Materials 1: Physics and Chemistry of the Inorganic Azides, (Eds H.D. Fair and R.F. Walker), Plenum, New York, 1977, Chap.7.
4. W.E. Garner, Chemistry of the Solid State, (Ed. W.E. Garner), Butterworths, London, 1955, Chap.9.
5. P. Gray and T.C. Waddington, Proc. R. Soc. (London), A235 (1956) 106, 481; A241 (1957) 110.
6. P. Gray, Q. Rev., Chem. Soc., 17 (1963) 441.
7. B.L. Evans, A.D. Yoffe and P. Gray, Chem. Rev., 59 (1959) 515.
8. P.W.M. Jacobs, J.G. Sheppard and F.C. Tompkins, Reactivity of Solids, (Ed. G.-M. Schwab), Elsevier, Amsterdam, 1965, p.509.
9. F.P. Bowden and A.D. Yoffe, Initiation and Growth of Explosion in Liquids and Solids, Cambridge University Press, Cambridge, 1952; Fast Reactions in Solids, Butterworths, London, 1958.
10. R.F. Walker, Energetic Materials 1: Physics and Chemistry of the Inorganic Azides, (Eds H.D. Fair and R.F. Walker), Plenum, New York, 1977, p.1.
11. D.A. Young, Decomposition of Solids, Pergamon, Oxford, 1966, p.174.
12. T.A. Richter, Energetic Materials 1: Physics and Chemistry of the Inorganic Azides, (Eds H.D. Fair and R.F. Walker), Plenum, New York, 1977, Chap.1.
13. G. Bertrand, J.P. Majoral and A. Baceiredo, Acc. Chem. Res., 19 (1986) 17.

14. C.S. Choi, Energetic Materials 1: Physics and Chemistry of the Inorganic Azides, (Eds H.D. Fair and R.F. Walker), Plenum, New York, 1977, Chap.3.
15. T.Gora, D.S. Downs, P. J. Kemmey and J. Sharma, Energetic Materials 1: Physics and Chemistry of the Inorganic Azides, (Eds H.D. Fair and R.F. Walker), Plenum, New York, 1977, Chap.5.
16. Z. Iqbal, H.J. Prask and S.F. Trevino, Energetic Materials 1: Physics and Chemistry of the Inorganic Azides, (Eds H.D. Fair and R.F. Walker), Plenum, New York, 1977, Chap.4.
17. F.C. Tompkins, Proc. Robert A. Welch Foundation Conf. on Chem. Res., XIV : Solid State Chemistry, Houston, Texas, 1970, p. 203.
18. P.G. Fox and R.W. Hutchinson, Energetic Materials 1: Physics and Chemistry of the Inorganic Azides, (Eds H.D. Fair and R.F. Walker), Plenum, New York, 1977, Chap.6.
19. M.M. Chaudhri and J.E. Field, Energetic Materials 1: Physics and Chemistry of the Inorganic Azides, (Eds H.D. Fair and R.F. Walker), Plenum, New York, 1977, Chap.8.
20. J. Alster , D.S. Downs, T. Gora, Z. Iqbal, P.G. Fox and P. Mark, Energetic Materials 1: Physics and Chemistry of the Inorganic Azides, (Eds H.D. Fair and R.F. Walker), Plenum, New York, 1977, Chap.9.
21. A. Wischin, Proc. R. Soc. (London), A172 (1939) 314.
22. N.F. Mott, Proc. R. Soc. (London), A172 (1939) 325; N.F. Mott and R.W. Gurney, Electronic Processes in Ionic Crystals, Oxford University Press, Oxford, 1948, Chap.8.
23. J.G.N. Thomas and F.C. Tompkins, Proc. R. Soc. (London), A209 (1951) 550; A210 (1951) 111.
24. K. Torkar and H.T. Spath, Monatsh. Chem., 98 (1967) 1696, 1712, 1733, 2020, 2382; 99 (1968) 118, 773.
25. K. Torkar, H.T. Spath and K. Mayer, Monatsh. Chem., 98 (1967) 2362.
26. K. Torkar, H.T. Spath and G.W. Herzog, Reactivity of Solids, (Ed. J.W. Mitchell), Wiley, New York, 1969, p.287.
27. S. Salot and J.C. Warf, Inorg. Chem., 13 (1974) 1776.
28. V.R. Pai Verneker and L. Avrami, J. Phys. Chem., 72 (1968) 778.
29. E.G. Prout and D.J. Moore, J. Inorg. Nucl. Chem., 31 (1969) 1595.
30. E.G. Prout and E.G. Sheppard, J. Inorg. Nucl. Chem., 43 (1981) 1977.
31. F.C. Tompkins and D.A. Young, Trans. Faraday Soc., 61 (1965) 1470.
32. P.W.M. Jacobs and F.C. Tompkins, Chemistry of the Solid State, (Ed. W.E. Garner) Butterworths, London, 1955, Chap.7.

33. E.G. Prout and M.E. Brown, Nature (London), 205 (1965) 1314.
34. E.G. Prout and V.C. Liddiard, J. Phys. Chem., 72 (1968) 2281; J. Inorg. Nucl. Chem., 35 (1973) 2183, 3419.
35. R.F. Walker, N. Gane and F.P. Bowden, Proc. R. Soc. (London), A294 (1966) 417.
36. R.F. Walker, J. Phys. Chem. Solids, 29 (1968) 985.
37. P.W.M. Jacobs and A.R. Tariq Kureishy, J. Chem. Soc., (1964) 4718; Canad. J. Chem., 44 (1966) 703.
38. W.J. McGill, J. S. Afr. Chem. Inst., 22 (1969) 139, 143.
39. K. Torkar and G.W. Herzog, Monatsh. Chem., 97 (1966) 765, 1217; 98 (1967) 245, 298.
40. K. Torkar, G.W. Herzog, W. Schintlmeister and D. Fischer, Monatsh. Chem., 97 (1966) 1339.
41. P. Ellis and P.G. Hall, Trans. Faraday Soc., 64 (1968) 1034.
42. B.J. Miller, J. Chem. Phys., 33 (1960) 889.
43. P.W.M. Jacobs and F.C. Tompkins, Proc. R. Soc. (London), A215 (1952) 265.
44. A. de Panafieu, B.S.H. Royce and T. Russell, J. Chem. Phys., 64 (1976) 1473.
45. Y. Sears and M. Steinberg, J. Catal., 11 (1968) 25; Thermochim. Acta, 75 (1984) 233.
46. P.W.M. Jacobs and A.R. Tariq Kureishy, J. Chem. Soc., (1964) 4723.
47. J. Sawkill, Proc. R. Soc. (London), A229 (1955) 135.
48. B.E. Bartlett, F.C. Tompkins and D.A. Young, Proc. R. Soc. (London), A246 (1958) 206.
49. A.C. McLaren and G.T. Rogers, Proc. Roy. Soc. (London), A240 (1957) 484.
50. Yu.A. Zakharov, E.S. Kurochkin, G.G. Savel'ev and Yu.N. Rufov, Kinet. Katal., 7 (1966) 425.
51. E.S. Kurochkin, G.G. Savel'ev, Yu.A. Zakharov and O.I. Chizhikova, Zh. Vses. Khim. Obschest., 13 (1968) 466.
52. P.J. Herley and W. Jones, J. Chem. Soc., Faraday Trans., 88 (1992) 3213.
53. K. Singh, Trans. Faraday Soc., 55 (1959) 124.
54. A.R. Ubbelohde, Chemistry of the Solid State, (Ed. W.E. Garner), Butterworths, London, 1955, Chap.11; Phil. Trans. R. Soc. (London), A241 (1948) 281.
55. P.J.F. Griffiths and J.M. Groocock, J. Chem. Soc., (1957) 3380.
56. D.A. Young, J. Chem. Soc., (1964) 3141.

57. R.W. Hutchinson, S. Kleinberg and F.P. Stein, J. Phys. Chem., 77 (1973) 870.
58. R.W. Hutchinson and F.P. Stein, J. Phys. Chem., 78 (1974) 478.
59. G.G. Savel'ev, Yu.A. Zakharov, G.T. Shechkov and R.A. Vasyutkova, Izv. Tomsk. Politekh. Inst., 151 (1966) 40.
60. J. Haberman and T.C. Castorina, Thermochim. Acta, 5 (1972) 153.
61. E.S. Kurochkin, G.G. Savel'ev, Yu.A. Zakharov and O.I. Chizhikova, Izv. Tomsk. Politekh. Inst., 199 (1969) 95.
62. M. Stammler, J.E. Abel and J.V.R. Kaufman, Nature (London), 192 (1961) 626.
63. G.T. Shechkov, Izv. Vyssh. Ucheb. Zaved., Khim. Tekhnol., 16 (1973) 1794.
64. V.R. Pai Verneker and A.C. Forsyth, J. Phys. Chem., 71 (1967) 3736.
65. R.F. Walker, Trans. Faraday Soc., 65 (1969) 3324.
66. S.K. Deb and A.D. Yoffe, Proc. R. Soc. (London), A256 (1960) 514, 528.
67. S.K. Deb, Trans. Faraday Soc., 66 (1970) 1802.
68. M. Bernard and A. Busnot, Bull. Soc. Chim. Fr., (1968) 3938.
69. V.I. Siele and F.C. Tompkins, J. Chem. Soc., (1964) 6191.
70. B.E. Bartlett, F.C. Tompkins and D.A. Young, J. Chem. Soc., (1956) 3323.
71. K. Singh and G.D. Palkar, J. Chem. Soc., (1965) 1474.
72. Z. Iqbal and A.D. Yoffe, Proc. R. Soc. (London), A302 (1967) 35.
73. T. Boddington and Z. Iqbal, Trans. Faraday Soc., 65 (1969) 509.
74. P.G. Fox and J. Soria-Riuz, Proc. R. Soc. (London), A317 (1970) 79.
75. W.E. Garner and L.W. Reeves, Trans. Faraday Soc., 51 (1955) 694.
76. W.E. Garner and J. Maggs, Proc. R. Soc. (London), A172 (1939) 299.
77. R.F. Walker, Ph D Thesis, Cambridge University, 1965.
78. P.G. Fox, J. Solid State Chem., 2 (1970) 491.
79. A.D. Yoffe, Developments in Inorganic Nitrogen Chemistry, (Ed. C.B. Colburn) Elsevier, London, 1966, Chap.2.
80. B.V. L'vov, Thermochim. Acta, 291 (1997) 179.
81. B.V. L'vov, Spectrochim. Acta, B, 50 (1995) 1427, 1459; 51 (1996) 533; 52 (1997) 20.
82. R. Audubert, C.R. Acad. Sci., Paris, 204 (1937)1192; 206 (1938) 748; Trans. Faraday Soc., 35 (1939) 197.
83. R.W. Nicholls, J. Phys. Chem., 64 (1960) 1760.

Chapter 12

DECOMPOSITION OF CARBONATES

12.1. INTRODUCTION

On heating, metal carbonates undergo dissociation of the anion ($CO_3^{2-} \rightarrow O^{2-} + CO_2$) in reactions that are endothermic and usually reversible. Complete uptake of the volatile product, to restore the original composition of the stoichiometric reactant, does not always occur, however, in any reasonable time interval. Calculated values of E_a are often close to the reaction enthalpy. Not all reactions proceed to completion in a single kinetic step. These reactions show a general pattern of similarity [1] with the dehydrations of crystalline hydrates, surveyed in Chapter 7, and also were amongst the earliest reactants used in kinetic and mechanistic investigations of solid state decompositions.

12.2. CALCIUM CARBONATE

The thermal decomposition of calcium carbonate is one of the most intensively studied solid state reactions because of its great technological importance. The conclusions derived from this work have been found to be of value in formulating the mechanisms of dissociations of other metal carbonates. The body of information related to the partially reversible dissociation of calcium carbonate is also most confusing, because of the variations of kinetic behaviour described and the wide range of kinetic parameters reported. This is usually regarded as being due to the sensitivity of the overall dissociation rate to the form of the reactant (particle size, etc.) together with the influences of the reverse process (CO_2 uptake) and heat flow. These effects are discussed below.

An additional possible reason for differences in behaviour is that $CaCO_3$ exists in three crystalline forms. Calcite is the most stable and aragonite and vaterite undergo transitions to calcite at 728 K and between 623 and 673 K, respectively. Consequently, at decomposition temperatures (above 900 K) the reactant $CaCO_3$ would normally be in the calcite structure [2]. Salvador *et al.* [3] concluded that the decompositions of both calcite and aragonite are phase boundary controlled. The apparent magnitude of E_a for the aragonite dissociation (87 kJ mol^{-1}) was smaller than that for calcite (111 kJ mol^{-1}). The difference was said to be due to crystalline structure and/or the effects of impurities. (See also Section 2.4.2. and Figure 2.3.)

The kinetics of the reaction:

$$CaCO_3(s) \rightarrow CaO(s) + CO_2(g) \tag{12.1}$$

are notoriously sensitive to the prevailing conditions and especially to the availability of carbon dioxide which, during reaction, may be dependent on the ease of diffusion away from the reaction zone. There is strong evidence that this reaction occurs at an advancing interface between the calcite and oxide phases. No phase of intermediate composition has been identified, although there is some evidence that the initially formed product may be metastable [4], possibly owing to its highly divided state.

Beruto and Searcy [4] studied the kinetics of advance of the $CaO/CaCO_3$ interface into calcite. The reactant was prepared in the form of a thin cleaved crystal of calcite, sheathed to prevent the reversible reaction on all surfaces except the (1011) cleavage face, so that the effects of particle size and geometry could be eliminated. A constant rate of evolution of carbon dioxide was observed to α greater than 0.8 in samples of up to 1 mm thickness. The gas pressure maintained during the reaction was low (about 5 x 10^{-5} Torr). Under these conditions it was concluded that intraparticular diffusion could not have influenced the rate of CO_2 release and kinetics were independent of the reverse process. The activation energy found, 205 kJ mol^{-1}, was about 18% greater than the reaction enthalpy (174 kJ mol^{-1}). Microscopic examination revealed the presence of a thin layer (about 30 nm) of solid product interposed between the unreacted salt and the outer layer of oxide product. This intermediate phase was bounded by flat surfaces and, although originally thought to be a metastable phase, has now been identified [5] as very fine (particle sizes about 10 nm) CaO needles.

Maciejewski and Reller confirm [6] that (as yet) there is no experimental evidence for the existence of an amorphous or metastable intermediate in the dissociation of calcite. They show that the CaO formed by $CaCO_3$ dissociation under high vacuum reacts relatively rapidly at low temperatures (below 320 K) in the reverse process on CO_2 readmission. Carefully controlled conditions are required to isolate and to study any highly active phases that may be formed during solid state decompositions.

Thomas and Renshaw [7] studied the role of dislocations in nucleus formation and early growth in reaction (12.1). The large calcite crystals used were first subjected to deformation (at 970 K in CO_2), to establish identifiable alignments of dislocations, and were then decomposed in a vacuum microbalance at very low rates to a small extent (α about 0.02). From microscopic examinations of the distributions of nuclei formed, it was concluded that decomposition commenced preferentially at dislocations which glide in {100} and {211} planes, but less readily on those which glide in {111} planes. These observations highlight the need to provide precise descriptions of the specific dislocations concerned in decompositions, and their strength. Microscopic evidence also indicated that the growth of product CaO crystals was topotactic. Further work [8] showed that the value of E_a = 210 kJ mol^{-1} for the early stages of reaction ($0.002 < \alpha < 0.01$) was not perceptibly different from that found for intermediate ranges of α. The self-energy of the dislocation thus does not appear to contribute to the initial value of E_a for calcite decomposition.

Fox and Soria-Ruiz [9] found that the decomposition rate may be influenced by the mobility of dislocations. They concluded that the increased reactivity in the vicinity of a dislocation arises from changes in the stereochemical environment of the ions concerned and not from the strain field associated with the distorted lattice at the dislocation core. Impurities may also be preferentially incorporated in the disorganized regions of the lattice [10] and exert some influence on reactions occurring there.

Kinetic data measured for the decomposition of calcium carbonate under isothermal and under programmed-temperature conditions [11] and varied reaction environments influencing the ease of removal of the CO_2 product, show that the apparent values of the kinetic parameters k, A and E_a may be influenced by sample heating rate, reactant self-cooling, sample mass, geometry and particle size, which determine the rate because of the reversible nature of the decomposition [12]. These effects can lead to compensation behaviour [13].

In general, the reported apparent magnitudes [14] of Arrhenius parameters (E_a from 110 to 1600 kJ mol^{-1} and A from 10^2 to 10^{69} s^{-1}) extend over a much larger range of values than can be realistically attributed to the operation of a single rate-controlling process. Some of the extreme values were calculated from non-isothermal data. Criado *et al.* [15] carried out a thermogravimetric study at 12 K min^{-1} in 1.3, 2.7, 12.0 and 20.0 kPa CO_2. From plots of $Y = \{(d\alpha/dt)(1 - \alpha)^{-2/3}\}/(1 - p/p^*)$ against $1/T$, where p is the partial pressure of CO_2 and p^* is the equilibrium pressure, mean values of $E_a = 191 \pm 5$ kJ mol^{-1} and ln (A/min^{-1}) = 20.3 were obtained which were in satisfactory agreement with the isothermal values (E_a=187 kJ mol^{-1}, ln(A/min^{-1}) = 21.7). These results confirm the necessity of controlling, or allowing for the contributions from, the pressure of the gaseous products in reversible decompositions. If the correct procedures are observed, acceptable kinetic parameters can be obtained from non-isothermal data.

Beruto and Searcy [4] argue that when large reactant samples are used, approximately equilibrium pressures of carbon dioxide may be established at regions within the sample bulk and the rate of outward CO_2 diffusion measured in kinetic work is determined by local pressure gradients. The apparent value of E_a is then close to the enthalpy of dissociation. During decompositions at low pressures, however, the value of the activation energy is greater because local internal pseudo-equilibrium is not attained [4].

When carbon dioxide removal from the reaction zone is neither rapid nor complete, the rate of decomposition of $CaCO_3$ may be controlled by the transfer of thermal energy and/or gas diffusion within the sample. Draper [16] focused attention on the heat transfer process and showed that the α-time curves can be altered (in a predictable manner) by systematic variations of those factors which determine the balance, distribution and movement of thermal energy within the reacting mass. Self-cooling of calcite, in the form of relatively large blocks (about 20 mm edge), during decomposition has been measured by Auffredic and Vallet [17]. Hills [18] concluded that the rate of isothermal decomposition of calcite was determined by contributions from both interfacial and transport processes. Values of transport parameters deduced from the kinetic data were in satisfactory agreement with those listed in the literature [19-21]. Vosteen [22] has provided a mathematical representation of the reaction controlled by gas diffusion and heat transfer.

Samples of CaO formed [23] from $CaCO_3$ decomposition have large surface areas that decrease linearly with reaction temperature, from 127 $m^2 g^{-1}$ at 920 K to 60 $m^2 g^{-1}$ at 1170 K. Areas were not diminished significantly by vacuum annealing at 1170 K. Microscopic observations showed that the first product formed was an aggregate with relatively uniform pores, but the texture changed when the CaO layer thickness was between 10 and 50 μm. It was further shown [24] that, in the presence of added LiCℓ, the area of CaO formed from $CaCO_3$ decomposition 870 to 970 K was less than about 1 $m^2 g^{-1}$, probably because a LiCℓ-CaO eutectic provides a path for CaO recrystallization through the liquid phase. In a CO_2 atmosphere, LiCℓ doubled the rate of $CaCO_3$ decomposition at 970 K, again ascribed to the presence of the liquid phase.

Where conditions eliminate diffusion resistance to reaction (12.1) [25], the rate is proportional to the surface area of the undecomposed $CaCO_3$, consistent with the interface advance or shrinking core model. Calcite breakdown with greater than 90% efficiency is achieved in 2.5 s at 1270 K and the surface area of the product CaO is a maximum (about 100 $m^2 g^{-1}$) after reaction at 1270 K.

The influence of low pressures of CO_2 (between $10^{-5}p^*$ to $4 \times 10^{-2}p^*$) on the rate of calcite dissociation (893 to 1073 K) were studied by Darroudl and Searcy [26]. Rates are relatively sensitive to CO_2 pressure above 10^{-2} p^*, but do not fit the parabolic rate law. At lower pressures, the rates are independent of $p(CO_2)$ which implies that behaviour is the same as that in vacuum [4]. It is concluded that the rate limiting process, in low CO_2 pressures, is either condensed phase diffusion of CO_2 or a surface step prior to CO_2 desorption.

The decomposition of $CaCO_3$ has also been studied in an atmosphere of CO_2. Under these conditions, very large apparent values of E_a are found [27,28], up to about 1500 kJ mol^{-1}. Hashimoto [27] has shown that by suitable extrapolation of such data to zero pressure, the values of E_a obtained are closer to those found for reactions proceeding under heat and gas diffusion control.

In contrast with the high vacuum studies used to investigate the mechanism of $CaCO_3$ decomposition, the dissociation rates of large limestone particles (2 to 7 mm) were studied [29] under near isothermal conditions, as in a commercial calcining process. Kinetic data fitted the contracting volume equation with apparent values of E_a of about 150 kJ mol^{-1} between 973 and 1173 K. Predictably, CO_2 decreased the reaction rate.

12.3. OTHER GROUP IIA METAL CARBONATES

12.3.1. Magnesium carbonate

$MgCO_3$ has the calcite structure [2]. Britton *et al.* [30] concluded that the decomposition of magnesite ($MgCO_3$) was an interface process, initiated at boundary surfaces and thereafter advancing inwards. The value of E_a found (150 kJ mol^{-1} between 813 and 873 K) was appreciably greater than the enthalpy of dissociation (101 kJ mol^{-1}). They considered the possible influences on the reaction rates of factors such as self-cooling, the recombination process at the reaction interface, the restriction of escape of carbon dioxide, and the rate of the nucleation step.

Values of E_a reported by Kubas *et al.* [31], 140 to 175 kJ mol^{-1}, were comparable with those cited above, but (as with calcite) in the presence of CO_2, E_a was increased to 234 kJ mol^{-1}. Addition of ZnO or NiO to $MgCO_3$ decreased the value of E_a slightly. The much greater decrease (to $E_a = 54$ kJ mol^{-1}) which resulted from the addition of CaO was ascribed to the formation of $CaCO_3$ as an intermediate. Doi and Kato [32] identified $MgO.MgCO_3$ as an intermediate in the decomposition of magnesite in N_2 or CO_2. The effects of procedural variables on the non-isothermal decompositions of magnesite and of dolomite have been reported by Sharp *et al.* [12].

12.3.2. Dolomite (Mg,Ca)CO_3

In a kinetic study of the decompositions of several specimens of natural dolomite, Britton *et al.* [30] related their observations to the behaviours of $MgCO_3$ and of $CaCO_3$ under comparable conditions. The value of E_a for the mixed carbonate, 220 kJ mol^{-1}, was appreciably larger than values found for the individual constituents. It was concluded that the initial reaction produced the mixed oxide (solid solution) (Mg,Ca)O which rapidly rearranges to yield MgO and CaO. The CaO present reacts to yield calcite if sufficient CO_2 remains available in the wake of the advancing interface. Carbon dioxide absorption by dolomite, which had been decomposed in vacuum, was somewhat less than the quantity required to convert all the calcium oxide present to calcite. These conclusions were supported by infrared [33] and by X-ray diffraction measurements [34]. The primary solid product was shown [34] to be a homogeneous rock-salt oxide equicomposition phase. This then unmixes to give two phases: Mg rich ($Ca_\epsilon Mg_{1-\epsilon}O$, ϵ about 0.1) and Ca rich ($Mg_\delta Ca_{1-\delta}O$, δ about 0.2). The paper discusses the mechanism of spinodal decomposition and its applications in preparing non-equilibrium oxide solid solutions.

MacKenzie and Meinhold [35], however, using ^{25}Mg NMR studies have shown that during decomposition of dolomite, amorphous MgO is formed much more readily and extensively than previously suggested, with no evidence of the participation of $MgCO_3$ as a discrete phase or solid solution, or of the involvement of a mixed oxide.

12.3.3. Strontium and barium carbonates

Judd and Pope [36] conclude that because the activation energies for decompositions of $CaCO_3$, $SrCO_3$ and $BaCO_3$ are all close to the corresponding enthalpies of dissociation (apparent values of E_a are 180, 222 and 283 kJ mol^{-1} and ΔH, 178, 235 and 269 kJ mol^{-1}, respectively) the mechanisms of decomposition in all three substances are the same as that proposed by Hills [18] for calcite. Strontium carbonate [37] generally resembles the calcium salt in that an increase in sample size results in a decrease in reaction rate. Differences in behaviour were ascribed [37] to the occurrence of a crystallographic transformation and to fusion.

The decomposition of barium carbonate requires higher temperatures [38] and the reaction in the solid state below 1300 K is influenced by the prevailing pressure of CO_2, while above 1300 K decomposition proceeds in a melt.

L'vov and Novichikhin [39] conclude that the decomposition of barium carbonate at 1200 K can be expressed as:

$BaCO_3(s) \rightarrow BaO(g) + CO_2(g)$

This was based on a comparison between evaporation rates calculated from thermodynamic data with those measured in low pressure experiments by Basu and Searcy [40]. The value of E_a determined experimentally (226 kJ mol^{-1}) was lower than that calculated theoretically (314 kJ mol^{-1}). The difference was ascribed to the opposition to product escape offered by an accumulating solid BaO layer as:

$BaO(g) \rightarrow BaO(s)$

The theoretical value was much closer to the experimental value of E_a (283 kJ mol^{-1}) reported by Judd and Pope [36], see above.

12.4. GROUP IA METAL CARBONATES AND BICARBONATES

12.4.1. Sodium bicarbonate

Reported [41-44] values of E_a for the thermal decomposition of $NaHCO_3$ were in the range 42 to 65 kJ mol^{-1}. The initial reaction results [42] in the formation and growth of Na_2CO_3 and subsequently the effects of heat and mass transfer become rate limiting.

Tanaka [45] used mass loss measurements to study the isothermal decomposition of $NaHCO_3$ between 383 and 397 K. Results fitted the Avrami-Erofeev equation, $n = 1.41$. The value of E_a was 109 ± 5 kJ mol^{-1}, which was greater than the reaction enthalpy of 71.8 ± 0.8 kJ mol^{-1}. It was concluded that this is a nucleation and growth process.

The thermal decomposition of $Na_2CO_3.NaHCO_3.2H_2O$ has been studied [46] between 350 and 487 K. In nitrogen, the Avrami-Erofeev equation ($n = 2$) applied and the value of E_a was 58 kJ mol^{-1}, below 390 K, decreasing to 24 kJ mol^{-1} above this temperature. At low temperatures, in a carbon dioxide atmosphere, $Na_2CO_3.3NaHCO_3$ and $Na_2CO_3.H_2O$ were formed by a first-order process with E_a = 67 kJ mol^{-1}. Kinetic and mechanistic interpretations are supported by, and discussed with reference to, a set of excellent electron micrographs of the nucleation and growth processes. All the reactions considered involve the decomposition of pairs of hydrogen carbonate ions ($2HCO_3^- \rightarrow CO_3^{2-} + H_2O + CO_2$). Aspects of the mechanism and the behaviour revealed by the micrographs remain unresolved.

A kinetic and microscopic study [47] of the thermal decomposition of solid sodium bicarbonate between 360 and 500 K:

$$2NaHCO_3 \rightarrow Na_2CO_3 + H_2O + CO_2$$

showed that, above 440 K, the reaction fits the contracting volume equation with relatively low values of the Arrhenius parameters (E_a = 32 kJ mol^{-1} and A = 10^1 s^{-1}). H_2O or CO_2 exert little effect on the reaction. The first-order equation applies below 390 K with E_a = 64 kJ mol^{-1} and A = 10^5 s^{-1}. Water vapour does not significantly influence this reaction, but CO_2 increases E_a to 130 kJ mol^{-1} and A to 10$^{13.5}$ s^{-1}.

Guarini *et al.* [48] reported that the decomposition of $NaHCO_3$ is (unusually) initiated within the crystal bulk, causing structure collapse and follows first-order kinetics. The apparent value of E_a (93 to 140 kJ mol^{-1}) was identified as being influenced by the presence of a surface layer that was poorly permeable to gaseous reaction products, formed on original crystal boundaries during storage in wet air. The role of outer boundaries in controlling the chemical properties of solids is emphasized. As shown by microscopic examination, reaction is not accompanied by modifications of surface textures that are indicative of chemical changes within the bulk. First-order behaviour is attributable to dependence of rate on the number of residual reactant crystallites.

12.4.2. Potassium bicarbonate

Tanaka [45] showed that the kinetics of decomposition of $KHCO_3$ (394 to 409 K) were closely similar to the behaviour of $NaHCO_3$ described above. The value of n in the Avrami-Erofeev equation was 1.43, E_a = 92.8 ± 5.7 kJ mol^{-1} and ΔH = 79.0 ± 1.3 kJ mol^{-1}. At 396 K, the rate of decomposition of $KHCO_3$ was about one-sixth that of $NaHCO_3$ and the reaction mechanisms are believed to be closely similar.

Hisatsune and Adl [49] studied the decomposition of $KHCO_3$ incorporated in a pressed KBr disc and the progress of reaction was determined from infrared measurements. In such a system the environment of the reactant ion may be significantly different from that in a crystal of the pure compound. $KHCO_3$ formed a bicarbonate cyclic dimer and reacted in the interval 410 to 490 K to yield two monomeric anions, E_a = 58 kJ mol^{-1}. Further reactions at higher temperatures (above 700 K) yielded formate.

12.5. SILVER CARBONATE

The variations in reactivity between different preparations of Ag_2CO_3 are so great that samples have been described [50] as 'active' or 'inactive'. The different forms cannot be distinguished on the basis of chemical analysis or X-ray diffraction. These variations of reactivity are attributed to changes in the concentrations of water incorporated in the structure. This impurity is believed to occupy a key role in decomposition by providing paths for migration of CO_2 through the crystal. Water is represented as occupying lattice sites in the form of impurity anion defects (OH^- and HCO_3^-) through:

$$O^{2-} + CO_2 + H_2O \rightarrow OH^- + HCO_3^-$$

together with cation vacancies as represented below:

CO_3^{2-}	Ag^+	CO_3^{2-}	Ag^+	CO_3^{2-}	Ag^+	CO_3^{2-}
Ag^+		□		Ag^+		Ag^+
CO_3^{2-}	Ag^+	$\mathbf{HCO_3^-}$	Ag^+	CO_3^{2-}	Ag^+	CO_3^{2-}
Ag^+		□		Ag^+		Ag^+
CO_3^{2-}	Ag^+	$\mathbf{OH^-}$	Ag^+	CO_3^{2-}	Ag^+	CO_3^{2-}
Ag^+		Ag^+		Ag^+		Ag^+
CO_3^{2-}	Ag^+	CO_3^{2-}	Ag^+	CO_3^{2-}	Ag^+	CO_3^{2-}

If Ag^+ vacancies (□) are located between the accommodated impurities, as indicated here, CO_2 may migrate from HCO_3^- to OH^- sites with small steric hindrance and without the requirement of the concurrent movement of a cation. In a solid containing a high proportion of such dissociated water, therefore, carbon dioxide may be readily transported across the crystal to the developing system of pores, which arise as a result of the volume change due to reaction. The migration of Ag^+ ions through cation vacancies aids the crystal reorganization involved in the formation of silver oxide.

There is no induction period to reaction, which is deceleratory throughout, and so it was concluded that the nucleation step is not difficult. The rates of decomposition, and of the reverse reaction, are largely controlled by the impurity (water) concentration, and hence reactivity is unusually sensitive to details of the method of preparation.

Consistent [50] with this model is the observation [51] that incorporation of Cd^{2+} as an impurity in the Ag_2CO_3 lattice increases the reaction rate, because this introduction of divalent ions must be accompanied by formation of cation vacancies. Wydeven *et al.* [52-54] decomposed pure Ag_2CO_3 and compared behaviour with that of the salt doped with Y^{3+} or Gd^{3+}. They also studied the effect of water vapour. The kinetic observations fitted the Polanyi-Wigner equation and it was concluded that decomposition proceeds by an interface mechanism.

12.6. LEAD CARBONATE

The decomposition of $PbCO_3$ proceeds to completion in stages. Maciejewski *et al.* [55], using TG, DTA, infrared and X-ray methods to study the decomposition of $PbCO_3$ in air and in CO_2, concluded that intermediate compounds having the ratios ($PbCO_3$:PbO) 5:1, 2:1, 7:5, 1:1 and 1:2 were formed. Shafrinskii *et al.* [56] describe the intermediates 5:3, 1:1 and 2:1. Doi and Chuzo [57] mention the 1:1 and 2:1 compounds and attribute some of the apparent complexity of behaviour to influences of reactant crystallinity and gas pressure. Komada *et al.* [58] identified the 1:1 intermediate and, from kinetic data, concluded that this compound is formed by an interface process (E_a = 104 kJ mol^{-1}), whereas its subsequent decomposition is controlled by transport of the gaseous product (E_a = 126 kJ mol^{-1}). $2PbCO_3.PbO$ was the only intermediate identified [58] during the decomposition of basic lead carbonate, $PbCO_3.Pb(OH)_2$. Ball and Casson [59] supplemented TG curves for the decomposition of $PbCO_3$ with X-ray diffraction measurements and found no evidence for the production of compounds other than those having 1:1 and 1:2 ratios.

12.7. ZINC CARBONATE

Decomposition of $ZnCO_3$ has been described [60,61] as occurring at an advancing interface (E_a = 96 kJ mol^{-1}) and the kinetic parameters were dependent on the CO_2 pressure. The reaction of basic zinc carbonate ($2ZnCO_3.3Zn(OH)_2.H_2O$) is believed to produce both CO_2 and H_2O at the same interface and to form no intermediates during decomposition. Wiedemann *et al.* [62] determined activation energies for the decomposition of $Zn_5(OH)_6(CO_3)_2$ under dry and moist conditions by thermogravimetry between 450 and 530 K. The value of E_a for the reaction at α = 0.5 in vacuum (184 kJ mol^{-1}) was greater than that in moist air (126 kJ mol^{-1}). Water vapour increased the reaction rate. Water retained on the solid within the zone of chemical change is suggested to promote recrystallization of the solid products.

12.8. CADMIUM CARBONATE

The decomposition of $CdCO_3$ is reversible and without side reactions [63]. The rate and activation energy are influenced by the gases present [64] especially carbon dioxide [65]. Mikhail *et al.* [66] measured a value of E_a (96 kJ mol^{-1}) which was very close to the dissociation enthalpy (99 kJ mol^{-1}). Doping with $A\ell^{3+}$, Li^+, Zn^{2+}, Ca^{2+} or Ba^{2+} increased the rate of decomposition, but resulted in the production of curved Arrhenius plots in which the apparent value of E_a decreased with increasing temperature. These impurities may change the mechanism of decomposition of $CdCO_3$ and also cause variations in the surface areas and structures of the solid products.

12.9. OTHER CARBONATES

12.9.1. Manganese carbonate

Although it has been maintained [67,68] that the decomposition of $MnCO_3$ in an inert and static atmosphere directly yields MnO, Westerdahl and Leader [69] concluded that the initial process is:

$$3MnCO_3 \rightarrow Mn_3O_4 + CO + 2CO_2$$

and that the oxide is reduced if the CO is not removed. When oxygen is available the residue contains Mn_3O_4, Mn_2O_3 and/or MnO_2.

The decompositions of $MnCO_3$-$CaCO_3$ solid solutions, which are isomorphous with calcite across the whole composition range $CaCO_3$ to $MnCO_3$, were studied [70] as a route to the preparation of finely divided $Mn_xCa_{1-x}O$ between x = 1 and x = 0. Results gave evidence that reactions other than CO_2 elimination occurred. It was suggested that CO formation may be accompanied by the production of Mn^{3+} with some carbonate breakdown through:

$$CO_3^{2-} + Mn^{2+} \rightarrow CO + O^- + O^{2-} + Mn^{3+}$$

At higher temperatures $2O^-$ will combine, release electrons ($Mn^{3+} + e \rightarrow Mn^{2+}$) and oxygen will oxidize CO, to give MnO - CaO and CO_2 as final products.

12.9.2. Basic nickel carbonate

Although there have been several studies [71-73] of the decompositions of the various basic salts, the information provided is limited. The decomposition [71]:

$$NiCO_3.2Ni(OH)_2.4.5H_2O(s) \rightarrow 3NiO(s) + CO_2(g) + 6.5H_2O(g)$$

probably involved several distinct rate processes. The surface area of the reactant (140 $m^2\ g^{-1}$) increased during decomposition to 240 $m^2\ g^{-1}$ at 573 K. On further heating to 973 K there was a rapid decrease in surface area of the product to 8 $m^2\ g^{-1}$, attributed to structure collapse and particle sintering. Nickel metal has been identified in the residual products [73].

12.9.3. Cobalt carbonate

The thermal decomposition of cobalt carbonate fitted the Avrami-Erofeev equation with $n = 1.37$ at 493 K and 2.54 at 543 K [74]. These data were later reanalyzed by Engberg [75] who allowed for an initial reaction, $\alpha_0 = 0.02$ and found $E_a = 95$ kJ mol^{-1}, which is close to the Co-O bond dissociation energy (91 kJ mol^{-1}).

12.9.4. Copper carbonates

Malachite ($CuCO_3.Cu(OH)_2$) is transformed by heating in air at 523 K to azurite ($2CuCO_3.Cu(OH)_2$) with $E_a = 165$ kJ mol^{-1} [76]. Further reactions occur at 583 K ($\rightarrow 2CuO + Cu(OH)_2 + 2CO_2$, $E_a = 200$ kJ mol^{-1}) and at 723 K to give CuO ($E_a = 227$ kJ mol^{-1}) as the final residual product. The textural changes accompanying these reactions are recorded.

12.9.5. The $Na_2CO_3 + CaCO_3$ reaction and the Hedvall effect

The 'Hedvall effect' [77] ascribes enhanced reactivity within particular temperature intervals in reactions of solids as being due to increased ionic mobility during a condition of crystalline disorder or recrystallization. Garn and Habash [78] selected the reaction:

$$Na_2CO_3 + CaCO_3 \rightarrow Na_2Ca(CO_3)_2$$

as a suitable system for investigation of the magnitude of the effect, because the product exhibits phase transitions (the necessary condition for crystalline disorder) within the the temperature range that the reaction occurs, 600 to 700 K. No Hedvall effect was detected under these most favourable reaction conditions, so the authors concluded that the explanation of the effect is not valid. An alternative explanation for the enhanced reactivity could be the intervention of a liquid phase.

12.10. KINETIC SUMMARY

Selected kinetic characteristics for the decompositions of metal carbonates are summarized in Table 12.1.

12.11. CONCLUSIONS

The decompositions of most metal carbonates are endothermic and at least partially reversible. To obtain reliable kinetic conclusions for reversible reactions it is necessary to obtain more extensive rate measurements and interpret these data with great care. This increased complexity is not always acknowledged. The unusually wide and diverse kinetic behaviour reported in the extensive literature for $CaCO_3$ amply illustrates the variations which may arise when reaction conditions are not comparable. The reported values of A and E_a extend over a range which is probably wider than for any other solid state decomposition and exhibit compensation behaviour [13]. This pattern of behaviour is believed to arise from the sensitivity of reaction rate to procedural variables such as CO_2 pressures and heat flows within typical reactant assemblages of small particles [12]. Apparent magnitudes of A and E_a measured by non-isothermal methods can be exceptionally large, particularly in CO_2 atmospheres [11,14,27]. Quantitative allowance for the CO_2 pressure can lead to more realistic calculated values for the Arrhenius parameters [15]. Beruto and Searcy [4], however, have shown that by careful design of experiments, meaningful information about the kinetics of the interface step can be obtained. While the carbonates of other Group IIA cations also show significant variations of kinetic characteristics with reaction conditions, the range of reported values is less than for $CaCO_3$.

The reported variations in reactivity of different samples of Ag_2CO_3 are ascribed [50] to crystal impurities and defects. The influence of doping on the rates of dissociation of Ag_2CO_3 and of $CdCO_3$ [66] has been investigated and doping could be used with advantage in mechanistic studies of other carbonate decompositions.

The establishment of the stoichiometries for the dissociations of carbonates other than those of Group IIA cations is less straightforward. Some of the intermediate phases recognized in the decomposition of lead carbonate may result from the

Table 12.1.
Selected kinetic characteristics for the decompositions of metal carbonates

Compound	Temperature range /K	Atmosphere	ΔH kJ mol^{-1}	E_a kJ mol^{-1}	Remarks	Ref.
$CaCO_3$	934-1013	Vacuum	178	205	Single (1011) face	[4]
$MgCO_3$	813-873	Vacuum	116.9	150	Contracting volume equation	[30]
$(Mg,Ca)CO_3$	850-940	Vacuum		208-230	Contracting volume equation	[30]
$SrCO_3$	1023-1293	Vacuum	234.5	222	Non-isothermal analysis based on contracting volume equation	[36]
$BaCO_3$	1103-1373	Vacuum	269.2	283	Non-isothermal analysis based on contracting volume equation	[36]
$BaCO_3$	1160-1210	Vacuum	252.1	225.9	Single (001) face	[40]
Ag_2CO_3					Contracting volume equation	
(active)	420-500	Vacuum	81.3	96		[79]
(inactive)	384-421	He	81.3	75-84	Pellets (IR) $\alpha < 0.1$	[52]

Table 12.1.(continued)
Selected kinetic characteristics for the decompositions of metal carbonates

Compound	Temperature range /K	Atmosphere	ΔH kJ mol^{-1}	E_a kJ mol^{-1}	Remarks	Ref.
$PbCO_3$	463-473	N_2	94.0	105	First stage	[58]
$PbCO_3.PbO$	463-473	N_2	-6.4	126	Second stage	[58]
$PbCO_3$	509-608	Vacuum	87.6	168-174	Contracting volume equation	[80,81]
$MnCO_3$	611-661	N_2	116.3	123	Prout-Tompkins equation	[82]
$CdCO_3$	573-773	Vacuum	99.0	151	Nucleation and growth	[63]
			99.0	96		[66]
$NaHCO_3$	383-397	N_2	71.8	109	Avrami-Erofeev	[45]
$KHCO_3$	394-409	N_2	79.0	93	Avrami-Erofeev	[45]

influence of gaseous CO_2 [59]. The range of oxide products from the $MnCO_3$ decomposition probably arises through secondary reactions involving cation oxidation.

The dissociations described above have mostly been identified as interface advance reactions for which the nucleation step occurs relatively readily. The dominant kinetic feature is the progress of the reaction zone inwards to the particle centres. The Polanyi-Wigner reaction model (Chapter 7) was developed to account for the rates of such processes. Shannon [83] identified 29 different chemical changes of this type and found that only one-third of the reported kinetic parameters were within an order of magnitude of the theoretically expected values. From these, the dissociations of $CaCO_3$ and $MgCO_3$ were selected for more quantitative application of the absolute reaction rate theory. The known crystal structure and physical properties of the participating bonds were used to represent the transition state as follows:

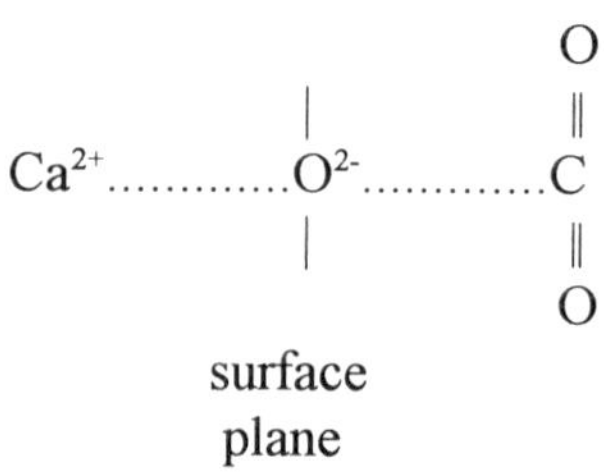

(with CO_2 rotating parallel to the surface)

Rates of reaction predicted by the Polanyi-Wigner theory for this reaction model were in good agreement with published data [30] (E_a for $CaCO_3$ = 166 kJ mol^{-1} and for $MgCO_3$ = 150 kJ mol^{-1}).

Subsequent work demonstrated [4] that E_a for $CaCO_3$ decomposition is significantly larger (+18%) than the reaction enthalpy, thereby invalidating the above conclusion. The influence of retention of gaseous products on the kinetics had been ignored. The forms and functions of interfaces are still insufficiently understood for successful application of absolute reaction rate theory. Even some of the apparent agreements with the predictions of Polanyi-Wigner theory [83] may be coincidental rather than significant.

The nucleation step in the decomposition of $CaCO_3$ appears to be determined more by stereochemistry [9] than by energetics [7,8]. Problems encountered in correlating the dispositions of growth nuclei with reactant dislocation structure [7] could arise from the rapid and comprehensive surface modifications accompanying the establishment of reaction conditions. Initial changes that influence surface reactivities have been established as a feature of many dehydrations [84] and similar behaviour might be expected to occur in other solid state decompositions.

Nucleus formation during $CaCO_3$ decomposition is accompanied by relatively restricted cracking of the outer (perhaps already deactivated) boundary layer covering the reaction zones (see plates in [84,85]). The boundary layers across nuclei appear to have undergone little change during subsurface reactions, while channel development within nuclei is similar to that observed during alum dehydrations [84]. Investigation of prereaction surface modifications [48] is important in characterizing nucleation processes.

Nucleation appears to occur readily in most carbonate dissociations. Product recrystallization may occur only after CO_2 release as shown for $CaCO_3$ [3,4] and basic nickel carbonate [71]. There is little evidence for promotion of the interface reaction by the solid product, instead the solid product is an inhibitor through increasing the contribution of the reverese (recombination) process. Measured values of E_a are usually comparable with reaction enthalpies (Table 12.1.). Together these features suggest that the rates of most carbonate dissociations are controlled by anion breakdown at the boundaries of reactant particles.

REFERENCES

1. W.E. Garner, Chemistry of the Solid State, (Ed. W.E. Garner), Butterworths, London, 1955, Chap.8.
2. R.C. Evans, An Introduction to Crystal Chemistry, Cambridge University Press, Cambridge, 2nd Edn, 1966, p.220.
3. A.R. Salvador, G.C. Calvo and A.B. Aparicio, Thermochim. Acta, 143 (1989) 339.
4. D. Beruto and A.W. Searcy, J. Chem. Soc., Faraday Trans. I, 70 (1974) 2145.
5. K. Towe, Nature (London), 274 (1978) 239.
6. M. Maciejewski and A. Reller, Thermochim. Acta, 142 (1989) 175.
7. J.M. Thomas and G.D. Renshaw, J. Chem. Soc. A, (1967) 2058.
8. J.M. Thomas and G.D. Renshaw, Chem. Commun., (1968) 1247.
9. P.G. Fox and J. Soria-Ruiz, Proc. R. Soc. (London), A314 (1970) 429.
10. E.N. Kovalenko and I.B. Cutler, Mater. Sci. Res., 3 (1966) 485.
11. P.K. Gallagher and D.W. Johnson, Thermochim. Acta, 6 (1973) 67.
12. F.W. Wilburn, J.H. Sharp, D.M. Tinsley and R.M. McIntosh, J. Thermal Anal., 37 (1991) 2003; J.H. Sharp, F.W. Wilburn and R.M. McIntosh, J. Thermal Anal., 37 (1991) 2021.
13. A.K. Galwey and M.E. Brown, Thermochim. Acta, 300 (1997) 107.
14. J. Zsakó and H.E. Arz, J. Thermal Anal., 6 (1974) 651.
15. J.M. Criado, M. Gonzalez, J. Malek and A. Ortega, Thermochim. Acta, 254 (1995) 121.

16. A.L. Draper, Proc. Robert A. Welch Foundation: Conferences on Chemical Research. XIV Solid State Chemistry, Houston, Texas, (1970) p.214.
17. J-P. Auffredic and P. Vallet, C.R. Acad. Sci. Paris, C, 263 (1966) 457, 652,1093; 264 (1967) 815; 267 (1968) 192; 271 (1970) 178; 275 (1972) 245.
18. A.W.D. Hills, Chem. Eng. Sci., 23 (1968) 297.
19. F.R. Campbell, A.W.D. Hills and A. Paulin, Chem. Eng. Sci., 25 (1970) 929.
20. T.F. Anderson, J. Geophys. Res., 74 (1969) 3918; 77 (1972) 857.
21. R.A.W. Haul and L.H. Stein, Trans. Faraday Soc., 51 (1955) 1280.
22. B. Vosteen, Chem. Ing. Tech., 43 (1971) 478.
23. E.K. Powell and A.W. Searcy, Commun. Amer. Ceram. Soc., (1982) C42.
24. G.F. Knutsen, A.W. Searcy and D. Beruto, J. Amer. Ceram. Soc., 65 (1982) 219.
25. Q. Zhong and I. Bjerle, Thermochim. Acta, 223 (1993) 109.
26. T.Darroudl and A.W. Searcy, J. Phys. Chem., 85 (1981) 3971.
27. H. Hashimoto, Kogyo Kagaku Zasshi, 64 (1961) 250, 1162,1166.
28. H. Mauras, Bull. Soc. Chim. France, (1960) 260.
29. J.T. Lee, T.C. Keener, M. Knoderer and S.J. Khang, Thermochim. Acta, 213 (1993) 223.
30. H.T.S. Britton, S.J. Gregg and G.W. Winsor, Trans. Faraday Soc., 48 (1952) 63, 70.
31. Z. Kubas, M.S. Szalkowicz and W. Rosnowski, Chem. Abs., 69 (1968) 30545y,110348b; 71 (1969) 63668k.
32. A. Doi and C. Kato, Kogyo Kagaku Zasshi, 74 (1971) 1597.
33. H. Jacobs and M.J.D. Low, J. Coll.Interface Sci., 46 (1974) 165.
34. G. Spinolo and U. Anselmi-Tamburini, J. Phys. Chem., 93 (1989) 6837.
35. K.J.D. MacKenzie and R.H. Meinhold, Thermochim. Acta, 230 (1993) 331.
36. M.D. Judd and M.I. Pope, J. Thermal Anal., 4 (1972) 31.
37. Z.N. Zemtsova, M.M. Pavlyuchenko and E.A. Prodan, Vestsi. Akad. Navuk Belaruss. SSR., Ser. Khim. Navuk, (1970)(6) 16; (1971)(6) 11; (1972)(1) 5.
38. W. Trzebiatowski and K. Skudlarski, Rocz. Chem., 36 (1962) 1279, 1427.
39. B.V. L'vov and A.V. Novichikhin, Thermochim. Acta, 290 (1997) 239.
40. T.K. Basu and A.W. Searcy, J. Chem. Soc., Faraday Trans. I, 72 (1976) 1889.
41. M. Kito and S. Sugiyama, Kagaku Kogaku, 30 (1966) 904.
42. C. Calistru and L. Ifrim, Bull. Inst. Politek Iasi, 14 (1968) 209.

43. B.R. Maiti, A.K. Biswas, P. Sengupta and N.C. Roy, Indian Chem. Eng., 12 (1970) 17.
44. I.P. Ishkin and E.S. Dubil, Zh. Prikl. Khim., 41 (1968) 52.
45. H. Tanaka, J. Thermal Anal., 32 (1987) 521.
46. M.C. Ball, C.M. Snelling, A.N. Strachan and R.M. Strachan, J. Chem. Soc., Faraday Trans., 88 (1992) 631.
47. M.C. Ball, C.M. Snelling, A.N. Strachan and R.M. Strachan, J. Chem. Soc., Faraday Trans. I, 82 (1986) 3709.
48. G.G.T. Guarini, L. Dei and G. Sarti, J. Thermal Anal., 44 (1995) 31.
49. I.C. Hisatsune and T. Adl, J. Phys. Chem., 74 (1970) 2875.
50. P.A. Barnes and F.S. Stone, Reactivity of Solids, (Ed. J.W. Mitchell), Wiley, New York, 1969, p.261; Thermochim. Acta, 4 (1972) 105.
51. Yu.A. Zakharov, V.V. Boldyrev and A.A. Alekseenko, Kinet. Katal., 2 (1961) 365.
52. T. Wydeven and M. Leban, Anal. Chem., 40 (1968) 363.
53. T. Wydeven and E. Rand, J. Catal., 12 (1968) 271.
54. T. Wydeven, J. Catal., 16 (1970) 82.
55. M. Maciejewski, J. Leyko and J. Werezynski, Bull. Acad. Pol. Sci., Ser. Sci. Chem., 18 (1970) 81, 205,711.
56. Y.S. Shafrinskii, Y.V. Rumyantsev and E.K. Vasil'ev, Chem. Abs., 69 (1968) 13301y.
57. A. Doi and K. Chuzo, Kogyo Kagaku Zasshi, 73 (1970) 488.
58. R. Komoda, Y. Nishi and M. Kano, Nippon Kagaku Zasshi, 88 (1967) 1038; 89 (1968) 478; 90 (1969) 30.
59. M.C. Ball and M.J. Casson, J. Inorg. Nucl. Chem., 37 (1975) 2253.
60. S. Bretsznajder and E. Cibor, Bull. Acad. Pol. Sci., Ser. Sci. Chim., 13 (1965) 709, 713, 765; 14 (1966) 119.
61. J. Pysiak, S.A. Slyshkina, L.I. Egortseva and M.M. Pavlyuchenko, Rocz. Chem., 45 (1971) 263.
62. H.G. Wiedemann, A.V. Tets and R. Giavanoli, Thermochim. Acta, 203 (1992) 241.
63. E.A. Prodan and M.M. Pavlyuchenko, Chem. Abs., 64 (1966) 11924d-g, 12173c.
64. K. Muraishi, Nippon Kagaku Zasshi, 91 (1970) 122.
65. M. Kazuo, Yamagata Diagaku Kiyo, Kogaku, 10 (1969) 459..
66. R. Sh. Mikhail, D. Dollimore, A.M. Kamel and N.R. El-Nazer, J. Appl. Chem. Biotechnol., 23 (1973) 419, 431.
67. A.J. Hegedus and K. Martin, Magy. Kem. Foly., 72 (1966) 404.
68. A.J. Hegedus and B.A. Kiss, Magy. Kem. Foly., 73 (1967) 41.
69. R.P. Westerdahl and P.J. Leader, Inorg. Nucl. Chem. Lett., 5 (1969) 199.

70. B. Fubini and F.S. Stone, J. Chem. Soc., Faraday Trans. I, 79 (1983) 1215.
71. D. Dollimore and J. Pearce, J. Thermal Anal., 6 (1974) 321.
72. R.M. Mallya and A.R. Vasudeva Murthy, J. Indian Inst. Sci., 43 (1961) 131.
73. Yu. N. Rufov and Z.A. Markova, Kinet. Katal., 6 (1965) 731.
74. L.K. Avramov and I.M. Yanachkova, Z. Phys. Chem. (Leipzig), 241 (1969) 244.
75. A. Engberg, Acta Chem. Scand., 27 (1973) 3929.
76. S.A.A. Mansour, J. Thermal Anal., 42 (1994) 1251.
77. J.A. Hedvall, Solid State Chemistry, Elsevier,Amsterdam, 1966.
78. P.D. Garn and T.S. Habash, J. Phys. Chem., 83 (1979) 229.
79. W.D. Spencer and B. Topley, J. Chem. Soc., (1929) 2633.
80. O. Kadlets and M.M. Dubinin, Izv. Akad. Nauk SSSR, Otd. Khim. Nauk., 3 (1961) 390.
81. G.I. Samal, Geterorg. Khim. Reakt., (1961) 92.
82. C. Iltis, Rev. Chim. Miner., 4 (1967) 233.
83. R.D. Shannon, Trans. Faraday Soc., 60 (1964) 1902.
84. M.E. Brown, A.K. Galwey and G.G.T. Guarini, J. Thermal Anal., 49 (1997) 1135.
85. D.C. Anderson, P. Anderson and A.K. Galwey, Fuel, 74 (1995) 1024,1018,1031.

Chapter 13

DECOMPOSITION OF METAL PERHALATES, HALATES AND HALITES

13.1. INTRODUCTION

The technological uses of these compounds as explosives and the importance of ammonium perchlorate as a propellant have resulted in the appearance of a large number of reports on the kinetics and mechanisms of their decompositions. Many undergo at least partial melting during anion breakdown. Products usually include anions containing a smaller number of oxygen atoms. Perchlorates have attracted the widest interest and perbromates and periodates have been generally less intensively studied. Some investigations have been concerned with the effects of anion variation, while others compare salts containing different cations but a common anion.

A comprehensive review of the pre-1976 literature on the decompositions of the salts of halogen oxyacids has been provided by Solymosi [1]. Another important review (1974) by Stern [2] is concerned with the high temperature properties of oxyhalides.

In this chapter, salts containing a common metal cation and various anions, XO_n^-, (where X = Cℓ, Br and I), are discussed in sequence from the most highly oxidized to those with lower values of n. Ammonium (and related) compounds are discussed in Chapter 15.

13.2. METAL PERCHLORATES

13.2.1. Group IA metal perchlorates

The kinetics of the initial stages of the thermal decompositions of the Group IA metal (Na, K, Rb, Cs) perchlorates between 623 and 773 K were studied by Cordes and Smith [3]. The major reaction products were the corresponding chlorates and oxygen. Because the rates of oxygen evolution below 683 K from all four reactants were almost equal, it was suggested that the rate-controlling factor was a property of the $C\ell O_4^-$ ion, rather than the band structure of the solid. At low α, the rate of gas evolution was almost constant and did not correlate with such parameters as particle size or sample mass. It was concluded that decomposition proceeded through a homogeneous bimolecular interaction between neighbouring perchlorate ions in the crystal. The value of A, estimated through application of the transition state theory to this model of the activated complex, is in reasonably close agreement with the mean value from experiment.

$LiC\ell O_4$ and $NaC\ell O_4$ melt before anion breakdown. The induction periods to the decompositions of K, Rb and $CsC\ell O_4$ are attributed [4] to slow anion breakdown within the solid, followed by a more rapid reaction in the melt formed as a eutectic between reactant and products. Because the activation energies for these reactions are about 270 kJ mol^{-1}, it was concluded from consideration of bond strengths that the rate-limiting step is the rupture of a $C\ell$-O bond. A sequence of further comparable disruption or dissociation reactions yields oxygen and the metal chloride as the final residual product.

$NaC\ell O_4.H_2O$ loses water [5] in two processes below 473 K, there is a phase transition between 568 and 588 K and melting accompanied by decomposition commences above 723 K and continues to 843 K. $KC\ell O_4$ undergoes an endothermic phase transformation, orthorhombic to cubic, between 553 and 603 K. Melting accompanied by decomposition to $KC\ell$ occurred between 858 and 903 K. Reactions, including the rapid decomposition accompanying fusion, were investigated using thermomicroscopy and a newly developed apparatus combining accoustic emission and DTA. Photomicrographs of the reactants illustrate the textural changes that appear during these decompositions.

The α-time curves for the decomposition of $KC\ell O_4$ are sigmoid and, in this and related systems, the pattern of kinetic behaviour is not identified with any nucleation and growth processes. The acceleratory stage arises because of an increasing contribution from the relatively more rapid reactions that occur during progressive fusion of the reactant in the molten eutectic formed between $KC\ell O_4$ and product anions containing less oxygen. During the deceleratory period, resolidification occurs. The chemical transformations are a complicated [6] combination of consecutive, concurrent and reverse reactions, including:

$$2KC\ell O_4 \rightarrow 2KC\ell O_3 + O_2 \rightarrow 2KC\ell + 4O_2$$
$$4KC\ell O_3 \rightarrow 3KC\ell O_4 + KC\ell$$

In addition, $KC\ell O_4$ reacts with both $KC\ell O_3$ and $KC\ell$. The gradual onset of melting and later solidification occur without discontinuity in the α-time plots.

Khorunzhii and Il'in [7] investigated the influence of MnO_2 on the decompositions of K and Na perchlorates and chlorates and concluded that both surface and grain-boundary diffusion played an essential role in control of the decomposition rate.

Hisatsune and Linnehan [8] used infrared spectra to measure the thermal decomposition of the $C\ell O_4^-$ ion in a $KC\ell$ matrix. Although the environment of the reactant ion may be different from that in crystalline $KC\ell O_4$, the kinetic behaviour is very similar to that reported by Cordes and Smith [3]. Reaction was second-order and $E_a = 185$ kJ mol^{-1}. The reaction of $C\ell O_3^-$ in $KC\ell$ was similar, but the value of E_a was lower (close to 125 kJ mol^{-1}). When both $C\ell O_4^-$ and $C\ell O_3^-$ were present, reaction was approximately first-order.

13.2.2. Group IIA metal perchlorates

Acheson and Jacobs [9] concluded that the decomposition of anhydrous $Mg(C\ell O_4)_2$, in vacuum, yielding residual MgO, involved the stepwise removal of oxygen from the anion. The first step in $C\ell O_4^-$ breakdown yields the chlorate anion and is believed to proceed through a chain mechanism:

$$C\ell O_4^- + O \rightarrow C\ell O_3^- + O_2$$

where the oxygen atoms are formed by breakdown of the chlorate ion ($C\ell O_3^- \rightarrow C\ell O_2^- + O$) and subsequently the chlorite ion to the hypochlorite ($C\ell O_2^- \rightarrow C\ell O^- + O$). The reaction undergoes a significant decrease in rate at $\alpha = 0.525$ and this was attributed to decomposition of magnesium hypochlorite. Kinetic behaviour is controlled by the diffusion of gas from the reaction zone, the value of E_a is small and the reaction is zero-order. An increase in rate of reaction resulting from the addition of $NH_4C\ell O_4$ was attributed to melt formation. Borisova *et al.* [10] agree that the decomposition occurs in two kinetic stages, but suggest that the second stage is the decomposition of a compound represented as $MgO.Mg(C\ell O_4)_2$.

Jacobs *et al.* [11] examined the data available on the decomposition of $Ba(C\ell O_4)_2$. The α-time curves consist of an acceleratory process followed by a discontinuous decrease in rate at α about 0.52. They proposed that the acceleratory stage is associated with chlorate formation:

$$C\ell O_4^- \rightleftarrows C\ell O_3^- + O$$
$$C\ell O_4^- + O \rightarrow C\ell O_3^- + O_2$$

and, because $C\ell O_3^-$ is less stable, it decomposes to yield an increasing concentration of oxygen atoms, so that reaction becomes acceleratory. As chloride accumulates in the reactant an alternative equilibration process becomes important:

$$C\ell^- + O \rightleftarrows C\ell O^-$$

Thus after α reaches about 0.5 the molar ratio $C\ell O_4^-$: $C\ell^-$ is approximately unity and the rate of reaction becomes dependent on the slow decomposition of the hypochlorite. While the occurrence of melting during these reactions cannot be definitely excluded, it is thought to be improbable.

The decomposition of $Ca(C\ell O_4)_2$ proceeds [12] in the solid state between 666 and 678 K with $E_a = 260$ kJ mol^{-1}, but at higher temperatures a melt containing $CaC\ell_2$ is formed and $E_a = 226$ kJ mol^{-1}.

The dehydration of $Sr(C\ell O_4)_2.4H_2O$ occurs [13] at about 470 K, possibly with aqueous fusion. Decomposition becomes apparent at 688 K and reaction is vigorous by about 720 K, but difficulties were experienced in analyzing the thermal data due to overlap of the exothermic reaction with melting.

13.2.3. Other metal perchlorates

The thermal stabilities of the three silver salts decrease [14] in the sequence $AgC\ell O_4 > AgC\ell O_3 > AgC\ell O_2$, which is also the order of decreasing ionic character. The fusion of $AgC\ell O_4$ well below its melting point is attributed to eutectic formation with $AgC\ell$ during an autocatalytic reaction. The chlorate decomposed at about 80 K above the melting point through disproportionation accompanied by some decomposition. In contrast, the disproportionation of silver chlorite:

$$3AgC\ell O_2 \rightarrow 2AgC\ell O_3 + AgC\ell$$

was not accompanied by gas evolution. Oxygen redistribution between the anions evidently does not involve the production of molecular oxygen, and presumably proceeds through exciton formation. The evolution of $C\ell O_2$ at higher temperatures is attributed to decomposition after electron transfer.

The thallium salts were less stable [14] than the silver analogues. $T\ell C\ell O_2$ decomposes at about 300 K and was, therefore, too labile for study. The reaction of $T\ell C\ell O_3$ differed from the behaviour of other chlorates in that decomposition resulted in cation oxidation following:

$$2C\ell O_3^- \rightarrow 2C\ell O_2 + 2O^-$$

and perchlorate was not formed. The decomposition of $T\ell C\ell O_4$, the most stable salt of the triad, was autocatalytic above 680 K and was accompanied by some sublimation. From the magnitude of E_a (227 kJ mol^{-1}) it was concluded that the primary step is rupture of a $C\ell$–O bond:

$$C\ell O_4^- \rightarrow C\ell O_3^- + O$$

and the cation is oxidized to $T\ell_2O_3$. Giridharan *et al.* [15] showed that the relative yields of $T\ell_2O_3$ and the volatile $T\ell C\ell$ depended on temperature and the prevailing atmosphere.

13.2.4. Irradiated Group IA and Group IIA metal perchlorates

Irradiation of alkaline and alkaline earth metal perchlorates [16] (^{60}Co source at ambient temperature) yielded $C\ell O_3^-$, $C\ell^-$, $C\ell O_2^-$, $C\ell O_2$, O_2 and metal oxide as primary products with $C\ell O^-$ formed by secondary reactions. The order of decreasing product yields was Cs > Rb > Mg > Sr > Na > Ca > K > Li > Ba.

13.3. METAL PERBROMATES AND PERIODATES

Only a few perbromates have so far been prepared [2]. The stability of BrO_4^- is about the same [17] as $C\ell O_4^-$ and IO_4^-. $KBrO_4$ decomposes exothermically between 548 and 553 K to $KBrO_3$ which is stable [18].

Potassium metaperiodate decomposes [19] in the range 529 to 561 K:

$$KIO_4 \rightarrow KIO_3 + \tfrac{1}{2}O_2$$

The initial deceleratory process ($\alpha < 0.003$) and subsequent constant rate of slow reaction ($0.003 < \alpha < 0.006$) were attributed to surface decomposition followed by advance into sub-grains. It was suggested that during the acceleratory stage of the main reaction ($\alpha > 0.006$) there is a chain-branching (crack propagation) process of the Prout-Tompkins type resulting in disintegration of the crystals. The subsequent deceleratory process is described by the contracting volume equation. The value of E_a (192 ± 17 kJ mol^{-1}) was identified with I-O bond rupture.

The products of decomposition of KIO_4 and $NaIO_4$ are unusual in undergoing efficient diffusion so that each partly reacted periodate crystal becomes surrounded by a fragile layer of almost pure iodate [20]. α-time curves fit the Avrami-Erofeev equation and values of n increase with temperature: n =1 to 2 for KIO_4 and n = 2 to 3 for $NaIO_4$. Further work is required to elucidate these complicated reactions.

When heated at 5 K min^{-1}, $CsIO_4$ decomposes ($\rightarrow CsIO_3 + ½O_2$) between 573 and 623 K by a two-stage reaction [21]. During the first stage ($\alpha < 0.2$) nuclei form according to the power law and $E_a = 226$ kJ mol^{-1}, from isothermal data. Later the contracting area equation fitted the data and $E_a = 138$ kJ mol^{-1} for nucleus growth.

Pacesová *et al.* [22] studied the thermal reactions of several silver periodates. The following exothermic reaction was identified as occurring between 448 and 473 K:

$$4Ag_2H_3IO_6 \rightarrow 2Ag_3IO_5 + 2AgIO_3 + 6H_2O + O_2$$

13.4. METAL CHLORATES

13.4.1. Group IA metal chlorates

The Group IA metal chlorates melt before decomposition [2]. The catalytic properties of Co_3O_4 in promoting [23] the solid state decomposition of $NaC\ell O_3$ were attributed to the ability of the oxide to donate an electron to an oxygen atom, temporarily accepted at the surface from a chlorate ion, prior to the formation and desorption of molecular oxygen. The systematic increase in E_a during reaction (from 120 to 200 kJ mol^{-1}) was ascribed to the progressive deactivation of the surface.

A small amount of decomposition occurs [24] in single $NaC\ell O_3$ crystals (0.1 to 1.0% mass loss) between 450 K and the melting point, about 535 K. X-ray topographic studies confirmed that 'grown-in' dislocations do not significantly participate in intracrystalline decomposition of this salt. The detailed photographic evidence of dislocation generation and distribution given in this paper is used to establish the role of line defects in the onset and development of reaction. It is concluded that growth dislocations may participate in surface decomposition but appear to be inactive in promoting reaction in the bulk. Significantly more reaction occurred in borax-doped $NaC\ell O_3$.

The catalytic activities of a number of substances (oxides, chlorides, sulphates, nitrates, etc. of various ionic cations) for $KC\ell O_3$ decomposition were compared [25]. The identity of the metal cation and its electronic configuration are suggested to be the factors that determine catalytic activity. The key to these catalytic processes is identified as the formation of a coordination bond between ClO_3^- and the cation. Cations with d^{1-9} configurations are highly active (including Cu^{2+}, Ni^{2+}, etc), d° configurations are moderately active (La^{3+}, Gd^{3+}), and d^{10} or noble gas configurations have little activity (Mg^{2+}, Ca^{2+}, etc.).

13.4.2. Group IIA metal chlorates

Between 650 K and melting at 669 K, $Ba(C\ell O_3)_2$ disproportionates [26] to $Ba(C\ell O_4)_2$ and $BaC\ell_2$ accompanied by some direct decomposition to $BaC\ell_2$.

13.4.3. Other metal chlorates

The decomposition of $Pb(C\ell O_3)_2$ [27] in the interval 467 to 489 K is complicated, yielding both $PbC\ell_2$ and PbO. Following an initial rapid deceleratory process, the sigmoid α-time curve for the main reaction fits the Prout-Tompkins equation ($0.35 < \alpha < 0.85$) with $E_a = 180$ kJ mol^{-1}. $AgC\ell O_3$ decomposes [14] in the melt between 589 and 612 K when α reaches about 0.8 as the $AgC\ell$ product accumulates (E_a = 239 kJ mol^{-1}). The decomposition [28] of thallium(I) chlorate (423 to 443 K, E_a = 150 kJ mol^{-1}) may be represented by:

$$7T\ell C\ell O_3 \rightarrow 3T\ell_2O_3 + T\ell C\ell + 6C\ell O_2$$

and again the mechanism is complicated.

13.5. METAL BROMATES AND IODATES

13.5.1. Group IA metal bromates and iodates

The decomposition of potassium bromate [29]:

$$KBrO_3 \rightarrow KBr + 1\tfrac{1}{2}O_2$$

proceeds by a first-order process in two temperature ranges, separated by a transition region. E_a values for reactions between 615 and 640 K and from 652 to 685 K (where melting occurs) were 260 and 221 kJ mol^{-1}, respectively. It was suggested that reaction involved a nucleation and growth mechanism accompanied by crystallite cracking. Acceleration of decomposition during the later stages is ascribed to more rapid reaction in an increasing amount of melt that results from eutectic formation. Similar results were obtained for $NaBrO_3$ [29]. Pre-irradiation enhanced melting, with a consequent increase in the rate of decomposition.

Decomposition of $CsBrO_3$ proceeds [30] in the molten or semi-molten state of a eutectic formed with the CsBr product. Kinetics were fitted by the Prout-Tompkins and Avrami-Erofeev equations. The reaction rate (673 K) was accelerated significantly both by γ-irradiation damage (which leads to rupture of Br-O bonds) and by the presence of added Ba^{2+} ions which introduce local strain into the crystal and thereby promote BrO_3^- ion breakdown.

Although alkali metal iodates decompose on heating at 750 K, in high vacuum (10^{-6} mbar), mass spectrometric studies [31] detected volatilization of the molecular species KIO_3, $RbIO_3$ and $CsIO_3$. Infrared spectra are consistent with C_{3v} symmetry involving tridentate coordination. On heating, KIO_3 underwent extensive decomposition and $LiIO_3$ and $NaIO_3$ yielded oxygen and metal iodide.

13.5.2. Group IIA metal bromates and iodates

The reactions of $Ba(BrO_3)_2$, proceeding in the range 563 to 653 K in air or 538 to 588 K in vacuum, result in the formation of BrO_2^- and BrO_4^-. E_a values reported [26,32] were 209 and 218 kJ mol^{-1}. Calcium and strontium bromates yield the bromides, but the decomposition of $Mg(BrO_3)_2$ gives MgO as the predominant residual product [33].

The solid state decomposition of $Ba(BrO_3)_2$ has been described [34] as proceeding through (i) an initial fast process; followed by (ii) a short induction period; (iii) a slow linear reaction; (iv) an acceleratory stage; and finally (v) a deceleration to completion. The activation energy during nucleation (159 kJ mol^{-1}) was smaller than that for the main reaction (191 kJ mol^{-1}). Irradiation increased the values of the rate coefficients.

The decomposition of $Ba(BrO_3)_2 \rightarrow BaBr_2 + 3O_2$ (573 to 583 K) gave [35] a sigmoid curve following a small initial gas evolution (to $\alpha = 0.06$). Data were well descibed by the Avrami-Erofeev equation with $n = 2$, ascribed to two-dimensional growth of nuclei. Values of E_a for the acceleratory and decay processes were 185 and 172 ± 5 kJ mol^{-1}. Reaction rates were accelerated and E_a values diminished by the addition of 10% w/w of Gd_2O_3 and Dy_2O_3. This was ascribed to participation of the dopant in the rate limiting step, identified as electron transfer :

$$O^-_{ADS} \rightarrow O_{ADS} + e \text{ (solid oxide)}$$

The decomposition of barium iodate [36,37]:

$$10Ba(IO_3)_2 \rightarrow 2Ba_5(IO_6)_2 + 8I_2 + 18O_2$$

yields solid $Ba_5(IO_6)_2$ which is stable to about 770 K in vacuum and to about 1220 K in air, and its decomposition to BaO, I_2 and O_2 is reversible. Calcium [38] and strontium [36] iodates behave similarly, while $Mg(IO_3)_2$ reacts [39] above 848 K to give MgO. Non-isothermal (5 K min^{-1}) DSC studies [40] of the dehydration and decomposition of $Mg(IO_3)_2.4H_2O$ identified the first step as water loss (- $4H_2O$ at 435 K) and then anion breakdown in two steps :

$$5Mg(IO_3)_2 \rightarrow Mg_5(IO_6)_2 + 4I_2 + 9O_2 \qquad (878\ K)$$

followed by:

$$2Mg_5(IO_6)_2 \rightarrow 10MgO + 2I_2 + 7O_2 \qquad (928\ K)$$

13.5.3. Other metal bromates and iodates

The decomposition [41] of $T\ell BrO_3$ is accompanied by cation oxidation to yield product $T\ell_2O_3$. This reaction proceeds in two stages through an initial ($\alpha < 0.05$) rapid first-order process ($E_a = 150$ kJ mol^{-1}) followed by a sigmoid α-time curve. The acceleratory and deceleratory stages of the main reaction were fitted by the Prout-Tompkins equation ($\alpha < 0.35$ and $E_a = 209$ kJ mol^{-1}) and the contracting volume ($E_a = 150$ kJ mol^{-1}) equation, respectively. Rapid initial decomposition at sub-grain boundaries was followed by advance of the reaction interface into the reactant particles.

The decompositions of cobalt and nickel iodates [37] at 688 K and above 763 K, respectively, yield the metal oxide and iodine. In hydrogen, NiI_2 is reduced to metal and HI at about 770 K, but the reaction of CoI_2 is slower. $Ni(IO_3)_2.2H_2O$ and $Zn(IO_3)_2.2H_2O$ dehydrate in a single stage, followed by decomposition to the oxides [42]. The deuterates were more stable because of structural changes.

The thermal stabilities of the lanthanide bromates and iodates [43] decrease with increase in cationic charge density, presumably as a consequence of increased anionic polarization. Metallic lead reacts [44] with K, Ca and Ba iodates to yield the iodites at about 700 K:

$$MIO_3 + Pb \rightarrow MIO_2 + PbO$$

Bancroft and Gesser [33] concluded that kinetic rather than equilibrium factors are dominant in determining whether decomposition of a metal bromate yields residual bromide or oxide.

13.6. METAL HALITES

The most frequently encountered reaction on heating metal chlorites [1,14,27] is disproportionation:

$$3MC\ell O_2 \rightarrow 2MC\ell O_3 + MC\ell$$

Barium and lead chlorites explode on rapid heating at about 460 K and 385 K, respectively. The decompositions of $NaC\ell O_2$ (above 413 K) and of $AgC\ell O_2$ (above 358 K) are first-order reactions and the values of E_a are 227 and 109 kJ mol^{-1}, respectively. The major products of $AgC\ell O_2$ decomposition above 373 K are Ag, Ag_2O and $C\ell O_2$. The thermal stabilities of these four chlorites decrease [27,28] in the same sequence (Na, Ba, Ag, Pb) as the ionic character of these solids, covalent character is associated with greater reactivity.

$Ba(BrO_2)_2$ is formed [45] during decomposition of the bromate in oxygen at 523 K.

13.7. RADIOLYSIS OF OXYHALIDES

Boyd *et al.* [46] have studied the γ-radiolysis of several oxyhalides. The extent of reaction in the Group IA and thallium bromates increased linearly with dose, yielding BrO_2^-, BrO^-, Br^- and O_2 in proportions which depended on the cation present. Later work extended the reactants to $KC\ell O_3$ (products $C\ell O^-$, $C\ell O_2^-$ and O_3^-), $CsBrO_3$ (products BrO^- and O_3^-) and $CsIO_3$ (increased absorption was ascribed to molecular iodine in the crystal). It was concluded that reaction within the crystal is initiated by ionization, with excitation of the oxyhalide ion, followed by rupture of the halogen-oxygen bond:

$$3C\ell O_2^- \rightarrow 2C\ell O_3^- + C\ell^-$$

The exothermic decomposition of $KC\ell O_2$ between 410 and 440 K gave the same stoichiometry as for radiolysis. The mechanism suggested was the formation of O atoms in the crystal which add to $C\ell O_2^- + O \rightarrow C\ell O_3^-$. The iodate is relatively very stable, yielding only small quantities of I^- and I_2.

Levy *et al.* [47] related the kinetics of decomposition of $NaBrO_3$ to the concentration of colour centres present. Rate expressions, derived from models involving defect formation followed by charge transfer, accurately described the experimental α-time results.

13.8. HALIDES AND HYDROXYHALIDES

Most simple halides, ionic or covalent, melt unchanged on heating. Crystal defects have been identified as being important in the decompositions of CoF_3 [48] and CdI_2 [49]. Barret [50] studied the decomposition of $CuBr_2 \rightarrow CuBr + \frac{1}{2}Br_2$ and discussed the role of diffusion within the mass of reactant particles on the overall kinetic behaviour. HF.LiF and HF.NaF decompose [51] before melting, in a single step, whereas the other Group IA hydrogen fluorides melt. The thermal stabilities of these and the Group IIA analogues correlate with the strength of hydrogen bonding within the structure.

Lanthanum oxychloride is very stable, reacting [52] between 1570 and 1870 K:

$$3LaOC\ell(s) \rightarrow La_2O_3(s) + LaC\ell_3(g)$$

Stabilities, measured by the effusion method, decreased [53] in the lanthanide sequence.

Secco, Ramamurthy *et al.* [54-57] have studied the decompositions of several metal (notably Zn, Cd, Cu and Pb) hydroxyhalides. Each metal is capable of forming several compounds and comparative studies, including consideration of structural aspects of the solids concerned, have sought correlations between constitution and reactivity. Although such trends have been identified amongst the large number of reactants studied, examples of unusual or apparently anomolous behaviour are also noted. The kinetic analyses of thermogravimetric data showed that the decompositions can be described by first-order, contracting volume (or area) or nucleation and growth equations, but these kinetic interpretations were not supported by observational studies. Arrhenius parameters were compared with the predictions of absolute reaction rate theory, and vibration frequencies were obtained from infrared measurements [54].

The presence of the halogen is identified [54-57] as stabilizing these hydroxycompounds and the ease of water loss often decreases in the sequence fluoride to iodide. Copper hydroxychloride, bromide and iodide undergo dehydroxylation in a similar way to the corresponding Cd and Zn salts, but the fluoride is hydrolyzed and yields water and HF. It was concluded from the kinetic analyses, which indicated that data fitted the contracting area, or volume, or nucleation and growth equations, that these reactions are interface processes. The decompositions of lead hydroxychloride, bromide and iodide differ in that they are better described by rate equations based on diffusion control, ascribed to the reaction of a mobile proton with OH^-, but said not to be due to the migration of water.

Comparable studies were made of the decompositions of a range of the double salts $xMe(OH)_2.yMC\ell_2$, where Me was Cd or Cu, and M was Mn, Co, Ni, Cu, Zn, Cd, Mg or Ca. Dehydroxylation was accompanied by a metathetical reaction. Decompositions of the salts of the Cd (= Me) series were apparently first-order processes, while the kinetics of decomposition of salts containing Cu (= Me) were more diverse, because examples of the applicability of both phase boundary advance and Avrami-Erofeev reaction models were found.

Ball and Coultard [58] have also studied the decompositions of basic copper halides, though the kinetic observations obtained from dynamic measurements were not in agreement with those referred to above [54]. Walter-Levy and Goreaud [59] found that, in the presence of air, the basic copper chlorides are oxidized to CuO with the release of free halogen. Sharkey and Lewin [60] compared the stabilities of various copper hydroxyhalides and found the sequence to be the same as that of the hydrogen bond strengths. Walter-Levy and Groult [61] found an increase in stability of six cadmium hydroxy and iodide compounds with increase in the halogen content in the order $Cd(OH)_2$, $Cd_4(OH)_6I_2$, $Cd_7(OH)_{10}I_4$, $Cd_3(OH)_4I_2$, Cd(OH)I, CdI_2.

13.9. CONCLUSIONS

The high values of E_a that are generally characteristic of the decomposition reactions of metal oxyhalides are widely interpreted [1] as evidence that the initial step in anion breakdown is the rupture of an X - O bond. The energy barriers to these reactions are not very sensitive to the properties of the cation present. Information of use in the formulation of reaction mechanisms has been obtained from radiolytic studies of oxyhalogen salts [16,46,47]. The tendency to melt, through eutectic formation, with faster reaction in the liquid phase, explains the sigmoid α-time curves in a number of systems, for example in the decompositions of $KC\ell O_4$, $RbC\ell O_4$ and $CsC\ell O_4$. It could be interesting to investigate quantitatively the participation of melting in some of these reactions.

Stability of the most highly oxidised species (XO_4^-) is generally greatest and this species is sometimes found as a product of disproportionation of XO_3^- and XO_2^-, so that a more highly oxidised intermediate may be formed during a sequence of reactions that ultimately give the halide (or, depending on thermodynamic controls [62], the oxide, e.g., magnesium salts) as the final residual product.

Ionic oxysalts are the most stable. Where anionic-cationic interactions introduce some covalent character, the decrease in anionic symmetry is accompanied by a marked decrease in stability. Transition metal perchlorates, and perchloric acid, tend to be explosives with unpredictable and often dangerous properties.

The many discussions of the mechanisms of decompositions of salts of oxyacids have included relatively few kinetic conclusions based on microscopic observations. Such textural investigations could provide information on the extent, and possibly the role of, melting during these reactions.

A second unresolved general feature of many decompositions in this group of reactants is the uncertainty regarding the chemical nature of the intermediate oxygen (radicals, O atoms ?) postulated in many of the decomposition mechanisms

[1,9,11,14,23,42]. Some information has been obtained from comparisons of thermal decompositions and radiolysis. Further spectroscopic studies, the introduction of additives capable of reacting with possible intermediates, analysis of quenched samples of partially decomposed salts, and rate measurements in a gas flow containing O atoms, could all provide valuable mechanistic information.

REFERENCES

1. F. Solymosi, Structure and Stability of Salts of Halogen Oxyacids in the Solid Phase, Wiley, New York, 1977.
2. K.H. Stern, J. Phys. Chem. Ref. Data, 3 (1974) 481.
3. H.F. Cordes and S.R. Smith, J. Phys. Chem., 72 (1968) 2189.
4. F. Solymosi, Acta Chim. Acad. Sci. Hung., 57 (1968) 11,35.
5. S. Shimada, J. Thermal Anal., 40 (1993) 1063.
6. A.E. Simchen, J. Phys. Chem., 65 (1961) 1093; J. Chim. Phys., (1961) 596.
7. B.I. Khorunzhii and K.G. Il'in, Izv. Vyssh. Ucheb. Zaved., Khim. Khim. Tekhnol., 15 (1972) 1886.
8. I.C. Hisatsune and D.G. Linnehan, J. Phys. Chem., 74 (1970) 4091.
9. R.J. Acheson and P.W.M. Jacobs, J. Phys. Chem., 74 (1970) 281; A.I.A.A.,J., 8 (1970) 1483.
10. S.I. Borisova, L.G. Berezkina, N.S. Tamm and A.V. Novoselova, Kinet. Katal., 16 (1975) 68.
11. P.W.M. Jacobs, F. Solymosi and J. Rasko, Combust. Flame, 17 (1971) 125.
12. L.G. Berezkina, S.I. Borisova and N.S. Tamm, Kinet. Katal., 13 (1972) 1162.
13. L.I. Chudinova, Zh. Neorg. Khim., 7 (1962) 431.
14. F. Solymosi, Z. Phys. Chem. N.F. (Frankfurt), 57 (1968) 1.
15. A.S. Giridharan, M.R. Udupa and G. Aravamudan, J. Thermal Anal., 7 (1975) 65.
16. L.A. Prince and E.R. Johnson, J. Phys. Chem., 69 (1965) 359,377.
17. M.M. Cox and J.W. Moore, J. Phys. Chem., 74 (1970) 627.
18. E.H. Appelman, Inorg. Chem., 8 (1969) 223.
19. B.R. Phillips and D. Taylor, J. Chem. Soc., (1963) 5583.
20. S. Takriti and G. Duplâtre, J. Chem. Soc., Faraday Trans. I, 84 (1988) 2831.
21. C.A. Strydom and L.C. Prinsloo, Thermochim. Acta, 241 (1994) 43.
22. L. Pacesová, H. Rathanová and J. Rosicky, Collect. Czech. Chem. Commun., 33 (1968) 3670.
23. T. Wydeven, J. Catal., 19 (1970) 162.

24. I.D. Begg, P.J. Halfpenny, R.M. Hooper, R.S. Narang, K.J. Roberts and J.N. Sherwood, Proc. R. Soc. (London), A386 (1983) 431.
25. J.C. Cannon and Y.C. Zhang, J. Thermal Anal., 41 (1994) 981; Thermochim. Acta, 228 (1993) 147.
26. P. Lumme and H. Lumme, Suom. Kemi., 35 (1962) 129.
27. F. Solymosi and T. Bánsági, Acta Chim. Acad. Sci. Hung., 56 (1968) 251,337.
28. F. Solymosi and T. Bánsági, Magy. Kem. Foly., 74 (1968) 32.
29. J. Jach, Reactivity of Solids, (Ed. J.H. de Boer), Elsevier, Amsterdam, 1960, p.334; J. Phys. Chem. Solids, 24 (1963) 63.
30. K.R. Biggs, R.A. Gomme, J.T. Graham and J.S. Ogden, J. Phys. Chem., 96 (1992) 9738.
31. D. Bhatta and G. Padhee, J. Thermal Anal., 37 (1991) 2693.
32. T. Dupuis and C. Rocchiccioli-Deltcheff, Anal. Chim. Acta, 44 (1969) 165.
33. G.M. Bancroft and H.D. Gesser, J. Inorg. Nucl. Chem., 27 (1965) 1537,1545.
34. S.D. Bhattamisra and S.R. Mohanty, J. Inorg. Nucl. Chem., 39 (1977) 2103.
35. D. Bhatta, S. Mishra and K.K. Sahu, J. Thermal Anal., 39 (1993) 275.
36. C. Várhelyi and E. Kékedy, Stud. Univ. Babes-Bolyai Chem., 1 (1962) 11.
37. V.V. Pechkovskii and A.V. Sofronova, Zh. Neorg. Khim., 11 (1966) 828.
38. J. Bousquet and P. Vermande, Bull. Soc. Chim. Fr., (1964) 214.
39. T. Unezawa, Y. Ninomiya and S. Tatuoka, J. Appl. Cryst., 3 (1970) 417.
40. M. Maneva and V. Koleva, J. Thermal Anal., 42 (1994) 1017.
41. M. Maneva and V. Koleva, J. Thermal Anal., 41 (1994) 817.
42. J. Simpson, D. Taylor, R.S. Fanshawe, J.M. Norbury and W.J. Watson, J. Chem. Soc., (1968) 3323.
43. G.R. Levi and E. Garrini, Gazz. Chim. Ital., 90 (1956) 447; G.R. Levi and C. Cipollone, Gazz. Chim. Ital., 53 (1923) 200; G.R. Levi, Sel. Chim., 13 (1953) 1091.
44. M. Harmelin and C. Duval, C.R. Acad. Sci. Paris, 260 (1965) 2461.
45. B. Tanguy, B. Frit, G. Turrell and P. Hagenmuller, C.R. Acad. Sci. Paris, C, 264 (1967) 301.
46. G.E. Boyd, E.W. Graham and Q.V. Larson, J. Phys. Chem., 66 (1962) 300; J.W. Chase and G.E. Boyd, J. Phys. Chem., 70 (1966) 1031; G.E. Boyd and L.C. Brown, J. Phys. Chem., 74 (1970) 1691, 3490.
47. P.W. Levy and P.J. Herley, Reactivity of Solids, (Ed. J.W. Mitchell), Wiley Interscience, 1969, p.75; P.W. Levy, M. Goldberg and P.J. Herley, J. Phys. Chem., 76 (1972) 3751.

48. I.V. Dmitriev and Zh.L. Vert, Chem. Abs., 80 (1974) 31250n.
49. J.R. Bristow and B.L. Rees, Nature (London), 198 (1963) 178.
50. P. Barret, Reactivity of Solids, (Ed. J.H. de Boer), Elsevier, Amsterdam, 1961, p.178.
51. A.A. Opalovskii, V.E. Fedorov and T.D. Fedorova, J. Thermal Anal., 1 (1969) 301; 2 (1970) 373; Thermal Analysis, (Ed. H.G. Wiedemann), Birkhauser, Basel, 2 (1971) 193.
52. E.I. Smagina, P.I. Ozhegov and N.P. Mikhailova, Zh. Fiz. Khim., 47 (1973) 1062.
53. Yu.B. Patrikeev, G.I. Novikov and V.V. Badovskii, Zh. Fiz. Khim., 47 (1973) 2454.
54. P. Ramamurthy and E.A. Secco, Canad. J. Chem., 47 (1969) 2181, 2185, 3915; 48 (1970) 1619, 3510.
55. O.K. Srivastava and E.A. Secco, Canad. J. Chem., 45 (1967) 579, 585,1375.
56. P. Ramamurthy, E.A. Secco and M. Badri, Canad. J. Chem., 48 (1970) 2617.
57. A-D. Leu, P. Ramamurthy and E.A. Secco, Canad. J. Chem., 51 (1973) 3882.
58. M.C. Ball and R.F.M. Coultard, J. Chem. Soc. A, (1968) 1417.
59. L. Walter-Levy and M. Goreaud, Bull. Soc. Chim. Fr., (1971) 444.
60. J.B. Sharkey and S.Z. Lewin, Thermochim. Acta, 3 (1972) 189.
61. L. Walter-Levy and D. Groult, C.R. Acad. Sci. Paris, C, 267 (1968) 310.
62. M.M. Markowitz, J. Inorg. Nucl. Chem., 25 (1963) 407.

Chapter 14

DECOMPOSITION OF METAL SALTS OF VARIOUS OXYACIDS

14.1. INTRODUCTION

This chapter considers together the thermal decompositions of metal salts of inorganic oxyacids. The decompositions of ammonium salts of these oxyacids (and other anions) are discussed in Chapter 15. The decomposition of potassium permanganate is considered first because it represents one of the earliest solid state reactions that was subjected to detailed mechanistic investigation. Substantial interest in the reaction has continued and has been extended to other metal permanganates.

14.2. PERMANGANATES

14.2.1. Potassium permanganate

The early study (1944) of the thermal decomposition of $KMnO_4$ by Prout and Tompkins [1] was significant [2] in providing a mechanistic explanation of the fit of α-time data to the kinetic expression:

$$\ln [\alpha/(1 - \alpha)] = k(t - t_o) \qquad (14.1)$$

(now widely referred to as the Prout-Tompkins equation). This important paper stimulated interest in this, and other, solid state reactions. Publications on the $KMnO_4$ reaction continue to appear. The autocatalytic nature of the $KMnO_4$ decomposition implied by equation (14.1) could not be ascribed [1] to an energy chain-branching model because the transfer step would have to be rapidly achieved (less than 10^{-12} s) to avoid energy dissipation as lattice vibrations [3]. Accordingly, the model proposed was that the initial reaction resulted in the development of a layer of solid products across surfaces. Subsequent reorganization of these product phases (incompletely characterized, but including MnO_2) resulted in surface strain that locally promoted reaction. Relaxation of this strain by cracking within the reactant crystal resulted in exposure of new surfaces at which reaction could continue. The characteristic acceleratory stage of the sigmoid α-time curves is, therefore, a consequence of the propagation of strain-induced cracks in divergent directions, accompanied by promoted anion disintegration within the strained surface zone. After the maximum reaction rate has been reached, the deceleratory stage is consistent with the absence of further crystallite disintegration, which was confirmed microscopically.

Evidence in support of chain branching during the decomposition of $KMnO_4$ was provided by Oates and Todd [4] using microscopy. They detected product formation along dislocation lines, with extensive branching, in sectioned crystals of partially reacted salt.

Hill *et al.* [5] extended measurements at the lower end of the temperature range down to 383 K to investigate the early (nucleation) stage of reaction, using plots of $d\alpha/dt$ against α to detect small effects. They concluded that nuclei were not randomly distributed, but were separated by a minimum distance related to the Burgers vector of the dislocations, and that the shapes of the α-time curves varied detectably with temperature. The relative significance of the acceleratory process diminished with temperature and below 423 K nucleation within grains was very slow compared with reactions at surfaces. From the rate of the surface process on grains of known size, the rate of interface advance was quantitatively expressed by the absolute reaction rate theory. Marked differences in the decomposition kinetics of fresh and aged $KMnO_4$ were attributed [6] to the onset of decomposition at surfaces and grain boundaries of the older material during storage. The variations in reported kinetic models applicable during the acceleratory period, i.e. the Prout-Tompkins equation [1], the power law [7], and the Avrami-Erofeev equation [8], with $n = 2$, can be traced to sample age and pretreatment. From measurements for eighteen preparations of $KMnO_4$ with different initial specific surface areas, Protashchik [9] found that the rate of reaction during the acceleratory period increased with increase in surface area, but the maximum decomposition rate decreased. From these results, he concluded that the sites of initial decomposition are within, rather than at, the surfaces of the crystals. Neutron irradiation decreased the decomposition temperature by about 50 K.

Phillips and Taylor [10] studied the decomposition of $KMnO_4$ in solid solutions of $KMnO_4$ and $KClO_4$ between 460 and 510 K. The first of the four stages of reaction ($\alpha < 0.01$) was identified as rapid decomposition at surface sites and was more strongly inhibited by $KClO_4$ than the other three stages which were: a constant rate process, followed by a stage fitted by the power law ($n = 2$), and then onset of the main acceleratory process. In dilute solid solution in $KClO_4$, the decomposition of

$KMnO_4$ at 508 K is incomplete and the extent of reaction is related [11] to the proportion of MnO_4^- present in the host lattice as other than isolated units [12]. Groupings of MnO_4^- ions, such as dimers or more extensive associations, are believed to be required to permit decomposition to proceed through an electron transfer step represented as:

$$2MnO_4^- \underset{k_{-1}}{\overset{k_1}{\rightleftharpoons}} MnO_4^{2-} + MnO_4 \xrightarrow{k_2} MnO_2 + O_2$$

Using a steady-state argument, Boldyrev [11] derived the rate expression:

$$\text{Rate} = k_1 k_2 [MnO_4^-]/k_{-1}[MnO_4^{2-}] + k_2$$

where $[MnO_4^{2-}]$ is a measure of the number of electrons present in the crystal and $[MnO_4^-]$ is the concentration per unit area of interface.

Additives either catalyze or inhibit decomposition according to whether their electronic work function is greater than or less than that of the reactant $KMnO_4$, as has been shown [11] for a range of different MnO_2 preparations. Kabanov and Zharova [13] concluded from conductivity studies in the early stages of decomposition ($\alpha < 0.16$) that K_2MnO_4 is formed in solid solution with $KMnO_4$. The double salt, $K_3(MnO_4)_2$, prepared by co-precipitation of $KMnO_4$ and K_2MnO_4, is more stable than $KMnO_4$. Thus MnO_2, one of the solid decomposition products of $KMnO_4$, catalyzes the reaction, while the other, MnO_4^{2-}, is an inhibitor.

$K_3(MnO_4)_2$ was identified by X-ray diffraction measurements in the residual phase during the early stages of reaction [14,15]. The intensities of peaks for this solid diminished as reaction advanced and the yield of product K_2MnO_4 increased. Erenburg *et al.* [16] identified manganates as the first products of decomposition of $RbMnO_4$ and $CsMnO_4$, behaviour which is contrasted with that found for $KMnO_4$.

Progress towards determining the detailed mechanism of this reaction has not fully realised the promise of the earlier work. Current understanding of the $KMnO_4$ decomposition reaction has been critically reviewed by Herbstein *et al.* [17] who provide a realistic perspective of the present situation against the background of an extensive publication list. The conclusions that are acceptable and the problems that remain unresolved are discussed.

Proposed decomposition mechanisms have included consideration of chain reactions, strain-induced autocatalysis [1], a diffusion-chain theory [5] and a nucleation and growth process [18]. These models must now be reinterpreted or replaced by a mechanism that takes into account the intermediate formation of $K_3(MnO_4)_2$ [15]. The predominant overall decomposition mechanism must be represented as two consecutive rate processes:

$$3KMnO_4 \rightarrow K_3(MnO_4)_2 + \delta\text{-}MnO_2 + O_2$$

$$2K_3(MnO_4)_2 \rightarrow 3K_2MnO_4 + \delta\text{-}MnO_2 + O_2$$

with a possible concurrent contribution from:

$$2KMnO_4 \rightarrow K_2MnO_4 + \delta\text{-}MnO_2 + O_2$$

A complete analysis of the reaction would require measurements of the variations with time of all the phases participating. The product giving unusual textures identified by Brown *et al.* [18] may, perhaps, be $K_3(MnO_4)_2$. The fit of kinetic data to the Avrami-Erofeev equation ($n = 2$) [18], together with the appearance of nuclei, illustrated in Figure 14.1., can be regarded now [17] as only an incomplete representation of this more complicated reaction.

Analytical measurements showed that the overall reaction could not be expressed by a simple stoichiometric equation [19,20] but was more satisfactorily represented by:

$$10KMnO_4 \rightarrow 2.6K_2MnO_4 + \text{"}K_4Mn_7O_{16}\text{"} + 6O_2$$

The product referred to above as δ-MnO_2 contains potassium (intercalated between $(MnO_2)_n$ layers) and the composition approximates to K_xMnO_{2+y} where x is about 0.5 (this the "$K_4Mn_7O_{16}$" phase in the above equation). The structures and compositions of the residual product, or products, other than K_2MnO_4, remain as yet incompletely characterised. This is consistent with the evidence that many reactions of manganese oxides require high temperature [21] and that the reaction of manganese oxide with KOH occurs between 573 and 673 K [22]. It must be concluded, therefore, that this residual product is a poorly crystallised, perhaps non-stoichiometric and inhomogeneous oxide mixture, higher temperatures are required to form recognisable crystalline products [18]. There may also be amorphous material, perhaps approaching a vitreous or fluid state, present at the reaction interface. (This represents an interesting parallel with the dehydrations discussed in

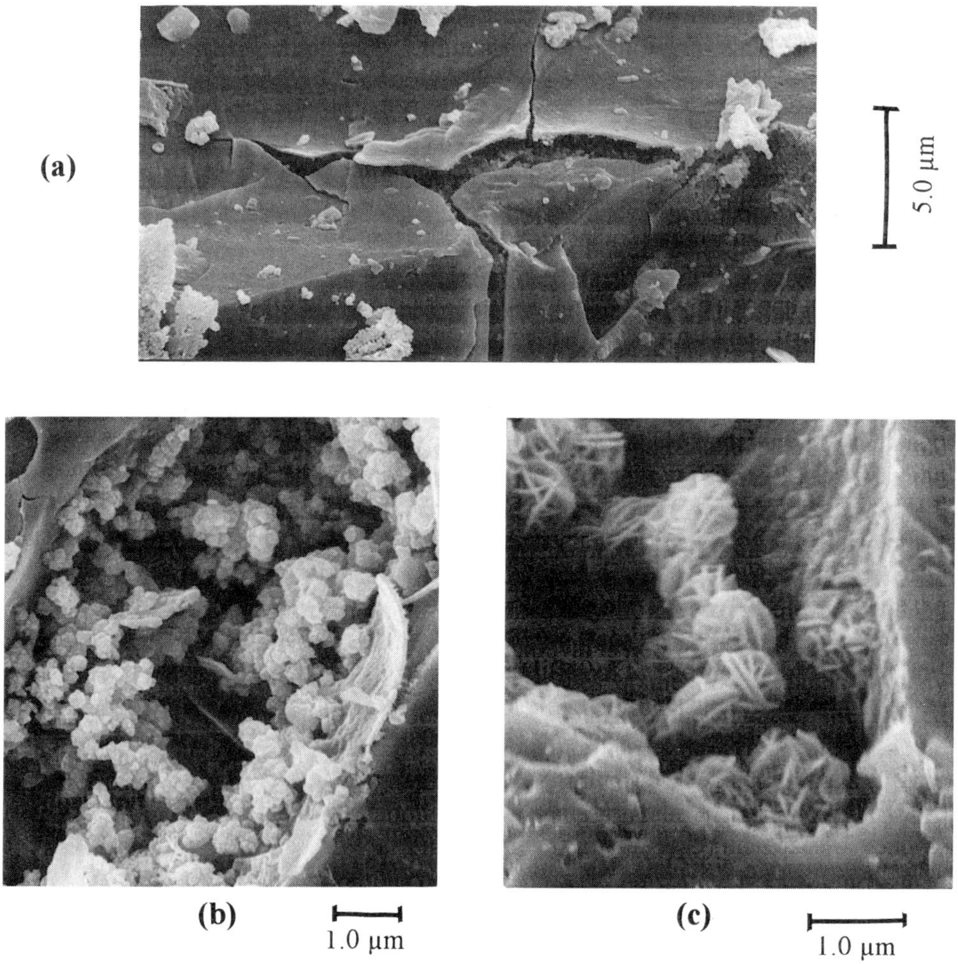

Figure 14.1.
Scanning electron micrographs of the surfaces of crystals of potassium permanganate [18]. (a) Crystals decomposed to completion ($\alpha = 1.00$) (scale bar = 5 μm). Local detachment of a thin layer of unretextured surface is evident. Such layers tend to obscure internal textural changes, but can be removed by light crushing. (b) Intensive local retexturing observed in partially decomposed ($\alpha = 0.10$ at 500 K) crystals (scale bar = 1.0 μm). Reaction has generated large numbers of irregularly shaped particles. Another view of similar particles is shown in (c) from a repeat experiment (scale bar = 1.0 μm).

Chapter 7, where product recrystallisation is the difficult step.) This possibility may be considered in interpreting the interfacial chemistry of $KMnO_4$ decomposition.

The decomposition of $KMnO_4$ is sensitive to pre-irradiation [23] by UV [24], X-rays, γ-rays, protons and neutrons. The effects of such pretreatments, which increase with dose, are to decrease the induction period and to increase the maximum reaction rate, although in general there is no significant change in the kinetic model or in the magnitude of E_a. (Comparable behaviour has been described for the other alkali permanganates [16,25,26]). There is some controversy about the mechanism whereby irradiation influences reactivity. Prout [23] favoured a displacement model in which multiple ionization of MnO_4 groups leads to random formation of interstitial cations. During thermal decomposition, these defects are annealed in the vicinity of dislocations and the release of stored (Wigner) energy (up to about 100 kJ mol^{-1}) may then rupture bonds in an adjacent MnO_4^- ion. The products formed cause structural deformation and enhance the annealing process through the development of decomposition spikes which eventually lead to fracture of the crystal. Irradiation nuclei behave differently from thermal nuclei. Boldyrev *et al.* [27,28] believe that the radiolysis products catalyze decomposition. Protashchik and Erofeev [29] take an intermediate view.

Listed E_a values for $KMnO_4$ decomposition are in the range 130 to 210 kJ mol^{-1}, and the energy for charge transfer measured spectroscopically [30] was somewhat higher. An X-ray diffraction study [31] of the structure of stoichiometric $KMnO_4.K_2MnO_4$ identified the discrete MnO_4^- and MnO_4^{2-} ions in the monoclinic cell. It was shown that there was a high rate of electron transfer, between about 270 and 330 K, with $E_a = 49 \pm 5$ kJ mol^{-1}.

14.2.2. Other Group IA metal permanganates

All the alkali-metal permanganates decompose on heating with sigmoid α-time curves [25,26]. For $LiMnO_4$ (above 380 K) [25], an initial linear process (E_a = 95 kJ mol^{-1}) is followed by reaction according to the Prout-Tompkins model (E_a = 138 kJ mol^{-1}). The decay period could also be described by a first-order equation with E_a = 130 kJ mol^{-1}. The Prout-Tompkins model described virtually the complete decomposition of $NaMnO_4$ (> 400 K) [25], with two rate coefficients but a single value of E_a (128 kJ mol^{-1}). The decomposition of the rubidium salt between 488 and 528 K [20] resembled that of the lithium salt, but with $E_a = 167 \pm 7$ kJ mol^{-1}. The decomposition of $CsMnO_4$ between 513 and 558 K [25] involved an initial slow process, described by the power law with $n = 2$ and E_a = 141 kJ mol^{-1}, and, thereafter, data fitted the Prout-Tompkins equation with E_a = 167 and 141 kJ

mol^{-1} in the acceleratory and deceleratory periods, respectively. Manganates have been identified [16] in the decomposition products of $RbMnO_4$ and $CsMnO_4$.

Recent studies of the thermal decompositions of $KMnO_4$ [18], $RbMnO_4$ [32] and $CsMnO_4$ [33] generally confirm the previous kinetic observations [25,26] that yield-time curves for all these reactants were sigmoid but had somewhat greater acceleratory character than is expressed by the solid state rate equations based on geometric models. It was concluded [32] that the influence of melting at the reaction interface increased in the sequence K < Rb < Cs, resulting in the relatively acceleratory behaviour in the early stages. Values of activation energies for these reactions were similar, 160 to 170 kJ mol^{-1}. Reactivities of these salts, measured from maximum reaction rates at 520 K [32] decreased in the sequence K > Rb > Cs. This is perhaps a consequence of increase in lattice spacing with cation size.

It seems probable that the reactivities of the alkali metal permanganates are predominantly controlled by the ease of anion disintegration. The values of E_a are similar. Subsequent reactions, however, vary with the cation present. No analogue of $K_3(MnO_4)_2$ is formed from the Cs salt [16] and that of Rb is less stable. The residual products (other than M_2MnO_4) are difficult to characterise, apparently containing some uncrystallized material, consistent with the possible intervention of an amorphous or molten interface phase.

Microscopic examinations of the textural changes that occur during these reactions, although often mentioned [1,25,26,34], have yielded remarkably little information of value in formulating reaction mechanisms. Three of these reactants form a superficial, smooth outer surface layer [18,32,33] of product that obscures changes proceeding beneath. Even when this layer is removed, the textures revealed are not obviously related to the kinetic characteristics. The evidence available has detected the formation of a crystalline product (possibly $K_3(MnO_4)_2$) during the early stages of $KMnO_4$ decomposition. No evidence of extensive reactant disintegration, as discussed in the early mechanistic proposals [1], was obtained. The rounded textures of the product are consistent with some sintering or fusion to form coherent aggregates of the small particles of the residual phases. The sigmoid α-time plots for the decompositions of both $LiMnO_4$ and $NaMnO_4$ were well represented (again) by the Prout-Tompkins equation. Activation energy values were somewhat lower (mean value of about 140 kJ mol^{-1} from values between 128 and 150 kJ mol^{-1} for $LiMnO_4$ and $NaMnO_4$) than those for the K, Rb and Cs salts [25,26]. The influence of pre-irradiation on the decomposition has been investigated [27].

14.2.3. Group IIA metal permanganates

Hardy [35] proposed that the following reactions participate in the decomposition of $Ba(MnO_4)_2$:

$$Ba(MnO_4)_2 \rightarrow BaMnO_4 + MnO_2 + O_2 \qquad (E_a = 142 \pm 4 \text{ kJ mol}^{-1})$$

$$2BaMnO_4 \rightarrow 2\ \alpha\text{-}BaMnO_3 + O_2 \qquad (E_a = 126 \pm 4 \text{ kJ mol}^{-1})$$

$$\downarrow$$

$$2\ \beta\text{-}BaMnO_3 \rightleftarrows 2BaMnO_3(\ell)$$

Rienaecker and Werner [36] suggested that $Ba_3(MnO_4)_2$ is the initial decomposition product, by analogy with results for $KMnO_4$. Isothermal α-time curves for vacuum decomposition at 413 to 463 K were sigmoid [37] following an initial ($\alpha = 0.04$) burst of gas . The acceleratory process was fitted by the Prout-Tompkins equation ($E_a = 151$ kJ mol^{-1}) and the power law with $n = 2$. Fragmentation into thin platelets was observed. The deceleratory period could be described by the first-order equation with $E_a = 124$ kJ mol^{-1}.

The effects of pre-irradiation [37], although still marked, were less for a given dose than those observed for the alkali metal permanganates. The other Group IIA permanganates are obtained as hydrates.

14.2.4. Other metal permanganates

The kinetics of decomposition of $AgMnO_4$ in vacuum between 378 and 393 K [38] differ from the behaviour of $KMnO_4$ in that the acceleratory periods of the α-time curves are described by the modified form of the Prout-Tompkins equation:

$$\ln[\alpha/(1 - \alpha)] = k \ln t + c$$

followed by a first-order decay with $E_a = 122 \pm 2$ kJ mol^{-1}. This difference of behaviour was attributed [38] to a decrease in the chain-branching coefficient, assumed to be inversely proportional to time. No disintegration of small crystals was observed but pre-irradiated crystals [39] shattered on completion of the induction period. X-ray diffraction studies [40] confirmed the existence of strain during the formation of the decomposition product. Addition of small amounts (5% by mass) of ZnO or ThO_2 accelerated the decomposition of $AgMnO_4$ at 388 K, TiO_2 decreased the rate, while NiO and Co_3O_4 had no effect [41].

Attempts to relate the kinetic characteristics of permanganate decompositions to the nature of the constituent cation have been discussed by Young [42].

The decompositions of $Ni(MnO_4)_2$ and $Cu(MnO_4)_2$ were studied [43-45] to investigate a possible low temperature method for the preparation of active mixed oxide catalysts. These reactants can be regarded as highly dispersed associations of two transition-metal oxides and, at the low temperatures of MnO_4^- breakdown (below 400 K), little recrystallisation is expected, giving high area, potentially highly active (probably non-stoichiometric) oxide catalysts. The structures, compositions and activities of these phases have only been partially characterised.

Nickel permanganate decomposed [44] ($\rightarrow NiMn_2O_5 + 1.5O_2$) between 356 and 400 K. The sigmoid α-time curves were well expressed by the Avrami-Erofeev equation ($n = 2$). An initial electron transfer step was identified as rate controlling, with $E_a = 100 \pm 5$ kJ mol^{-1}. The rate of the first half of reaction ($\alpha < 0.5$) was decreased by the presence of water vapour. The rate of this autocatalytic reaction also proceeded more rapidly in the solid state than the comparable reaction in aqueous solution.

Copper(II) permanganate [45] was slightly less stable than the nickel salt and kinetic studies were completed between 335 and 370 K. The sigmoid oxygen evolution curves (1.27 mol O_2 yield) again fitted the Avrami-Erofeev equation ($n = 2$) and $E_a = 115 \pm 15$ kJ mol^{-1}. Electron transfer was again suggested as the rate controlling step, possibly involving the intervention of Cu^+ with later reoxidation to Cu^{2+}. Reaction is identified as yielding a poorly crystallised product.

14.3. CHROMATES

Thermal analysis studies of chromates and isopolychromates have been reviewed by Charsley [46]. Darrie *et al.* [47,48] studied the decompositions of several chromates to ascertain whether values of E_a determined for these rate processes could be correlated with any feature of the band structure of the solid, investigated by spectral methods. During the decompositions [48] of the (Mg, La, Nd and Sm) chromates (VI), the oxidation state of the chromium was decreased to +3. From comparisons of the activation energies, obtained by isothermal gravimetric measurements, with charge transfer energies determined spectroscopically, it was concluded that the activation step involved a one electron transfer from oxygen to chromium. The decomposition of magnesium chromate:

$$4MgCrO_4 \rightarrow 2MgCr_2O_4 + 2MgO + 3O_2$$

was well represented by the contracting-volume equation. Decompositions of the other three trivalent salts took place in two stages:

$$4M_2(CrO_4)_3 \rightarrow 8MCrO_4 + 2Cr_2O_3 + 5O_2$$

$$2MCrO_4 \rightarrow 2MCrO_3 + O_2$$

From similar studies [47] of the reactions of the lanthanide chromates(V), which form the second step in the decompositions of the chromates(VI), it was concluded that the breakdown of $LaCrO_4$ and of $PrCrO_4$ involved the transfer of energy between the CrO_4 units. In the other lanthanide chromates, however, energy was transferred by exchange from an activated CrO_4 group to an adjacent cation. CrO_4 groups at a surface act as energy sinks by absorbing energy reaching the surface and then decompose.

Furusaki *et al.* [49] reported that the kinetic characteristics of the conversion of $LaCrO_4$ (monazite structure) to $LaCrO_3$ (perovskite structure) depended on $p(O_2)$ and that the rate decreased with increasing availability of oxygen. In N_2, the reaction (908 to 938 K) fitted the contracting volume expression with $E_a = 198$ kJ mol^{-1}. The rate limiting step for reaction in O_2 is the diffusion of oxygen across the product layer, the Jander rate equation applies with E_a about 506 kJ mol^{-1}.

14.4. NITRATES AND NITRITES

14.4.1. Introduction

Most nitrates and nitrites melt well below their decomposition temperatures so that, in spite of their technological importance, these reactions do not often contribute much to the understanding of solid state reactions. An extensive review of the high temperature properties and decompositions of nitrates has been given by Stern [50]. The temperatures, stoichiometries and enthalpies of these decompositions are considered for 36 cations, although for some the amount of available information is small. Particular attention is directed towards the chemistry of secondary reactions that are considered with reference to equilibria between N_2, O_2 and the oxides of nitrogen (thermodynamic data for the most important reactions are given). On heating, most ionic nitrates melt to a liquid, some of which are stable above the melting point. The enthalpies of decomposition of many nitrates ($\rightarrow MO + N_2O_5$) have been shown to be a linear function of $\Omega^{1/2}/Z$ where Ω is the cation radius and Z is the effective nuclear change.

Several nitrates containing monovalent cations (alkali metals, NH_4^+, Tl^+ and Ag^+) undergo one or more phase transformations [51] below their melting points. Overall decompositions are often exothermic. Gaseous products may include O_2, N_2, N_2O, NO, NO_2, N_2O_3, N_2O_4 and N_2O_5 [50], and depend largely on the reaction temperature and the possibility of secondary chemical changes between primary products. N_2O_5 is unstable above 298 K and reacts to form N_2O_4 or N_2O_3 and oxygen. N_2O_4 dissociates to NO_2 with increasing temperature and dissociation is complete above 400 K. Similarly N_2O_3 dissociates to NO_2 and NO above 400 K. Below 500 K, NO will react completely with any O_2 to form NO_2, but above 500 K significant amounts of NO remain unreacted. NO and NO_2 do not react with N_2, nor do they decompose significantly to N_2 and O_2 below 1000 K. N_2O is stable to about 800 K.

Nitrites generally decompose at lower temperatures than the corresponding nitrates. The formation of nitrites as intermediates during the decomposition of nitrates has been suggested and may be accompanied by further changes that result in the regeneration of a proportion of the nitrate. The decomposition of nitrites is suggested to take place according to the scheme:

$$2MNO_2(s) \rightarrow M_2O(s) + NO(g) + NO_2(g)$$

Further reactions amongst the accumulated gaseous products may result [50] in partial oxidation of the nitrite to the nitrate.

Nitrates may retain water strongly and special preparative methods may be required to obtain the anhydrous salts. Residual water in nitrates and nitrites can hydrolyze the reactant on heating with the formation of basic salts and nitric acid. Studies of the radiolysis of nitrates [52-55] have provided additional information on decomposition mechanisms.

Alkali and alkaline earth metals, silver and cadmium form ionic nitrates but other metals, e.g. copper and zinc, give covalent nitrates. Nitrate breakdown may proceed through the intervention of the less stable nitrite and because this may be reoxidised by NO_2 to NO_3^- the mechanism may appear to be complicated and the behaviour observed depends on conditions locally prevailing.

The peroxonitrite anion, $ONOO^-$, has been identified [56] as a primary product of UV-irradiation of solid nitrates at 254 nm. This has been identified as a principal path for nitrate breakdown.

A knowledge of the thermal chemistry of nitrates is of value in considering several rate processes that are or have been regarded as proceeding in the solid state. These include the decompositions of NH_4ClO_4, $(NH_4)_2Cr_2O_7$ (Chapter 15) and more complex salts, where oxidized nitrogen may be present.

14.4.2. Group IA and Group IIA metal nitrates

The alkali, alkaline earth and ammonium nitrates all melt before decomposing. The homogeneous decomposition of the molten sodium salt proceeds [57] through intermediate formation of the nitrite. Johnson and Forten [52] concluded from studies of the radiation-induced decompositions of (Na^+, K^+, Cs^+ and Pb^{2+}) nitrates that the mechanisms are complex and depend on the relative importance of the reactions:

$$NO_3^- + O \rightarrow NO_2^- + O_2 \quad \text{and} \quad NO_2^- + O \rightarrow NO_3^-$$

A thermodynamic study of the condensed phases of $NaNO_3$, KNO_3 and $CsNO_3$ and their transitions has been reported [58].

Calcium nitrate melts below the temperature of onset of the principal decomposition reactions which Doumeng [59] has identified as follows :

$$Ca(NO_3)_2 \rightarrow Ca(NO_2)_2 + O_2$$
$$Ca(NO_2)_2 \rightarrow CaO + 2NO + \tfrac{1}{2}O_2$$
$$Ca(NO_2)_2 \rightarrow CaO + NO + NO_2$$
$$Ca(NO_2)_2 + 2NO_2 \rightleftarrows Ca(NO_3)_2 + 2NO$$
$$Ca(NO_2)_2 + 2NO \rightleftarrows Ca(NO_3)_2 + N_2$$

Ettarh and Galwey [60] measured the melting point of anhydrous $Ca(NO_3)_2$ as 836 $\pm$ 2 K. Any water present was evolved, without accompanying anion breakdown, between 420 and 460 K. Fusion was accompanied by some reaction but the main decomposition, detected by rising temperature experiments, occurred at 932 or 949 K for salt heated at 5 and 10 K min $^{-1}$, respectively. Two complementary kinetic studies were undertaken. Rate data measured from gas evolved in a constant volume pre-evacuated system, without a cold trap, gave sigmoid α-time curves with an approximately constant rate ($0.3 < \alpha < 0.7$). Products measured by a method of titrating the acid formed on reaction of NO or NO_2 with water, gave two linear intervals of reaction ($0.1 < \alpha < 0.5$) and a relatively more rapid reaction ($0.7 < \alpha < 0.9$). The overall reaction stoichiometry was :

$$4Ca(NO_3)_2 \rightarrow 4CaO + 4O_2 + 6NO_2 + N_2$$

These results are approximately expressed by a mechanism involving two dominant consecutive reactions, in which the proportion of acid gases is greater at higher α values :

$$Ca(NO_3)_2 \rightarrow Ca(NO_2)_2 + O_2$$
$$Ca(NO_2)_2 \rightarrow CaO + NO_2 + NO$$

Values of E_a for these steps were 220 ± 10 and 298 ± 10 kJ mol^{-1}. These magnitudes correspond approximately to the enthalpies of these highly endothermic decompositions.

The $Ca(NO_3)_2$ decomposition is catalyzed by metal oxides. Comparative studies [61] of the relative activities of 16 selected metal oxides in promoting nitrate ion breakdown in $Ca(NO_3)_2$ led to the conclusion that all of the reactions were accompanied by melting. Most of the mixtures containing oxides of metals of the first transition period underwent reaction at temperatures below or close to the melting point of $Ca(NO_3)_2$, 836 K, and well below the decomposition temperature, 949 K. Cations with only partially filled *d* shells were the most active, Mn^{2+} and Fe^{3+}, followed by Cr^{3+} and Co^{2+}. It was concluded that these reactions were promoted by the electron transfer steps:

$$M^{n+} + {}^{-}O - NO_2 \rightarrow M^{(n+1)+} + O^{2-} + NO_2$$
$$M^{(n+1)+} + {}^{-}O - NO_2 \rightarrow M^{n+} + NO_3 \rightarrow M^{n+} + NO_2 + \tfrac{1}{2}O_2$$

Interactions at oxide/salt contacts were suggested [61] to result in formation of a melt, which, after the initiation of decomposition, contained the ions: Ca^{2+}, M^{n+}, NO_3^-, O^{2-} and probably NO_2^-. Reaction proceeds more rapidly in the liquid and the melting step was identified as exerting an important control on reactivity. The overall abilities of metal oxides to promote $Ca(NO_3)_2$ decomposition decreased in the sequence:

$MnO > Fe_2O_3 > Cr_2O_3 > CoO > Cu_2O = CuO > PbO = PbO_2 > MnO_2 > NiO > V_2O_5 > TiO_2 > ZnO > PtO$.

The low activities of Al_2O_3 and MgO were not explained, but may arise from water retained in the reaction mixture up to the decomposition temperature.

14.4.3. Other metal nitrates

Several hydrated transition metal nitrates (Co, Cu, Cr, Zn and Ni) undergo aqueous fusion [62] prior to hydrolysis which yields basic salts. Vratny and Gugliotta [63] reported sigmoid α-time curves for the decomposition of $Pb(NO_3)_2$ between 523 and 708 K, and concluded that the cation was not further oxidized. Margulis *et al.* [64] identified both dissociation and autoxidation steps during the decomposition of lead nitrate. The first step involved melting of the eutectic formed from the reactant and the initial product, $2PbO.Pb(NO_3)_2$.

The influences of the ions Ca^{2+}, Mg^{2+} and Ba^{2+} and of MnO_2 crystals on the decomposition of $Mn(NO_3)_2$ in the presence of water have been investigated [65].

From a FTIR product analysis coupled with TG, it was concluded [66] that basic copper nitrate, $Cu_2(OH)_3NO_3$, decomposes in one step between 430 and 500 K when heated at 10 K min^{-1}. The reactant breaks down to yield CuO + H_2O + HNO_3, accompanied by reversible partial dissociation of the nitric acid ($\rightleftarrows$ 0.5 H_2O + NO_2 + 0.25O_2).

Two stages in the decomposition of $Ni(NO_3)_2.2H_2O$ were identified [67] by a thermogravimetric study (2.5 K min^{-1}). The first step (at 440 K) is:

$$2Ni(NO_3)_2.2H_2O \rightarrow Ni(NO_3)_2.Ni(OH)_2.H_2O + 2H_2O + (NO,NO_2,O_2)$$

followed by decomposition of the solid intermediate (at 545 K):

$$Ni(NO_3)_2 \rightarrow NiO + (NO, NO_2, O_2)$$

$$Ni(OH)_2.H_2O \rightarrow NiO + 2H_2O$$

The residual NiO is non-stoichiometric with less than 1% Ni^{3+}.

To explain mass spectroscopic detection of the gaseous molecules (CoO, CuO, Cu_2O, NiO, PbO and $Mg(OH)_2$) during low temperature decompositions of the anhydrous and hydrated nitrates of these metals, L'vov [68] has proposed a mechanism for the thermal decompositions of solids based on congruent gasification of all reaction products followed by subsequent condensation of the species of low volatility.

14.4.4. Nitrites

Barium and lead nitrites [69] decompose to yield the nitrates [70]:

$$2Ba(NO_2)_2 \rightarrow Ba(NO_3)_2 + BaO + NO + \tfrac{1}{2}N_2$$

$$3Pb(NO_2)_2 \rightarrow Pb(NO_3)_2 + 2PbO + 4NO$$

The behaviour [71] of $AgNO_2$ is closer to that expected from the breakdown of a nitro compound than for a nitrite because decomposition between 308 and 363 K yields Ag metal and NO_2. α-time curves are sigmoid with a prominent linear region ($0.15 < \alpha < 0.45$), but the Arrhenius plot was curved above 333 K. This was attributed to inadequate removal of gaseous products. In contrast to the behaviour observed for most other solids, pre-irradiation with γ-rays inhibits the subsequent thermal decomposition [71].

Amorphous sodium *trans*-hyponitrite ($Na_2N_2O_2$), heated in a DSC in the absence of oxygen, decomposed [72] suddenly at various temperatures between 633 and 663 K with the formation of a viscous froth that later changed to a white residual solid. The reaction ($Na_2N_2O_2 \rightarrow Na_2O + N_2O$) may be accompanied by the formation of nitrite ($Na_2N_2O_2 + O_2 \rightarrow 2NaNO_2$) and further oxidation to nitrate due to breakdown of N_2O to oxygen by an unidentified mechanism. When $Na_2N_2O_2$ is heated in oxygen the decomposition temperature is lowered by oxidation of $N_2O_2^{2-}$ to NO_2^- and subsequently to NO_3^-.

14.5. PHOSPHATES

14.5.1. Introduction

The chemistry of the thermal decomposition of metal phosphates is of considerable intrinsic and technological interest. There are many reactants in this group and they exhibit a wide diversity of properties. The decompositions involve several separate steps and progress towards understanding the theoretical principles and the chemical controls of the reactions of solid phosphates has been slow. The products of decomposition include water and residual solids that contain compounds with more highly condensed anions. On heating, phosphate groups tend to polymerize with loss of water, perhaps in several steps, with hydrogen bonding being replaced by P-O-P bonds. The mechanisms of decomposition of phosphates are often discussed with reference to the mobility of protons and vacancies in the reactant structure, the bonding of water and the recrystallization of the various participating phases.

The trivalent orthophosphate anion (PO_4^{3-}) readily forms double salts, so that the number of reactants available is very large. Studies have included the decompositions of many acid salts, and acid salts may also be generated during decomposition of ammonium salts following the release of ammonia gas. Comparisons between the decomposition behaviour of related compounds (e.g. metal and acid salts) can yield useful mechanistic information. Removal of water often yields pyrophosphates or metaphosphates. Some higher molecular mass substances form glassy phases and these crystallize only with difficulty. The decompositions of ammonium phosphates are considered in Chapter 15.

14.5.2. Group IA metal phosphates

The salts MH_2PO_4, where M = Li, Na, K and Cs, decompose initially to $M_2H_2P_2O_7$, for example:

$$2NaH_2PO_4 \rightarrow Na_2H_2P_2O_7 + H_2O$$

The rate-determining step for the sodium salt has been identified [73] as a surface process below 470 K (E_a = 158 kJ mol^{-1}) and, above 470 K, the diffusion of H_2O becomes controlling. Sodium metaphosphate, $NaPO_3$, is formed only above 510 K.

Dombrovskii [74] studied the influence of the additives H_3PO_4 or Na_3PO_4 (< 2% molar) on the decomposition of Na_2HPO_4. The rate of condensation and polycondensation is controlled by the concentration of mobile protons in both surface and bulk and depends on the structure, number of defects and the O:H ratio. The added acid acts as a proton donor, so increasing the rate of water formation, while the basic additive introduces anionic defects and so stabilizes the Na_2HPO_4. The ability of ammonium oxalate or ammonium chloride to accelerate the dehydration of $Na_2HPO_4.12H_2O$ to the pyrophosphate was similarly attributed to catalysis resulting from proton donation. As the concentration of mobile protons is increased, there is more rapid water evolution, whereas the addition of the corresponding sodium salts results in some inhibition.

Prodan *et al.* [75] studied the low pressure (10^{-4} Torr), low temperature (from 273 to 373 K) dehydration of $Na_5P_3O_{10}.6H_2O$ in the form of fine crystals. Reaction occurred in two stages (with E_a = 56 and 84 kJ mol^{-1}) both of which were diffusion controlled. The activation energy increased with extent of reaction. The rate of reaction of this salt was enhanced [76] by water vapour, attributed to its ability to reorganize the diffusion layer. This effect (Smith-Topley behaviour) has been noted in many dehydration reactions (Chapter 7).

The dehydration of KH_2PO_4 to the metaphosphate becomes [77] detectable at 470 K, whereas the reaction of K_2HPO_4 commences at 555 K yielding $2K_2HPO_4.K_4P_2O_7$ as an intermediate prior to formation of the pyrophosphate at 670 K. The changes occurring on heating mixtures of the same reactants were also studied. Products mentioned include $K_3HP_2O_7$ and $K_5P_3O_{10}$.

A generally similar pattern of changes was found [78] for the decompositions of lithium and caesium orthophosphates. This comparability of reactivity is ascribed to the similarities of the three-dimensional hydrogen bonding in the anion structure with those of the sodium salts. Reaction transforms these condensed hydrogen-bonded systems to the alternative oxygen-bonded structures. A trend towards lower reaction temperatures in the solids containing the smaller cations (which vary in radius from 0.068 to 0.134 nm) was attributed to the minor effect of increased polarization with reduction in ionic size.

14.5.3. Group IIA metal phosphates

From infrared and thermogravimetric studies, Pechkovskii *et al.* [79] identified the following steps in the decomposition of $Mg(H_2PO_4)_2.2H_2O$:

$Mg(H_2PO_4)_2.2H_2O$ (388 to 523 K) $\rightarrow$ $Mg(H_2PO_4)_2$ (433 to 723 K) $\rightarrow$ $MgH_2P_2O_7$

(633 to 873 K) $\rightarrow$ $[Mg(PO_3)_2]$ (amorphous) (873 to 1273 K) $\rightarrow$

$[Mg(PO_3)_2]$(crystalline)

The reaction of $Ca(H_2PO_4)_2.H_2O$ is even more complicated [79], involving the formation of $Ca_2P_2O_7$ and $Ca_5(P_3O_{10})_2$. Anhydrous $CaHPO_4$ (monetite) decomposes [80] to γ-$Ca_2P_2O_7$ at about 770 K with $E_a = 213$ kJ mol^{-1}, and is transformed to the β-modification at higher temperatures.

14.5.4. Other metal phosphates

The water elimination reactions of $Co_3(PO_4)_2.8H_2O$ [81], zirconium phosphate [82] and both acid and basic gallium phosphates [83] are too complicated to make kinetic studies of more than empirical value. The decomposition of the double salt, $Na_3NiP_3O_{10}.12H_2O$, has been shown [84] to obey a composite rate equation comprised of two processes, one purely chemical and the other involving diffusion control, for which $E_a = 38$ and 49 kJ mol^{-1}, respectively.

14.5.5. Metal phosphites

In vacuum above 323 K the decomposition of copper(II) hypophosphite can be expressed [85]:

$$2Cu(H_2PO_2)_2 \rightarrow 2Cu + 2H_3PO_2 + H_2O + P_2O_3$$

Reaction is characterised by an appreciable induction period and values of E_a for the sigmoid α-time curves were about 100 ± 5 kJ mol^{-1}. The mechanism proposed involved initial proton transfer and the formation of the copper(I) intermediate, $CuHPO_2$, which decomposes to the products shown above.

14.6. SULFATES

14.6.1. Introduction

The temperatures of onset of anion decomposition of most simple metal sulfates are appreciably above those for the dehydrations of the crystal hydrates often formed from aqueous solution, so that decomposition studies are usually concerned with the anhydrous salts. Water loss from prepared hydrates may proceed to completion in more than a single stage (see Chapter 7). If product water is not removed, the sulfate may dissolve in the condensed liquid and reactions are complicated by hydrolysis.

Comparatively few detailed studies of the kinetics and mechanisms of sulfate decompositions have been described. The relatively high reaction temperatures and the corrosive properties of the gaseous products cause practical problems in studying these reactions, although their technological importance has been a stimulus. The available observations indicate that sulfate decompositions are often reversible [86] and there are similarities with the behaviour of carbonates, described in Chapter 12. For example, the magnitude of E_a for the decomposition of $MgSO_4$ is approximately equal to the enthalpy of dissociation [87]. For some reactants, e.g. $ZnSO_4$ [88], the rate of decomposition is decreased by the presence of SO_3, which may be strongly adsorbed on the residual solid products [87-89]. Diffusion within the reactant mass [90] may thus be rate controlling. The equilibria involved are more complicated than the single dissociation occurring in carbonates, because two steps may contribute:

$$MSO_4(s) \rightleftharpoons MO(s) + SO_3(g)$$

$$SO_3(g) \rightleftharpoons SO_2(g) + \tfrac{1}{2}O_2(g)$$

Oxygen in the reaction vessel may influence kinetic behaviour by participating in these equilibria. There is also the possibility that changes in oxidation state of the cation may occur during the decomposition (e.g. $FeSO_4$ and $CrSO_4$). A number of sulfates (e.g. $CaSO_4$ and rare earth sulfates) decompose to yield intermediate oxysulfates. The possibility of melt formation before or during reaction cannot always be excluded and is a further factor that complicates the kinetic interpretations.

14.6.2. Group IA metal sulfates

Anhydrous Group IA metal sulfates melt before decomposition. The anhydrous sodium salt undergoes several phase transformations on heating:

Na_2SO_4 (ℓ) melts $\rightarrow$ Na_2SO_4 (g) and $2Na + SO_2 + O_2$ 1153 - 1473 K [91]

Several studies have been made of the decomposition of ammonium sulfate. Ammonia is evolved at 383 K:

$$(NH_4)_2SO_4 \rightleftharpoons NH_4HSO_4 + NH_3$$

The bisulfate melts [92] at 420 K and subsequently yields the pyrosulfate or $(NH_4)_3H(SO_4)_2$. The low temperature reactions are probably diffusion-controlled.

14.6.3. Group IIA metal sulfates

Mohazzabi and Searcy [93] have made a very precise kinetic investigation of the decomposition of barium sulfate. Effects of changes in the area of the reaction interface during progress of reaction were eliminated through the use of a single cleaved (001) surface of a $BaSO_4$ crystal. The decomposition rate was unchanged during reaction which extended to the formation of a product layer of 1 mm thickness, and this layer was shown not to constitute a barrier to the diffusive escape of the volatile products. From the observed close similarity between the values of E_a and the dissociation enthalpy (both were about 580 kJ mol^{-1}), it was concluded that all the steps in the reaction are at equilibrium except for the desorption of SO_2 and O_2, so that the controlling reaction is:

$$BaSO_4(s) \rightarrow BaO(s) + SO_2(g) + \tfrac{1}{2}O_2(g)$$

The effective area at which this reaction can occur is estimated to represent about 1.4% of the total exposed surface of the crystal.

The enthalpies of formation of $MgSO_4$ and $CaSO_4$ have been measured [86] from dissociation equilibria 1173 to 1638 K. There was no evidence for the formation of basic sulfates. Hulbert [87] concluded that magnesium sulfate decomposed by a contracting volume phase-boundary reaction in the range 1193 to 1353 K and, as in the barium salt, E_a was approximately equal to the enthalpy of dissociation (312 kJ mol^{-1}). The rate of reaction was decreased by an increase in particle size and by the retention of SO_3 in the product layer.

The isothermal decomposition of $BeSO_4$ between 875 and 990 K fitted [94] the contracting area equation, $E_a = 217$ kJ mol^{-1}. These observations were interpreted as evidence of the existence of a connecting columnar structure within the reactant.

14.6.4. Transition-metal sulfates

A feature of the literature on sulfate decompositions is the diversity of reported values of E_a. Mihalik and Horvath [95] found that the decompositions of Zn, Cu, Mg, Fe, Ni, Co and Mn sulfates could be described by the zero-order kinetic equation and that E_a was about 250 kJ mol^{-1}. Kolta and Askar [96], however, described the decompositions of Zn, Fe, Co, Ni and Cu sulfates as being slightly deceleratory with E_a in the range 96 to 125 kJ mol^{-1}.

In hydrogen, $CrSO_4$ decomposes [97] directly to the oxide (Cr_2O_3) at 648 K, but, in oxygen, the reaction is complete by about 770 K and yields both Cr_2O_3 and $Cr_2(SO_4)_3$. At 1170 K, $MnSO_4$ yields both Mn_2O_3 and Mn_3O_4 by reactions for which the values of E_a are 259 and 213 kJ mol^{-1}, respectively. Pechkovskii *et al.* [98] reported a much greater value of E_a (660 kJ mol^{-1}).

The decomposition of $FeSO_4$ occurs [99,100] in two steps via the formation of $Fe_2O_2(SO_4)$. Kinetic behaviour approximated to the contracting geometry models, with E_a in the range 188 to 255 kJ mol^{-1}. Values of E_a for the reactions in oxygen were lower than those in nitrogen. A more detailed kinetic analysis [101] of these consecutive reactions confirmed the conclusions of the earlier study. Kamel *et al.* [102] reported that $FeSO_4.H_2O$ undergoes concurrent oxidation and hydrolysis to yield $Fe_2O(SO_4)_2$ and $Fe(OH)SO_4$ and subsequent reactions yield $Fe_2(SO_4)_3$ which decomposes in two steps to give residual α- Fe_2O_3. The Mössbauer effect has been used to determine the changes in oxidation state of the cation during these reactions. Oxidation [103] is completed at 618 K. In non-oxidizing atmospheres the anhydrous salt is formed between 453 and 643 K and in air there is evidence [104] of oxysulfate intermediates.

The decomposition of $Fe_2(SO_4)_3$ [$\rightarrow Fe_2O_3 + 3SO_3(\rightleftharpoons 3SO_2 + 1\frac{1}{2}O_2)$] was studied between 823 and 923 K [105]. Reaction was zero order and $E_a = 219$ kJ mol^{-1}, in air or in argon. FeS as additive did not change the behaviour. These conclusions differ somewhat from previous reports for this reaction.

Thermogravimetric studies [106] of the rates of the isothermal decompositions of five transition metal sulfates in flowing nitrogen used compressed pellets of finely ground reactant. Each sample was dehydrated for 20 min at 570 K before being heated to reaction temperature. Reaction occurred at an interface that advanced inwards from the pellet surfaces. The thickness of the oxide product layer for decompositions of $Fe_2(SO_4)_3$, $CoSO_4$ and $NiSO_4$ was directly proportional to time and E_a values were 212, 257 and 211 kJ mol^{-1} respectively. Two reaction rates were observed in the decompositions of $CuSO_4$ and $ZnSO_4$, the first was decomposition of sulfate ($E_a = 217$ and 238 kJ mol^{-1}) followed by the breakdown of the oxysulfate (221 and 275 kJ mol^{-1}). Values of E_a are closely related to the enthalpies of decomposition of the sulfates.

Reche and Landsberg [107] measured the value of E_a for the decomposition of $NiSO_4$ as 270 kJ mol^{-1}. The decomposition of $CuSO_4$ occurs in two steps [98,107] via the intermediate $Cu_2O(SO_4)$. Ingraham and Marier [89] concluded that reaction in the pelleted salt proceeded at a well-defined interface which advanced at a constant rate and was inhibited by SO_3 in the prevailing atmosphere. Values of E_a for the two steps were 238 and 280 kJ mol^{-1}. These values were in excellent agreement with those reported by Shah and Khalafalla [108] for the same processes, who observed a linear rate of interface advance and inhibition of the reaction by gaseous oxygen.

Ingraham and Marier [88] identified the intermediate $ZnO.2ZnSO_4$ in the decomposition of $ZnSO_4$. This is a topochemical process, following linear kinetics and the rate is decreased by the adsorption of SO_3 on the product oxide. The reverse reaction ($ZnO + SO_3 \rightarrow ZnSO_4$) is very slow if the solid is not finely divided.

14.6.5. Other sulfates

Aluminium sulfate is dehydrated [109] below 670 K and decomposes in oxygen between 920 and 1160 K, but some 50 K lower in nitrogen. Reactions fitted the contracting area (using freeze-dried reactant) and the contracting volume (for the reagent grade commercial sample) equations with E_a between 284 and 374 kJ mol^{-1}.

Two equilibria were involved:

$$Al_2(SO_4)_3(s) \rightleftharpoons Al_2O_3(s) + 3SO_3(g)$$

$$SO_3(g) \rightleftharpoons SO_2(g) + \tfrac{1}{2}O_2(g)$$

This reaction model has been disputed by Papazian *et al.* [110].

From a thermogravimetric investigation of the decomposition of UO_2SO_4, Notz and Jaffe [111] concluded that U_3O_8 is formed in the range 912 to 994 K by a nucleation and growth process with E_a = 245 kJ mol^{-1}. A phase transition precedes decomposition.

Selected kinetic characteristics for the decompositions of metal sulfates are collected in Table 14.1.

14.6.6. Decomposition of sulfates by effusion methods

Brittain *et al.* [112] have reported studies of the decompositions of several sulfates using a torsion effusion detection technique to identify the primary gaseous decomposition products. $ZnSO_4$ and $ZnO.2ZnSO_4$ yielded SO_3 as the sole volatile product between 800 and 900 K, together with finely divided residual solids. $MgSO_4$ reacted similarly between 900 and 1000 K, but in the presence of additives expected to promote breakdown of SO_3 (platinum group metals and transition metal oxides) the products were consistent with the equilibrium mixture:

$$MgSO_4 \rightleftharpoons MgO + SO_2 + \tfrac{1}{2}O_2$$

$CuSO_4$ and $CuO.CuSO_4$ decomposed between 700 and 780 K to give only SO_3. These reactions were not catalyzed by platinum powder, which did, however, promote dissociation of SO_3 ($\rightleftharpoons SO_2 + \tfrac{1}{2}O_2$).

14.7. SULFITES

The endothermic, reversible dissociations of sulfites exhibit a superficial resemblance to the reactions of carbonates ($MSO_3 \rightleftharpoons MO + SO_2$). However, the possibility of anion oxidation to the thermally more stable sulfate makes investigations of the simple reactions of these salts difficult. Some of the problems are exemplified by the many studies of the reactivity of $CaSO_3$. Calcium sulfite is a probable intermediate in the reactions of $CaCO_3$ used to desulphurize the flue gases emitted from coal-fired power stations. On heating $CaSO_3$ two alternative reactions are possible [113]:

$$CaSO_3 \rightleftharpoons CaO + SO_2 \qquad \text{or} \qquad 4CaSO_3 \rightarrow 3CaSO_4 + CaS$$

It has recently been shown [114] that, between 823 and 973 K, the early part of the dissociation reaction could be studied without significant influence of the disproportionation step. The evolution of SO_2 was deceleratory, without any induction period, ascribed to the early commencement of reaction across all the irregular surfaces of the dehydrated reactant. A full kinetic analysis was not possible because the gas yield corresponding to complete dissociation could not be measured later when there was significant $CaSO_4$ formation. The value of E_a for the dissociation was 210 ±10 kJ mol^{-1} and, subtracting the reaction enthalpy (ΔH=150 kJ mol^{-1}), the energy barrier to anion breakdown was estimated to be about 60 kJ mol^{-1} (approximately twice the value for $CaCO_3$ breakdown [115]).

Verma [116] has provided a comprehensive review of the thermal behaviour of metal sulfites and an analogous review covering the metal selenites is in preparation [117].

14.8. THIOSULFATES

Spectroscopic methods were used [118] to identify intermediates in the decompositions of $Na_2S_2O_3$ and $K_2S_2O_3$. The sodium salt gave the polysulfides Na_2S_2 and γ- Na_2S_5. The potassium salt gave K_2S_3, K_2S_4 and K_2S_5 which, in air, are oxidized to the final sulfate product. It was suggested that small amounts of sulfite participate as a reaction intermediate.

14.9. CONCLUSIONS

The common feature of the decompositions of the diverse group of compounds described in this chapter is anion breakdown. Decompositions in which the cation may have a dominant influence are considered in the chapters on hydrates (Chapter 7), ammonium salts (Chapter 15) and coordination compounds (Chapter 17).

In spite of the wide variety of kinetic behaviour observed for individual compounds, it is possible to classify the reactions represented in this chapter into two broad groups as *irreversible* and *reversible* decompositions.

Irreversible decompositions

The irreversible decompositions of permanganates and chromates release a proportion of the anionic constituent oxygen and form a non-volatile metal oxide product. The activation energy for the decomposition of $KMnO_4$ is similar to values for the decompositions of the rubidium and caesium salts, which suggests a common rate controlling step [32]. This is probably electron transfer within the anionic sub-lattice [11]. The O_2 released is not extensively adsorbed on the solid products and is not expected to participate further in the anion breakdown step and, hence, kinetic characteristics are found to be insensitive to reaction conditions and magnitudes of E_a reported by different workers are similar.

Unresolved features of the decomposition of $KMnO_4$ are the role of $K_3(MnO_4)_2$ [17], for which measurements of amounts present during the reaction are lacking, the possible intervention of melting, and the identities of the residual solids. The reaction temperature is probably too low to permit recrystallization of the finely-divided and/or amorphous manganese oxides (with K_2O) into recognizable crystalline products [18]. The long history of investigations of the $KMnO_4$ decomposition [17] is a good example of the difficulties inherent in characterizing in detail the mechanism of a solid state process.

For the decompositions of some chromates it has been possible to relate values of E_a to spectroscopically measured charge transfer energies. Such studies could be of value in identifying the bonding situations ("transition state") which arise during the course of solid state decompositions. The possibility that electron-accommodating energy levels at surfaces or interfaces may be involved has been discussed [119].

Reversible decompositions

These include the decompositions of many sulfates and nitrates. Measurements of rates and other kinetic parameters for reversible reactions may be subject to the same problems as those described for $CaCO_3$ (Chapter 12) unless experimental conditions are carefully designed to ensure that influence of the reverse reaction is minimised, as described for the dissociation of $BaSO_4$ [93].

Table 14.1.
Selected kinetic characteristics for the decompositions of metal sulfates

Reactant	Solid product	Temperature range / K	E_a /kJ mol^{-1}	$\Delta H_{decomp,298\ K}$ /kJ (mol reactant)$^{-1}$	Remarks	Ref.
$BeSO_4$	BeO	875-99	217	191	Contracting area	94
$MgSO_4$	MgO	1193-1353	312	285	Contracting volume	87
			343		Constant rate (α<0.25)	120
		1293-1333	661		Constant rate and Contracting volume	121,98
$BaSO_4$	BaO	1422-1550	575 (588 at 1500 K)	512	(001) face, constant rate	93
$Al_2(SO_4)_3$	η-Al_2O_3	923-1223	268	580	Contracting volume	125
		910-1040	310	(526 at 1000 K)		109
			289		Freeze dried, Contracting area	
$In_2(SO_4)_3$	In_2O_3	1073-1273		792	Avrami-Erofeev, n = 1.0-1.6	122
$MnSO_4$	Mn_3O_4 (in N_2)	1073-1173	213	305 *	Initial rates	123
	Mn_2O_3 (in O_2)	1123-1173	259	289 *		

(continued)

Table 14.1.(continued)
Selected kinetic characteristics for the decompositions of metal sulfates

Reactant	Solid product	Temperature range / K	E_a /kJ mol^{-1}	$\Delta H_{decomp,298 K}$ /kJ (mol reactant)$^{-1}$	Remarks	Ref.
$FeSO_4$	$Fe_2O_2SO_4$	748- 848	260		Contracting area	101
$FeSO_4$	Fe_2O_3	953-1003	253	230 *	Contracting volume	124
$Fe_2O_2SO_4$	Fe_2O_3	748- 848	188-255		N_2 or vacuum	100
$Fe_2(SO_4)_3$	Fe_2O_3	973-1073	83	526(1000 K)	N_2, contracting volume	125
			154		Vacuum, contracting volume	126
		823- 923	107		Air, contracting volume	96
		823- 923	219		Air, Ar, zero order	105
$CoSO_4$	CoO (in N_2)	1008-1098	222	209(1000 K)	Contracting geometry	127
	(in O_2)	1088-1178	222	252	Contracting geometry	
		1113-1153	318	252	Contracting volume	98
[Co_3O_4	CoO	1083-1133	356	351(1000 K)		127]
$NiSO_4$	NiO	968-1083	269	236	Constant rate	107
		1273-1423	255			121

(continued)

Table 14.1.(continued)
Selected kinetic characteristics for the decompositions of metal sulfates

Reactant	Solid product	Temperature range / K	E_a /kJ mol^{-1}	$\Delta H_{decomp,298\ K}$ /kJ (mol reactant)$^{-1}$	Remarks	Ref.
$CuSO_4$	$Cu_2O(SO_4)$	1053-1123	212			128
			238			98
			238		Contracting volume	108,123
		963-1083	281 ± 36		Constant rate	107
$Cu_2O(SO_4)$	CuO	1133-1193	346			128
		1033-1073	268			98
			280		Contracting volume	108,123
		963-1083	262		Parabolic rate law	107
$CuSO_4$	CuO			219	(Overall value)	121
$ZnSO_4$	$Zn_3O(SO_4)_2$	923-1023	357 ± 10		Parabolic rate law	107
		1023-1123	174 ± 20		Constant rate	107
$Zn_3O(SO_4)_2$	ZnO	1023-1123	267 ± 9		Parabolic rate law	107
$ZnSO_4$	$Zn_5O_2(SO_4)_3$	1113-1173	212		Non-isothermal	126

(continued)

Table 14.1.(continued)
Selected kinetic characteristics for the decompositions of metal sulfates

Reactant	Solid product	Temperature range / K	E_a /kJ mol^{-1}	$\Delta H_{decomp,298\ K}$ /kJ (mol reactant)$^{-1}$	Remarks	Ref.
$Zn_5O_2(SO_4)_3$	ZnO	1233-1293	77		Non-isothermal	126
$ZnSO_4$		1148-1163	271		Unspecified intermediate	98
	ZnO		385		Unspecified intermediate	98
$ZnSO_4$	ZnO			235	(Overall value)	121
$PbSO_4$	PbO	1133-1573	385	305	Constant rate	126
UO_2SO_4	U_3O_8	912- 994	245	400	Nucleation and growth	111

Enthalpies of decomposition, $\Delta H_{decomp,298\ K}$, refer to 1 mole of the reactant specified (crystalline) with evolution of SO_3 gas, except where marked *, where the product is SO_2 gas and an appropriate yield of oxygen.
$\Delta H^o_{298\ K}$ for the reaction $SO_3(g) \rightarrow SO_2(g) + \frac{1}{2}O_2$ is 99 kJ mol^{-1}.

Under less-rigorously controlled conditions, contributions from secondary reactions between primary products cannot be ruled out (such as $SO_3(g) \rightleftharpoons SO_2(g) + \frac{1}{2}O_2(g)$ and the more complicated reactions of nitrogen oxides). Further complicating factors are the presence of reactive gases such as oxygen in the reaction vessel, the formation and secondary reactions of intermediates such as sulfites and nitrites (with possible melting), and changes in the oxidation state of the cation.

General comment

The thermal behaviour of the metal phosphates differs from those of the two groups above in that elimination of water results in the (irreversible) condensation of the anionic groups. Rates of many of these dehydrations are controlled by diffusion within the residual solids.

REFERENCES

1. E.G. Prout and F.C. Tompkins, Trans. Faraday Soc., 40 (1944) 488.
2. M.E. Brown, Thermochim. Acta, 300 (1997) 93.
3. J.Y. Macdonald, J. Chem. Soc., (1936) 832,839; Trans. Faraday Soc., 34 (1938) 979.
4. W.A. Oates and D.D. Todd, Proc. 1st Australian Conf. Electrochem., Sydney, Hobart, 1965, p.88.
5. R.A.W. Hill, R.T. Richardson and B.W. Rodger, Proc. R. Soc. (London), A291 (1966) 208.
6. R.A.W. Hill and J.N. Welsh, Trans. Faraday Soc., 56 (1960) 1059.
7. S. Roginskii, Trans. Faraday Soc., 34 (1938) 959.
8. B.V. Erofeev and N.I. Mitskevich, Reactivity of Solids, (Ed. J.H. de Boer), Elsevier, Amsterdam, 1961, p.273.
9. V.A. Protashchik, Vesti Akad. Navuk Beloruss SSR, Ser. Khim. Navuk, (4) (1966) 46; Dokl. Akad. Nauk Beloruss. SSR, 16 (1972) 623.
10. B.R. Phillips and D. Taylor, J. Chem. Soc., (1962) 4242.
11. V.V. Boldyrev, J. Phys. Chem. Solids, 30 (1969) 1215.
12. J. Berlinger, J. Chem. Phys., 29 (1958) 537.
13. A.A. Kabanov and L.A. Zharova, Zh. Fiz. Khim., 49 (1975) 2310.
14. V.V. Boldyrev, Z.G. Vinokurova, L.N. Senchenko, G.P. Shchetinina and B.G. Erenburg, Zh. Neorg. Khim., 15 (1970) 2585.
15. V.V. Boldyrev, A.P. Voronin, T.A. Nevolina and V.V. Marusin, J. Solid State Chem., 20 (1977) 327.
16. B.G. Erenburg, L.N. Senchenko, V.V. Boldyrev and A.V. Malysh, Zh. Neorg. Khim., 17 (1972) 1121, 2154.

17. F.H. Herbstein, M. Kapon and A. Weissman, J. Thermal Anal., 41 (1991) 303.
18. M.E. Brown, A.K. Galwey, M.A. Mohamed and H. Tanaka, Thermochim. Acta, 235 (1994) 255.
19. F.H. Herbstein, G. Ron and A. Weissman, J. Chem. Soc. A, (1971) 1821.
20. F.H. Herbstein and A. Weissman, J. Chem. Soc., Dalton Trans., (1973) 1701.
21. T.E. Moore, M. Ellis and P.W. Selwood, J. Amer. Chem. Soc., 72 (1950) 852.
22. G.A. El-Shobaky, K.A. El-Barawy and A.A. Ibrahim, Thermochim. Acta, 102 (1986) 21.
23. E.G. Prout, J. Inorg. Nucl. Chem., 7 (1958) 368.
24. E.G. Prout and C.M. Lownds, Inorg. Nucl. Chem. Lett., 9 (1973) 617.
25. E.G. Prout and P.J. Herley, J. Phys. Chem., 64 (1960) 675; 66 (1962) 961; J. Chem. Soc., (1959) 3300.
26. P.J. Herley and E.G. Prout, J. Inorg. Nucl. Chem., 16 (1960) 16.
27. V.V. Boldyrev and A.N. Oblivantsev, Kinet. Katal., 3 (1962) 887.
28. V.V. Boldyrev and L.I. Bystrykh, Russ. Chem. Rev., 32 (1963) 426.
29. V.A. Protashchik and B.V. Erofeev, Dokl. Akad. Nauk Beloruss SSR, 10 (1966) 658; Dokl. Akad. Nauk SSSR, 172 (1967) 1129.
30. M. Cieslak-Golonka, E. Ingier-Stocka and A. Barteri, J. Thermal Anal., 43 (1995) 157.
31. M.B. Hursthouse, K.C. Quillin and D.R. Rosseinsky, J. Chem. Soc., Faraday Trans., 88 (1992) 3071.
32. A.K. Galwey, S.A. Lyle and S.A.A. Mansour, Thermochim. Acta, 235 (1994) 239.
33. A.K. Galwey and S.A.A. Mansour, Thermochim. Acta, 228 (1993) 397.
34. M.E. Brown, C.S. Sole and M.W. Beck, Thermochim. Acta, 89 (1985) 27; 92 (1985) 149.
35. A. Hardy, Bull. Soc. Chim. Fr., (1961) 1329.
36. G. Rienaecker and K. Werner, Z. Anorg. Allg. Chem., 327 (1964) 281.
37. E.G. Prout and P.J. Herley, J. Phys. Chem., 65 (1961) 208.
38. E.G. Prout and F.C. Tompkins, Trans. Faraday Soc., 42 (1946) 482.
39. E.G. Prout and M.J. Sole, J. Inorg. Nucl. Chem., 9 (1959) 232.
40. E.G. Prout and G.S. Woods, Am. Soc. Test. Mater. Spec. Tech. Publ., 359 (1964) 50.
41. V.F. Komarov, A.A. Kabanov and V.T. Shipatov, Kinet. Katal., 8 (1967) 550.
42. D.A. Young, Decomposition of Solids, Pergamon Press, Oxford, 1966.
43. M.C. Ball, P. Dowden and A.K. Galwey, J. Mater. Chem., 7 (1997) 315.

44. A.K. Galwey, S.A.A. Fakiha and K.M. Abd El-Salaam, Thermochim. Acta, 206 (1992) 297.
45. A.K. Galwey, S.A.A. Fakiha and K.M. Abd El-Salaam, Thermochim. Acta, 239 (1994) 225.
46. E.L. Charsley, Differential Thermal Analysis, Vol.1, (Ed. R.C. Mackenzie), Academic Press, London, 1970, p.423.
47. R.G. Darrie and W.P. Doyle, Reactivity of Solids, (Ed. J.W. Mitchell), Wiley-Interscience, 1969, p.281.
48. R.G. Darrie, W.P. Doyle and I. Kirkpatrick, J. Inorg. Nucl. Chem., 29 (1967) 979.
49. A. Furusaki, H. Konno and R. Furuichi, Thermochim. Acta, 253 (1995) 253.
50. K.H. Stern, J. Phys. Chem. Ref. Data, 1 (1972) 747.
51. A.C. McLaren, Rev. Pure Appl. Chem., 12 (1962) 54.
52. E.R. Johnson and J. Forten, Discuss. Faraday Soc., 31 (1961) 238.
53. E.R. Johnson, Am. Soc. Test. Mater. Spec. Tech. Publ., 359 (1963) 71.
54. V.V. Boldyrev, V.M. Lykhin, A.N. Oblivantsev and K.M. Salikhov, Kinet. Katal., 7 (1966) 432; Khim. Vys. Energ., 2 (1968) 566.
55. K.C. Patil, R.K. Gosavi and C.N.R. Rao, Inorg. Chim. Acta, 1 (1967) 155.
56. R.C. Plumb and J.O. Edwards, J. Phys. Chem., 96 (1992) 3245.
57. B.D. Bond and P.W.M. Jacobs, J. Chem. Soc. A, (1966) 1265.
58. T. Jriri, J. Rojez, C. Bergman and J.C. Mathieu, Thermochim. Acta, 266 (1995) 147.
59. M. Doumeng, Rev. Chim. Miner., 7 (1970) 876.
60. C. Ettarh and A.K. Galwey, Thermochim.Acta, 288 (1996) 203.
61. C. Ettarh and A.K. Galwey, Thermochim.Acta, 261 (1995) 125.
62. A.M. Sirina, I.I. Kalinichenko and A.I. Purtov, Zh. Neorg. Khim., 15 (1970) 2430; 13 (1968) 2931; 19 (1974) 314, 1547.
63. F. Vratny and F. Gugliotta, J. Inorg. Nucl. Chem., 25 (1963) 1129.
64. E.V. Margulis, M.M. Shokarev, L.A. Savchenko, L.I. Beisekeeva and F.I. Bershinina, Zh. Neorg. Khim., 17 (1972) 43.
65. J. Pelovski, O. Matova and St. Shoumkov, Thermochim. Acta, 196 (1992) 503.
66. I. Schildermans, J. Mullens, B.J. Van der Veken, J. Yperman, D. Franco and L.C. Van Poucke, Thermochim. Acta, 224 (1993) 227.
67. A. Malacki, R. Gajerski, S. Labus, B. Prochowska-Klisch and K. Wojciechowski, J. Thermal Anal., 39 (1993) 545.
68. B.V. L'vov, Mikrochim.Acta [Wien] II (1991) 299; B.V. L'vov and A.V. Novichikhin, Spectrochim. Acta, 50 (1995) 1427, 1459.
69. K.C. Patil, R.K. Gosavi and C.N.R. Rao, Inorg. Chim. Acta, 1 (1967) 155.
70. S.R. Mohanty and M.N. Ray, Inorg. Nucl. Chem. Lett., 2 (1966) 387.

71. J.W. Schneller and T.B. Flanagan, Nature (London), 215 (1967) 729.
72. J.D. Abata, M.P. Dziobak, M. Nachbor and G.D. Mendenhall, J.Phys.Chem., 93 (1989) 3368.
73. E. Kowalska and W. Kowalski, Przem. Chem., 41 (1962) 734.
74. N.M. Dombrovskii, Kinet. Katal., 9 (1968) 250.
75. E.A. Prodan, M.M. Pavlyuchenko and V.A. Budnikova, Vestsi Akad. Navuk Beloruss. SSR, Ser. Khim. Khim. Navuk, 4 (1968) 53.
76. E.A. Prodan, M.M. Pavlyuchenko, L.A. Lesnikovich and V.A. Sotnikova-Yuzhik, Dokl.Adad. Nauk Beloruss. SSR, 14 (1970) 905; E.A. Prodan, React. Solids, 8 (1990) 299.
77. R.K. Osterheld and L.F. Audrieth, J. Phys. Chem., 56 (1952) 38.
78. R.K. Osterheld and M.M. Markowitz, J. Phys. Chem., 60 (1956) 863.
79. V.V. Pechkovskii, L.N. Shchegrov, R.Ya. Mel'nikova, A.S. Shul'man and S.S. Gusev, Vestsi Akad. Navuk Beloruss SSR, Ser. Khim. Navuk, (1) (1968) 36.
80. N.W. Wikholm, R.A. Beebe and J.S. Kittelberger, J. Phys. Chem., 79 (1975) 853.
81. L.N. Shchegrov, V.V. Pechkovskii, R.Ya. Mel'nikova and S.S. Gusev, Zh. Fiz. Khim., 44 (1970) 64.
82. S.E. Horsley, D.V. Nowell and D.T. Stewart, Spectrochim. Acta, A30 (1974) 535.
83. Y.Y. Kharitonov, N.N. Chudinova and I.V. Tananaev, Izv. Akad. Nauk SSR Neorg. Mater., (1) (1965) 525, 539.
84. J. Pysiak, Dokl. Akad. Nauk Beloruss. SSR, 14 (1970) 131.
85. O.I. Lomovsky, Yu. I. Mikhailov, V.V. Boldyrev and V.M. Mastikhin, Thermochim. Acta, 43 (1981) 135.
86. E.W. Dewing and F.D. Richardson, Trans. Faraday Soc., 55 (1959) 611.
87. S.F. Hulbert, Mater. Sci. Eng., 2 (1967/8) 262.
88. T.R. Ingraham and P. Marier, Can. Met. Quart., 6 (1967) 249.
89. T.R. Ingraham and P. Marier, Trans. A.I.M.E., 233 (1965) 359, 363.
90. D.G. Ivanov and S.P. Kosev, see Chem. Abs., 69 (1968) 5580v.
91. J.C. Halle and K.H. Stern J. Phys. Chem., 84 (1980) 1699.
92. R. Kiyourda and K. Urano, Ind. Eng. Chem., 9 (1970) 489.
93. P. Mohazzabi and A.W. Searcy, J. Chem. Soc., Faraday Trans. I, 72 (1976) 290.
94. D.W. Johnson and P.K. Gallagher, J. Amer. Ceram. Soc., 55 (1972) 232.
95. A. Mihalik and Z. Horvath, see Chem. Abs., 71 (1969) 64637m.
96. G.A. Kolta and M.H. Askar, Thermochim. Acta, 11 (1975) 65.
97. A. Noel, D. Thomas, J. Tudo and G. Tridot, Bull. Soc. Chim. Fr., (1968) 95.

98. V.V. Pechkovskii, A.G. Zvezdin and T.I. Beresneva, Kinet. Katal., 4 (1963) 208.
99. D.W. Johnson and P.K. Gallagher, J. Phys. Chem., 76 (1972) 1474.
100. D.W. Johnson and P.K. Gallagher, J. Phys. Chem., 75 (1971) 1179.
101. D.W. Johnson and P.K. Gallagher, Thermochim. Acta, 5 (1973) 455.
102. A.H. Kamel, Z. Sawires, H. Khalifa, S.A. Saleh and A.M. Abdallah, J. Appl. Chem. Biotechnol., 22 (1972) 591, 599.
103. A. Bristoti, J.I. Kunrath, P.J. Viccaro and L. Bergter, J. Inorg. Nucl. Chem., 37 (1975) 1149.
104. K. Skeff Neto and V.K. Garg, J. Inorg. Nucl. Chem., 37 (1975) 2287.
105. P.G. Coombs and Z.A. Munir, J. Thermal Anal., 35 (1989) 967.
106. H. Tagawa and H. Saijo, Thermochim. Acta, 91 (1985) 67.
107. W. Reche and R. Landsberg, Z.Anorg. Allgem. Chem., 375 (1970) 128.
108. I.D. Shah and S.E. Khalafalla, see Chem. Abs., 77 (1972) 142650a, 156920a.
109. D.W. Johnson and P.K. Gallagher, J. Amer. Ceram. Soc., 54 (1971) 461.
110. H.A. Papazian, P.J. Pizzolato and R.R. Orrel, Thermochim. Acta, 4 (1972) 97; 6 (1973) 337.
111. K.J. Notz and H.H. Jaffe, J. Inorg. Nucl. Chem., 25 (1963) 851.
112. R.D. Brittain, K.H. Lau, D.R. Knittel and D.L. Hildenbrand, J. Phys. Chem., 90 (1986) 2259; see also 84 (1980) 1890; 87 (1983) 3713; 93 (1989) 5316.
113. R. Matsuzaki, H. Masumizo, N. Murakami and Y. Saeki, Bull. Chem. Soc. Japan, 51 (1978) 121.
114. D.C. Anderson and A.K. Galwey, Proc. R. Soc. (London), A452 (1996) 585, 603.
115. D. Beruto and A.W. Searcy, J.Chem.Soc., Faraday Trans. I, 70 (1974) 2145.
116. V.P. Verma, Thermochim. Acta, 89 (1985) 363.
117. V.P. Verma, to be published.
118. K.D. Cleaver and J.E.D. Davies, J.Chem.Soc., Dalton Trans., (1980) 245.
119. A.K. Galwey and M.E. Brown, Proc.R.Soc. (London), A450 (1995) 501.
120. W.E. Brownwell, J. Amer. Ceram. Soc., 46 (1963) 374.
121. K.H. Stern and E.L. Weise, High Temperature Properties and Decomposition of Inorganic Salts. Part 1. Sulphates. Natl. Stand. Ref. Data Ser., Natl. Bur. Stand. (U.S.), No. 7, 1966.
122. N.A. Khovorostukhina, Yu.V. Rumyantsev and I.K. Skobeev, Tr. Vost. Sib. Fil. Akad. Nauk SSSR, Sib. Otd., 41 (1962) 83.
123. T.R. Ingraham and P. Marier, Trans. Metall. Soc. AIME, 233 (1965) 359; 236 (1966) 1064; 242 (1968) 2039.

124. V.V. Pechkovskii, A.G. Zvezdin and S.V. Ostrovskii, J. Appl. Chem. USSR, 35 (1963) 1454; 36 (1963) 1454.
125. N.A. Warner and T.R. Ingraham, Can. J. Chem. Eng., 40 (1962) 263.
126. H.N. Terem and S. Akalan, C. R. Acad. Sci. Paris, 232 (1951) 973.
127. T.R. Ingraham and P. Marier, Thermochim. Acta, 1 (1970) 39.
128. H.N. Terem and M. Tugtepe, see Chem. Abs., 52 (1958) 1878.

Chapter 15

DECOMPOSITION OF AMMONIUM SALTS

15.1. INTRODUCTION

Ammonium salts have been grouped in a separate chapter to emphasize the similarities of behaviour in reactions involving the NH_4^+ ion, or yielding the volatile NH_3 molecule, following a proton transfer step. Detailed kinetic studies have been published for the solid-state decompositions of many ammonium salts. Comparisons with the metal salts containing the same anion are often useful. The reactions of solid coordination compounds in which ammonia is present as a ligand are discussed in Chapter 17.

Because the decompositions of many ammonium salts evolve NH_3 and H_2O simultaneously (or consecutively) in a 2 : 1 molar ratio, it is often convenient to represent the formula of such salts in the form $\{m(NH_4)_2O.n(\text{metal oxide}).xH_2O\}$, where m, n and x are integers (see ammonium chromates, vanadates, tungstates and molybdates below).

Thermal analyses of 26 ammonium salts have been reported by Erdey *et al.* [1]. Of all the ammonium salts, ammonium perchlorate (AP), on account of its importance as a solid propellant, and also for its intrinsic interest due to the apparently unique pattern of thermal behaviour, has received exceptional attention.

15.2. AMMONIUM PERCHLORATE

The behaviour of ammonium perchlorate (AP) on heating is unusually complicated. Two detailed reviews [2,3] of the extensive literature to mid-1968 have appeared. Relatively fewer articles on the thermal properties of AP have appeared in recent years perhaps because of a decline in funding for rocket research.

The thermal decomposition of AP proceeds to completion in two distinct stages. The *low temperature decomposition* (420 to 600 K) is an autocatalytic nucleation and growth process. Kinetic behaviour is influenced by a crystallographic transformation (orthorhombic to cubic) at 513 K. Reaction rates just above 513 K are significantly slower than rates immediately below this temperature. The residual solid product of this low temperature reaction (about 70% of the initial reactant mass) is chemically and crystallographically identical with the reactant, but has an increased surface area. This residue is stable in this temperature range unless it is subjected to textural change through recrystallization or mechanical treatment, and it sublimes in a vacuum. The *high temperature decomposition* (above about 600 K) leaves no solid residue in the heated zone. This reaction is also accompanied by sublimation if the prevailing pressure is low.

α-time curves for the low temperature isothermal decomposition of AP are sigmoid [2-6]. These are well expressed by the Avrami-Erofeev equation, although the value of n is influenced by the physical form of the reactant (crystal, powder or pellet) and the occurrence of the crystallographic transformation at 513 K. Values of E_a reported [2] for the low temperature reaction are in the range 100 to 140 kJ mol^{-1}. The Avrami-Erofeev model is an acceptable description of the kinetic characteristics because Krautle [7] has confirmed that the rate of advance of the reactant-product interface during growth of nuclei is independent of α and the prevailing gas pressure. Analytical evidence [2,4,8] has confirmed that the proportion of non-condensable (78K) products ($N_2 + O_2$) does not change during the progress of reaction. Measurements of product yields from the pressures of permanent gases generated can therefore be used to determine α.

Jacobs and Ng [9,10] fitted the α-time curves for the low temperature reaction (below 513 K) ($0.01 < \alpha < 1.0$) by the *extended* form [11] of the Avrami-Erofeev equation, rather than the more usually used approximate forms of this equation. The onset of this main reaction was preceded by three distinguishable evolutions of gas. The first was an almost instantaneous release of a small quantity of gas (when $\alpha < 10^{-5}$), probably desorbed from the surface of the salt. This was followed, to α about 10^{-4}, by a linear rate of reaction identified as surface nucleation [5]. When $10^{-4} < \alpha < 4 \times 10^{-3}$ there was an exponential dependence of product yield on time. This acceleratory process can be termed `branching', though it is more probable that during this stage dislocations are generated as a consequence of local strains in the vicinity of existing nuclei and that these act as centres for new nucleation. Values of A and E_a reported [9,10] for the nucleation, growth and branching processes in the decomposition of both orthorhombic and cubic forms of $NH_4C\ell O_4$ are given in Table 15.1. These values show compensation behaviour [12].

Microscopic observations confirm the existence of compact nuclei which grow in three dimensions [4,7]. The texture of the residual material, the solid "product" of reaction has also been characterized [13]. Because the nucleation step has been associated with crystal imperfections, the distribution of dislocations at the surface of the reactant has been studied [14] by an etch technique. This work was later extended [15] to investigation of the mechanism whereby pre-irradiation accelerated the onset of reaction in $NH_4C\ell O_4$. It was concluded that the gaseous products from irradiation-induced decomposition imposed strains within the crystal which caused an increase in the density of dislocations. This generated additional reactive sites so that the rate of nucleation and, therefore, of reaction, was increased.

Table 15.1.
Arrhenius parameters for the low-temperature (420 to 600 K) thermal decomposition of ammonium perchlorate [9,10]

Crystal Form		E_a/kJ mol^{-1}	ln (A /min^{-1})
Orthorhombic:	nucleation	118 ± 10	24.4 ± 2.3
(< 513 K)	growth	129 ± 5	27.6 ± 1.2
	branching	148 ± 4	32.5 ± 0.9
Cubic:	nucleation	103 ± 5	21.2 ± 1.4
(> 513 K)	growth	99 ± 6	20.5 ± 1.4
	branching	130 ± 7	28.6 ± 1.6

The rate of the high temperature decomposition of $NH_4C\ell O_4$ is influenced by the experimental conditions prevailing and most significantly the pressure of gases present. Sublimation accompanies reaction and overall behaviour is sensitive to the rate at which material escapes from the hot zone within which there are secondary reactions amongst the gaseous products. The "Unified Reaction Scheme" proposed by Jacobs *et al.* [2,16,17] to account for the several possible reactions is reproduced in Figure 15.1. After proton transfer, the breakdown of $HC\ell O_4$ undoubtedly involves several species and rate processes, some of which result in ammonia oxidation. The compositions of evolved product mixtures vary with reaction temperatures, and clearly depend on the relative importance of the contributing processes.

Many other workers [18-24] have similarly concluded that the first step in decomposition of $NH_4C\ell O_4$ is the transfer of a proton from the cation to the anion. This explains [4] the increase in induction period to reaction following exposure of salt to ammonia and the more rapid onset of reaction of reactant containing small amounts of $HC\ell O_4$. The inhibiting effect of NH_3 changes across the crystallographic transformation [25]. The stability of the residual phase in the low temperature interval is ascribed to the formation of an outer layer of adsorbed NH_3 (resulting from incomplete oxidation) and/or superficial NH_4^+ ions which acts as a protective surface, suppressing the formation of perchloric acid which is the unstable entity [2,16]. At higher temperatures both of the adsorbed products of proton transfer (NH_3(a) and $HC\ell O_4$(a)) are volatilized. If rapid diffusion from the heated zone is prevented (by the presence of gases), secondary reactions occur, probably proceeding through initial breakdown of the acid [2,16,17]. The pressure insensitive low-temperature reaction, which is a heterogeneous decomposition proceeding at the surface of the reactant salt, may be contrasted with the pressure sensitive high-temperature decomposition, which is homogeneous, occurring predominantly in the gas phase, and persists until the dissociated material leaves the hot zone.

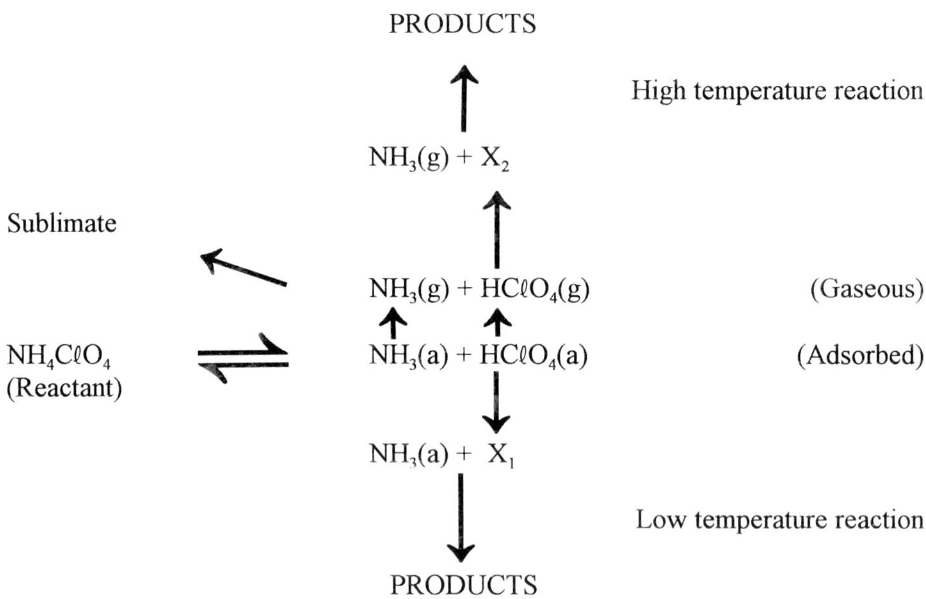

Figure 15.1.
The "Unified Reaction Scheme" proposed by Jacobs *et al.* [2,16,17] to account for the several possible thermal reactions of ammonium perchlorate (X_1 and X_2 represent mixtures of intermediates).

At low pressures, the products of proton transfer (NH_3 + $HC\ell O_4$) escape unchanged from the site of reaction and recombine on a cool surface elsewhere as sublimate. A kinetic study of the rate of $NH_4C\ell O_4$ sublimation has been reported by Jacobs and Russell-Jones [26]. They developed a kinetic equation from consideration of the diffusion processes on the surface and in the gas phase. The theoretical foundations of the assumptions in this treatment have subsequently been reexamined by Guirao and Williams [27].

Although proton transfer has often been identified as the initial step in the decomposition of ammonium perchlorate, alternatives have been suggested. Maycock and Pai Verneker [28,29], for example, discussed the role of point defects in this decomposition and, from electrical conductivity measurements, suggested that the first step may be the transfer of either an electron or a proton. Boldyrev *et al.* [18], while accepting that proton transfer is the first step in the low temperature reaction, suggested that the $HC\ell O_4$ so formed may migrate rapidly within the reactant. Decomposition of $HC\ell O_4$ occurs within the pore structure of the partially eroded crystal so that regions of distorted lattice are removed. Jacobs *et al.* [30] accounted for AC conductivity data by a proton transfer mechanism that does not involve vacancies. Owen *et al.* [31] concluded that there was no evidence for proton conduction at or near the crystal surfaces, which is the zone in which such transfer is to be expected to participate in sublimation. These workers favoured the view that electron transfer is the initial step in the decomposition of $NH_4C\ell O_4$.

The rate of the low temperature reaction is influenced by the particle size of the reactant [4,29], ageing [29] and the presence of volatile decomposition products [29]. Measurements have been made of the surface temperature of $NH_4C\ell O_4$ during deflagration and other aspects of the combustion reactions [32,33].

There have also been many studies of the influences of additives on the decomposition of $NH_4C\ell O_4$. Jacobs and Whitehead [2] distinguish the following three types of catalytic behaviour. (i) Additives have been identified which catalyze the high temperature reaction more efficiently than the low temperature decomposition. Because the oxides NiO, ZnO, Cr_2O_3, Co_3O_4 and CoO catalyze [33] the decomposition of crystals of $NH_4C\ell O_4$ with which they have no physical contact, it is concluded that volatilized $HC\ell O_4$ diffuses to the active surface of the oxide on which it undergoes catalytic heterogeneous decomposition. Accordingly studies have been made of the activities of various oxides in this reaction [34,35]. (ii) Cadmium and silver ions influence the surface concentration of ammonia, through amine formation, and thus promote proton transfer [2]. Copper and strontium ions accelerate the nucleation reaction. (iii) Cd, Mg and Zn perchlorates, as well as CdO and ZnO, catalyze the decomposition of $NH_4C\ell O_4$ and reduce the deflagration temperature. These additives yield a melt containing metal perchlorate in which, presumably, proton transfer occurs more easily and there may also be amine formation. There have been more detailed studies [36,37] of the reactions of mixtures of $NH_4C\ell O_4$ with $Mg(C\ell O_4)_2$ or $Zn(C\ell O_4)_2$. A further mechanism of reaction (iv) may operate in the carbon catalyzed decomposition of $NH_4C\ell O_4$, where changes in rate and extent of reaction may result from displacement of surface equilibria due to adsorption of mobile participants on the high-area additive. Such an effect resembles that given in (ii) above, but here subsequent steps in the reaction could be modified.

Discussions in several of the papers concerned with the decomposition of AP [38] have been principally concerned with determining whether the first step in reaction is proton transfer [17], or electron transfer [31]. Galwey and Mohamed [38], considering the unexpected result that several nitrates strongly promote the low temperature decomposition of $NH_4C\ell O_4$, developed a different and more general explanation of the unusual features of this reaction. This proposal also included the observation that the kinetics of $NH_4C\ell O_4$ decomposition are similar to the breakdown of $NO_2C\ell O_4$: both reactions occur above about 470 K and E_a values are 140 kJ mol^{-1}. Furthermore, chemical analysis of samples of partly reacted $NH_4C\ell O_4$ ($\alpha = 0.5$) detected small amounts of oxidised nitrogen (corresponding to 0.08 % $NO_2C\ell O_4$). Recognising the strong oxidising and acidic properties of perchloric acid, the intervention of $NO_2C\ell O_4$ as an intermediate provides an explanation of the unusual (perhaps unique) features of the $NH_4C\ell O_4$ reaction.

The porous product is reactant crystal penetrated by a labyrinthine system of often parallel rounded pores (about 0.3 μm diameter) connecting with larger, various shaped holes up to about 5 μm diameter. This texture is explained by the occurrence of reaction inside froth-like droplets of a fluid medium composed largely of molten $NO_2C\ell O_4$ within which $C\ell O_4^-$ undergoes breakdown leading to oxidation of NH_4^+ ($\rightarrow NO_2^+$). The droplets advance into undecomposed material. Chance encounters between droplets result in locally enhanced reaction to form the larger holes and the remaining fluid particles advance further into other zones of unreacted material. The essential active reaction intermediate is identified as $NO_2C\ell O_4$ which decomposes to yield products that are sufficiently strong oxidising agents to replenish the active, mobile droplet that is the seat of chemical change. The mechanism has been described in greater detail [38]. Further support for this mechanism has been provided by microscopic and chemical studies [39,40] (see Figure 15.2.).

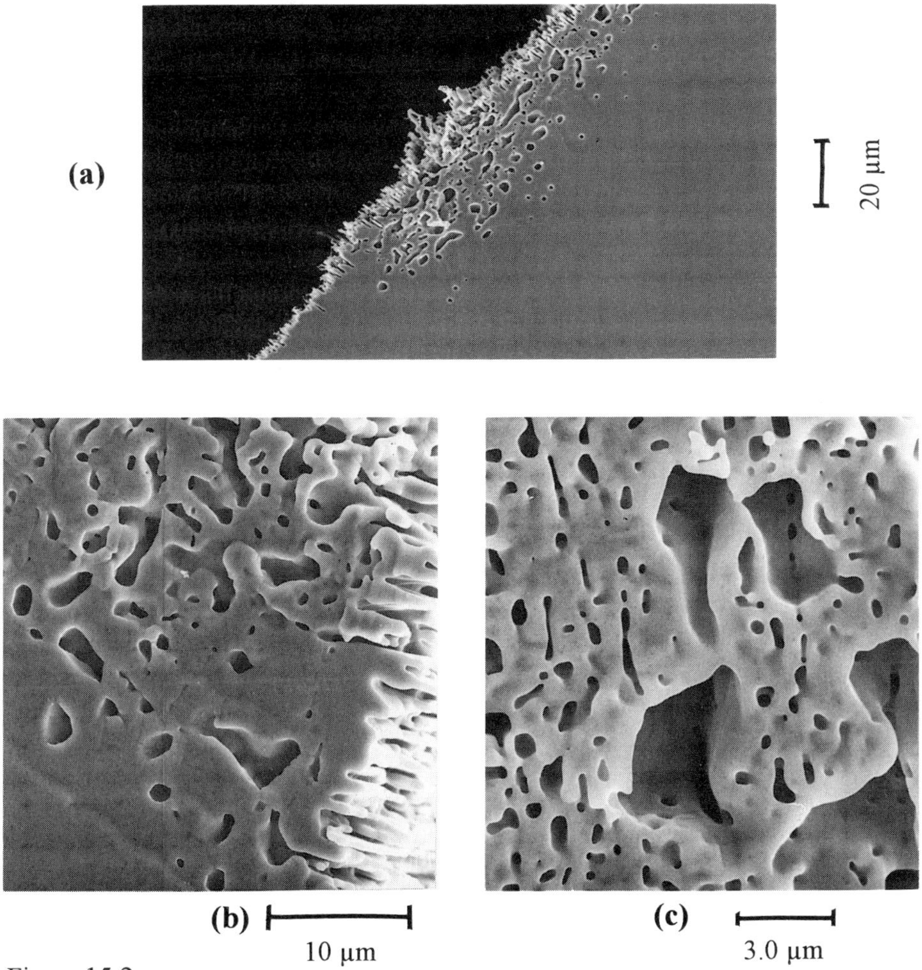

Figure 15.2.
Micrographs of ammonium perchlorate crystals [38]. (a) Cleaved section across the reacted zone during early stages of growth of a nucleus near the crystal surface (scale bar = 20 μm). (b) Section of growth nucleus adjoining the original crystal face. No sharp reaction interface is identified (scale bar = 10 μm). (c) Section across a large growth nucleus. The appearance is typical of product from the low temperature reaction. The various sized holes represent about 30% reactant decomposed (scale bar = 3.0 μm). (Reproduced with permission of the Royal Society, London.)

15.3. NITRONIUM AND NITROSONIUM PERCHLORATES

The decompositions of $NO_2C\ell O_4$ and $NOC\ell O_4$ are considered together because each one is at least partially converted into the other during breakdown. Maycock and Pai Verneker [41] identified three distinct stages in the decomposition of $NO_2C\ell O_4$, carried out in helium to suppress sublimation:

$$NO_2C\ell O_4 \rightarrow NOC\ell O_4 + \tfrac{1}{2}O_2 \qquad (350\ K)$$
$$2NOC\ell O_4 \rightarrow NO_2C\ell O_4 + NO_2 + C\ell O_2 \qquad (395\ K)$$
$$NO_2C\ell O_4 \rightarrow NO_2 + O_2 + C\ell O_2 \qquad (420\ K)$$

The first two reactions give sigmoid α-time curves that are well expressed by the Prout-Tompkins equation. These are believed to involve a branching-chain mechanism. The third process is fitted by the contracting volume equation, possibly due to decomposition through ion-pair evaporation. The thermal stability of $NO_2C\ell O_4$ is decreased [42] by the incorporation of impurities which increase the number of cation vacancies, whereas the creation of anion vacancies has the opposite effect.

Glasner *et al.* [43] suggested that, irrespective of which of the two salts is heated, during reaction a quasi-equilibrium involving both will occur. Decomposition involves the intermediate production of the relatively stable oxide, $C\ell O_3$, and the third step (420 K), above, was identified as controlling the overall rate of reaction.

On heating to between 400 and 600 K the mono-, di- and trimethylamine perchlorates dissociate to yield the amine and perchloric acid, the latter undergoing limited breakdown [44,45]. Decomposition temperatures increase with the extent of cation substitution, but the values of E_a show the opposite trend. Using the cold-matrix isolation technique, it has been shown that hydroxylamine, methylamine and hydrazine (also ammonium) perchlorates vaporize, following proton transfer, to yield $HC\ell O_4$. The same initial step is probably important in reactions of hydrazine diperchlorate [46] and guanidinium perchlorate [47]. The significance of melting in these reactions has not been established.

15.4. AMMONIUM HALATES

The sequence of thermal stabilities of the ammonium halates [48] is: $NH_4BrO_3 < NH_4C\ell O_3 < NH_4IO_3$, and again it is proposed that the first step in reaction is proton transfer. Reactions proceed at interfaces by nucleation and growth mechanisms.

From microscopic measurements is was shown that the ratio of the rate of nucleation to that of growth increased in the sequence $NH_4C\ell O_3 > NH_4BrO_3 > NH_4IO_3$.

Unlike $NH_4C\ell O_4$, the residual product is not chemically unchanged reactant. The decomposition of $NH_4C\ell O_3$ yields [49] residual NH_4NO_3 and a mixture of volatile products. Guillory *et al.* [50] concluded that decomposition following proton transfer gives predominantly NH_3, H_2O and $C\ell O_2$. When reaction occurs at or near lattice voids, however, the sequence of subsequent processes results in the formation of NH_4NO_3, $C\ell_2$, H_2O and N_2. The sigmoid α-time curves for the decomposition of $NH_4C\ell O_3$ were fitted by [49] the Avrami-Erofeev equation, $n = 2$ or 3, and E_a was about 96 kJ mol^{-1}, but this relation is inapplicable in the temperature range immediately below the onset of deflagration at about 360 K. The salt is stabilized by the presence of NH_3, but stability is decreased by ageing or the presence of $HC\ell O_3$. Added oxides cause little acceleration of reaction. This is believed to result from the absence of participation of such oxides in the slow step (proton transfer), and decomposition of the acid is rapid even in the absence of a catalyst.

Ammonium bromate is very unstable [51], and decomposition at about 270 K may be expressed as:

$$4NH_4BrO_3 \rightarrow NH_4NO_3 + 2Br_2 + O_2 + N_2O + 6H_2O$$

while explosion occurs above 327 K and yields no residual solid product:

$$2NH_4BrO_3 \rightarrow N_2 + Br_2 + O_2 + 4H_2O$$

Isothermal α-time curves were sigmoid. The early stages fitted the exponential law and this was followed by an extended ($0.2 < \alpha < 0.8$) linear period, with $E_a = 120$ kJ mol^{-1}. Proton transfer was again proposed as the rate-limiting step, consistent with the observed inhibition by NH_3 and promoting effect of acid.

The isothermal decomposition of NH_4IO_3, the last stable member of this group, was studied (418 to 438 K) by volumetric and gravimetric techniques [52]. Explosion occurred above 453 K. Unlike the behaviour of the related substances, this reaction was deceleratory throughout, fitted by either the contracting volume or first-order equations ($E_a = 276$ kJ mol^{-1}) during a reaction which yielded residual HI_3O_8 and I_2O_5. The proposed mechanism was, again, proton transfer followed by oxidation of NH_3 by the product oxides. This explains the absence of NH_3 in the evolved gases, in contrast with the decompositions of $NH_4C\ell O_3$ and NH_4BrO_3. Catalytic effects were noted when NH_3 was introduced to partially decomposed reactant.

15.5. AMMONIUM NITRATE

Non-isothermal investigations [53] and solid-state ^{15}N NMR studies [54] of the several phase transitions of NH_4NO_3 have been reported. Following three phase transformations [55] above 298 K, decomposition of solid NH_4NO_3 begins [1,56] above 423 K, but only becomes extensive well above the melting point (440 K). Decomposition with the evolution of N_2O and H_2O from the melt is first-order [57,58] (E_a = 153 to 163 kJ mol^{-1}). The proposed mechanism involves formation of nitramide as an intermediate. Other proposed schemes [59,60] have identified NO_2^+ (see $NH_4C\ell O_4$ [38] above) or the radical $NH_2NO\cdot$ (below 473 K) as possible participants. The influence of additives on NH_4NO_3 decomposition has been studied [61,62].

Brower *et al.* [63] proposed that the thermal decomposition of NH_4NO_3 proceeds by an ionic mechanism in the melt below about 570 K. After proton transfer:

$2HNO_3 \rightarrow H_2NO_3^+ + NO_3^-$ and

$H_2NO_3^+ \rightleftarrows H_2O + NO_2^+$, then

$NH_3 + NO_2^+ \rightarrow NH_3NO_2^+ \rightarrow N_2O + H_3O^+$

An increase in E_a from 118 kJ mol^{-1} to 193 kJ mol^{-1} at higher temperatures is believed to result from a change of mechanism in which rate control is by HNO_3 breakdown. The same products were formed at the same rate from vapour or liquid reactants.

Szabo *et al.* [64] used NH_4HSO_4 as a solvent to enable reproducible kinetic data to be obtained for NH_4NO_3 decomposition. The first step is identified as thermal dissociation followed by heterolytic breakdown of the nitric acid formed:

$2HNO_3 \rightleftarrows NO_2^+ + NO_3^- + H_2O$

followed by:

$[H_3N\text{-}NO_2]^+ \rightarrow H^+ + H_2N\text{-}NO_2$

Proton transfer results in isomerization of the nitramine to HNNO(OH) which breaks down to $N_2O + H_2O$. This mixture is identified as a potentially useful reagent for nitration.

Kinetic studies [65] of the decomposition $NH_4NO_3(\rightarrow N_2O + 2H_2O)$ between 441 and 484 K showed that reaction proceeded at a constant rate of interface advance with evolution of the products from the surface with $E_a = 84$ to 88 kJ mol^{-1}.

In general, ammonia is evolved before anion breakdown in the decompositions of ammonium metal nitrates [66,67]. Data for the decomposition of hydrazonium nitrate [68] fit second-order kinetics and reaction presumably occurs in a melt.

15.6. AMMONIUM PHOSPHATES

Kubasova [69] has reviewed the chemistry of the ammonium salts of polyphosphoric acids. Much interest in these compounds derives from their agricultural uses as fertilizers. Both NH_3 and H_2O tend to be eliminated simultaneously on heating, but dehydration alone may be achieved [69] in an atmosphere of NH_3 (548 to 598 K), for example:

$$nNH_4H_2PO_4 \rightarrow (NH_4PO_3)_n + nH_2O$$

Thermal analyses [1,70-72], supported by chemical analyses and X-ray diffraction [71] measurements, show that $(NH_4)_3PO_4$ and $(NH_4)_2HPO_4$ both lose NH_3 in a manner analogous to the dehydration of hydrates to form $NH_4H_2PO_4$. All three salts are isostructural. Above about 420 K, water elimination is accompanied by anion condensation to form $(NH_4)_2H_2P_2O_7$ and $(NH_4)_3H_2P_3O_{10}$, proceeding to NH_4PO_3 at higher temperatures. A glassy melt is formed after an exothermic transition at 553 K. The relative thermal stabilities of condensed ammonium phosphates have been studied [73]. Most fuse to a glass-like substance but some polymerize slowly at about 450 K in the solid state [73,74].

Liteanu *et al.* [75] studied the isothermal dehydration of $(NH_4)_2HPO_4$ in a fluidized bed. The α-time curves were fitted by the contracting volume equation, and the rate of dehydration in air increased with decreasing particle size. For a fixed particle size, the reaction rates and apparent values of E_a increased with variation in carrier gas in the sequence: air < methane < hydrogen. The apparent values of E_a were 80 $\pm$ 4, 93 $\pm$ 3 and 104 $\pm$ 3 kJ mol^{-1}, respectively. The reaction rate also increased with water vapour pressure in the air, but E_a decreased from 84.5 $\pm$ 0.5 in dry air to 64.0 $\pm$ 0.5 kJ mol^{-1} when the water vapour pressure was 17.7 Torr. Water vapour on the grain surfaces promoted dissociation:

$$(NH_4)_2HPO_4 + H_2O \rightleftarrows NH_4H_2PO_4 + NH_3 + H_2O$$

Small amounts (< 3.5%) of ammonium salts markedly accelerated [76] the dehydration of $Na_2HPO_4.12H_2O$ to $Na_4P_2O_7$. This was attributed to an increase in the concentration of delocalized protons in the structure, as a consequence of the proton donor properties of NH_4^+.

15.7. AMMONIUM SULFATES

Ammonium sulfate releases [1] ammonia above 523 K in an endothermic reaction:

$$(NH_4)_2SO_4 \rightarrow NH_4HSO_4 + NH_3$$

This temperature is higher than those given in other reports, 383 K [77] and 413 to 513 K [78] indicating a possible influence of retained water on reactivity. Although ammonium sulfate does not melt, the product NH_4HSO_4 does and decomposes to NH_3 and H_2SO_4 between 623 and 723 K. The intermediate formation of the double salt $(NH_4)_3H(SO_4)_2$ at 433 K was detected by X-ray diffraction during the decomposition [79] of $(NH_4)_2SO_4$ between 413 and 513 K. $(NH_4)_3H(SO_4)_2$ decomposed to NH_4HSO_4 at 453 K. Decomposition of ammonium hydrogen sulfate at 473 K proceeded through the formation of molten sulfamic acid:

$$NH_4HSO_4 \rightleftarrows NH_2SO_3H + H_2O$$

which then reacted to form the pyrosulfate:

$$NH_2SO_3H + NH_4HSO_4 \rightarrow (NH_4)_2S_2O_7$$

with the evolution of SO_3 above 523 K. Values of E_a for the first-order deammination and dehydration processes were [78] 67 and 54 kJ mol^{-1}, respectively, indicating possible diffusion control. Decomposition was generally inhibited by water vapour.

The thermal decomposition of NH_4HSO_4 ($\rightleftarrows NH_3 + H_2O + SO_3$) between 618 and 650 K was well described [80] by the zero-order equation, identified as a one-dimensional phase boundary reaction, $E_a = 141$ kJ mol^{-1}. This reaction was studied as a potential energy-storage system.

Basic aluminium ammonium sulfate [81], $(NH_4)_2O.3Al_2O_3.4SO_3.xH_2O$ ($x = 6$ to 8), loses water at about 473 K. Deammination and complete dehydration commences above 673 K, and SO_3 evolution starts at about 873 K to yield residual Al_2O_3 which retains traces of SO_3. α-time data fitted the contracting volume equation.

15.8. AMMONIUM VANADATES

Decomposition of ammonium metavanadate, NH_4VO_3, (AMV) can be represented [82-84] as stepwise decreases in the $(NH_4)_2O : V_2O_5$ ratio. The first step (from 1:1 to 1:2) between 432 and 453 K yields $(NH_4)_2O.2V_2O_5$ which decomposes further between 453 and 483 K to $(NH_4)_2O.3V_2O_5$. Each step occurs as a single reaction in moist ammonia. Between 533 and 573 K, decomposition yields V_2O_5. The overall reaction is readily reversible [84]:

$$2NH_4VO_3 \rightleftarrows V_2O_5 + 2NH_3 + H_2O$$

The kinetics of all the above reactions were described [84] by the contracting volume equation, sometimes preceded by an approximately linear process. Values of E_a, for the isothermal reactions in air, were 175, 133 and 143 kJ mol^{-1}. The rate-limiting step for decomposition in inert atmospheres was identified as [84] the evolution of NH_3, whereas in oxidizing atmospheres it is the release of H_2O. Thermal analyses for the decomposition in air [85,86] revealed only the hexavanadate intermediate and values of E_a for the two steps detected were 180 and 163 kJ mol^{-1}.

The mechanisms of decomposition have been discussed [84] with reference to the crystal structures of the reactants, intermediates and products, the thermodynamics of the contributory steps, and the surface textures of the reacting particles. From the reversibility of the several reactions and the lack of evidence of extensive recrystallization, it was concluded [84] that the decompositions involve only minor structural changes and that crystal structure imposes a high degree of control on the chemical changes occurring. Interfaces advance inwards from all surfaces, there is no indication of a difficult nucleation step or the extensive structural reorganisation characteristic of many nucleation and growth processes. Later studies using DSC and TG [87] and DTA, TG, IR and XRD complemented by electron microscopy [88] supported the proposed mechanisms [84] and the influence of retained structural order.

Von Sacken and Dahn [89], using TG-MS, showed that below 523 K the thermal decomposition of AMV proceeds by simultaneous release of ammonia and water in a constant proportion, regardless of whether the atmosphere is inert or oxidizing. Above 523 K, residual ammonia reduces the solid products so that the stoichiometry of the final residue is dependent upon the sample size, gas flow, heating rate and reactor design.

De Waal *et al.* [90] have used Raman spectroscopy to measure the isothermal kinetics (423 to 443 K) of decomposition of AMV by monitoring the $\upsilon(VO_2)_s$ band at 927 cm^{-1} against time. Reaction was approximately first-order with $E_a = 48.6$ kJ mol^{-1}.

The complicated effects on the decomposition of AMV of doping with metal oxides have been described [91,92]. Reactions vary widely with added cations, the role of defects when iron, cobalt or nickel is present has been discussed [91]. Baran *et al.* [93] examined the detailed stoichiometry of the thermal decomposition of the mixed-valence vanadate, $(NH_4)_2V_3O_8$.

The decomposition of ammonium peroxovanadate $(NH_4)_4V_2O_{11}$ in N_2 was studied [94] by TG at 0.1 to 4.0 K min^{-1}. The first step yields water and $2NH_4[VO(O_2)_2NH_3]$ by a reaction that fits the Avrami-Erofeev equation, $n = 2$, with $E_a = 144$ kJ mol^{-1}. Isothermal studies gave $E_a = 148.8$ kJ mol^{-1}. The product of this reaction decomposes as follows:

$$NH_4[VO(O_2)_2NH_3](s) \rightarrow NH_4[VO_3.NH_3](s) + O_2(g)$$

$$\downarrow$$

$$\tfrac{1}{2}V_2O_5(s) + NH_3(g) + \tfrac{1}{2}H_2O \leftarrow NH_4VO_3(s) + NH_3(g)$$

15.9. AMMONIUM CHROMATES

The first step in the thermal decomposition of ammonium chromate is dissociation to the dichromate [95] at about 375 K. The rate of this process is decreased by the presence of oxygen [96]. The deammination and dehydration reaction:

$$2(NH_4)_2CrO_4 \rightarrow (NH_4)_2Cr_2O_7 + 2NH_3 + H_2O$$

is well represented by the contracting area equation [97]. This interface advance reaction model is supported by microscopic observations which reveal, in sections of partly reacted lath-shaped crystals, the formation of an outer product layer. Arrhenius parameters for this reaction were $E_a = 97 \pm 5$ kJ mol^{-1} and $\ln A /s^{-1} = 13.8$ from rate studies between 334 and 383 K. At the lower end of the temperature interval investigated there was a relative decrease of reaction rate in the presence of the gaseous products. It was concluded that the first step in the dissociation is reversible proton transfer ($\rightleftarrows NH_3 + HOCrO_3^-$) followed by the irreversible dehydration step ($\rightarrow Cr_2O_7^{2-} + H_2O$).

De Waal *et al.* [98] used Raman spectroscopy to measure the decomposition kinetics from the isothermal time-dependence of the totally symmetric Cr - O vibration mode in $(NH_4)_2CrO_4$ between 343 and 363 K. The results were fitted to the Avrami-Erofeev equation with $n = 2$. E_a for microcrystals was 97 ± 10 kJ mol^{-1} and 49 ± 1 kJ mol^{-1} for powdered samples.

Thermoanalytical measurements distinguished four stages in the decomposition [99] of ammonium dichromate. At 508 K:

$$3(NH_4)_2Cr_2O_7 \rightarrow 3(Cr_2O_5.NH_3) + N_2 + 6H_2O + NH_3$$

followed at 533 K by:

$$3(Cr_2O_5.NH_3) \rightarrow [(CrO_2)_6.H_2O] + N_2 + 2H_2O + NH_3$$

The hydrated oxide loses water at about 570 K and oxygen at higher temperatures, giving Cr_2O_3. The final solid product is active [100,101] as a catalyst.

Three consecutive stages were identified [102] for the decomposition of single crystals of $(NH_4)_2Cr_2O_7$ between 453 and 488 K. These stages were: (i) an induction period identified as nucleation; (ii) the main reaction with evolution of $N_2(+N_2O)$, which was fitted by the Prout-Tompkins equation with a single rate coefficient (E_a = 138 kJ mol^{-1}); and finally (iii) a zero-order process which was ascribed to a constant area of interface advance, for which E_a was 159 to 172 kJ mol^{-1}, depending on particle size. The induction period decreased with ageing of the sample. The evolution of NH_3 and H_2O ceased after completion of the acceleratory stage. These initial products inhibited reaction and it was concluded (by analogy with other ammonium compounds) that the first step is proton transfer. The ammonia released was oxidized, accounting for the formation of NO_2^- and NO_3^- in the residual phase, detected by infrared examination. The maximum concentrations of these intermediates occurred at the end of the acceleratory period, though the reaction temperature was above the stability range of NH_4NO_3. Fischbeck and Spingler [103] analyzed the sigmoid α-time curves (461 to 491 K) with the power law (n = 4) during the acceleratory period, followed by a later fit to the contracting volume equation (E_a = 205 kJ mol^{-1}).

Particle size was found [104] to influence both the kinetics of decomposition of $(NH_4)_2Cr_2O_7$ and the composition of the gaseous products. There was one maximum only in the isothermal (473 K) α-time curves for samples of small surface area, but on increasing the specific surface area of the reactant by grinding, a second and earlier maximum appeared. The role of dislocations in promoting reaction has been studied [105].

Microscopic observations [106] showed that reactions commenced locally with the generation of froth-like material, recognised as involving melting (fusion nuclei) because of the rounded surfaces of all bubbles in the coherent groups. Closely similar textures were shown to be generated during the reactions of CrO_3 in NH_3 + H_2O vapour and on heating $CrO_3 + (NH_4)_2CO_3$ mixtures. The melting point of CrO_3 (468.5 K) is close to the lower end of the temperature range of the present study. It is concluded that $(NH_4)_2Cr_2O_7$ first dissociates to yield CrO_3 and reaction proceeds in the melt containing this strong oxidising agent. Kinetic characteristics change, the reaction rate decreases, after the constituent water and ammonia are volatilized or oxidised. Several rate processes contribute to the overall chemical changes, probably concurrently. More detailed characterization of the reaction mechanism would require measurements of the changes in concentrations of the several participants reacting in the condensed phase, or parallel studies of model mixtures. Measurements of the gaseous products alone are probably insufficient to formulate a quantitative reaction model for $(NH_4)_2Cr_2O_7$ decomposition in the melt.

In a later paper, de Waal and Heyns [107] used IR, Raman and electronic spectroscopy, as well as XRD, to reexamine the decomposition stoichiometry. The products observed were found to depend greatly on the conditions during heating.

15.10. AMMONIUM PERMANGANATE

The sensitivity of NH_4MnO_4 to explosion makes study difficult, but Bircumshaw and Tayler [108] decreased self-heating by immersing the reactant crystals in an inert oil. The slow reaction (343 to 384 K) gave a residue containing MnO_2, Mn_2O_3 and NH_4NO_3, and the gaseous products were H_2O, N_2, O_2, NO_2 and N_2O. The kinetics of the decomposition are unusual in that the rate accelerated continuously almost to $\alpha = 1.00$. This behaviour may be a consequence of several reactions occurring in the solid state in which only a small proportion of reactant is converted to gaseous products, nitrogen and oxygen being retained in the condensed phase as NH_4NO_3.

Pavlyuchenko *et al.* [109] described the decomposition of NH_4MnO_4 as slow between 343 K and 363 K, very fast above 369 K and explosive above 373 K. They observed a maximum rate in their α-time curves and proposed a radical-chain mechanism. The rate of decomposition was decreased in the presence of H_2O and NH_3, but increased by O_2. The activation energies reported in both these studies [108,109] were in agreement at 109 kJ mol^{-1}.

Radwan *et al.* [110] studied the solid products of decomposition in air as potential catalysts.

The rate of decomposition of NH_4MnO_4 at 351 K increased at first with dose of UV pre-irradiation [111], but above a particular exposure there was a decrease in rate. A similar rate maximum was observed for samples which had been aged for various times. These effects were ascribed to partial decomposition. Low concentrations of products accelerated decomposition, but higher concentrations effectively decreased the rate of reactant breakdown by decreasing self-heating.

15.11. OTHER AMMONIUM SALTS

15.11.1. Ammonium azide

NH_4N_3 sublimes in an inert atmosphere below 523 K [112]. At higher temperatures and/or pressures there is slow decomposition, increasing to explosion as conditions become more severe, for example 723 K at 10 Torr, or 583 K at 150 Torr. Products are N_2, H_2, NH_3 and HN_3.

15.11.2. Ammonium carbonate and ammonium bicarbonate

$(NH_4)_2CO_3$ decomposes [1] without melting, but with some sublimation between 293 and 373 K. Microscopic observations [113] during the initial stages ($\alpha < 0.15$) of decomposition of NH_4HCO_3 showed that the originally transparent crystals became opaque and cavities were formed which grew at constant rate of interface advance until coalescence. Cavities were regarded as analogous to nuclei because the reaction yields no solid product. Kinetic studies in the range 293 to 343 K showed that the rate of decomposition decreased when reaction proceeded in different environments, in the sequence vacuum, water vapour, carbon dioxide and ammonia, but the influence of gases present on the magnitude of E_a was relatively small, all values were between 68 and 80 kJ mol^{-1}. When reaction occurred in a mixture of all three product gases, the inhibition was not large, but E_a was increased to 140 kJ mol^{-1} in the higher temperature interval between 360 and 387 K. It was concluded that decomposition proceeds through the intermediate formation of ammonia and carbonic acid and, again, the initial step was identified as proton transfer.

15.11.3. Ammonium carboxylates

Many of these salts melt or sublime either before or accompanying anion breakdown, or other reaction. Decomposition temperatures generally increase with molar mass. Thermal analyses for several ammonium carboxylates have been given by Erdey *et al.* [1], who concluded that the base strength of the anion increases with temperature until it reaches that of NH_3. Decompositions of ammonium acetate (above 333 K) and ammonium oxalate (above 473 K) proceed through amide formation. Ammonium benzoate and ammonium salicylate sublime (above 373 K) without decomposition, but ammonium citrate decomposes (above 423 K) to yield some residual carbon.

Hájek *et al.* [114] studied the thermal reactions of some ammonium salts of terephthalic and isophthalic acids in an inert-gas fluidized bed between 373 and 473 K. Simultaneous release of both NH_3 molecules occurred from the diammonium salts, without dehydration or amide formation, during which the reactant crystallites maintained their external shapes and sizes. The α-time curves were fitted by the contracting volume equation. Rate coefficients for decompositions of diammonium terephthalate, monoammonium terephthalate and diammonium isophthalate at 423 K were in the ratio 7.4 : 1.0 : 134 and values of E_a (in the same sequence) were 87, 108 and 99 kJ mol^{-1}.

The decomposition of $(NH_4)_3[Cr(C_2O_4)_3]$ at 623 K in air [115] yields non-stoichiometric chromium oxides of composition close to $CrO_{1.9}$ with high surface areas. At higher temperatures these decomposed to α-Cr_2O_3 by reactions that yielded oxygen in two stages. The kinetics of O_2 evolution varied with particle morphology and could be related to the path length for oxygen diffusion.

15.11.4. Ammonium molybdates and tungstates

The stoichiometries of decomposition of these more complex salts have not been clearly established and there is little reliable kinetic or mechanistic information available. On heating the complex compound $(NH_4)_{10}H_2W_{12}O_{42}.7H_2O$ [116] the first thermal event (390 K) is the loss (1.5%) of impurities CO_3^{2-} and HCO_3^-. At about 470 K some NH_3 and H_2O are evolved with the formation of ammonium metatungstate: $(NH_4)_5H_2W_{12}O_{40}.7H_2O$ (E_a = 41 kJ mol^{-1}). Collapse of the polytungstate structure occurs between 513 and 560 K, to give a dehydrated amorphous material that sometimes contains some NH_4^+ (6.5% mass loss and E_a = 45 kJ mol^{-1}). WO_3 is formed between 613 and 643 K and there is recrystallization at 659 K.

15.11.5. Ammonium perrhenates

Decomposition of NH_4ReO_4 occurs in two steps [117]. Below 633 K, in dry nitrogen the first step yields $ReO_2N_{0.5}$(+¼N_2 + 2H_2O). The solid product oxynitride is poorly crystalline and decomposes above about 633 K in a second step to monoclinic ReO_2 (+¼N_2). At higher temperatures (about 873 K), there is a crystallographic transformation to orthorhombic ReO_2 which disproportionates to Re_2O_7 and Re above about 1120 K.

α-time curves for the decomposition of NH_4ReO_4 are deceleratory and may be described by either the contracting volume or the Avrami-Erofeev equation with n = 1.0 to 1.3 (this is close to first-order behaviour), when $0.2 < \alpha < 0.9$. Values of E_a obtained from these rate equations were 151 and 113 kJ mol^{-1}, respectively. Surface nuclei were detected in electron microscopic observations.

Similar kinetics were found for the decomposition of the intermediate $ReO_2N_{0.5}$. The contracting volume equation applied between $0.08 < \alpha < 0.6$ with E_a = 180 kJ mol^{-1}. The Avrami-Erofeev equation, now with n = 1.7, alternatively fitted the data within the similar range $0.1 < \alpha < 0.7$ with E_a = 255 kJ mol^{-1}, and with n = 1.0 thereafter (E_a = 176 kJ mol^{-1}).

The disproportionation of ReO_2($\rightarrow$ Re_2O_7 and Re) was also deceleratory and the contracting volume equation applied from $0 < \alpha < 0.5$ with E_a = 423 kJ mol^{-1} and (somewhat surprisingly) the Prout-Tompkins equation was applied to the rather flat decay period (E_a = 318 kJ mol^{-1}).

Ratner *et al.* [118,119] used samples of defined particle sizes and surface areas and reported the rapid formation of nuclei of a new phase on NH_4ReO_4 surfaces at 503 to 523 K. Decomposition was also accompanied by an increase in surface area and a marked decrease in conductivity (x 10^{-8}) near the onset of reaction. Structures and compositions of the product phases are discussed.

α-time curves for the decomposition of NH_4ReO_4 in hydrogen [118,119]:

$$2NH_4ReO_4 + 7H_2 \rightarrow 2Re + 2NH_3 + 8H_2O$$

were deceleratory and between 473 and 623 K were fitted by the contracting volume equation ($\alpha < 0.75$). The value of E_a (80 kJ mol^{-1}) was less than that found for the reaction in N_2. The reaction rate varied with the pressure of water vapour in the hydrogen, being a maximum at about 5% H_2O. Above 548 K, intermediates (ReO_3, etc.) were volatilized and reduced in the gas phase.

15.11.6. Ammonium uranates

Ammonium diuranate, identified as $UO_3.NH_3.2H_2O$, decomposes in two stages [120]. The first stage, completed at about 500 K, was identified as a one-dimensional diffusion process, by its conformity to the parabolic rate law ($\alpha^2 = kt$) with $E_a = 42 \pm 4$ kJ mol^{-1}. The product, $UO_3.2H_2O$, is dehydrated at 500 to 600 K to β-UO_3. This reaction is fitted by the contracting volume equation, with $E_a = 75 \pm 8$ kJ mol^{-1} and there is an increase of surface area. UO_3 decomposes to U_3O_8

above about 600 K, and again is described by the contracting volume model, but E_a is larger, 92 ± 8 kJ mol^{-1}. Price [121] investigated the influence of salt preparation and reaction atmosphere on the decomposition of ammonium diuranate.

15.12. CONCLUSIONS

Ammonium salts resemble hydrates in several respects because, like H_2O, the NH_3 grouping may occupy a variety of bond situations from which it may be displaced or react on heating. Both ammonia and water act as ligands in coordination compounds (Chapters 7 and 17) and such metal-ligand bonds are usually dissociated at higher temperatures. Water is almost invariably released unchanged, whereas release of ammonia may be followed by reactions with the coordinating ion or with other ligands, usually involving oxidation of the nitrogen.

A common feature of the decompositions of many ammonium compounds is the identification of the first step in reaction as proton transfer. The consequent accumulation of protons with the residual oxy-anions may be followed by the elimination of water, accompanied by condensation (or continued condensation) of the anions, see chromates and phosphates above, ultimately leading to residual oxide formation. Alternatively the acid may be volatilized, many ammonium salts sublime ($NH_4Cl(s) \rightleftarrows NH_3(g) + HCl(g)$).

The deamminations and dehydrations of ammonium salts, which are generally reversible processes, are often complicated by the occurrence of solid-phase transformations, sublimation (for example NH_4ClO_4 [2]) and/or melting. It is, therefore, difficult to determine experimentally the effects of crystal structure, defects and surface properties on the chemical changes occurring.

The ability of strongly oxidizing anions to react with ammonia released on dissociation results in the thermal behaviour of ammonium salts being very different from those of other salts of the same anion. NH_4ClO_4, for example, decomposes at temperatures well below those characteristic of the alkali metal perchlorates [122]. In an early comment, Dodé [123] suggested that NH_4NO_3 was the expected product of decomposition of ammonium salts of strongly oxidizing anions. This has been shown to hold for the decompositions of NH_4MnO_4 [108] and NH_4ClO_3 [49], and NO_3^- and NO_2^- have been detected during the decomposition of $(NH_4)_2Cr_2O_7$ [102]. NH_4ClO_4 was an apparent exception to this generalization [4,5], until NH_4NO_3 was shown to be a catalyst for the decomposition [38,39]. NO_2^+ was identified as an essential participant in the mechanism, being formed as an intermediate from the breakdown of NH_4NO_3 [64]. The decomposition of NH_4ReO_4 occurs at 633 K, well above the temperature of decomposition of NH_4NO_3, and N_2 is formed because oxygen is retained as ReO_2. Some properties of the NH_4^+ ion in solid ammonium salts have been discussed by Johnson [124].

Decompositions in which there is oxidation of the ammonia formed initially yield an increased number of intermediates and products, by concurrent and/or consecutive rate processes. Kinetic analysis of the overall reaction has to take these complications, together with those introduced by possible melting or sublimation, into account.

REFERENCES

1. L. Erdey, S. Gál and G. Liptay, Talanta, 11 (1964) 913.
2. P.W.M. Jacobs and H.M. Whitehead, Chem. Rev., 69 (1969) 551.
3. A.G. Keenan and R.F. Siegmund, Q. Rev. Chem. Soc., 23 (1969) 430.
4. L.L. Bircumshaw and B.H. Newman, Proc. R. Soc. (London), A227 (1954) 115.
5. L.L. Bircumshaw and B.H. Newman, Proc. R. Soc. (London), A227 (1955) 228.
6. A.K. Galwey and P.W.M. Jacobs, Proc. R. Soc. (London), A254 (1960) 455.
7. K.J. Kraut1e, J. Phys. Chem., 74 (1970) 1350.
8. G.A. Heath and J.R. Majer, Trans. Faraday Soc., 60 (1964) 1783.
9. P.W.M. Jacobs and W.L. Ng, Reactivity of Solids, (Eds J.S. Anderson, M.W. Roberts and F.S. Stone), Chapman and Hall, London, 1972, p. 398.
10. P.W.M. Jacobs and W.L. Ng, J. Solid State Chem., 9 (1974) 315.
11. P.W.M. Jacobs and F.C. Tompkins, Chemistry of the Solid State, (Ed. W.E. Garner), Butterworths, London, 1955, Chap.7.
12. M.E. Brown, D. Dollimore and A.K. Galwey, Reactions in the Solid State, Comprehensive Chemical Kinetics, (Eds C.H. Bamford and C.F.H. Tipper),Elsevier, 1980, Vol. 22, p.197.
13. P.J. Herley, P.W.M. Jacobs and P.W. Levy, Proc. R. Soc. (London), A318 (1970) 197.
14. P.J. Herley, P.W.M. Jacobs and P.W. Levy, J. Chem. Soc. A, (1971) 434.
15. P.W. Levy and P.J. Herley, J. Phys. Chem., 75 (1971) 191.
16. P.W.M. Jacobs and A. Russell-Jones, A.I.A.A.J., 5 (1967) 829.
17. J.V. Davies, P.W.M. Jacobs and A. Russell-Jones, Trans. Faraday Soc., 63 (1967) 1737.

18. V.V. Boldyrev, Yu.P. Savintsev and T.V. Moolina, Reactivity of Solids, (Eds J.S. Anderson, M.W. Roberts and F.S. Stone), Chapman and Hall, London, 1972, p.421.
19. W.A. Rosser, S.H. Inami and H. Wise, Combust. Flame, 12 (1968) 427.
20. P.O. Korobeinichev, V.V. Boldyrev, Y.Y. Karpenko and N.Z. Lyakhov, Izv.Akad. Nauk SSSR, Ser.Khim., (1969) 778.
21. R.N. Isaev, Y.A. Zakharov and V.V. Bordachev, Zh. Fiz. Khim., 44 (1970) 302.
22. G.L. Pellett, A.I.A.A.J., 8 (1970) 1560.
23. A.G. Keenan and R.F. Siegmund, J. Solid State Chem., 4 (1972) 362.
24. V.V. Boldyrev, V.V. Alexandrou, E.V. Boldyreva, V.I. Gritsan, Yu.Ya. Karpenko, O.P. Korobeinitchev, V.N. Panfilov and E.F. Khairetdinov, Combust. Flame, 15 (1970) 71.
25. B.I. Kaidymov and V.S. Gavazova, J. Inorg. Nucl. Chem., 36 (1974) 3848.
26. P.W.M. Jacobs and A. Russell-Jones, J. Phys. Chem., 72 (1968) 202.
27. C. Guirao and F.A. Williams, J. Phys. Chem., 73 (1969) 4302.
28. J.N. Maycock and V.R. Pai Verneker, Proc. R. Soc. (London), A307 (1968) 303.
29. V.R. Pai Verneker and J.N. Maycock, J. Inorg. Nucl. Chem., 29 (1967) 2723.
30. P.W.M. Jacobs, F.E. Lovatt and W.L. Ng, Canad. J. Chem., 50 (1972) 3154.
31. G.P. Owen, J.M. Thomas and J.O. Williams, J. Chem. Soc., Faraday Trans. I, 68 (1972) 2356; 70 (1974) 1934.
32. M.W. Beckstead and J.D. Hightower, A.I.A.A.J., 5 (1967) 1785.
33. E.V. Boldyreva, B.N. Bezrukov and V.V. Boldyrev, Kinet. Katal., 8 (1967) 249.
34. R. Gilbert and P.W.M. Jacobs, Canad. J. Chem., 49 (1971) 2827; Combust. Flame, 17 (1971) 343.
35. F. Solymosi, L. Gera and S. Börcsök, Thirteenth Symposium (International) on Combustion, The Combustion Institute, Pittsburgh, 1971, p.1009.
36. R.J. Acheson and P.W.M. Jacobs, A.I.A.A.J., 8 (1970) 1483; J. Phys. Chem., 74 (1970) 281.
37. F. Solymosi and J. Rasko, Z. Physik. Chem., 67 (1969) 76.
38. A.K. Galwey and M.A. Mohamed, Proc. R. Soc. (London), A396 (1984) 425.
39. A.K. Galwey, P.J. Herley and M.A. Mohamed, React. Solids, 6 (1988) 205; Thermochim. Acta, 132 (1988) 205.

40. A.K. Galwey, M.A. Mohamed and D.S. Cromie, React. Solids, 1 (1986) 235.
41. J.N. Maycock and V.R. Pai Verneker, Anal. Chem., 40 (1968) 1935; J. Phys. Chem., 71 (1967) 4077; 72 (1968) 4004.
42. J.N. Maycock, V.R. Pai Verneker and C.S. Gorzynski, J. Phys. Chem., 72 (1968) 4015.
43. A. Glasner, I. Pelly and M. Steinberg, J. Inorg. Nucl. Chem., 31 (1969) 3395; 32 (1970) 33.
44. W.A. Guillory and M. King, J. Phys. Chem., 73 (1969) 4367.
45. P.R. Nambiar, V.R. Pai Verneker and S.R. Jain, J. Thermal Anal., 7 (1975) 587; 8 (1975) 15.
46. C.J. Grelecki and W. Cruice, Advan. Chem. Ser., 54 (1965) 73.
47. A. Glasner and A. Makovky, J. Chem. Soc., (1953) 182; (1954) 1606.
48. V.V. Boldyrev, T. Bánsági, O.P. Korobeinichev, Yu.P. Savintsev, T.V. Mulina, Yu.Ya. Karpenko and F. Solymosi, Mag.Kem.Foly., 78 (1972) 215; Kinet.Katal., 13 (1972) 80.
49. F. Solymosi and T. Bánsági, Combust.Flame, 13 (1969) 262; Acta Chim. Hung., 74 (1972) 9.
50. W.A. Guillory, M. King and J.L. Mack, J. Phys. Chem., 73 (1969) 4370.
51. F. Solymosi and T. Bánsági, J. Phys. Chem., 74 (1970) 15.
52. F. Solymosi and K. Jaky, J. Inorg. Nucl. Chem., 33 (1971) 2829.
53. E. Jóna, T. Sramko and D. Nagy, J. Thermal Anal., 27 (1983) 37.
54. K.L. Anderson-Altmann and D.M. Grant, J. Phys. Chem., 97 (1993) 11096.
55. R.N. Brown and A.C. McLaren, Proc. R. Soc. (London), A266 (1962) 329.
56. N.S. Shah and T.N. Oza, J. Chem. Soc., (1932) 725.
57. G. Guiochon, Ann. Chim., 5 (1960) 295.
58. A.M. Kerdivarenko and R.L. Davidovich, see Chem. Abs., 54 (1960) 10474i.
59. W.A. Rosser, S.H. Inami and H. Wise, J. Phys. Chem., 67 (1963) 1753.
60. B. Yu. Rozman, Zh. Prikl. Khim., 33 (1960) 1052.
61. K.S. Barclay and J.M. Crewe, J. Appl. Chem., 17 (1967) 21.
62. A.G. Keenan and I.J. Ferrer, J. Phys. Chem., 76 (1972) 2844.
63. K.R. Brower, J.C. Oxley and M. Tewari, J. Phys. Chem., 93 (1989) 4029.
64. Z.G. Szabo, E. Hollós and J. Trompler, Z. Physikal. Chem. N.F., 144 (1985) 187.
65. N.Koga and H. Tanaka, Thermochim. Acta, 209 (1992) 127.
66. H.P. Chung, H. Ku and E.J. Chi, see Chem. Abs., 64 (1966) 13726f.
67. M. Harmelin, C.R. Acad. Sci. Paris, C, 264 (1967) 1009.

68. Yu.I. Rubtsov and G.B. Manelis, Zh. Fiz. Khim., 44 (1970) 396.
69. L.V. Kubasova, Russ. Chem. Rev., 40 (1971) 1.
70. F. Kasparek, see Chem. Abs., 60 (1964) 2536g.
71. E.V. Margulis, L.I. Beisekeeva, N.I. Kopylov and M.A. Fishman, Zh. Prikl. Khim., 39 (1966) 2364.
72. M.N. Nabiev, M.T. Saibova, I.A. Borukhov and N.A. Parpiev, Zh. Neorg. Khim., 14 (1969) 2950.
73. R.V. Coates and G.D. Woodward, J. Chem. Soc., (1964) 1780.
74. C.Y. Shen, N.E. Stahlheber and D.R. Dyroff, J. Am. Chem. Soc., 91 (1969) 62.
75. C. Liteanu, F. Margineanu and P. Krobel, J. Thermal Anal., 2 (1970) 119, 267; 4 (1972) 53.
76. N.M. Dombrovskii and R.N. Dombrovskaya, Izv. Vyssh. Uchebn. Zaved. Khim. Khim. Tekhnol., 11 (1968) 748.
77. D.G. Ivanov and S.P. Kosev, see Chem. Abs., 69 (1968) 5580y.
78. R. Kiyoura and K. Urano, Ind. Eng. Chem. Process. Des. Dev., 9 (1970) 489.
79. C. Castellani-Bisi and E.R. Garrini, Gazz. Chim. Ital., 93 (1963) 1252.
80. J. Li, G. Zhang and J. Wang, Thermochim. Acta, 207 (1992) 219.
81. S. Bretsznajder and J. Pysiak, Bull. Acad. Pol. Sci., 12 (1964) 177, 201, 315.
82. M.E. Brown, L. Glasser and B.V. Stewart, Prog. Vac. Microbalance Tech., 2 (1973) 125.
83. M.E. Brown and B.V. Stewart, J. Thermal Anal., 2 (1970) 287; Thermal Analysis, (Ed. H.G. Wiedemann), Birkhauser, Basel, 1972, p. 313.
84. M.E. Brown, L. Glasser and B.V. Stewart, J. Thermal Anal., 6 (1974) 529; 7 (1975) 125.
85. I.K. Bhatnagar, D.K. Chakrabarty and A.B. Biswas, Indian J. Chem., 10 (1972) 1025.
86. V.V. Subba Rao and V.N. Mulay, Indian J. Chem., 8 (1970) 750.
87. F. Stander and C.P.J. van Vuuren, Thermochim. Acta, 157 (1990) 357.
88. A. Legrouri, T. Baird and J.R. Fryer, React. Solids, 5 (1988) 53.
89. U. von Sacken and J.R. Dahn, J. Power Sources, 26 (1989) 461.
90. D. de Waal, A.M. Heyns and K.-J. Range, Mat. Res. Bull., 25 (1990) 43.
91. A.A. Said, J. Thermal Anal., 37 (1991) 849.
92. R.M. Gabr, A.M. El-Awad and M.M. Girgis, J. Thermal Anal., 37 (1991) 249.

93. E.J. Baran, M.B. Vassallo and D. De Marco, Thermochim. Acta, 237 (1994) 143.
94. C.A. Strydom, J.Thermal Anal., 40 (1993) 1069.
95. I.H. Park, Bull. Chem. Soc. Jpn., 45 (1972) 2749, 2753.
96. E.S. Osinovik and A.F. Yanchuk, Dokl. Akad. Nauk BSSR, 10 (1966) 29.
97. S. Rajam and A.K. Galwey, J. Chem. Soc., Faraday Trans. I, 78 (1982) 2553.
98. D. de Waal, A.M. Heyns and K.-J. Range, J. Solid State Chem., 80 (1989) 170.
99. B. Mathieu, D.J. Apers and P.C. Capron, J. Inorg. Nucl. Chem., 33 (1971) 2857.
100. J. Burzyk, J. Deren and J. Haber, Reactivity of Solids, (Ed. G.M. Schwab), Elsevier, Amsterdam, 1965, p.486.
101. S.R. Rao and V. Ramakrishna, Indian J. Chem., 4 (1966) 43.
102. J. Simpson, D. Taylor and D.M.W. Anderson, J. Chem. Soc., (1958) 2378.
103. K. Fischbeck and H. Spingler, Z. Anorg. Allg. Chem., 235 (1938) 183; 241 (1939) 209.
104. B.V. Erofeev, Reactivity of Solids, (Ed. J.H. de Boer), Elsevier, Amsterdam, 1961, p.273.
105. B.V. Erofeev and I.I. Gorelkin, Dokl. Akad. Nauk Beloruss., 19 (1975) 800.
106. A.K. Galwey, L. Pöppl and S. Rajam, J. Chem. Soc., Faraday Trans. I, 79 (1983) 2143.
107. D. de Waal and A.M. Heyns, J. Alloys and Compounds, 187 (1992) 171.
108. L.L. Bircumshaw and F.M. Tayler, J. Chem. Soc., (1950) 3674.
109. M.M. Pavlyuchenko, N.G. Rafal'skii and I.A. Isybul'ko, Geterog. Khim. Reakt., (1961) 99.
110. F.M. Radwan, A.M. Abd El-Hameed, M.R. Mahmoud and R.B. Fahim, J. Thermal Anal., 32 (1987) 883.
111. V.V. Boldyrev, see Chem. Abs., 55 (1961) 11043h.
112. P. Gray and T.C. Waddington, Research Corresp., 8 (11)(1955) S56.
113. M.M. Pavlyuchenko, E.A. Prodan and T.N. Samoseiko, Kinet. Katal., 15 (1974) 796; E.A. Prodan, T.N. Samoseiko and M.M. Pavlyuchenko, Zh. Fiz. Khim., 46 (1972) 2653; Vestsi Akad. Navuk Belaruss, SSR., Ser. Khim. Navuk, (1972) (6) 18; (1973)(1) 14.
114. M. Hájek, J. Málek and V. Bazant, Collect. Czech. Chem. Commun., 36 (1971) 84.
115. A. Lerch and A. Rousset, Thermochim. Acta, 232 (1994) 233.

116. M.A. Mohamed, S.A.A. Mansour and M.I. Zaki, Thermochim. Acta, 138 (1989) 309.
117. P. Gibart, Bull. Soc. Chim. Fr., (1964) 70.
118. Yu.E. Ratner, L.G. Berezkina and Yu.V. Tsvetkov, Met. Izv. Redk. Met., Acad. Nauk SSSR, Inst. Met., (1967) 96; Zh. Neorg. Khim., 13 (1967) 1516.
119. Yu.E. Ratner, D.M. Chizhikov and Yu.V. Tsvetkov, Izv. Akad. Nauk SSSR Met., (1968) 100.
120. M.C. Ball, C.R.H. Birkett, D.S. Brown and M.J. Jaycock, J. Inorg. Nucl. Chem., 36 (1974) 1527.
121. G.H. Price, J. Inorg. Nucl. Chem., 33 (1971) 4085.
122. F. Solymosi, Structure and Stability of Halogen Oxyacids in the Solid State, Wiley, London, 1977.
123. M. Dodé, C. R. Acad. Sci. Paris, 200 (1935) 63; Bull. Soc. Chim. Fr., 5 (1938) 170.
124. D.A. Johnson, J. Chem. Soc., Dalton Trans., (1988) 445.

Chapter 16

THE THERMAL DECOMPOSITIONS OF METAL SALTS OF ORGANIC ACIDS

16.1. INTRODUCTION

Most molecular organic compounds melt unchanged. The weak intermolecular links (usually Van der Waals and hydrogen bonds) undergo thermal breakdown at lower temperatures than most other possible bond rearrangements. If the crystal structure is particularly stable and the molecule is unstable, containing strained structures and/or highly reactive groups, decomposition may precede or accompany fusion. For (predominantly) ionic compounds containing a metal cation and an organic anion there may be a true heterogeneous reaction with metal (or oxide) promoted anion breakdown at an advancing interface. The thermal decompositions of metal carboxylates, notably formates and oxalates, have been extensively and intensively studied and such studies have contributed significantly towards the advancement of the subject. These reactions yield simple mixtures of gaseous products and are particularly suitable for kinetic studies. Many of these reactants do not melt and comparative investigations of rates with systematic variations of cation have been made.

There appears to be, as yet, no general agreement concerning the chemical parameters that control the thermal reactivities of metal carboxylates or the initial step in these reactions. The numerous kinetic studies of these salts have not led to the acceptance of any common pattern of reaction mechanisms. Groups of related reactants, selected to include diverse chemical features which may help major trends to be identified, are surveyed below.

16.2. METAL FORMATES

16.2.1. Introduction

Nickel formate has been the subject of particular attention partly because of the suggestion that surface-bonded formate is formed during the nickel catalysed decomposition of formic acid. The role of this intermediate has been discussed in detail [1-5]. The decomposition of copper formate has also received considerable attention for similar reasons, but also to determine the effect of alternative reactant structures on the kinetic behaviour. In contrast, the decomposition of silver formate has been almost ignored. This omission is surprising, because the kinetics of the decomposition of formic acid on silver metal, and the decomposition of silver oxalate have both been studied in detail.

16.2.2. Nickel formate

The kinetics of decomposition of nickel formate [6,7] are sensitive both to the experimental conditions [8] and the reactant structure. α-time curves for the isothermal decomposition (about 450 K) are usually [8], though not invariably [9], sigmoid and there is microscopic evidence [6] that reaction proceeds through nucleation and growth. The induction period [6] and the shape of the subsequent acceleratory process [8] are influenced by the rapidity with which product water vapour is removed from the vicinity of the reactant. Data fit the Prout-Tompkins equation with E_a about 100 kJ mol^{-1}.

Electron micrographs of the surfaces of salt crystallites decomposed to small extents (generally $\alpha < 0.09$) revealed a very marked inhomogeneity in the density of nuclei [7]. Some surfaces contained large numbers of closely-spaced product crystallites, while others remained apparently unreacted. No dependence of nucleation density on reaction temperature could be recognized.

The rate of salt decomposition was also shown to vary with sample disposition (crystallites packed in a fine tube, loosely aggregated or dispersed on glass wool). Inspection of partially-reacted salt revealed that nucleation was not homogeneous within the reactant mass. This confirms the sensitivity of onset of reaction to the local availability of water vapour [8] and it is concluded that nucleation is strongly inhibited by water, presumably adsorbed at potential sites of product formation. (Dehydration of $Ni(HCOO)_2.2H_2O$ precedes decomposition.) On heating, the quantity of water within the sample is gradually decreased until nucleation becomes possible, but the effectiveness of removal will vary locally with ease of withdrawal along intercrystallite channels. The rate of nucleation may thus be expected to depend on several terms:

$$dN/dt = k\, N_o\, f\,[p(H_2O)]\, f\,[x]$$

where f $[p(H_2O)]$ expresses the dependence of the probability of nucleation on the prevailing pressure of water vapour. This term may include kinetic control of the rate of dehydration of the nucleation sites N_o and also due allowance for the temperature dependent equilibrium between surface and volatile water. f [x] expresses the variation of the f [$p(H_2O)$] term within the reactant mass. The forms of both terms are not known quantitatively.

Decomposition of nickel formate may be represented by the concurrent reactions [6,8,9]:

$$Ni(HCOO)_2(s) \begin{cases} \nearrow Ni(s) + H_2O(g) + CO(g) + CO_2(g) \\ \searrow Ni(s) + H_2(g) + 2CO_2(g) \end{cases}$$

where the formate ion breaks down on the surface of the metal product. Erofeev *et al.* [10], however, concluded from the influence of doping on the rate of decomposition of nickel formate that the primary step involved electron transfer:

$$HCOO^- \rightarrow HCOO + e$$

A similar conclusion is given [10] for decomposition of the cobalt salt.

The volume occupied by each approximately spherical Ni metal product particle, 20 to 50 nm diameter [7], is considerably less than that of the salt from which it was derived. The growth of nuclei, interface advance, is less strongly influenced by the availability of water vapour. The factors which determine the limiting size of each small particle of the product metal have not been established.

16.2.3. Copper(II) formate

Three different crystallographic modifications of the anhydrous salt can be prepared. The decompositions of these reactants (430 to 506 K) exhibit sigmoid α-time curves which have been attributed [11] to a chain-type process through the formation of the volatile and unstable copper(I) salt as intermediate. This was indicated by the formation of a copper mirror on the containing vessel. Schuffenecker *et al.* [11] identified the rate limiting step as electron transfer. Erofeev *et al.* [10,12] showed that the kinetic behaviours of two forms of the salt were different. This was ascribed to variations in the dispositions of copper ions in the crystal structures of the reactants.

Galwey *et al.* [13] supplemented kinetic measurements for the three salts with electron microscopic examination of partially and fully decomposed material. The significant differences in kinetic behaviour were attributed to the dissimilar lattice structures, variations in sizes and shapes of crystallites and disintegration of the reactant particles. The volatile and unstable copper(I) formate dimer [14] is identified as the probable intermediate responsible for the evident mobility of the metal. This is formed at the salt-metal reaction interface and its diffusion outwards and decomposition elsewhere yields the copper sublimate on the walls of the reaction vessel. This metal dispersal (usually) acts in opposition to the growth of an existing nucleus. Microscopic observations indicate that (again usually) there is preferred internal nucleation which may be followed by decomposition of the short lifetime species in the vicinity of the site of generation. The product metal, if deposited on a surface of unreacted salt, may appear as effective nucleation or result in the formation of further copper(I) formate by a chain type process. Alternatively, if product metal is deposited upon existing nuclei, the particles may aggregate to give the appearance of sintering, although this arises through the growth of existing nuclei. The main features of these concurrent and interacting processes [13] are summarized in Figure 16.1.

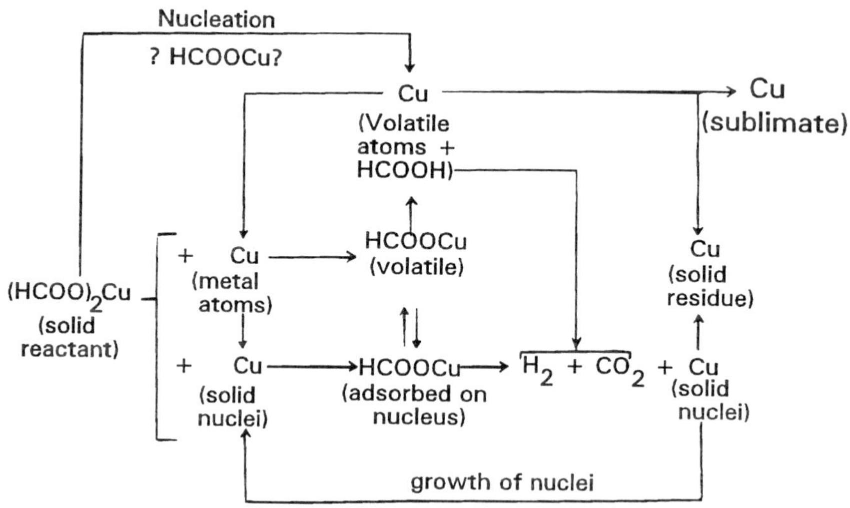

Figure 16.1. The main features of the concurrent and interacting processes involved in the decomposition of copper(II) formate. (Based on Scheme 1 from Ref. [13]).

A consequence of this mobility of the intermediate is that neither the nucleation nor the growth contributions can be correctly represented by rate expressions based on irreversible nucleation processes and the growth of existing nuclei. The autocatalytic behaviour is subject to the redistribution of metal and a dependence of rate upon reaction conditions. Kinetics are influenced by the form of the sample, e.g. by pelleting, which changes the sample surface/volume ratio and, therefore, the ease of sublimation. Activation energies for the acceleratory processes during the decompositions of the three different reactant structures studied [13] were close to 115 kJ mol^{-1}, but increased during the deceleratory stage. The largest value of E_a was 146 kJ mol^{-1}. The α - time data fitted the Avrami-Erofeev equation with $n = 2$ or 3, depending upon the crystal structure.

Kinetic measurements for the overall process do not provide sufficient information for the rate coefficients for the individual steps in Figure 16.1. to be determined separately, or for a single rate-limiting process to be identified from the several interdependent reactions mentioned. As with the nickel salt, several connected contributory rate processes participate in the overall reaction.

16.2.4. Zinc formate

Dollimore and Tonge [15] ascribed the deceleratory decomposition of zinc formate in air ($0 < \alpha < 0.3$) to an initial instantaneous and extensive nucleation of reactant crystallite surfaces with product zinc oxide and the operation of a contracting sphere mechanism. For $0.3 < \alpha < 0.8$ the reaction rate is almost constant, probably as a result of reactant cracking. E_a for both processes is 67 kJ mol^{-1}. During the course of reaction the yields of hydrogen and carbon monoxide increased, while that of carbon dioxide decreased. This was attributed to a decrease in the catalytic activity of the product oxide, possibly as a result of sintering. The formation of higher molecular mass products was mentioned.

Djega-Mariadassou *et al.* [16] reported the occurrence of incongruent melting and the formation of an unidentified phase during the decomposition of zinc formate. Additional products identified were methane, ethane, acetone and methyl acetate. From the results of qualitative analyses, it was concluded that reaction proceeds by several routes involving radical formation, shown [bracketed]:

$$\nearrow ZnO(s) + [HCOO](g) + [HCO](g)$$

$$Zn(HCOO)_2(s) \rightarrow ZnO(s) + 2[H](g) + CO(g) + CO_2(g)$$

$$\searrow ZnO(s) + [H](g) + [HCO](g) + CO_2(g)$$

Subsequent reactions between these species yielded the products observed. Studies were made of surface properties of the product zinc oxide.

16.2.5. Thorium(IV) formate

The kinetics of the decomposition [17] of thorium tetraformate to ThO_2 can be described by the Prout-Tompkins equation with $E_a = 150$ kJ mol^{-1} from 498 to 553 K. The autocatalytic process was ascribed to participation of the oxide in breakdown of the carboxyl groups at the reaction interface to yield ThO_2, formaldehyde and carbon dioxide as the primary products of reaction. The volatile products could, however, react further on the surface of the active solid to yield a number of secondary products amongst which the following gases were identified: H_2, CO, H_2O, CH_3OH, $HCOOCH_3$, HCOOH and $(CH_3)_2O$. Addition of nickel formate to the reactant not only accelerated decomposition but also influenced the composition of the gases evolved, yielding predominantly CO, CO_2 and H_2 (which are the main products of nickel formate decomposition).

16.2.6. Uranium(IV) and uranyl(VI) formates

The kinetics of decomposition [18] of anhydrous uranyl(VI) formate in vacuum (538 to 557 K) were described by the Prout-Tompkins equation ($E_a = 169$ kJ mol^{-1}). At 553 K the stoichiometry is:

$$3UO_2(HCOO)_2(s) \rightarrow 2HCO_2H(g) + 4CO(g) + H_2O(g) + 3\alpha\text{-}UO_3(s)$$

The decomposition of uranium(IV) formate in a marginally lower temperature interval could be described by the linear law with $E_a = 119$ kJ mol^{-1} and at 553 K:

$$2U(HCOO)_4(s) \rightarrow 3HCOOH(g) + 4CO(g) + H_2(g) + CO_2(g) + 2UO_2(s)$$

From consideration of the reaction products it was suggested that the most probable first step is breakdown of the formate ion as follows:

$$HCO_2^- \rightarrow H^+ + CO_2^{2-}$$

with further reactions of these first-formed species.

On heating uranium(IV) formate in air at 553 K, uranyl(VI) formate was formed by a slightly deceleratory process which approximately fitted the contracting area equation with $E_a = 35$ kJ mol^{-1}.

16.2.7. Other formates

When the cation present is sufficiently electropositive, the residue from formate decomposition may be the carbonate, as has been described [19,20] for the thermal reactions of Group IIA metal formates. The decomposition of calcium formate dispersed in a potassium bromide disc also yields residual carbonate [21]. The lanthanide formates decompose with the formation of basic carbonates.

Behaviour depends on the cation present. Studies of the decompositions of some mixed formates, e.g. Ni - Mg [22], Zn - Ba [23] and Cu - Ba [23], have been reported. Sodium and potassium formates melt prior to anion decomposition and reaction is believed [24] to proceed through the intermediate $CO_2^{\cdot}$. In the absence of oxygen, this entity dimerizes to yield the oxalate, whereas in the presence of oxygen the dismutation reaction, which produces carbon monoxide and carbonate, is preferred.

16.2.8. Formates as intermediates in the decomposition of formic acid on metals and oxides

Formates have often been proposed as intermediates in the heterogeneous catalytic decomposition of formic acid on metals or oxides . While there is strong evidence that surface formate can exist under the conditions prevailing during the catalytic reaction [1-4], there is not general agreement as to whether this is a necessary intermediate [3-4]. Furthermore, the identification of a surface participant is but one factor in the quantitative kinetic representation of surface processes [25]. The complete formulation of a reaction mechanism necessitates the determinations of surface concentrations, reactivities and interactions of all participants. The reaction occurring at the salt-metal interface during the decomposition of, for example, nickel formate is not necessarily identical with the heterogeneous catalytic process, because the reactant consists of carboxylate anions, and the composition differs from the acid molecule which includes an additional proton. At the reaction interface the metal surface is also in close proximity with the salt and the specialised environment includes charge imbalance and steric strain. Consequently, while the solid phase decomposition and heterogeneous catalytic reaction may include common intermediates, the reactivities of these and the rate-controlling factors are not necessarily identical. Mechanistic conclusions from catalytic studies (see, for example [26]) may, however, be useful in formulating mechanisms of decomposition of solid formates.

The role of the formate ion as an essential intermediate in the decomposition of formic acid on nickel metal [1-5] has not been quantitatively characterized, but its participation is consistent with the observations. It is probable, therefore, that the rate limiting steps for the catalytic reaction and the solid state decomposition of nickel formate are comparable. Further work is required to clarify the details of both reaction mechanisms.

During the heterogeneous decomposition of formic acid on copper, the active metallic phase undergoes sintering, sublimation and modifications of surface texture [25]. By comparison with the behaviour of copper(II) formate [13], it is concluded that copper(I) formate is formed. Thus variations in the values of A and E_a reported for the catalytic process may be attributed to a dependence of kinetic behaviour on the lifetime of the volatile participant, and thus upon metal mobility and reaction conditions [5]. The decompositions of the copper formates and the catalytic decomposition of formic acid on copper metal thus include the participation of common intermediates, but these different reactions each consist of a sequence of several interdependent processes.

16.3. METAL ACETATES

16.3.1. Nickel acetate

Though salt dehydration was not accompanied [27] by particle disintegration, the anhydrous pseudomorph was shown by X-ray diffraction measurements to be very poorly crystallized (a characteristic feature of many nickel carboxylates). Decomposition in air (554 to 631 K) proceeded at a constant rate ($0.1 < \alpha < 0.8$ and $E_a = 96$ kJ mol^{-1}), ascribed to the operation of an autocatalytic mechanism. The reaction in vacuum (562 to 610 K) gave a sigmoid α-time curve which was fitted by the Prout-Tompkins equation. Because the activation energy was the same as that for reaction in air, it was concluded that the same mechanism operated. The reaction in air yielded residual nickel oxide, while reaction in vacuum gave the carbide with excess carbon and some oxide. In addition to carbon dioxide, the volatile products of decomposition included water and acetic acid.

Manabe and Kubo [28] reported that sample disposition influenced kinetic behaviour and concluded that reaction proceeded to completion in two stages, the first of which yielded a basic nickel acetate. Dorémieux [29] identified this product of partial hydrolysis during water removal in nitrogen as $0.9Ni(CH_3COO)_2.0.1Ni(OH)_2$. The course of the subsequent decomposition depends on reaction temperature. Below 593 K reaction occurs without an induction period to yield nickel oxide, acetone, carbon dioxide and particles of nickel metal. Acetate radicals are also generated which rearrange to acetic acid and the anhydride. Some of the oxide is reduced and the metal is carbided ($E_a = 310$ kJ mol^{-1}). Above 593 K a different mechanism dominates. Rupture of two C–C bonds occurs in the acetate groups and nickel oxide is the residual product ($E_a = 176$ kJ mol^{-1}).

A further isothermal study [30] confirmed the autocatalytic character of the nickel acetate decomposition. The variation with temperature of the rate of decomposition in vacuum gave apparent values of E_a of 150 kJ mol^{-1} between 548 and 578 K and 80 kJ mol^{-1} in the higher range, 578 to 618 K. No melting was detected during reaction and textures of the reaction interface were below the limits of microscopic resolution available. The predominant solid product, Ni_3C, was identified as promoting the nucleation and growth process. The pattern of rate behaviour was explained by some decomposition of nickel carbide, particularly later in the reaction and at the higher temperatures, to give nickel metal which is a more effective catalyst for anion breakdown.

16.3.2. Cobalt acetate

Partial melting was reported [27] to take place during decomposition of this salt in air. Kinetic behaviour included a decrease in rate at α about 0.3. In vacuum the reaction was accompanied by sublimation. Kinetic behaviour showed some irreproducibility and the decrease in rate occurred at α about 0.1. The residue contained metal and oxide, but, unlike the nickel salt, not the metal carbide.

The following steps were proposed [31] for the decomposition of cobalt acetate in nitrogen. Dehydration of the tetrahydrate is accompanied by some hydrolysis (as for the nickel salt) to yield a basic anhydrous salt of composition $Co(CH_3COO)_{1.88}(OH)_{0.12}$. This salt decomposes on heating, with the evolution of acetic acid and acetic anhydride, to yield two well-crystallized products $Co_6O(CH_3COO)_{10}$ (blue) and $Co_3O(CH_3COO)_4$ (rose). The blue salt is transformed into the rose salt before reacting to yield a green powder of approximate composition $40CoO.Co(CH_3COO)_2$ of the wurzite (zinc blende) structure. The final step is reaction to cobalt oxide (CoO, fcc) with trace quantities of the metal.

When the products of reaction are confined, additional intermediates can be identified. Initial dehydration yields the anhydride which hydrolyzes to yield, successively, salts represented as: $Co_4(CH_3COO)_7(OH).2H_2O$; $Co_3(CH_3COO)_5(OH)$ and $Co_3O(CH_3COO)_4$. This last compound decomposes as described above for the reactions in nitrogen.

16.3.3. Copper(II) acetate

The decomposition of copper(II) acetate (473 to 563 K) is accompanied [32] by significant reactant volatilization and, as with the formate, a deposit of copper metal is formed on the container walls. Copper(I) acetate was proposed as the mobile intermediate [14,33,34].

16.3.4. Silver acetate

The overall decomposition of silver acetate (453 to 553 K) may be represented [32] as:

$$2Ag(CH_3COO)(s) \rightarrow 2Ag(s) + C(s) + CH_3COOH(g) + CO_2(g) + H_2(g)$$

Acetone was not produced and no mechanism of reaction was provided.

16.3.5. Group IIA metal acetates

Bernard and Busnot [35] have identified the solid phases formed during the decompositions of magnesium, calcium and strontium acetates in air. At 510 K the magnesium salt is partially transformed into the basic salt $4Mg(CH_3COO)_2.Mg(OH)_2$. At higher temperatures there may be melting as the reactant decomposes to yield the oxide. Anhydrous calcium acetate decomposes to the carbonate at 710 K, accompanied by [32] melting and evolution of acetone. Following water removal, strontium acetate undergoes [35] a phase transformation at 593 K and at about 650 K yields the carbonate. Barium acetate decomposes [36] (740 K) by a topochemical reaction (E_a = 161 kJ mol^{-1}).

Inconsistencies in the literature on the decompositions of Group IIA metal acetates have been attributed [37] to the sensitivity of thermogravimetric measurements to sample configuration, and the influence of product accumulation. The decompositions of magnesium and cadmium acetates involve formation of amorphous intermediates and partial melting.

16.3.6. Other acetates

Decomposition of the acetate ion isolated in a potassium iodide matrix was studied [38] (753 to 847 K) by infrared measurements of the quantities of acetate or formate (product) ions present. Reaction, suggested to be controlled by the step:

$$CH_3COO^- \rightarrow \cdot CH_3 + \cdot CO_2^-$$

followed first-order kinetics with E_a = 195 kJ mol^{-1}. Hydrogen abstraction, largely from water present in the matrix, follows to give the formate ion. A similar reaction yields methane, though dimerisation may occur to form ethane. In different matrices ($KC\ell$ and KBr) the reaction appeared to be diffusion controlled and yielded formate and carbonate ions.

The solid products obtained from the decomposition of uranyl acetate [39] in nitrogen and in air were UO_{2+x} (with some 5% carbon) and U_3O_8, respectively. The gaseous products contained carbon monoxide, carbon dioxide, acetic acid and acetone. Following a short acceleratory process, the isothermal decomposition of this salt in nitrogen proceeded at an approximately linear rate with E_a = 151 kJ mol^{-1}. The reaction in air between 520 and 573 K consisted of an initial approximately linear α-time process (E_a = 107 kJ mol^{-1}). This was followed, after an appreciable increase in rate which occurred at a value of α which increased with temperature, by a second approximately linear process (E_a = 165 kJ mol^{-1}). The initial step in reaction is suggested to be the fission of the UO_2 to acetate bonds, followed by rearrangements of the acetate radical and oxidation of the residual material if oxygen is present. The constant reaction rate was attributed to the inward advance of the interface predominantly at the larger surfaces of the plate-like crystals.

The thermal decomposition of manganese(II) acetate tetrahydrate [40] commenced with the loss of water in two stages. Hydrolysis with acetic acid evolution was detected at 393 K. The decomposition resembled that of cobalt(II) acetate [41] in that both yield metal acetyl acetate and metal acetate hydroxide (e.g. $Mn(CH_3COO)_2.OH$) intermediates. The product from reaction in N_2 or H_2 was MnO and in air Mn_3O_4 was obtained.

Mohamed *et al.* [42] have reported non-isothermal kinetic studies of the decompositions of nickel and lead acetate hydrates. They review previous studies and report analyses of reaction products.

A survey of the temperatures of dehydrations and decompositions of thirteen hydrated acetates has been given by Baraldi [43] who used spectroscopic methods to identify intermediates and products.

Studies of the decompositions of reactants yielding several different volatile products require monitoring of more than one evolved compound. Barnes *et al.* [44] describe the use of TG-GLC-MS as a technique for evolved gas analysis during the decomposition of calcium propionate. Products detected included H_2O, CO_2, C_2H_5CHO, $C_2H_5COCH_3$, $C_2H_5COC_2H_5$ and $CaCO_3$. The reactant melted at 673 K. α-time curves for decompositions between 647 and 665 K were sigmoidal and data fitted the Avrami-Erofeev equation (n = 2), identified as nucleation followed by two dimensional growth with E_a = 315 kJ mol^{-1} and A = 4.8 x 10^{21} s^{-1}. A radical mechanism involving the intervention of C_2H_5 and C_2H_5CO is discussed.

Metal propiolates (prop-2-ynoates) (Na, K, Rb, Co, Ni or Zn) [45] containing the anion $(O_2CC \equiv CH)^-$ and, in some salts, $(HO_2CC \equiv CH)$, decomposed usually between 430 and 470 K by decarbonylation ($\rightarrow M_2CO_3$, CO, C and C_2H_2), decarboxylation ($\rightarrow M_2CO_3$, CO_2, C and C_2H_4) and/or coupling to yield an acetylenedicarboxylate salt. The coupling reaction was favoured where the reactants were slowly decomposed under confined conditions between two BaF_2 plates.

Lower reaction temperatures favoured decarboxylation and the formation of coupled products (Rb^+, Co^{2+}, Zn^{2+} salts), while decarbonylation tended to occur at higher temperatures (Na^+, K^+, Ni^{2+} salts).

16.4. METAL OXALATES

16.4.1. Introduction

The major products from most metal oxalate decompositions can be predicted from thermodynamic data [46,47] (Chapter 2). Interpretation of observations must allow for the possibility that the identifiable solid phases may not be those initially formed, but arise as stable products of a secondary process. Secondary reactions may involve adsorption-desorption steps at the surfaces of finely-divided, reactive and perhaps non-stoichiometric solids. The composition of product gases may also vary within the mass of reactant, through chemical interactions between primary products.

16.4.2. Nickel oxalate

Jacobs and Tariq Kureishy [48] identified three stages in the decomposition of nickel oxalate: a surface reaction, a period of slow development of nuclei and finally the growth of these nuclei at constant rate during which the interface advances through the bulk of the reactant crystallites.

The initial process corresponds to about 1% reaction, is deceleratory throughout, and is fitted by the contracting area equation (E_a = 210 kJ mol^{-1}). Reaction is confined to surfaces of the original particles and is initiated at edges and corners [48]. After this, a long induction period preceded the main reaction, which was a nucleation and growth process fitted by the Avrami-Erofeev equation with $n = 2$ ($0.1 < \alpha < 0.8$). The induction period corresponds to an interval of slow growth after which the rate of interface advance increases when nuclei have attained a critical size. This model has been discussed [48] quantitatively. The activation energy for reaction under differential conditions (130 kJ mol^{-1}) is less than that for reaction under accumulatory conditions (210 kJ mol^{-1}), where product removal was less rapid. The difference in E_a (about 80 kJ mol^{-1}) is close to that found for the precursor dehydration reaction (77 kJ mol^{-1}), and it was concluded that under accumulatory conditions, water could be retained at the reactive nickel ions. The activation energy requirement thus includes the energy necessary for water vapour removal, in addition to the electron transfer step which is identified as being rate limiting. Initiation of this reaction may involve the transference of an electron to a metallic product particle, or to a vacancy at the interface. Grinding the salt increased the rate of reaction slightly, but the incorporation of nickel product or noble metal in the original solid reactant did not significantly catalyse the reaction.

Dominey *et al.* [49] used the reduced-time method of kinetic analysis. The initial reaction was again identified as a surface decomposition process for which E_a was 159 kJ mol^{-1}. This reaction was regarded as yielding a monolayer of product. Because the kinetic behaviour provides no evidence for a contracting volume model, it was suggested that the induction period represents the time required for the migration of nickel atoms, formed in the surface reaction, to aggregate at preferred sites on crystallite faces. Such aggregation yielded metal nuclei which were not randomly dispersed but disposed at linear imperfections [50]. The subsequent growth of nuclei was expected to follow a cubic rate law, but, following coalescence at linear arrays, cylindrical nuclei were formed and the rate law exponent would then decrease from $n = 3$ to $n = 2$. The experimental data support this model, $E_a = 150$ kJ mol^{-1} (503 to 533 K).

Support for the nucleation and growth model was provided by electron microscopy which showed that nucleation was not confined to surfaces, but occurred at lines of internal dislocation. Overlap, following growth, resulted in the formation of approximately cylindrical particles of product. The presence of gaseous hydrogen exerted [51] only a small influence on the kinetics of decomposition.

Residual water from incomplete dehydration reacts to produce nickel oxide [52]. The rate of salt decomposition was shown to be sensitive to the prevailing conditions. The presence of hydrogen, which reduced any oxide present, markedly increased the rate of reaction, in contrast with the conclusion [51] above. The total area of the reactant increased during the early stages to a maximum value at α about 0.15 and thereafter decreased. The area of metallic product similarly increased to a maximum at around $\alpha = 0.5$ for reaction in vacuum, but when hydrogen was present again the nickel area increased.

16.4.3. Iron oxalates

Iron(II) oxalate dihydrate loses water above 420 K to yield the anhydrous salt [53] in an inert atmosphere, but in the presence of oxygen Fe_2O_3 may be formed [54]. At about 590 K the anhydrous oxalate decomposed slowly to yield magnetite and iron but no wüstite was detected [53]. Kadlec and Danes [55] showed that the α-time curves are sigmoidal and $E_a = 167$ kJ mol^{-1}. During decomposition of the mixed (Fe-Mg) oxalates, magnesium may stabilize the wüstite product against either oxidation or reduction [56].

The decomposition of iron(II) oxalate in hydrogen is an autocatalytic reaction which fits the Avrami-Erofeev equation with $n = 2$ ($0.02 < \alpha < 0.25$) and $n = 3$ (0.3

< α < 0.85). The composition of the gaseous products varies throughout the course of reaction. Two chemical processes are involved: the reduction of the salt to iron, and the decomposition to yield both iron and magnetite.

Iron(III) oxalate decomposed between 410 and 450 K to give CO_2 and iron(II) oxalate which retained about 10% of the iron(III) salt [57]. Melting was not detected. The sigmoid α - time curves were identified as being due to a nucleation and growth process. The first half of the reaction was well represented by the Avrami - Erofeev equation (n = 2) and the latter half by the contracting volume equation. Values of E_a were relatively low, 107 to 120 kJ mol^{-1}, and rate control was ascribed to either electron transfer or C - C bond rupture.

The more stable iron(II) oxalate decomposed [58] between 596 and 638 K to yield FeO, as the initial solid product, which then disproportionated to Fe and Fe_3O_4, together with CO_2 and CO (in a 3:2 ratio). α - time data again fitted the Avrami - Erofeev equation (n = 2) with E_a = 175 kJ mol^{-1}. The reaction was proposed to proceed by a nucleation and growth process without melting. The behaviour was comparable with the decompositions of other metal oxalates.

16.4.4. Cobalt oxalate

The α-time curves for the decomposition of anhydrous cobalt oxalate (570 to 590 K) were [59] sigmoid, following an initial deceleratory process to α about 0.02. The kinetic behaviour was, however, influenced by the temperature of dehydration. For salt pretreated at 420 K, the exponential acceleratory process extended to α = 0.5 and was followed by an approximately constant reaction rate to α = 0.92, the slope of which was almost independent of temperature. In contrast, the decomposition of salt previously dehydrated at 470 K was best described by the Prout-Tompkins equation (0.24 < α < 0.97) with E_a = 165 kJ mol^{-1}. This difference in behaviour was attributed to differences in reactant texture. Decomposition of the highly porous material obtained from low temperature dehydration was believed to proceed outwards from internal pores, and inwards from external surfaces in a region of highly strained lattice. This geometry results in zero-order kinetic behaviour. Dehydration at 470 K, however, yielded non-porous material in which the strain had been relieved and the decomposition behaviour was broadly comparable with that of the nickel salt. Kadlec and Danes [55] also obtained sigmoid α-time curves which fitted the Avrami-Erofeev equation with n = 2.4 and E_a = 184 kJ mol^{-1}. The kinetic behaviour of cobalt oxalate [60] may be influenced by the disposition of the sample in the reaction vessel.

On decomposition, cobalt oxalate shows a greater tendency than the nickel salt to form oxide. In vacuum, or in an inert atmosphere, reaction proceeds in two steps [61,62]:

$$CoC_2O_4(s) \rightarrow CoO(s) + CO(g) + CO_2(g)$$

$$CoO(s) + CO(g) \rightarrow Co(s) + CO_2(g)$$

The composition of the products depends on the relative significances of these processes during reaction. When oxygen is present the residue consists of oxides, whereas the metal is formed in hydrogen.

16.4.5. Manganese(II) oxalate

The dehydration that precedes the decomposition of manganese(II) oxalate exhibits the Smith-Topley effect. The residual product of decomposition, MnO, is readily oxidized and the temperature of reaction is decreased in the presence of oxygen. The decomposition of anhydrous manganese(II) oxalate in vacuum [63] can be represented as:

$$MnC_2O_4(s) \rightarrow MnO(s) + CO(g) + CO_2(g)$$

During the early stages of decomposition, the CO/CO_2 ratio in the gaseous products was about 1.3, evidence for the initial formation of a higher oxide, including presumably, the Mn^{3+} ion. In oxygen, the salt decomposed according to:

$$4MnC_2O_4(s) + 3O_2(g) \rightarrow 2Mn_2O_3(s) + 8CO_2(g)$$

α-time curves for both reactions were sigmoid though not symmetrical. The induction period was short and the acceleratory reaction was fitted by the power law with n between 2 and 3 when $\alpha < 0.2$. Thereafter the reaction was best described by the contracting volume equation, after surface nuclei had covered all external boundaries of the reactant crystallites.

Studies of the decomposition of the anhydrous salt (600 to 670 K) [63] gave activation energies of from 140 to 180 kJ mol^{-1}, and some variations in the form of the $g(\alpha)$-time functions were found. These differences in behaviour were attributed to the use of different reactant preparations (effect of crystallite size, impurities, etc.) and the possible presence of traces of oxygen during the onset of reaction. The kinetics of reaction in oxygen have been studied less fully, but it is known that decomposition occurs in a lower temperature interval, 500 to 590 K, and E_a is also lower (100 kJ mol^{-1}).

The reactivity of manganese(II) oxalate is significantly influenced by the conditions during dehydration. Rate coefficients for decomposition of reactant previously held (2 h) at 460 K in a maintained vacuum were about 2.5 times greater than those found for a similarly treated sample heated at 460 K in air. This difference was apparent in both decomposition and oxidation reactions. These data, together with electron microscopic observations, showed that nucleation in the vacuum dehydrated porous salt occurred at internal surfaces, while in the air dehydrated salt some reorganization of the crystal structure was possible and there was more pronounced product formation at the external surfaces.

Decomposition is suggested [63] to involve the formation of Mn^{3+} ions at the salt-oxide interface and subsequent bond redistributions to involve electron transfer steps, resembling those in solution, thereby regenerating active Mn^{3+} ions at the interface. In the absence of oxygen, the carbon monoxide evolved reduces the oxide product to a finely divided and very reactive form of MnO. Decomposition is more rapid when oxygen gas is present, due to increased availability of the intermediate Mn^{3+} at the nucleus surface and this species is abundant in the residual oxide phase.

16.4.6. Silver oxalate

Many kinetic studies of the thermal decomposition of silver oxalate have been reported. Some α-time data have been satisfactorily described by the cube law during the acceleratory period ascribed to the three-dimensional growth of nuclei. Other results were fitted by the exponential law which was taken as evidence of a chain-branching reaction. Results of both types are mentioned in a report [64] which attempted to resolve some of the differences through consideration of the ionic and photoconductivities of silver oxalate. Conductivity measurements ruled out the growth of discrete silver nuclei by a cationic transport mechanism and this was accepted as evidence that the interface reaction is the more probable. A mobile exciton in the crystal is trapped at an anion vacancy (see barium azide, Chapter 11) and if this is further excited by light absorption before decay, then decomposition yields two molecules of carbon dioxide:

$$[C_2O_4]^{2-} \rightarrow 2CO_2 + V_{anion} + 2e$$

where V_{anion} is an anion vacancy which traps the two electrons.

The initial acceleratory stage of the reaction was attributed to the rapid formation of nuclei at surface lattice defects, identified [64] as slip lines. The applicability of the exponential law was explained by growth, through branching, of two-dimensional plate-like nuclei. The important role of dislocations and impurities both in the initial nucleation step and in the subsequent branching process was emphasized. Pre-irradiation of the reactant had a relatively small influence on the rate of subsequent salt decomposition.

The difficulties of interpreting the observations reported for the thermal decomposition of silver oxalate have been discussed [65]. Isothermal data for freshly prepared salt at 399 K were sufficiently irreproducible for different experiments to fit different rate expressions during the acceleratory process, for example, either the power law, or the exponential law. In the second half of reaction (i.e. $\alpha > 0.5$), however, data were more reproducible and results were satisfactorily described by the contracting volume equation.

Leiga [66] supplemented kinetic studies with microscopic examination of partially and fully decomposed salt and also investigated the influence of additives. The conclusions of the earlier studies above were confirmed by the demonstration that surface nucleation yielded compact particles which grew in three dimensions during decomposition or photolysis in vacuum. Comparable behaviour was observed for reaction in air of suitably doped salt. When no suitable additive was present, compact nuclei were not seen, which indicates that air inhibits the development of nuclei.

The kinetics of thermal decomposition (388 to 400 K) of pure silver oxalate and samples of salt containing organic dyes, chosen for their electron donating or electron withdrawing properties [67], were fitted by the power law. The values of n obtained for samples containing those dyes with donor properties were not much influenced by the additive and were generally between 3 and 4, but, in material doped with dyes having acceptor properties, the values of n were larger, up to 5. Reaction may involve the trapping of a mobile exciton at a site adjacent to a crystal impurity or imperfection, probably at a surface, followed by decomposition of the oxalate anion. Electron transfer is facilitated at those active sites occupied by a donor molecule, but is retarded by the presence of acceptors, because these locally raise the energy barrier to the transfer step.

Certain acid dyes [67] stabilize silver oxalate by forming surface compounds, while other dyestuffs accelerate the decomposition because their redox properties enhance the ease of electron transfer from the oxalate ion to the silver. The influences of incorporated cadmium, copper and other ions on the rate of thermal decomposition, and on the concentration and mobility of interstitial silver ions, have been reviewed [46,68].

16.4.7. Copper(II) oxalate

Broadbent *et al.* [69] showed that α-time curves for the decomposition of copper(II) oxalate (503 to 533 K) were sigmoidal and that data for the vacuum reaction fitted the Avrami-Erofeev equation with values of $n = 2.9$ initially and later $n = 3.5$ ($E_a = 136$ kJ mol^{-1}). Electron transfer was identified as the step controlling the reaction. There was no evidence from X-ray diffraction studies for the intervention of the Cu^+ salt : the orthorhombic structure was present until disappearance of the reactant and product copper metal was detected. However, many metal carboxylates, chilled after dehydration, yield anhydrous salts that are amorphous to X-rays or poorly crystalline, see, for example [70].

Mohamed and Galwey [71], in contrast, obtained evidence that copper underwent stepwise reduction when α was less than 0.5, from titration measurements of the copper(II) content ($Cu^{2+} + KI \rightarrow I_2$) in samples of partially decomposed salt. This is consistent with the behaviour found for other copper carboxylates [72]. Moreover, it was shown that the reactivities of both (Cu^{2+} and Cu^+) oxalates were similar and reactions overlapped. In agreement with the previous study [69], α-time curves for the overall decomposition of copper(II) oxalate were sigmoidal and, in this slightly higher temperature interval (515 to 550 K), E_a was increased to 180 ± 7 kJ mol^{-1}.

The thermal decomposition of copper oxalate in N_2 has been studied recently by Coetzee *et al.* [73]. Decomposition started at about 533 K and took place in one stage. The mass of the solid product from the decomposition corresponded to the formation of Cu_2O. Evolved gas analysis showed that both CO_2 and CO are evolved during the decomposition. The proportion of CO in the evolved gas was greatest during the initial period and declined steadily as the reaction proceeded. The DSC results showed that (unlike the oxalates $[M(C_2O_4)(H_2O)_2]_n$) the decomposition of copper oxalate in N_2 is exothermic with an enthalpy change of -33 kJ mol^{-1}.

16.4.8. Zinc oxalate

The decomposition of anhydrous zinc oxalate [74] (573 to 643 K) yields residual oxide:

$$ZnC_2O_4(s) \rightarrow ZnO(s) + CO(g) + CO_2(g)$$

Carbon dioxide is formed in a slight excess (the CO_2/CO ratio is about 1.09), but no zinc metal product was detected. Although there was an acceleratory process ($\alpha <$ 0.4), the isothermal α-time curves were predominantly deceleratory and $E_a = 210$ kJ mol^{-1}. During the acceleratory stage the original nuclei were complemented by the simultaneous random formation of new nuclei, so that kinetic behaviour was dominated by nucleation.

Danforth and Dix [75] compared the behaviours of both zinc and magnesium (see below) oxalates. Kinetic measurements were made by independent determinations of the yields of carbon monoxide and of carbon dioxide at appropriate reaction times. The kinetics of reaction of the zinc salt were fitted by the Prout-Tompkins equation (620 to 645 K) and $E_a = 196$ kJ mol^{-1}. The activation step was identified as the formation of the radical ion $C_2O_4^-$, following the transfer of an electron to a cation. The decomposition of this activated species was then accelerated by the product zinc oxide. The term 'accelerated' rather than 'catalysed' was preferred because the magnitude of E_a was not decreased by the presence of the oxide product.

16.4.9. Magnesium oxalate

Similarities were found between the patterns of behaviour reported for the zinc and magnesium oxalates. The stoichiometry of decomposition of magnesium oxalate (claimed to be accurate to 1%) (623 to 773 K) is [76]:

$$MgC_2O_4(s) \rightarrow MgO(s) + CO(g) + CO_2(g)$$

The initial acceleratory stage ($\alpha < 0.3$) was attributed primarily to the growth of nuclei formed during removal of water from the dihydrate, and it was suggested that similar elementary processes operate during the decay period. Kinetic observations [75] for the isothermal decomposition of magnesium oxalate (603 to 633 K) were represented by the two term rate equation:

$$d\alpha/dt = k_1 (1 - \alpha) + k_2 \alpha (1 - \alpha)$$

The activation energy determined (200 kJ mol^{-1}) was, within experimental error, the same value as that for decomposition of zinc oxalate. A further (slightly lower) value of E_a for MgC_2O_4 decomposition was reported by Kadlec and Danes [55] as 187 kJ mol^{-1}.

16.4.10. Uranyl(VI) oxalate

Dehydration of uranyl oxalate trihydrate [77] proceeded through the monohydrate to the anhydrous salt which then decomposed:

$$UO_2C_2O_4(s) \rightarrow UO_{2+x}(s) + xCO(g) + (2 - x) CO_2(g)$$

The residual oxide had a high surface area. If carbon dioxide was continually removed during reaction the product was stoichiometric UO_2. During decarboxylation in air, the residue is oxidized and contains the higher oxides UO_3 and U_3O_8, to an extent which depends on the availability of oxygen.

The rate of decomposition in vacuum at 575 K was increased [78] by preliminary reactor irradiation and, following such pretreatment, the α-time curves were fitted by the power law with $n = 2$. From the kinetic analysis, it was concluded that there is preferred nucleation along tracks of radiation damage in the crystal followed by cylindrical growth of product nuclei.

The kinetics of decarboxylation [79] of anhydrous uranyl oxalate in argon (576 to 595 K) fitted the Prout-Tompkins equation ($0.05 < \alpha < 0.99$) and $E_a = 260$ kJ mol^{-1}. The reaction in air (598 to 616 K) proceeded to completion in two stages. During the first stage, data again fitted the Prout-Tompkins equation and the second stage was better described by the Avrami-Erofeev equation with $n = 2$. Values of E_a found for both stages were similar to that for the decomposition in argon. The primary product was uranium dioxide, which is rapidly oxidized when air is present. Microscopic observations [78] indicated that reaction is initiated at the points of surface emergence of sub-grain boundaries, then extends across surfaces and later penetrates the particles.

Dollimore *et al.* [80] confirmed these kinetic observations between $0 < \alpha < 0.5$, for decomposition in air, though their value of E_a was slightly lower (240 kJ mol^{-1}). The surface area measured at intervals during reaction, reached a maximum at $\alpha = 0.55$, the point of inflection on the mass loss curve. The product became increasingly activated during the first part of decomposition, whereas during the latter stages sintering predominated. The initial kinetic characteristics were consistent with a growing pore structure and $n = 2$ is in accordance with the development of plates. When α is greater than 0.55, however, the porous structure rapidly collapsed and the subsequent rate was governed by growth in a non-porous structure.

16.4.11. Lanthanide oxalates

Isothermal (593 to 693 K) α-time curves for the decomposition of lanthanum oxalate [81] were sigmoid but asymmetrical with the decay period being the more pronounced. The Prout-Tompkins equation applied in two linear regions with $E_a =$ 132 kJ mol^{-1}. A branching chain mechanism was proposed during which channels from the surface penetrate the crystals as reaction proceeds. The predominant initial product was carbon monoxide, which disproportionated to yield carbon dioxide, and the residual solid contained carbonate and finely divided carbon.

The decompositions of the light lanthanide [81] oxalates (613 to 713 K), were generally similar to that described above for the lanthanum salt. The relative stabilities of these six salts, and also of the normal and basic carbonates, is: Gd > Sm > Nd > La > Pr > Ce. With few exceptions, the intermediates (identified by composition and infrared measurements) were:

$$M_2(C_2O_4)_3(s) \rightarrow M_2(C_2O_4)(CO_3)_2(s) \rightarrow M_2(CO_3)_3(s) \rightarrow M_2O_{0.6}(CO_3)_{2.4}(s)$$

$$\rightarrow M_2O(CO_3)_2(s) \rightarrow M_2O_3(s)$$

The initial gaseous product of decomposition, carbon monoxide, disproportionated. The extent was greatest for cerium(III) oxalate (72%) and least for the gadolinium salt (58%), following almost the reverse order to that given above.

Erbium and lutetium oxalates [81] decomposed after long induction periods required for complete dehydration. The carbonates which then formed were unstable and dissociated to yield oxides. The extents of carbon monoxide disproportionation during breakdown of these reactants were smaller (17% for Er and 6% for Lu).

The decomposition of europium oxalate [82] occurred at a significantly lower temperature (520 K). Results were interpreted as indicating cation reduction to the divalent state. In vacuum, the products of decomposition were europium(II) carbonate and finely-divided carbon. In carbon dioxide at 593 K, europium(II) oxalate is stabilized. In an oxidizing atmosphere, europium is reoxidised and Eu_2O_3 is formed at 663 K. A reaction mechanism was proposed in which carbon monoxide reacted with the oxalate ion to form carbon dioxide and $C_2O_3^{2-}$. The reactions of ytterbium oxalate were similar.

The intermediate formation of europium(II) carbonate has been questioned by Gallagher *et al.* [83]. From Mössbauer spectra and other data, the alternative participant, $Eu_2O_2OCOCO_2$, was recognised, which loses carbon monoxide to form the dioxycarbonate $Eu_2O_2CO_3$. Carbon dioxide evolution then yields Eu_2O_3.

The behaviour of yttrium oxalate (573 K) was [82] generally similar to that of the lanthanide oxalates. No carbonate could be detected, suggesting that the carbonate is less stable than the oxalate. Carbon monoxide disproportionation proceeded to about 60%, accompanied by carbon deposition.

16.4.12. Lithium oxalate

Decomposition of lithium oxalate [84] (742 to 765 K) yielded the carbonate and was accompanied by an initial large increase in surface area, followed by extensive sintering. The α-time curve was sigmoid with no induction period and $E_a = 223$ kJ

mol^{-1}. An initial acceleratory process was followed by an approximately constant rate of reaction ($0.1 < \alpha < 0.8$) ascribed to a phase boundary mechanism. A quantitative model is given to account for the variation in area with α.

Girgis and El-Awad [85] reported an isothermal study of the decomposition of $Li_2C_2O_4$ between 829 and 858 K, using thermogravimetry. The sigmoid curves were fitted by the Avrami-Erofeev equation with $n = 2$, $E_a = 269$ kJ mol^{-1} and $\ln(A/s^{-1}) = 31.8$. Some ferrospinel additives, $Cd_{1-x}Co_xFe_2O_4$ (x = 0, 0.5 and 1.0), were crushed with the reactant and their activities as catalysts for the decomposition were in the above sequence. Values of E_a were 264, 245 and 213 kJ mol^{-1}, and of ln (A/s^{-1}) were 27.3, 28.6 and 23.9, respectively. This promotional activity is a consequence of the appreciable solubility of the additives in the host oxalate crystal. The decomposition mechanism involves the sequence of steps:

$$C_2O_4^{2-} \rightleftarrows 2CO_2^- \rightarrow [OCOCO_2]^{2-} \rightarrow CO_3^{2-} + CO$$

The additives cause lattice defects or structural deformation and may participate in electron transfer.

16.4.13. Other oxalates

A comparative DTA study [86] of some 25 oxalates in N_2 and in O_2 has been published. The onset temperatures of dehydration and decomposition reactions are given and the influence of atmosphere upon the temperatures and products of decomposition is discussed. Variations in the surface areas of the solid products from decomposition of a number of oxalates have been reported by Dollimore and Nicholson [87].

The kinetics of reactions of calcium, strontium and barium oxalates have been studied in less detail than might have been anticipated, particularly in view of the extensive use of calcium oxalate monohydrate as a standard in thermal analysis.

Anhydrous calcium oxalate decomposes at 745 K to yield calcite [88]. Freeberg *et al.* [89] reported $E_a = 284$ kJ mol^{-1} for the first-order reaction:

$$CaC_2O_4(s) \rightarrow CaCO_3(s) + CO(g)$$

Bose *et al.* [90] found that the thermal decomposition of calcium oxalate ($\rightarrow CaCO_3 + CO$ or $CO_2 + C$) between 693 and 733 K was well described by the contracting area equation, but with two linear ranges that together applied for α between 0.065 and 0.98. Values of E_a for the two stages were 245 and 128 kJ mol^{-1}, respectively. The reaction was catalyzed by $La_2O_3 > Sm_2O_3 > Gd_2O_3$, rate constants being almost doubled and the apparent values of E_a were appreciably diminished. It is concluded that reaction proceeds as follows:

$$C_2O_4^{2-} \rightarrow CO + O^{2-} + O = C^+ - O^- \rightarrow CO + CO_3^{2-}$$

The thermal decomposition of $BaC_2O_4.0.5H_2O$ in N_2 is reported [89,91,92] to take place in three major stages:

(i) $BaC_2O_4.0.5H_2O(s) \rightarrow BaC_2O_4(s) + 0.5H_2O(g)$ (323 to 383 K)

(ii) $BaC_2O_4(s) \rightarrow BaCO_3(s) + CO(g)$ (603 to 723 K)

(iii) $BaCO_3(s) \rightarrow BaO(s) + CO_2(g)$

The temperature range for the dehydration is significantly lower than those found for the other monooxalates. This might be due to the weaker electrostatic field generated by the barium ion. The decomposition was initially exothermic but later became strongly endothermic. Koga and Tanaka [93] have given a value of 41.8 kJ mol^{-1} for the enthalpy of decomposition.

Stage (ii) is accompanied by the formation of carbon, via the Boudouard reaction: $2CO(g) = CO_2(g) + C(s)$. Between 950 and 1500 K, the residual carbon reacts with the $BaCO_3$ to form BaO and 2CO. The onset temperature for the carbonate decomposition was found to increase with the CO_2 partial pressure. In N_2, the carbonate decomposition started at 976 K, whereas in pure CO_2 it started at 1621 K.

The zero-order rate equation gave the best fit [93] for the isothermal decomposition of BaC_2O_4 to $BaCO_3$ in the temperature range, 666 to 694 K (E_a = 181 kJ mol^{-1}). The authors also studied the decompositions of CaC_2O_4 and SrC_2O_4 and reported that the Avrami-Erofeev equation (n = 1.5) gave the best fit for the decomposition of CaC_2O_4, and the contracting geometry equation (n = 1.54) gave the best fit for the decomposition of SrC_2O_4.

Thorium oxalate yielded [94] the carbonate below 593 K, but reactions at higher temperature yielded ThO_2 directly. The rate controlling process changed across the temperature interval studied (573 to 623 K). The decomposition of cadmium oxalate at 620 K may be represented [95] as:

$$CdC_2O_4(s) \rightarrow CdO(s) + CO(g) + CO_2(g)$$

but, at higher temperatures, the residue contained metal and the lower oxide (E_a = 144 kJ mol^{-1}).

The decomposition of mercury(II) oxalate [96] yields the metal, which is volatile under reaction conditions, mercury(II) oxide and both CO and CO_2. α-time curves were sigmoid (453 to 478 K). The reaction initially spread rapidly over the surfaces of crystallites (E_a = 107 kJ mol^{-1}) from a number of reaction centres. Subsequently the rate was controlled by the inward advance of a reaction interface (E_a = 155 kJ mol^{-1}). The reaction rate was not markedly influenced by the presence of gaseous or solid products. The initial process is accelerated by radiation, which is believed to yield the mercury(I) salt following electron transfer.

The stoichiometry of decomposition of lead oxalate [97] in air is analogous to that given for cadmium above. The sigmoid α-time curves are acceleratory to $\alpha = 0.6$ and the reported activation energy (142 kJ mol^{-1}) agrees closely with earlier values [98,99]. Bircumshaw and Harris [98] concluded that reaction proceeded by a branching chain mechanism. Yankwich and Copeland [100] discussed the relative yields of CO_2/CO in the gaseous products, with reference to the competing decompositions by oxygen abstraction and C-C bond rupture. The shapes of the α-time curves for isothermal decomposition of lead oxalate [68] vary with ageing, from a deceleratory process characteristic of freshly prepared salt, to the sigmoid curve found for the reactant after storage for several days. The change is attributed to variations in the number and distribution of dislocations, through mutual annihilation of line defects, or grouping through polygonization.

Selcuk and Price [101] reported that the thermal decomposition of PbC_2O_4 in N_2 between 611 and 707 K, studied by isothermal thermogravimetry, fitted the Avrami-Erofeev equation with n = 2, ascribed to random nucleation followed by two dimensional growth, $E_a = 119 \pm 7$ kJ mol^{-1}. Incorporation of 1% Zn^{2+} was found to increase E_a to 137 ± 5 kJ mol^{-1} but did not change the kinetic fit. It is suggested that zinc oxalate within the host crystal restricts the formation of Smekal cracks along which decomposition is favoured.

16.4.14. Mixed metal oxalates

The thermal decompositions of mixed magnesium-nickel oxalates [77] commenced at sites of incorporation of cations of the additive into the oxalate lattice, regardless of the relative stabilities of the oxalates concerned. A further study compared the behaviour of magnesium-cobalt and magnesium-nickel oxalates [55].

The thermal behaviour of the mixed metal copper oxalates, $MCu(C_2O_4)_2.3H_2O$, (M = Fe, Co, and Ni) [73,102] differs from that of the individual metal oxalates and of copper(II) oxalate. The dehydrations were single-stage deceleratory processes, as with the oxalates $[M(C_2O_4)(H_2O)_2]_n$. The onset temperatures for the dehydrations, from DSC experiments, increased in the following order: $FeCu(C_2O_4)_2.3H_2O$ (412 K), $CoCu(C_2O_4)_2.3H_2O$ (414 K), and $NiCu(C_2O_4)_2.3.5H_2O$ (475 ±10 K).

The decompositions of the mixed oxalates, $MCu(C_2O_4)_2.3H_2O$, in N_2, [73,102] were endothermic processes and took place in two overlapping stages, in contrast to the exothermic decomposition characteristic of copper(II) oxalate. The mass of the residue from the decomposition of $FeCu(C_2O_4)_2.3H_2O$ corresponded to the formation of FeO and Cu_2O by loss of CO and CO_2. The masses of the decomposition products of $CoCu(C_2O_4)_2.3H_2O$ and $NiCu(C_2O_4)_2.3H_2O$ were closest to the formation of metal products, Co and Cu, and Ni and Cu, respectively, with the evolution of four moles of CO_2 per mole of reactant. The onset temperatures for decomposition, from DSC experiments, increased in the order: $CoCu(C_2O_4)_2$ (618 K), $FeCu(C_2O_4)_2$ (624 K), $NiCu(C_2O_4)_2$ (635 K).

The isothermal kinetics of decomposition were complex, with at least two overlapping processes taking place. The shapes of the peaks indicated that both processes were initially acceleratory, and then deceleratory. The isothermal rate was assumed to be made up of weighted contributions from individual processes which could be described by the Avrami-Erofeev equation, with various values of n.

The thermal decompositions of nickel(II)-cobalt(II) oxalate solid solutions were studied using TG and TM [103]. A series of the mixed binary Ni(II)-Co(II) oxalate samples was prepared at 25% (atom) intervals across the system. Physical mixtures were also prepared by mixing the pure end members. The DTG and DTM curves showed that the decomposition proceeds to completion in two overlapping stages. The kinetics of the individual steps were not studied. From the DTG curves, the authors stated that the physical mixtures behaved as individual oxalates, while the coprecipitate decomposed as a single entity. The TM curves showed that the products formed from the physical mixture and the coprecipitate were distinctly different. The magnetic behaviour of the product from the coprecipitate was consistent with the behaviour predicted for a Ni-Co alloy, but the products from the physically mixed oxalate do not show the transition temperature predicted for an alloy. The kinetics of decomposition of iron-nickel mixed oxalates have been studied by Dorémieux *et al.* [104].

During the decomposition of Ni-Co mixed oxalates the production of nickel metal accelerates reaction, because the nickel constituent reacts about 60 K lower than cobalt oxalate (Sections 16.4.2. and 16.4.4. above). The significance of the relative yields of metal and oxide in the decompositions of the mixed oxalates Co-Ni; Co-Mg and Ni-Mg has been discussed [99].

The thermal decomposition of barium titanyl oxalate tetrahydrate, $BaTiO(C_2O_4)_2.4H_2O$, occurs in three stages [105]: (i) dehydration, (ii) decomposition of the anhydrous oxalate to the carbonate, and (iii) decomposition of the carbonate forming barium titanate. Isothermal α-time curves for stage (ii), 509 to 599 K in vacuum, derived from separate measurements of pressures of evolved CO and CO_2, were deceleratory and superimposable up to $\alpha = 0.3$. CO evolution was slower beyond $\alpha = 0.3$ and a diffusion mechanism was proposed, $E_a = 189$ kJ mol^{-1}.

The thermal behaviour of $Ba[Cu(C_2O_4)_2(H_2O)].5H_2O$ in N_2 has been examined [106,107] using TG and DSC. The dehydration starts at relatively low temperatures (353 K), but continues until the onset of the decomposition (553 K). The decomposition takes place in two major stages. The mass of the intermediate after the first stage corresponded to the formation of barium oxalate and copper metal and, after the second stage, to the formation of barium carbonate and copper metal. The dehydration enthalpy was 311 ± 30 kJ mol^{-1}, and the overall enthalpy change for the decomposition of $Ba[Cu(C_2O_4)_2]$ in N_2 was -347 kJ mol^{-1}.

The kinetics of the thermal dehydration and decomposition [106] were studied using isothermal TG. The dehydration was strongly deceleratory and the α-time curves could be described by the three-dimensional diffusion model. The values of the activation energy and the pre-exponential factor for the dehydration were 125 ± 4 kJ mol^{-1} and (1.38 ± 0.08) x 10^{15} min^{-1}, respectively. The decomposition was complex, and was analysed in terms of two overlapping deceleratory processes. The faster process could be described by an equation of the form of the contracting geometry model, but with $n = 5$ (for which no physical model is available, see Table 3.3.). The slow process could also be described by the contracting geometry model, but with $n = 2$. The values of E_a and A were 206 ± 23 kJ mol^{-1} and (2.2 ± 0.5) x 10^{19} min^{-1} for the fast process, and 259 ± 37 kJ mol^{-1} and (6.3 ± 1.8) x 10^{23} min^{-1} for the slow process.

Because the decompositions of the mixed metal oxalates $MCu(C_2O_4)_2$ and $Ba[Cu(C_2O_4)_2]$ could be described, at high values of α, by the contracting geometry equation, the Arrhenius parameters for the later stages of all these decompositions are comparable. Arrhenius plots for the second stage of the decompositions of the mixed metal oxalates, $MCu(C_2O_4)_2$, gave activation energies of 155 ± 40, 101 ±16 and 91 ± 2 kJ mol^{-1} for M = Fe, Co and Ni. The comparable deceleratory process for the decomposition of $Ba[Cu(C_2O_4)_2]$ gave an activation energy of 259 ± 37 kJ mol^{-1}. Here the removal of the bulkier gaseous products of the decomposition (CO_2 and CO) is suggested [106] to occur through channels made by the prior dehydration, and that the kinetic behaviour is on the borderline between control by geometrical progression of the reactant/product interface and control by diffusion.

16.5. METAL MELLITATES

The decompositions in vacuum of the mellitates [72,108-110] of metals in the first transition period (Cr^{3+}, Mn^{2+}, Fe^{3+} (basic), Co^{2+}, Ni^{2+}, Cu^{2+} and Zn^{2+}) were selected as being of interest to allow comparison of the reactivities of the mellitate anion ($C_{12}O_{12}^{6-}$) with the oxalate ion. Both anions contain no hydrogen but differ in structure and proportion of constituent carbon. In general the α-time curves for decompositions of mellitates were deceleratory, but comparisons of kinetics were complicated by the occurrence of two distinct rate processes during the decompositions of the Cr^{3+}, Mn^{2+} and Fe^{3+} salts. A plot of measured values of activation energies reported for the decomposition of oxalates and mellitates, against enthalpy of formation of the appropriate oxide, contained two approximately linear relations [108], see Figure 16.2. One was for those reactions which yield a metallic product and the other for reactions in which the residual (oxides) material is not catalytically active. The rate limiting steps in the decompositions of both series of salts were identified as the ruptures of two carboxyl group-to-metal, or to-cation bonds, respectively.

There was an increase in the rate of decomposition of cobalt-nickel mixed mellitates [111] with increase in the proportion of nickel present. Values of E_a measured during the early stages of reaction were close to that characteristic of nickel mellitate (184 kJ mol^{-1}), but in the later period of reaction increased towards the value found for the cobalt salt (250 kJ mol^{-1}). No discontinuous change in kinetic behaviour was apparent and it was concluded that reaction proceeded in a solid solution.

16.6. METAL SQUARATES

16.6.1. Introduction

The decompositions of selected transition metal squarates (MC_4O_4) were identified as being of interest because they contain a further organic anion that, like the oxalates and mellitates, contains no hydrogen, so decreasing the complexity of product mixtures and secondary reactions.

16.6.2. Nickel squarate

The decomposition kinetics of nickel squarate [112] were difficult to measure reliably due to the sensitivity to salt dispersion within the reaction vessel. This behaviour was a consequence of the overlap and interaction of the reversible salt dehydration and irreversible decomposition steps in the same temperature interval:

$NiC_4O_4.2H_2O(s) \rightleftarrows NiC_4O_4(s) + 2H_2O(g)$

$NiC_4O_4(s) \rightarrow Ni(s) + 4CO(g)$

The sigmoid α-time curves for decompositions between 490 and 540 K were well fitted by the Avrami-Erofeev equation, with $n = 2$, between $0.03 < \alpha < 0.95$, and the value of E_a (163 ± 10 kJ mol^{-1}) was found by extrapolation to zero reactant mass to eliminate the contribution from the reversible dehydration step. Scanning electron

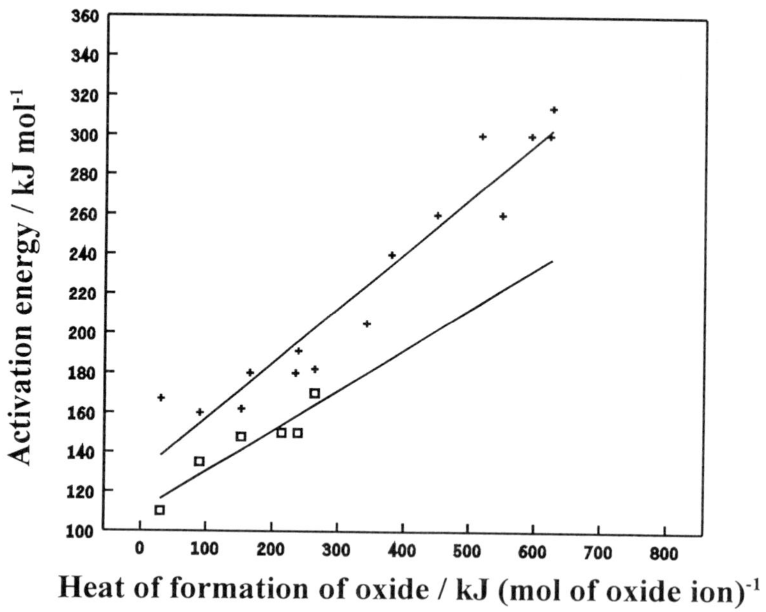

Figure 16.2.
Correlation between selected values of E_a for decompositions of oxalates and mellitates with the enthalpies of oxide formation [108] (some mean values of E_a are included). The magnitudes of E_a thus appear to be controlled by the strengths of the M-O bonds. Two trends may be discerned. Reactions where breakdown of the oxalate anion are believed to be catalytically promoted by the metal product, have slightly lower E_a values (□), compared to reactions in which the residual product does not promote decomposition, including the mellitates where carbon deposition inhibits catalysis (+).

microscopic observations gave evidence of rectilinear crack development (see Figure 16.3.) in the approximately cube-shaped reactant particles (10 - 20 μm edge) but the solid product could not be recognised. At higher magnifications, using replication techniques and a transmission electron microscope, product nuclei were identified as assemblages of small nickel particles, less than 50 nm in diameter, so reaction involves nucleation and growth. Anion breakdown on metal surfaces may proceed through the intervention of nickel carbonyl that is unstable and decomposes at reaction temperature.

16.6.3. Copper(II) squarate

The decomposition of copper(II) squarate, as with the copper(II) carboxylates, proceeded to completion in two distinct rate processes with stepwise cation reduction [113]. The first step ($\alpha < 0.5$) fitted zero-order kinetics with $E_a = 150 \pm 15$ kJ mol^{-1} between 530 and 590 K. The second step was approximately first-order with an increase in E_a to 210 ± 20 kJ mol^{-1} and reaction temperature, 590 to 670 K. No reaction interface could be identified in scanning electron microscopic studies.

Silver squarate decomposed [114] between 473 and 510 K without melting, by a predominantly deceleratory reaction with $E_a = 190 \pm 8$ kJ mol^{-1}.

16.7. METAL MALONATES, FUMARATES AND MALEATES

16.7.1. Cobalt(II) malonate

Dehydrated cobalt(II) malonate decomposed in vacuum (540 to 600 K) [115] to yield carbon dioxide as the predominant volatile product, although when α was greater than 0.5 a significant quantity of carbon monoxide (about 15%) was produced. Three phases were identified in the residual material (Co, CoO and Co_2O_3) but these were poorly crystallized and X-ray diffraction data were imprecise. Kinetic measurements identified an initial zero-order reaction ($0.06 < \alpha < 0.3$, $E_a = 179$ kJ mol^{-1}). There was a marked decrease in rate (x 0.2) at α from 0.35 to 0.4, followed by a second period of zero-order behaviour ($0.45 < \alpha < 0.72$, $E_a = 183$ kJ mol^{-1}). The reaction rates were sensitive to the conditions of removal of product gases. From electron microscopic examination of partially and fully decomposed salt, it was concluded that decomposition occurred at a salt-product interface, through the growth of poorly-crystallized nuclei initially established at internal surfaces. Such nuclei contain (at least) three phases and have a large area, but the structure is disordered and the mixture exhibits catalytic properties. Some similarities with the behaviour of the nickel salt [116] were pointed out.

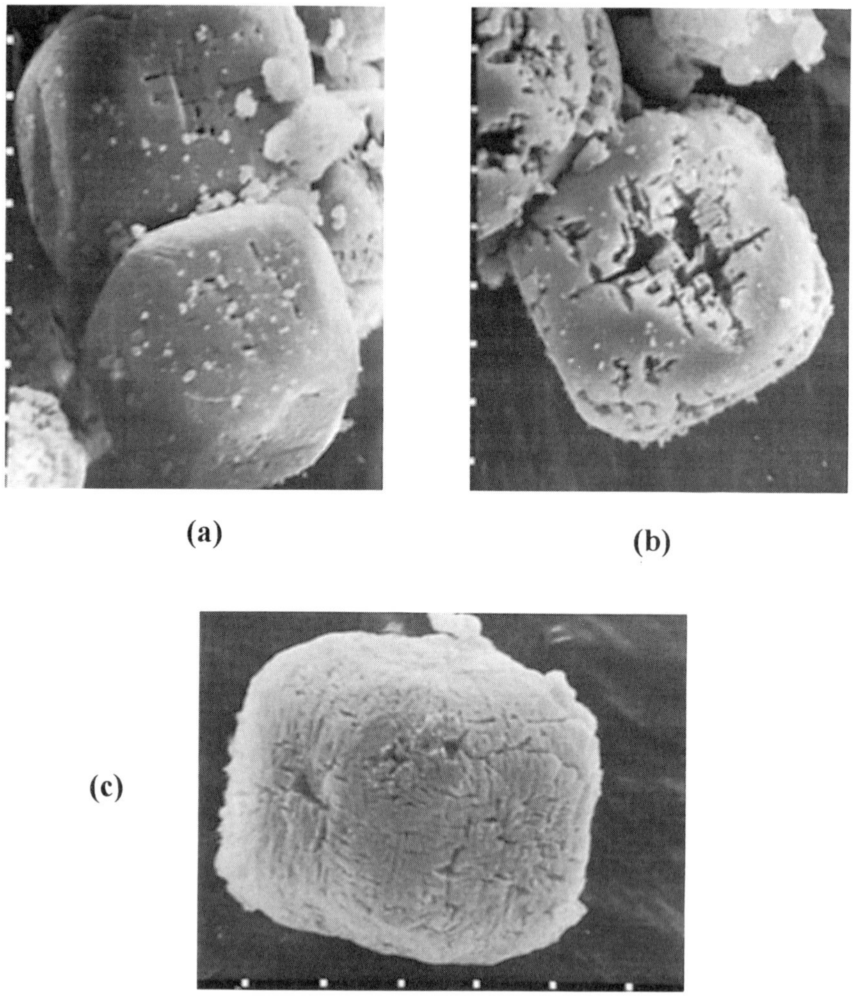

Figure 16.3.
Scanning electron micrographs of nickel squarate crystallites [112]. (Scale bars are 3.0 μm.) (a) During the early stages of reaction, holes appear in the cube faces and cracks develop parallel to the edges. (b) As decomposition proceeds, the holes widen and extend. (c) On completion of decomposition, the residual particles are pseudomorphic with the reactant crystallites, but are extensively cracked. (Reproduced with permission of the Royal Society of Chemistry, London.)

16.7.2. Nickel malonate, fumarate and maleate

The predominant products of decomposition of nickel malonate [116] are carbon dioxide and nickel carbide, though smaller yields of several other compounds (CO, CH_4 and acetic acid) were detected. From kinetic studies of the reaction in vacuum, 543 to 613 K, it was shown that reaction consisted of an initial deceleratory process to $\alpha = 0.12$, followed by a short acceleratory period (Avrami-Erofeev equation, $n = 3$) after which the rate was constant between $0.2 < \alpha < 0.9$ and $E_a = 179$ kJ mol^{-1}. Examination of partially decomposed salt by optical microscopy suggested that nuclei were formed as parallel lamellae which traversed the thickness of the hexagonal platelets of reactant. The acceleratory stage of the main reaction is ascribed to the early spherical growth of nuclei, but overlap yielded parallel laminae of product. Advance of the approximately planar surfaces of such assemblages of metal into the undecomposed reactant sustained a constant rate of product evolution until neighbouring systems impinged on each other. There was evidence from microscopy that the lamellae which provided sites for nucleation at internal surfaces are developed during dehydration. The anhydrous pseudomorph of the prepared dihydrate was almost amorphous to X-rays.

Le Van *et al.* [117] concluded, from X-ray studies of the decomposition of nickel malonate in an inert atmosphere, that finely divided nickel was the first formed residual product and this was subsequently carbonized by carbon monoxide. The shapes of the α-time curves for decomposition in vacuum and in the presence of water vapour [116] or oxygen [118] were closely comparable, although the values of E_a (and the reaction products) were significantly changed (179, 137 and 157 kJ mol^{-1}, respectively for the reactions mentioned).

A more recent study [119] confirmed that the sigmoid α-time curves characteristic of the decomposition of nickel malonate [116] were preceded by an initial deceleratory process, completed at $\alpha = 0.12$. The rate was constant between $0.3 < \alpha < 0.6$ and $E_a = 176 \pm 10$ kJ mol^{-1} (543 to 613 K). During the main reaction ($0.2 < \alpha < 0.6$), however, it was shown that about 20% acetate was present with the reactant and this proportion diminished during the final stages of decomposition. Electron microscopic observations gave no evidence of melting but otherwise provided little information of value in formulating the reaction mechanism. A parallel study [30] showed that the rates of decompositions of malonate and acetate were similar within this temperature interval. The main reaction of $CH_2(COO)_2Ni$ was reinterpreted and is now identified as the concurrent decompositions of both malonate and acetate (formed as an intermediate) anions on a nickel carbide product (perhaps also containing nickel metal) surface that exhibits some catalytic activity. This is a nucleation and growth process.

The decomposition of nickel fumarate (573 to 613 K, $E_a = 208 \pm 8$ kJ mol^{-1}) [120] gave sigmoid α-time curves but with a short initial reaction ($\alpha = 0.02$). There was no evidence of reactant melting. It was concluded that anion breakdown was promoted by the nickel carbide product, in a nucleation and growth reaction.

Nickel maleate decomposed [121] in a slightly lower temperature range (543 to 583 K, $E_a = 188 \pm 12$ kJ mol^{-1}) by a deceleratory reaction. It was concluded that the reaction rate was controlled by maintained nucleation involving the continual formation of nickel particles, the growth of which was inhibited by the deposition of carbonaceous residues.

A comparative study of the thermal reactions in oxygen of the above three reactants [118], together with nickel formate, showed that the shapes of the α-time curves were comparable with those for vacuum decompositions. The magnitudes of the Arrhenius parameters (with the exception of the decomposition of the formate which does not give carbide product) were diminished. It was concluded that the initiation of reaction was unaffected by the presence of oxygen, so that the geometry of each rate process was unchanged. In the presence of O_2, interfacial reactions on the nickel oxide product ($\rightarrow CO_2 + H_2O$) proceeded appreciably more rapidly by a common mechanism and all values of E_a were close to 150 kJ mol^{-1}.

16.7.3. Copper(II) malonate, fumarate and maleate

α-time curves for the thermal decomposition of copper(II) malonate [122,123] were composed of two recognizably distinct rate processes. The first reaction was acceleratory to almost $\alpha = 0.5$ after which there was a rapid, almost discontinuous, diminution in rate and the second slower process was deceleratory to $\alpha = 1.00$. Chemical analyses of reactant samples, partially decomposed to a range of different, known, α values showed that acetate was formed as an intermediate during the first reaction. From NMR data it was concluded that the amount of [CH_3COO^-] present increased linearly with α to $\alpha = 0.5$ and decreased rapidly thereafter so that only small amounts remained when α was greater than 0.7. The quantities of Cu^{2+} present in partially decomposed samples of reactant were measured by a titration method and it was shown that amounts progressively diminished as α increased, becoming negligible when α was greater than 0.5. Electron microscopic studies, again using reactant samples decomposed to known α values, showed that, during this reaction, the individualities of reactant crystallites were preserved to give product particles that were pseudomorphous with those of the reactant. There was some surface distortion with the development of bubble-like protuberances. Internal surfaces, exposed by gentle crushing of reacted material, revealed the progressive development of a froth-like texture as reaction proceeded to $\alpha = 0.5$. The rounded surfaces of all the (cooled) solid interbubble material is indicative of textural control by surface tension forces, identified as generated by gas evolution resulting from anion breakdown ($\rightarrow CO_2$) in a liquid system.

These observations are satisfactorily explained by a mechanism in which salt decomposition is completed through two consecutive reactions during which there is stepwise cation reduction $Cu^{2+} \rightarrow Cu^{+} \rightarrow Cu^{o}$. The appearance of a liquid phase during acetate formation, when α is less than 0.5 and $Cu^{2+} \rightarrow Cu^{+}$, accounts for the first acceleratory rate process which fits the autocatalytic exponential law ($d\alpha/dt = k\alpha$) with $E_a = 200 \pm 15$ kJ mol^{-1} between 468 and 505 K. Reaction occurs in an intracrystalline fused material but the maintenance of an outer unreactive surface prevents particle coalescence or comprehensive fusion. The product of this rate process, believed to be predominantly copper(I) malonate, then decomposes by a first order reaction for which $E_a = 188 \pm 6$ kJ mol^{-1} between 477 and 528 K. This probably occurs in the solid state to give a residue identified as a dispersion of copper metal crystallites on a coherent carbon substrate. Structures and properties, including the catalytic activity of this material, have been reported [124].

The thermal decompositions of copper(II) fumarate and of copper(II) maleate [125] showed some important chemical similarities with the reaction of the malonate. The most notable common feature was that the Cu^{2+} content in both salts diminished to 5% of the original value when $\alpha = 0.5$, so that all three decompositions proceed with stepwise cation reduction : $Cu^{2+} \rightarrow Cu^{+} \rightarrow Cu^{o}$. The first reaction in the copper(II) maleate decomposition was accompanied by melting and α-time values fitted the Prout-Tompkins equation with $E_a = 225 \pm 6$ kJ mol^{-1} between 473 and 508 K. During these reactions the maleate anion isomerized to the fumarate and the extent of the second, deceleratory, rate process ($E_a = 139 \pm 15$ kJ mol^{-1}) decreased as the reaction temperature was increased in the range 509 to 528 K.

Textural changes observed during the decomposition of copper(II) fumarate [125] were less extensive and no positive evidence of melting was obtained. No anion other than fumarate was detected during this reaction. Both rate processes in the overall decomposition were deceleratory and isothermal data were well expressed by the first-order equation during both stages. For the first reaction $E_a = 167 \pm 2$ kJ mol^{-1} ($0.12 < \alpha < 0.40$) between 533 and 597 K and for the subsequent slower reaction $E_a = 205 \pm 5$ kJ mol^{-1} ($0.54 < \alpha < 0.96$) between 576 and 620 K. The fumarate is significantly more stable to decarboxylation than the maleate.

16.7.4. Silver malonate, fumarate and maleate

The thermal decomposition of silver malonate [126,127] shows points of similarity with the decomposition of copper(II) malonate described above [123] and it is useful to compare and contrast the behaviour of these two reactants. Both kinetic studies were completed in comparable temperature intervals (silver : 453 to 485 K and copper : 468 to 505 K and 477 to 528 K). Sigmoid α-time curves for the decomposition of $CH_2(COOAg)_2$ fitted the Avrami-Erofeev equation, $n = 2$, and $E_a = 170 \pm 10$ kJ mol^{-1}. This was identified as a nucleation and growth reaction. Microscopic evidence confirmed the appearance of solid product nuclei composed of finely divided silver particles, diameters 0.04 to 0.16 μm. In contrast with the textural observations for the copper salt, there was no evidence of melting and it appeared that acetic acid (or ketene) was volatilized from the residual reacting phase. Later work [128] showed that γ-irradiation increased the rate of silver malonate reaction, in direct proportion to the radiation dose. This is ascribed to an increase in the number of nuclei participating in salt breakdown.

The contrasting reaction mechanisms identified in these comparative studies [123,126] of two chemically similar compounds, decomposing in comparable temperature intervals, is surprising and exemplifies the difficulties inherent in the formulation of reaction mechanisms for these rate processes. Silver metal evidently promotes anion breakdown and this is the more electropositive element. This reaction proceeds in a marginally lower temperature interval than the decomposition of the copper salt where there is no metal formed in the first step and reaction may be promoted by relaxation of the intracrystalline forces of the solid on melting.

Mohamed *et al.* [129] have reported non-isothermal studies of the decompositions of silver maleate and fumarate. Following some initial sublimation, both salts reacted to form maleic anhydride and CO_2 as dominant products, together with other organic compounds. The decomposition of silver maleate commenced at 443 K and the anion isomerized to fumarate at 518 K which decomposed at 557 K. The main decomposition of silver fumarate occurred at 566 K.

16.7.5. Manganese(II) malonate

The decomposition of the dihydrate ($C_3H_2O_4Mn.2H_2O$) yielded CO_2, acetone and traces of acetic acid as the volatile carbon-containing products, and MnO in N_2 or H_2 atmospheres, or Mn_3O_4 as the solid product from reaction in air [130]. Using infrared analyses, manganese acetate or the basic salt $(CH_3COO)_2MnOH$ were identified as intermediates. In this respect the manganese(II) salt resembles the other malonates studied here and, again, proton transfer is identified as an important step in anion breakdown.

16.7.6. Calcium malonate, fumarate and maleate

These reactions were studied to compare the observations above with reactions in which there is anion breakdown, but where cation reduction is not thermodynamically feasible. Calcium malonate, in common with the other malonates discussed above, forms [131] acetate intermediate during salt decomposition (612 to 653 K). The value of E_a was large, 456 kJ mol^{-1}. There was no evidence of melting. Following an initial brief acceleratory stage, completed at $\alpha = 0.2$, the reaction was predominantly deceleratory. Decomposition was proposed to involve proton transfer and the monovalent acetate ions formed to interrupt, and probably decrease the stability of, extended chains of bonded divalent anion to cation sequences. Calcium fumarate and calcium maleate [132] were significantly more stable (protons on unsaturated carbon only) and kinetic studies were undertaken above 733 K. E_a values were 195 to 215 kJ mol^{-1} for the fumarate and 340 to 430 kJ mol^{-1} for the maleate, for both deceleratory reactions. Electron microscopic observations gave no evidence that was of value in the formulation of the reaction mechanism. Anion breakdown in all three salts of the strongly electropositive cation occurred in a significantly higher temperature range than for the transition metal salts that yielded metal product. The retention of a large proportion of the volatile product of carboxylate breakdown (CO_2) in these reactions as $CaCO_3$ restricts the value of kinetic studies based on gas evolution.

16.8. OTHER METAL CARBOXYLATES

16.8.1. Other nickel carboxylates

Leicester and Redman [27] noted that nickel (and also cobalt) butyrates and caprylates decompose in the absence of air at 553 K, with liquefaction, to yield the corresponding ketone and a residue which is rich in metal. There was no systematic change in shape of the pyrolysis curve with chain length for nickel salts between propionate and stearate. Le Van and Périnet [133] studied the decomposition of nickel salts of C_1 to C_5 fatty acids and concluded that nickel oxide is the initial product, which is then reduced to the metal and subsequently carbided. (The differences in behaviour for the same group of copper and cobalt salts are also discussed). Nickel benzoate [134] melted before decomposition ($E_a = 188$ kJ mol^{-1}). The nickel salt of cyclohexane carboxylic acid did not melt, but the marked irreproducibility of kinetic behaviour was attributed to secondary interactions between acid and anhydride, which constituted the major volatile products, and the metallic product which apparently catalysed dehydrogenation, yielding benzene in the evolved gases. Studies of these reactions have been largely qualitative. The rates and relative product yields are sensitive to the ease of contact between the evolved gases and the surfaces of the catalytically active residual metal.

The decompositions in vacuum of nickel phthalate [135] (566 to 663 K) and of nickel mellitate [109] (591 to 668 K) were deceleratory throughout and proceeded, without melting, in non-crystalline matrices to yield nickel metal and carbon dioxide as the most abundant products (the phthalate also gave benzoic acid and other volatile compounds). The values of E_a were close to that of the benzoate (196 and 190 kJ mol^{-1}, respectively). The decomposition of nickel terephthalate [136] proceeded in a higher temperature range (638 to 703 K) and, again, the solid residue was nickel metal. This reaction was predominantly deceleratory, but with some evidence for the occurrence of an initial acceleratory process, and $E_a = 224$ kJ mol^{-1}. The deceleratory (second-order) reactions of the phthalate and mellitate are suggested to be rate controlled by the necessity to maintain nucleation. The growth of the particles of metal product is inhibited by the surface decomposition of aromatic nuclei which are immobile and stable after formation at these temperatures. Nickel terephthalate decomposes in a higher temperature range, which is comparable with that for carbide decomposition, and accordingly carbon possesses some mobility in the product. Some growth of nuclei is, therefore, possible and this explains the initial acceleratory behaviour and the less deceleratory reaction.

16.8.2. Lead citrate

The thermal decomposition of lead citrate [137] in vacuum, proceeds to completion in two stages. The kinetic behaviour is, however, influenced by which of the two hydrates was used to prepare the anhydrous reactant. The first stage in the decomposition of the anhydrous salt prepared from the tetrahydrate was a deceleratory process which was well represented by the first-order equation and was completed at $\alpha = 0.3$ ($E_a = 52$ kJ mol^{-1}, 488 to 509 K). The subsequent reaction, studied in the temperature interval 553 to 578 K, fitted the contracting volume equation and $E_a = 265$ kJ mol^{-1}. Reactant prepared by removal of water from the dihydrate also decomposed in two stages. The initial rate process ($\alpha < 0.3$) fitted the Avrami-Erofeev equation with n between 2 and 3 ($E_a = 150$ kJ mol^{-1}, 528 to 558 K). This behaviour contrasts with that of the other preparation, but again the later stages were satisfactorily expressed by the contracting volume equation ($E_a = 210$ kJ mol^{-1}, 578 to 598 K). The initial sigmoid curve, characteristic of the reaction of the more perfect crystallites obtained on dehydration of the dihydrate was shown by electron microscopy to correspond with the growth of spherical particles of lead. This reaction was restricted to crystallite surfaces and growth of each nucleus ceased on attaining a metal particle diameter of about 30 nm. Completion of the first stage of reaction is attributed to the exhaustion of all sites capable of reacting to form a nucleus and completion of the growth of all such nuclei. These product particles are identified as providing the nucleation sites for the second stage of reaction and the reactant/product interface advances into crystallites according to the contracting volume model.

The thermal decompositions of iron(II) carboxylates have been studied together with investigations of the electrical properties of the finely divided oxide product (γ - Fe_2O_3) which is potentially of commercial value in magnetic recording media [138].

16.8.3. Metal styphnates (salts of 2, 4, 6 - trinitroresorcinol)

Barium styphnate can be prepared either as a monohydrate or a trihydrate and the kinetics of the dehydrations and the decompositions of both forms have been studied by Tompkins and Young [139]. The decomposition of the dehydrated monohydrate (542 to 581 K) exhibited a sigmoid curve which initially fitted the power law and the induction period was very short (E_a = 153 kJ mol^{-1}). Reaction at a non-coherent interface proceeded along grain boundaries and into the sub-grains.

Water removal from the trihydrate resulted in a different reactant structure and, providing water was excluded, this was not converted to the monohydrate. Different kinetic behaviour was observed during subsequent decomposition. The temperature of reaction was lower, studied between 509 and 538 K, attributed to decomposition occurring in a less stable structure. In contrast with the monohydrate, α-time curves for this reactant were fitted by the exponential rate equation when $\alpha > 0.05$ and E_a = 140 kJ mol^{-1}. The mechanism proposed was that reaction advanced through the creation of cracks in a strained reactant crystallite assemblage.

Tompkins and Young [139] concluded that the reaction of lead styphnate monohydrate (468 to 502 K) was comparable with that of barium styphnate monohydrate. The α-time curves for both salts were not identical, but again there was nucleation at grain boundaries, followed by the advance of a non-coherent reaction interface into the reactant crystallites. The activation energy (138 kJ mol^{-1}) is slightly less than that for the barium salt, but was ascribed to the same primary process, activation of a surface styphnate ion.

Flanagan [140,141] has also studied the dehydration and decomposition of lead styphnate. The products of decomposition were identified as CO_2: N_2: NO in the ratio 3.1:1:0.4 and discrete nuclei could not be identified in microscopic examination of partially decomposed salt. The shapes of α-time curves for reaction in 0.8 Torr water vapour were significantly different from those in vacuum. This was attributed to differences in the physical structure of the reactant. In water vapour the rate of reaction was approximately constant up to α = 0.8, ascribed to an initial rapid reaction covering all crystallite faces, followed by interface penetration into blocks predominantly in one crystallographic direction. In vacuum there was an initial evolution of gas followed by a linear rate corresponding to penetration of blocks from two opposite faces. Strain results in the propagation of cracks from the interface and the exponential law is obeyed during the acceleratory phase. Following textural reorganization there was a decrease in rate to approximately zero order kinetics and decomposition proceeded predominantly in one direction.

16.9. GENERAL CONCLUSIONS

16.9.1. Dehydration

Many metal carboxylates are prepared in the form of hydrates and water is lost in a lower temperature range than that required for the onset of anion decomposition [86]. The removal of water from the structure must be accompanied by substantial bond redistribution within the anhydrous phase so formed, and this can sometimes lead to structural reorganization within the solid [142]. Nickel salts, prepared in the form of crystalline hydrates, however, show a low degree of lattice order after water removal [116,118,121]. The crystal structures of few dehydrated metal carboxylates are known in any detail, an omission that must be remembered in formulating reaction mechanisms.

The surface area of the anhydrous salt, the reactant in many decomposition studies, may be markedly dependent upon the conditions which obtained during water removal, including temperature [59] and prevailing water vapour pressure [52,143] (see also Section 7.7.). Temporarily retained water may promote reorganization of the very finely divided product [142]. Both the distribution and development of decomposition nuclei and the kinetic behaviour can, in some salts, be significantly determined by the crystallite reorganization which accompanies the precursor dehydration step [59,63] of the prepared reactant hydrate.

16.9.2. Hydrolysis

In addition to the physical reorganization of the reactant, any retained water present can react chemically with other crystal constituents. Partial reactant hydrolysis, prior to the onset of decomposition, has been described for the reactions of cobalt [31] and nickel [29] acetates. Tournayan *et al.* [52,143] suggested that chemisorbed oxygen, derived from water retained by the crystal, may poison the active surfaces of the product nickel (and, presumably, other metal) nuclei.

16.9.3. Secondary reactions

The decomposition products identified following reaction are not necessarily the primary compounds which result directly from the rate limiting step. Particularly reactive entities may rapidly rearrange before leaving the reaction interface and secondary processes may occur on the surfaces of the residual material which often possesses catalytic properties. The volatile products identified [144] from the decomposition of nickel formate were changed when these were rapidly removed from the site of reaction. The primary products of decomposition of thorium formate were identified [17] as formaldehyde and carbon dioxide, but secondary processes occurring on the residual thoria yielded several additional compounds. The oxide product similarly catalysed interactions between the primary products of decomposition of zinc acetate [145]. During the decomposition of rare earth oxalates, carbon monoxide disproportionates extensively to carbon dioxide and carbon [81,82].

The mass and disposition [37] of reactant crystallites in the reaction vessel may influence the rate of salt decomposition.

The residual reaction product identified during or on completion of reaction is not necessarily the first-formed phase. Kornienko [146], for example, reported that six metal carboxylates (Mn $\rightarrow$ Zn) yielded oxide as the initial product, but suggested that the oxide can be reduced to metal by the carbon monoxide concurrently evolved.

16.9.4. The interface reaction

The objective of a mechanistic study is to identify all contributory steps that are involved in the transformation of the reactants into products. There are difficulties of investigating behaviour in the highly localized laminar zones (reaction interfaces) within which the major activity is located. Identifications of rate limiting steps in carboxylate reactions have usually been based upon comparisons between related salts. Many factors must be considered and little is known of the character of bonding (i.e. covalent or ionic) in the anhydrous reactant structures. Acheson and Galwey [108] concluded that the activation step in the decompositions of oxalates and mellitates involved the rupture of the bond between the cation and anionic oxygen (see Figure 16.2.).

Boldyrev *et al.* [46] identified the C - C link as the least stable bond, which may be broken as the initial step during oxalate pyrolysis ($C_2O_4^{2-} \rightarrow 2CO_2^-$). These radicals ($CO_2^-$) may react in three ways : (i) revert to oxalate, (ii) form $(OCOCO_2)^{2-} \rightarrow CO_3^{2-} + CO$, or (iii) reduce the cation ($\rightarrow M + 2CO_2$). The common initial step explains the similar values of E_a for decompositions of several oxalates (often about 170 kJ mol^{-1}). This is lower than the energy of rupture of the C - C bond (approximately 270 kJ mol^{-1} [38]) due to local conditions that influence the chemical changes occurring at the reaction interface.

The rate limiting step in carboxylate breakdown has been variously identified [147] as rupture of each of the principal linkages, R - CO - O - M(-), where R is an organic group and M is the metal, which may be bonded to another group. To this may be added proton transfer. Such a step undoubtedly participates in the conversion of malonate to acetate, but this does not necessarily control the overall rate.

Carboxylate decompositions are generally irreversible and it is noteworthy that compensation behaviour, a pattern of behaviour frequently associated with reversible processes, is not a feature of the literature concerned with these reactants. More usually the E_a values for carboxylate decompositions are accepted as approximating to a constant value. For example, seven values of E_a reported for the decomposition of nickel oxalate (134, 135, 136, 150, 150, 168 and 177 kJ mol^{-1}) were (with one exception) all within 151 ± 17 kJ mol^{-1}, which is typical of the reproducibility found in this field. Greater variation may arise when the decomposition is influenced by other reactions, such as prior dehydration [112].

Steps which have been proposed as rate controlling in the decompositions of one or more carboxylates include the following.

(i) *Bond rupture*. One of three different bonds in the carboxylate group could be the first to rupture: (a) metal-oxygen (M-O) [39,108]; (b) oxygen-carbon (C-O) [148,149]; and (c) the carboxyl group link (C-C) [18,46,95].

(ii) *Electron transfer*. The reaction of nickel oxalate is believed [48] to involve electron transfer, followed by decomposition of the oxalate radical so formed.

(iii) *Catalysis by solid products* [6,13]. In zinc and magnesium oxalates [75], following an electron transfer step, the radical ion so formed ($C_2O_4^-$) may decompose thermally or at an oxide surface. Cation oxidation may be important [63,148].

(iv) *More than one reaction considered.* In the decomposition of lead oxalate it is presumed [100] that there is competition between oxygen abstraction and carbon-carbon bond rupture.

The unambiguous identification of one of these possibilities, in preference to all others, must be achieved in the context of the difficulties inherent in establishing reliable identifications of primary products, kinetic data and other relevant information for the system under investigation. Combinations of the above steps are also possible, together with other types of chemical changes (e.g. proton transfer in malonate decomposition).

16.9.5. Influence of the cation

Boldyrev *et al.* [46] pointed out that the products of oxalate breakdown depend on the electropositive character of the cation :

$$MC_2O_4 \nearrow MCO_3 + CO$$

$$MC_2O_4 \rightarrow MO + CO + CO_2$$

$$MC_2O_4 \searrow M + 2CO_2$$

Further reactions occur, including processes catalyzed by the solid products, where the different organic anions give various carbonaceous radicals. The participation of acetate in malonate breakdown has already been discussed above. Decompositions of calcium carboxylates have been used as a method of ketone preparation in organic chemistry.

The chemical properties of the residual solid products from metal carboxylate decomposition are largely controlled by the reactant cation present and may include one or more of the following: metal, carbide, oxide or oxides, carbonate.

Manganese in carboxylates is not readily reduced below the divalent state (MnO), although the intermediate formation of a higher oxide containing Mn^{3+} has been reported [149]. The behaviour of iron(II) salts is markedly influenced by the tendency to disproportionate to metal and Fe^{3+}, which may proceed concurrently with a carboxylate decomposition [58]. The residue in air is ultimately Fe_2O_3, whereas in vacuum the presence of excess carbon can result in carbide production. Decompositions of cobalt carboxylates [115] are anomalous in that metal (perhaps both cubic and hexagonal forms [109]) and oxide(s) (CoO and/or Co_3O_4) occur in the product mixture. The mechanisms of these reactions are thus not readily identified. Some cobalt salts are volatilized from the heated zone during reaction.

Nickel salts have been the most extensively studied. Reactions which yield only metal occur at a lower temperature than those resulting in deposition of carbon which may alternatively (depending on temperature) be incorporated in the metal as a carbide or inhibit the development of metal nuclei [30].

The behaviour of some copper salts is complicated by the concurrent volatilization of metal, attributed to the formation of unstable gaseous copper(I) intermediates. There is strong evidence that reactions proceed to completion through stepwise reduction of the cation ($Cu^{2+} \rightarrow Cu^{+} \rightarrow Cu^{0}$) [72].

Zinc salts yield oxide product and the non-stoichiometric nature of this compound could influence the chemistry of the interface reactions. Deposition of excess carbon can [108] result in oxide reduction and metal sublimation.

Reactions of carboxylates containing the more electropositive cations yield product carbonates, or sometimes the basic carbonates. Some of these salts, e.g., those of the alkali metals, melt before decomposition. The oxide products from decomposition of the lanthanide compounds may contain carbon deposited as a result of carbon monoxide disproportionation. Kinetic measurements must include due consideration of the possible retention of carbon dioxide by the product (as CO_3^{2-}) and the secondary reactions involved in carbon deposition.

16.9.6. Patterns of reactivity

The following discussion of reactivity patterns compares the behaviour within groups of salts containing either common cations or common anions. Reactivities are measured from the *temperature* limits of the ranges used in the determinations of the magnitudes of E_a (representing limiting magnitudes of reaction rates corresponding to completion in times ranging from about 10 min to 10 h) and the calculated magnitudes of E_a. The magnitudes of A are less reliable because they can seldom be related to an identifiable common physical characteristic of the rate processes concerned and are not considered here.

Nickel salts. The thermal decompositions of nickel salts of simple organic acids proceed by nucleation and growth mechanisms in which the product metal is identified as promoting anion breakdown [49], but this activity is diminished by the presence of carbon in the residue. Comparisons of published kinetic data show [30] that those salts that yield product metal only (formate, oxalate, squarate) react in a somewhat lower temperature range (approximately 495 to 530 K) and with lower values of E_a (148 to163 kJ mol^{-1}) than those reactants that yield nickel carbide and residual carbon (530 to 590 K) and E_a values are 177 to 208 kJ mol^{-1} (malonate, maleate, fumarate, mellitate: the acetate gives residual carbon and also reacts in the higher temperature range, but kinetic characteristics are more complicated). This is parallel behaviour with the known [5] lower catalytic activity of the carbide towards the reactions of hydrocarbons. When the anion contains sufficient hydrogen, decomposition yields acid (anhydride) or the appropriate ketone.

The vacuum decompositions of the two closely related salts [120,121] nickel maleate and nickel fumarate yielded similar mixtures of product gases (predominantly carbon dioxide with smaller quantities of CO and CH_4) but exhibited significantly different kinetic behaviour. The low proportion of constituent hydrogen in the unsaturated anions decreases the significance of secondary hydrocarbon reactions. Neither salt showed any evidence of melting. The decomposition of nickel maleate 543 to 583 K proceeded through an initial ($\alpha = 0.07$) deceleratory stage, identified as a surface process, followed by the main reaction which was strongly deceleratory. The residue was a finely divided nickel-carbon mixture and it was concluded that the development of nuclei was inhibited by the deposition of unsaturated carbonaceous residues. The kinetic behaviour is quantitatively expressed by an equation derived on the assumption of continued first-order nucleation at the surfaces of reactant particles, the size of which diminish progressively during the course of reaction. In contrast, the decomposition of the fumarate 573 to 613 K proceeded by a nucleation and growth process (a sigmoid α-time curve) ascribed to the advance of cylindrical reactant/nickel carbide interfaces. At the appreciably higher temperature of reaction of this isomer, carbon can migrate within or on the metallic product phase and the development of nuclei is not inhibited. Some evidence supporting these kinetic interpretations is provided by electron micrographs of the partially and fully decomposed reactant. Clearly the difference in anion structure exerts a significant influence on salt stability and results in a marked variation in the kinetics and mechanism of decomposition.

From the above comparisons it is evident that both structure and composition of the anion may influence the mechanism of decomposition of nickel carboxylates. The crystal structure of the reactant can probably be discounted as a rate controlling parameter because dehydration usually yields amorphous materials. Depending on temperature, carbon deposited on the surface of a germ metallic nucleus may effectively prevent or inhibit growth, it may be accommodated in the structure to yield carbide, or be deposited elsewhere (by carbide decomposition). These mechanistic interpretations are based on the relative reactivities of the nickel salt and of nickel carbide, for which the temperature of decomposition is known, 570 K [150].

A method of estimating the activation energy for nickel carboxylate decompositions has been proposed [110,136]. The rate limiting step in reaction is assumed to be the rupture of one or more Ni to carboxylate bonds with $E_a = 113$ kJ mol^{-1} for each such bond ruptured. This energy barrier to decomposition is decreased by stabilization of the transition-state complex where (i) the metal is incorporated in an existing nucleus and (ii) a resonating structure may be generated within the anion. While this model explains the relative magnitudes of the values of E_a for various nickel carboxylates, the application to nickel formate is difficult [6], because, for this reactant, E_a is sensitive to reaction conditions and is not identified with a specific chemical step.

Comparative studies have been made [118] of the reactions in oxygen of nickel formate, nickel malonate, nickel maleate and nickel fumarate (488 to 593 K). Each reactant yielded nickel oxide and carbon dioxide (also, presumably, water) only. The shapes of the α-time curves were similar to those for the decompositions of each salt in vacuum and there was no evidence to suggest that the oxide formed constituted a barrier to the continuation of reaction (as occurs in the oxidation of the metal). The initial disposition and subsequent advance of the reactant-oxide interface during reactions in vacuum and in oxygen were comparable. The magnitude of E_a during oxidation of all four reactants studied was 150 ± 10 kJ mol^{-1} and different from the values determined for vacuum decomposition of each individual substance. It is suggested that reactions of all four salts involve (i) reaction of oxygen with vacancy formation at the surface of nickel oxide; and (ii) subsequent reactions of anions with the positive holes and cations with the vacancies. The mechanism of breakdown of anions at the interface during all four reactions is identical and controlled by the surface and defect properties of the solid product, thus nucleation is apparently independent of the gaseous reactant, but growth is strongly dependent on the presence of oxygen.

Copper salts. The decompositions of copper(II) salts of organic acids proceed [72] with stepwise reduction of the cation, $Cu^{2+} \rightarrow Cu^{+} \rightarrow Cu$. There was no evidence that copper metal promotes carboxylate anion breakdown, unlike nickel, and it is improbable that there was an appreciable amount of copper metal present in these reactants when α was less than 0.5. Values of E_a for the decompositions of copper(II) salts (formate, oxalate, fumarate, squarate, mellitate) were 167 kJ mol^{-1}, or less. The reactions of the formate and mellitate occurred in lower temperature intervals than the other salts. Decompositions of the malonate and maleate, which were accompanied by melting, proceeded at relatively low temperatures (about 480 to 506 K), but E_a values were at the higher end of the range (200 and 225 kJ mol^{-1}, respectively). Except for the formate and the maleate (which decomposes to the fumarate) values of E_a for the decompositions of the copper(I) salts were generally greater than those of the copper(II) salts, being between 180 and 210 kJ mol^{-1} and reaction temperatures, during the second rate processes for decompositions of the malonate, fumarate and squarate, were higher. No simple explanation can be offered for these varied and diverse patterns of reactivity where behaviour is influenced by the different anions present [72]. Levels of reactivity and E_a values are generally similar to those of the nickel salts which yield product metal. This result is surprising because of the evidence that the nickel metal product is catalytically active in promoting anion breakdown.

Silver salts. Silver formate and silver oxalate readily decompose below about 420 K, but salts containing more complicated anions that yield carbon in the residue, breakdown at higher temperatures (above 440 K for the malonate, mellitate, squarate, maleate and fumarate [129]). The formate and oxalate of this less electropositive element are less stable than the corresponding nickel and copper salts, but the differences are less marked where the residual product retained carbon. The reaction of silver squarate was *not* a silver promoted nucleation and growth process, but proceeded with the intervention of silver oxide [114].

Calcium salts. The thermal decompositions of calcium oxalate, malonate, maleate and fumarate were studied in significantly higher temperature ranges (above 720, 612 to 653, 733 to 763 and 733 to 803 K, respectively) than those of the same salts of the transition metals. This is evidence of a stabilizing influence on these anions of the strong bond formed with this strongly electropositive cation.

Transition metal oxalates. From studies of the decompositions of six oxalates (Mn, Fe, Co, Ni, Cu and Zn) Kornienko [146] concluded that the initial reaction product is the oxide and that for four of these salts (Fe, Co, Ni and Cu) the oxides are subseqently reduced to the metal by carbon monoxide. The values of the activation energies for these decompositions are comparable (about 176 kJ mol^{-1}) and stability is attributed to the polarizing power of the cation. Extensions of this approach to include salts of other cations have been discussed by Dollimore *et al.* [47] and Kadlec and Danes [55].

Boldyrev *et al.* [46], from quantum mechanical calculations of bond strengths in the oxalate anion, and from observations [38] of the decomposition of this species in potassium bromide matrices, concluded that the most probable controlling step in the breakdown of the oxalate ion is rupture of the C-C bond. This model is (again) based on the observation that the magnitudes of the activation energies for decompositions of many metal salts of oxalic acid are comparable. This model was successfully applied [46,68] to the decompositions of many oxalates, with the possible exception of silver oxalate where the strengths of the C-C and Ag-O bonds are similar.

Zhabrova *et al.* [151] identified the reactions of nickel, cobalt and copper oxalates (E_a = 150, 159 and 129 kJ mol^{-1} respectively) as redox processes in which there is an autocatalytic effect by product metal on the electron transfer step. The decomposition rate was determined by the area of the reactant and results were fitted by the Prout-Tompkins equation. In contrast, the reactions of magnesium, manganese and iron oxalates (E_a = 200, 167 and 184 kJ mol^{-1}) are not autocatalytic and the area of product increased during reaction, but the area of reactant decreased linearly with α. The absence of a catalytic effect by the oxide products of decomposition of magnesium and zinc oxalates has similarly been emphasized by Danforth and Dix [75] (for which reported values of E_a are about 200 kJ mol^{-1}).

Using DTA measurements, Macklen [148] compared the temperatures of onset of decomposition of manganese, iron, cobalt and nickel oxalates for reactions in nitrogen and in air. The reactivities of these salts follows the sequence of ease of cation oxidation [152] ($M^{2+} \rightarrow M^{3+}$) and from this it was concluded that the first step for reaction in air is cation oxidation followed by rapid breakdown of the oxalate anion.

Ag, Cu, Ni and Ca carboxylates. Comparisons of the temperature intervals of the decompositions of the Ag, Cu, Ni and Ca salts of each of the anions considered (oxalate, malonate, maleate, fumarate, mellitate and squarate) showed that the silver salts were invariably least stable. Copper salts reacted in the next higher temperature range, with the exception of the squarate where the nickel salt decomposed at the lower temperature, and approximately equal reactivities were found for Cu and Ni oxalates. (For both these exceptions, the reactions were promoted by the more active product, nickel metal). Calcium salts were the most stable, although data for the mellitate and squarate were not available. This stability sequence is the same as that of the electrochemical series, emf values for cation reduction were: Ag, 0.80; Cu, 0.16; Ni, -0.23 and Ca, -3.02 V. This is consistent with control of reactivity by the strength of the metal-oxygen link. This conclusion is supported [108] by the correlation between values of E_a for decompositions of the oxalates and mellitates with the enthalpies of oxide formation (Figure 16.2.). Magnitudes of E_a, for the series of Cu and Ni salts compared here, were mostly between relatively narrow limits (170 to 200 kJ mol^{-1}) which also supports this controlling factor.

REFERENCES

1. G.C. Bond, Catalysis by Metals, Academic Press, London, 1962.
2. P. Mars, J.J.F. Scholten and P.Zwietering, Adv. Catal., 14 (1963) 35.
3. J.M. Trillo, G. Munuera and J.M. Criado, Catal. Rev., 7 (1972) 51.
4. K. Fukuda, S. Nagashima, Y. Noto, T. Onishi and K. Tamaru, Trans. Faraday Soc., 64 (1968) 522.
5. A.K. Galwey, Adv. Catal., 26 (1977) 247.

6. A.K. Galwey, M.J. McGinn and M.E. Brown, Reactivity of Solids, (Eds J.S. Anderson, M.W. Roberts and F.S. Stone), Chapman and Hall, London, 1972, p.431; J. Chim. Phys., 75 (1978) 147.
7. M.E. Brown, B. Delmon, A.K. Galwey and M.J. McGinn, Reactivity of Solids, (Eds J. Wood, O. Lindqvist, C. Helgesson and N-G. Vannerberg), Plenum, New York, 1976, p.221.
8. L.L. Bircumshaw and J. Edwards, J. Chem. Soc., (1950) 1800.
9. P.G. Fox, J. Ehretsmann and C.E. Brown, J. Catal., 20 (1971) 67.
10. B.V. Erofeev, M.V. Maiorova and L. Ya. Mostovaya, Vesti Akad. Navuk Bellarus. SSR, Ser. Khim. Navuk SSR, (1966) 23; (1965) 5.
11. R. Schuffenecker, Y. Trambouze and M. Prettre, Ann. Chim., 7 (1962) 127, 133.
12. B.V. Erofeev and L.S. Kravchuk, Dokl. Akad. Nauk Beloruss. SSR, 11 (1967) 516; Kinet.Katal., 10 (1969) 533.
13. A.K. Galwey, D.M. Jamieson and M.E. Brown, J. Phys. Chem., 78 (1974) 2664.
14. D.A. Edwards and R. Richards, Inorg. Nucl. Chem. Lett., 8 (1972) 783.
15. D. Dollimore and K.H. Tonge, J. Inorg. Nucl. Chem., 29 (1967) 621.
16. G. Djega-Mariadassou, A. Marques and G. Pannetier, Bull. Soc. Chim. Fr., (1971) 3166.
17. B. Mentzen, Ann. Chim., 3 (1968) 367.
18. G.D. Buttress and M.A. Hughes, J. Chem. Soc. A, (1968) 1272.
19. M.D. Judd, M.I. Pope and C.G. Wintrell, J. Appl. Chem. Biotechnol., 22 (1972) 679.
20. R. Canning and M.A. Hughes, Thermochim. Acta, 6 (1973) 399.
21. K.O.Hartman and I.C. Hisatsune, J. Phys. Chem., 69 (1965) 583.
22. G. Rienaecker and W. Toursel, Monatsber. Deut. Akad. Wiss. Berlin, 13 (1971) 340.
23. E.S. Osinovik, L.P. Kostyuk-Kul'gravchuk and Z.M. Grabovskaya, Vesti Akad.Navuk Belaruss. SSR, Ser. Khim. Navuk, (1966) 112; (1967) 111.
24. R. Sabbah, P. Bianco and J. Haladjian, Bull. Soc. Chim. Fr., (1969) 2304.
25. D.M. Jamieson and A.K. Galwey, J. Catal., 34 (1974) 156.
26. J. McCarty and R.J. Madix, J. Catal., 38 (1975) 402.
27. J. Leicester and M.J. Redman, J. Appl. Chem., 12 (1962) 357.
28. K. Manabe and T. Kubo, Kogyo Kagaku Zasshi, 69 (1966) 1727.
29. J-L. Dorémieux, C. R. Acad. Sci., Paris, C, 261 (1965) 4426; Bull. Soc. Chim. Fr., (1969) 1508.

30. A.K. Galwey, S.G. McKee, T.R.B. Mitchell, M.E. Brown and A.F. Bean, React. Solids, 6 (1988) 173.
31. J.L. Dorémieux, Bull. Soc. Chim. Fr., (1967) 4586, 4593.
32. M.D. Judd, B.A. Plunkett and M.I. Pope, J. Thermal Anal., 6 (1974) 555.
33. D.C.K. Lin and J.B. Westmore, Canad. J. Chem., 51 (1973) 2999.
34. D.A. Edwards and R.N. Hayward, Canad. J. Chem., 46 (1968) 3443.
35. M.A. Bernard and F. Busnot, Bull. Soc. Chim. Fr., (1968) 2000.
36. V.I. Yakerson and A.M. Rubinshtein, Kinet. Katal., 2 (1961) 172.
37. H.G. McAdie and J.M. Jervis, Thermochim. Acta, 1 (1970) 19.
38. K.O. Hartman and I.C. Hisatsune, J. Phys. Chem., 71 (1967) 392; F.E. Freeberg, K.O. Hartman and I.C. Hisatsune, J.Phys.Chem., 71 (1967) 397.
39. P.S. Clough, D. Dollimore and P. Grundy, J. Inorg. Nucl. Chem., 31 (1969) 361.
40. M.A. Mohamed and S.A. Halawy, Thermochim. Acta, 242 (1994) 173.
41. M.A. Mohamed, S.A. Halawy and M.M. Ebrahim, J. Thermal Anal., 41 (1994) 387.
42. M.A. Mohamed, S.A. Halawy and M.M. Ebrahim, J. Anal. Appl. Pyrolysis, 27 (1993) 109; Thermochim. Acta, 236 (1994) 249.
43. P. Baraldi, Spectrochim. Acta, 38A (1982) 51.
44. P.A. Barnes, G. Stephenson and S.B. Warrington, J. Thermal Anal., 25 (1982) 299.
45. C.E. Stoner and T.B. Brill, Inorg. Chem., 28 (1989) 4500.
46. V.V. Boldyrev, I.S. Nev'yantsev, Yu.I. Mikhailov and E.F. Khairetdinov, Kinet. Katal., 11 (1970) 367.
47. D. Dollimore, D.L. Griffiths and D. Nicholson, J. Chem. Soc., (1963) 2617.
48. P.W.M. Jacobs and A.R. Tariq Kureishy, Reactivity of Solids, (Ed. J.H. Boer), Elsevier, Amsterdam, 1961, p.352; Trans. Faraday Soc., 58 (1962) 551.
49. D.A. Dominey, H. Morley and D.A. Young, Trans. Faraday Soc., 61 (1965) 1246.
50. A. Engberg, Acta Chem. Scand., 27 (1973) 3929.
51. J-L. Dorémieux and P. Brissaud, C. R. Acad. Sci. Paris, C, 269 (1969) 1465.
52. L. Tournayan, H. Charcosset, Y. Trambouze and G. Blandenet, Bull. Soc. Chim. Fr., (1968) 925; (1969) 1553.
53. M.J. Halsey and A.M. Pritchard, J. Chem. Soc. A, (1968) 2878.
54. G.C. Nicholson, J. Inorg. Nucl. Chem., 29 (1967) 1599.

55. O. Kadlec and V. Danes, Collect. Czech. Chem. Comm., 32 (1967) 1871.
56. H. Foerster, K. Meyer and K. Nagorny, Z. Phys. Chem. (Frankfurt), 87 (1973) 64.
57. A.K. Galwey and M.A. Mohamed, Thermochim. Acta, 213 (1993) 279.
58. M.A. Mohamed and A.K. Galwey, Thermochim. Acta, 213 (1993) 269.
59. D. Broadbent, D. Dollimore and J. Dollimore, J. Chem. Soc. A, (1966) 1491.
60. T. Kubo and K. Manabe, Kogyo Kagaku Zasshi, 67 (1964) 1355.
61. J. Amiel and C. Paulmier, C. R. Acad. Sci. Paris, C, 255 (1962) 2443.
62. G. Périnet, M. Le Van and P. Bianco, C. R. Acad. Sci. Paris, C, 266 (1968) 1152.
63. M.E. Brown, D. Dollimore and A.K. Galwey, J. Chem. Soc., Faraday Trans. I, 70 (1974) 1316.
64. A. Finch, P.W.M. Jacobs and F.C. Tompkins, J. Chem. Soc., (1954) 2053.
65. R.M. Haynes and D.A. Young, Discuss. Faraday Soc., 31 (1961) 229.
66. A.G. Leiga, J. Phys. Chem., 70 (1966) 3254, 3260.
67. Yu.I. Mikhailov, V.V. Boldyrev and R.M. Belkina, Kinet. Katal., 11 (1970) 75, 614.
68. V.V. Boldyrev, Bull. Soc. Chim. Fr., (1969) 1054, and the extensive bibliography contained in this review.
69. D.Broadbent, J. Dollimore, D. Dollimore and T.A. Evans, J.Chem.Soc., Faraday Trans., 87 (1991) 161.
70. A.K. Galwey, J. Chem. Soc., (1965) 5433.
71. M.A.Mohamed and A.K. Galwey, Thermochim. Acta, 217 (1993) 263.
72. A.K. Galwey and M.A. Mohamed, Thermochim. Acta, 239 (1994) 211.
73. A. Coetzee, D.J. Eve, M.E. Brown and C.A. Strydom, J. Thermal Anal., 41 (1994) 357.
74. P.E. Yankwich and P.D. Zavitsanos, J. Phys. Chem., 68 (1964) 457.
75. J.D. Danforth and J. Dix, J. Amer. Chem. Soc., 93 (1971) 6843.
76. P.E. Yankwich and P.D. Zavitsanos, J. Phys. Chem., 69 (1965) 442.
77. V.G. Guslev, V.V. Boldyrev, I.S. Nev'yantsev and Yu.Ya. Karpenko, Kinet. Katal., 15 (1974) 53.
78. D.A. Young, J. Chem. Soc., (1960) 4533.
79. D. Dollimore, J. Dollimore and J. Little, J. Chem. Soc. A, (1969) 2946.
80. D. Dollimore, L.F. Jones, T. Nicklin and P. Spooner, J. Chem. Soc., Faraday Trans. I, 69 (1973) 1827.
81. A. Glasner and M. Steinberg, J. Inorg. Nucl. Chem., 16 (1961) 279; 22 (1961) 39, 156.

82. A. Glasner, E. Levy and M. Steinberg, J. Inorg. Nucl. Chem., 25(1963) 1119, 1415; 26 (1964) 1143.
83. P.K. Gallagher, F. Schrey and B. Prescott, Inorg. Chem., 9 (1972) 215.
84. D. Dollimore and D. Tinsley, J. Chem. Soc. A, (1971) 3043.
85. M.M. Girgis and A.M. El-Awad, Thermochim. Acta, 214 (1993) 291.
86. D. Dollimore and D.L. Griffiths, J. Thermal Anal., 2 (1970) 229.
87. D. Dollimore and D.Nicholson, J. Chem. Soc., (1962) 960.
88. K.P. Pribylov, D.Sh. Fazlullina and R.M. Cherchetkin, Zh. Neorg. Khim., 13 (1968) 3182.
89. F.E. Freeberg, K.O. Hartman, I.C. Hisatsune and J. Schempf, J. Phys. Chem., 71 (1967) 397.
90. S. Bose, K.K. Sahu and D. Bhatta, Thermochim. Acta, 175 (1995) 175.
91. J. Paulik, F. Paulik and L. Erdey, Anal. Chim. Acta, 44 (1969) 153.
92. A.H. Verdonk and A. Boersma, Thermochim. Acta, 6 (1973) 95.
93. N. Koga and H. Tanaka, J. Thermal Anal., 32 (1987) 1521.
94. E.S. Osinovik and A.K. Yanchuk, Vesti Akad. Navuk Belaruss. SSR, Ser. Khim. Navuk, (1966) (3) 131.
95. V.P. Kornienko, M.N. Dubrovskaya and T.I. Gritsan, Kinet. Katal., 9 (1968) 535.
96. E.G. Prout and F.C. Tompkins, Trans. Faraday Soc., 43 (1947) 148.
97. S.E. Morsi, J. Chem.U.A.R., 13 (1970) 113.
98. L.L. Bircumshaw and I. Harris, J. Chem. Soc., (1948) 1893.
99. O. Kadlec and J. Rosmusovà, Collect. Czech. Chem. Comm., 31 (1966) 4324; O. Kadlec and V. Danes, Collect. Czech. Chem. Comm., 33 (1968) 1174.
100. P.E. Yankwich and J.L. Copeland, J. Amer. Chem. Soc., 79 (1957) 2081.
101. A. Selcuk and D. Price, Thermochim. Acta, 209 (1992) 195.
102. A. Coetzee, D.J. Eve and M.E. Brown, J. Thermal Anal., 39 (1993) 947.
103. P.K. Gallagher, P.T. Chao, L. Zhong and J. Subramanian, J. Thermal Anal., 39 (1993) 975.
104. J-L. Dorémieux, Colloques Internal CNRS (France), No. 201 - Thermochimie; J-L. Dorémieux, P. Dugleux and G. Cizeron, Silicates Indust., 35 (1972) 125.
105. S. Otta and S.D. Bhattamisra, J. Thermal Anal., 41 (1994) 419.
106. M. Isausti, R. Cortes, M.I. Arriortua, T. Rojo and E.H. Bocanegra, Solid State Ionics, 63-65 (1993) 351.
107. J. Bacsa, D.J. Eve and M.E. Brown, J. Thermal Anal., 50 (1997) 33.
108. R.J. Acheson and A.K. Galwey, J. Chem. Soc. A, (1967) 1167.

109. A.K. Galwey, J. Chem. Soc. A, (1966) 87.
110. A.K. Galwey, Kinet. Katal., 10 (1969) 765.
111. R.J. Acheson and A.K. Galwey, J. Chem. Soc. A, (1968) 1125.
112. A.K. Galwey and M.E. Brown, J. Chem. Soc., Faraday Trans. I, 78 (1982) 411.
113. A.K. Galwey, M.A. Mohamed, S. Rajam and M.E. Brown, J. Chem. Soc., Faraday Trans. I, 84 (1988) 1349.
114. A.K. Galwey, M.A. Mohamed and M.E. Brown, J. Chem. Soc., Faraday Trans. I, 84 (1988) 57.
115. A.K. Galwey, D.M. Jamieson, M. Le Van and C. Berro, Thermochim. Acta, 10 (1974) 161.
116. K.A. Jones, R.J. Acheson, B.R. Wheeler and A.K. Galwey, Trans. Faraday Soc., 64 (1968) 1887.
117. M. Le Van, G. Périnet, F. Boubli and R. Lafont, C. R. Acad. Sci. Paris, C, 271 (1970) 69.
118. B.R. Wheeler and A.K. Galwey, J. Chem. Soc., Faraday Trans. I, 70 (1974) 661.
119. A.K. Galwey, S.G. McKee, T.R.B. Mitchell, M.A. Mohamed, M.E. Brown and A.F. Bean, React. Solids, 6 (1988) 187.
120. M.J. McGinn, B.R. Wheeler and A.K. Galwey, Trans. Faraday Soc., 66 (1970) 1809.
121. M.J. McGinn, B.R. Wheeler and A.K. Galwey, Trans. Faraday Soc., 67 (1971) 1480.
122. N.J. Carr and A.K. Galwey, Reactivity of Solids, (Eds P. Barret and L-C. Dufour), Elsevier, Amsterdam, 1985, p.697.
123. N.J. Carr and A.K. Galwey, Proc. R. Soc. (London), A404 (1986) 101.
124. J. Cunningham, M. O'Neill, G. Patrick, N. Hickey, Z. Wang, A.K. Galwey and J.L.G. Fierro, J. Thermal Anal., 41 (1994) 651.
125. N.J. Carr and A.K. Galwey, J. Chem. Soc., Faraday Trans. I, 84 (1988) 1357.
126. A.K. Galwey and M.A. Mohamed, J. Chem. Soc., Faraday Trans. I, 81 (1985) 2503.
127. M.A. Mohamed, J. Thermal Anal., 42 (1994) 1081.
128. A.K. Galwey, P.J. Herley and M.A. Mohamed, J. Chem. Soc., Faraday Trans. I, 84 (1988) 729.
129. M.A. Mohamed, S.A.A. Mansour and G.A.M. Hussien, J. Thermal Anal., 41 (1994) 405.
130. M.A. Mohamed, J. Anal. Appl. Pyrolysis, 30 (1994) 59.

131. A.K. Galwey and M.A. Mohamed, Solid State Ionics, 42 (1990) 135.
132. A.K. Galwey and M.A. Mohamed, Thermochim. Acta, 273 (1996) 239.
133. M. Le Van and G. Périnet, Bull. Soc. Chim. Fr., (1966) 3109; C. R. Acad. Sci. Paris, C, 266 (1968) 732.
134. A.K. Galwey, J. Chem. Soc., (1965) 6188.
135. A.K. Galwey, J. Catal., 4 (1965) 697.
136. R.J. Acheson and A.K. Galwey, J. Chem. Soc. A, (1967) 1174.
137. M.E. Brown, J. Chem. Soc., Faraday Trans. I, 69 (1973) 1202.
138. A. Venkataraman and A.J. Mukhedkar, J. Mater. Sci., 23 (1988) 3505; J. Thermal Anal., 36 (1990) 1495; Bull. Mater. Sci., 16 (1993) 51.
139. F.C. Tompkins and D.A. Young, Trans. Faraday Soc., 52 (1956) 1245; J. Chem. Soc., (1956) 3331; (1957) 4281.
140. T.B. Flanagan, Trans. Faraday Soc., 57 (1961) 797.
141. T.B. Flanagan, J. Phys. Chem., 66 (1962) 416.
142. W.E. Garner, Chemistry of the Solid State, (Ed. W.E. Garner), Butterworths, London, 1955, Chap.8.
143. L. Tournayan, H. Charcosset, B.R. Wheeler, M.J. McGinn and A.K. Galwey, J. Chem. Soc. A, (1971) 868.
144. J. Fahrenfort, L.L. Van Reyen and W.M.H. Sachtler, Heterogeneous Catalysis, (Ed. J.H. de Boer), Elsevier, Amsterdam, 1960, p.23.
145. G. Djega-Mariadassou, E. Kerboub and G. Pannetier, Bull. Soc. Chim. Fr., (1970) 1353.
146. V.P. Kornienko, Ukrain. Khim. Zhur., 23 (1957) 159.
147. M.E. Brown, D.Dollimore and A.K. Galwey, Reactions in the Solid State, Comprehensive Chemical Kinetics, (Eds C.H. Bamford and C.F.H. Tipper),Vol.22, Elsevier, Amsterdam, 1980, p.210.
148. E.D. Macklen, J. Inorg. Nucl. Chem., 30 (1968) 2689.
149. K. Nagase, Chem. Letters, (1972) 205.
150. L.J.E. Hofer, E.M. Cohn and W.C. Peebles, J. Phys. Colloid Chem., 54 (1950) 1161.
151. G.M. Zhabrova, B.M. Kadenatsi and A.V.Shkarin, Zh. Fiz. Khim., 45 (1971) 1702; G.M. Zhabrova and B.M. Kadenatsi, Probl. Kinet. Katal., 15 (1973) 175.
152. G. Bakcsy and A.J. Hegedus, Thermochim. Acta, 10 (1974) 399.

Chapter 17

DECOMPOSITIONS OF COORDINATION COMPOUNDS

17.1. INTRODUCTION

So many diverse substances have been referred to in the literature as *complexes* that the term may be regarded as a non-specific description of any compound which contains three or more identifiable groups. This term will not be used here and this chapter is primarily concerned with the thermal decompositions of coordination compounds. Because the literature is so extensive [1] and the number of compounds in this class is so large and diverse, only a selection of substances, representing different types of characteristic behaviour, is included in this account. The preparations and structures (electronic, stereochemical and crystallographic) of this wide class of compounds are to be found in texts of inorganic chemistry, or specialized monographs [1-3].

A *coordination compound* consists of a *central atom* which is bonded to species known as *ligands*. The ligands may be monatomic or polyatomic and are capable of independent existence in the absence of coordinate bonding to the central atom. Ligands may be formally derived from neutral molecules (such as H_2O, NH_3, ethylenediamine (en), propylenediamine (pn), pyridine (py), CO, NO, etc.), or from ions (e.g., anions such as Cl^-, Br^-, NO_3^-, SO_4^{2-}, etc., or cations such as NH_4^+, RNH_3^+, etc.). Polyatomic ligands may bond to the central atom through a single atom in the ligand (*monodentate ligand*) or may form rings (*chelate*) by bonding through two or more separate atoms in the ligand (*polydentate ligands*). The *coordination number* of the central atom is defined by the number of sigma bonds formed between ligands and the central atom. Coordination compounds may contain only one coordinated central atom (*mononuclear*) or two or more coordinated central atoms (*polynuclear*). Polydentate ligands may act as *bridges* in polynuclear compounds. Compounds in which the number of bonds between the central atom and separate carbon atoms is at least half the coordination number of the central atom may be arbitrarily distinguished [1] as *organometallic compounds*. The

IUPAC definition is broader, including any chemical species containing a carbon to metal bond.

The geometric shape of the fixed positions occupied by the ligating atoms defines the *coordination polyhedron*. Irregular polyhedra may occur when the ligands are not identical, or impose additional stereochemical constraints.

The hydrates undoubtedly constitute the most completely studied group of coordination compounds. The central atom is bonded to water molecules, e.g. $[Mg(H_2O)_6]^{2+}$, $[Ni(H_2O)_6]^{2+}$ and $[Cu(H_2O)_4]^{2+}$. Because of the large amount of information available, together with the widely accepted convention of regarding hydrates as a distinct group, dehydrations, principally of this type of reactant, have been considered separately in Chapter 7. An important characteristic of hydrates is that the neutral H_2O ligand is particularly unreactive and is almost invariably released unchanged on heating. Hydrates are also strongly stabilized by hydrogen bonding in the crystal structure. Water molecules, after dissociation of the coordinate link, may remain chemisorbed for a time and participate in the reorganization of the solid product. Many examples are known of mixed coordination compounds containing both water molecules and other ligands, e.g. $[Co(NH_3)_5H_2O]Cl_3$, or in which water molecules are not specifically held in the coordination shell of the central atom, e.g. $[Co(NH_3)_6]_2(SO_4)_3.2.5\ H_2O$.

The simplest class of coordination compounds is that of mononuclear compounds with monodentate ligands only. Then follow, in order of increasing complexity, mononuclear compounds with polydentate ligands, binuclear compounds with monodentate ligands only, binuclear compounds with non-bridging ligands, and binuclear compounds with bridging ligands, etc.

The kinetic behaviour, the rate controlling processes and the mechanisms of thermal decomposition of most coordination compounds are generally similar to those that characterize the decompositions of simple salts. However, the increased number and variety of constituent groupings accommodated in the reactant structures introduces diversity and complications into the chemistry of the decomposition mechanisms. Ligand release, through dissociation of the coordinate linkage, may be accompanied by, or rapidly followed by, secondary reactions, so that mechanisms generally involve several different contributory processes. As a result of this complexity, relatively few detailed studies of the kinetics and mechanisms of the solid state decompositions of coordination compounds have been reported. Where ligands other than H_2O occupy the coordination shell, the complexity of the chemical changes has often made it necessary to limit the objectives of investigations to comparison of the thermal behaviour of members of a series of related substances. Thermal analysis, supplemented by spectral or other measurements, has been [4-6] extensively used. Measurements of enthalpy changes and mass losses enable the individual steps in a sequence of successive rate processes to be characterized (at least qualitatively). The influences of specific changes in particular substituents within the reactant compounds on the progress of the decompositions can be determined. Differences in behaviour resulting from ligand substitutions or variation of ion structure are small, so that comparisons require accurate and carefully controlled measurements.

Complete formulation of the overall reaction mechanism requires, in addition to the considerations discussed in Chapter 6, the identification of all changes in bonding, disposition and structure of the ligands remaining, as well as any changes in the oxidation state of the central atom. Complete determination of the reaction stoichiometry is an essential part of any mechanistic study.

Decompositions of representative coordination compounds are described below. Reactants have been grouped together on the basis of a common central atom, comparable compositions and ligand structures. Reviews of the complicated, dispersed and often inconsistent literature concerned with the decompositions of coordination compounds include those by Dollimore [5] (concerned with DTA), Wendlandt and Smith [4] (concerned with the thermal properties of transition metal ammines), Carp and Segal [7], Hill [8], D'Ascenzo *et al.* [9] and Liptay [10].

Most of the studies of decompositions of coordination compounds have used a flow of inert gas or a vacuum to remove the volatile products. Occasionally the influences of specified additives or reactive gases have been examined. Lack of any mention of the occurrence of phase changes, recrystallization or melting (in most reports) has been taken to imply that the reaction of interest occurs in the crystalline phase of the original reactant.

The reactions of many coordination compounds in solution are well documented [1,2]. Comparisons of chemical processes occurring under homogeneous and heterogeneous conditions have potential value in determining the factors which control the reactivities of 'free' ions and the influences of crystal structure on thermal stability.

17.2. MONONUCLEAR COORDINATION COMPOUNDS WITH MONODENTATE LIGANDS

17.2.1. Cobalt(III) ammines

Wendlandt [11] has shown that the thermal stabilities of salts containing the $[Co(NH_3)_6]^{3+}$ ion (heated in helium at 5 K min^{-1}) vary with the anion present (Cl^-, Br^-, NO_3^- and (hydrated) SO_4^{2-}) and that these anions also influence the chemical changes which occur. The onset of reaction for both halides became detectable at about 473 K. DTA endotherms were detected at 553 K, followed by two broad peaks between 573 and 648 K for the chloride and between 513 and 598 K for the bromide. The nitrate showed the onset of an endotherm at 500 K, but the development of this process was obscured by a sharp exotherm at 513 K. The sulfate, $[Co(NH_3)_6]_2(SO_4)_3.2.5H_2O$, was shown by TG to be stable to about 493 K, but a sequence of at least seven endotherms (each accompanied by gas evolution) was detected thereafter (an early endotherm at about 433 K may be due to desorbed water). Representative DTA traces are shown in Figure 17.1.

Although all four salts commence reaction within the narrow temperature interval between 470 and 500 K, the properties of the anions influence the stability and determine the subsequent steps of salt breakdown.

The exotherm in the DTA trace of the nitrate is attributed to the decomposition of this anion which gave Co_3O_4 as the residue (after heating to 873 K). NH_4NO_3 was also formed and the gaseous products were, in order of abundance: H_2O, N_2, NH_3 and N_2O. Immediately prior to the rapid decomposition there was perceptible transfer of NO_3^- to the coordination shell. The reaction of the sulfate yielded $CoSO_4$, $(NH_4)_2SO_4$, NH_3, H_2O and a small quantity of nitrogen resulting from the reduction of Co^{3+} to Co^{2+}. The sequence of reactions occurring between 440 and 620 K was not identified.

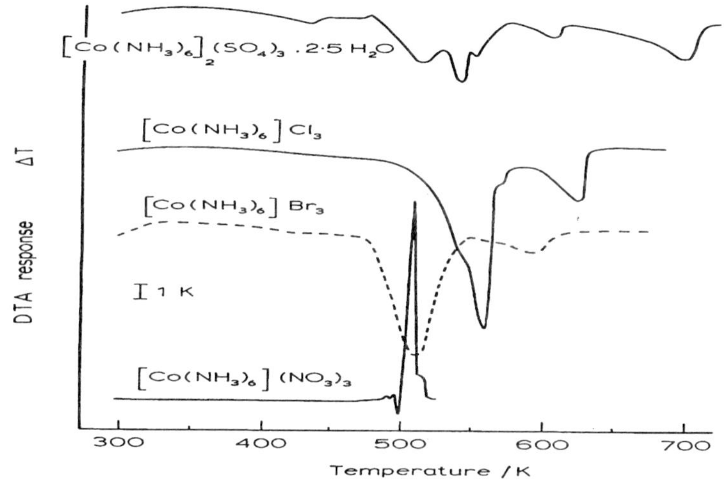

Figure 17.1.
DTA curves for hexaamminecobalt(III) salts [11].

Magnetic susceptibility measurements [12] for the decompositions of the chloride and the bromide supported the reduction of Co^{3+} to Co^{2+} between 473 and 593 K. Consideration of these observations together with TG data showed [12] that little, if any, intermediate production of a different ammine compound occurred during heating at 5 K min^{-1}. Watt [13], using a slower heating rate, identified the process at 553 K as conversion of the hexaammine to $[Co(NH_3)_5Cl]Cl_2$. Rapid decomposition was apparent at the temperature of incipient melting (562 K). The initial step in decomposition was, therefore, identified as electron transfer [14] giving the less stable cobalt(II) coordination compound which rapidly reacts further to yield the products summarized in Figure 17.2.

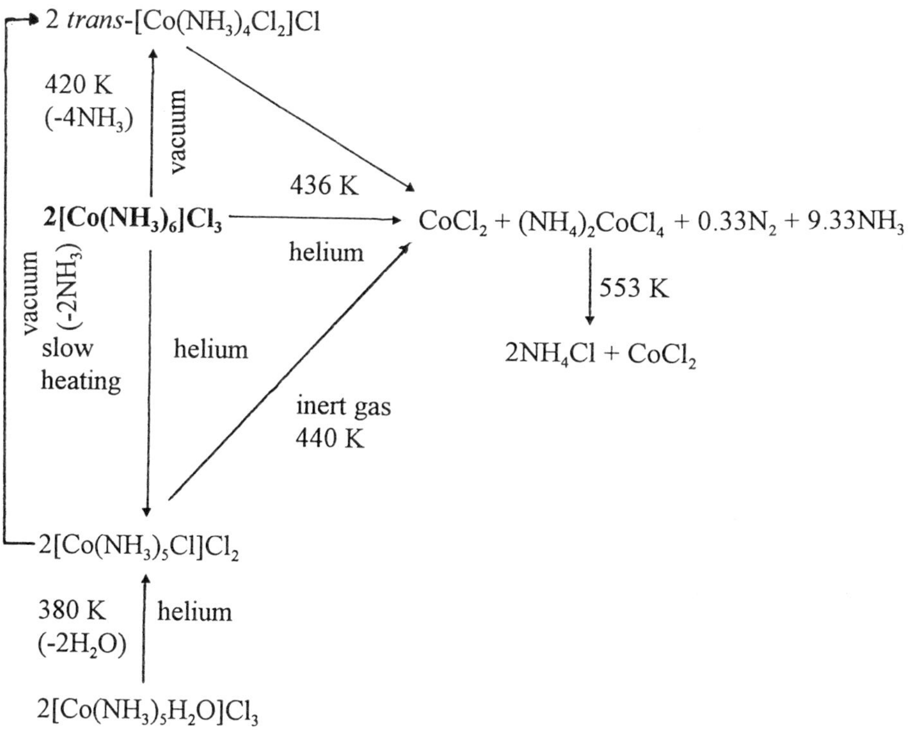

Figure 17.2. Thermal decomposition of hexaamminecobalt(III) chloride.

While deammination may be partially reversible, the possibility of concurrent cation reduction (to form Co^{2+} and N_2) distinguishes these rate processes from the truly reversible dehydration reactions. The reactivity of ammonia is greater than that of water, so that chemical involvement here extends beyond the simple dissociation of a coordinate linkage. The identification [12] (see also [13]) of the intermediate $(NH_4)_2CoCl_4$ represents a novel alternative to other proposed mechanisms [4]. On further heating, the intermediate dissociates to leave a residue of $CoCl_2$. The decompositions of the hexaammine and several related compounds are summarized in Figure 17.2.

Although the hexaammine compound can be decomposed without extensive substitution of the ammonia within the coordination shell by the chloride [4], substitution occurs when the reactant is maintained for a time at temperatures immediately below the range within which decomposition becomes significant. Thus during slow heating $[Co(NH_3)_5Cl]Cl_2$ is formed [13]. In vacuum, where presumably deammination is enhanced, the reaction temperature is slightly decreased and substitution proceeds [15] a step further to *trans*-$[Co(NH_3)_4Cl_2]Cl$. In general, the possibility of concurrent deammination and decomposition must be considered.

Zivkovic [16] using heating rates between 2.5 and 20 K min^{-1} found E_a values for the decomposition of $[Co(NH_3)_6]Cl_3$ to $CoCl_2$ and $(NH_4)_2CoCl_4$ (453 to 553 K) and the further decomposition of $(NH_4)_2CoCl_4$ to NH_4Cl and $CoCl_2$ (623 K) to be 132 and 54 to 67 kJ mol^{-1}, respectively.

The behaviour of the bromides is analogous to that given for the chlorides in Figure 17.2., though reaction temperatures may be slightly different and, in vacuum, $[Co(NH_3)_5Br]Br_2$ probably [15] does not yield *trans*-$[Co(NH_3)_4Br_2]Br$.

Decompositions of pentaamminehalocobalt(III) halides, $[Co(NH_3)_5X]Y_2$ (where X and Y are Cl^-, Br^- or I^-) have been studied by Wendlandt and Smith [17], and several of their observations are summarized in Figure 17.3.

Reactions in vacuum (TG) became detectable at lower temperatures than in helium (DTA), where the release of ammonia was slower, although the sequence of relative stability was almost identical. The minimum temperatures [17] of reduction of Co^{3+} are included in Figure 17.3. Most values were close to the corresponding DTA values. Ingier-Stocka [18] confirmed this sequence of changes and added textural evidence from scanning electron micrographs. From a non-isothermal kinetic study, it was concluded that an isothermal study is required to obtain reliable kinetic parameters. Values of E_a and A varied markedly with rate equations and conditions.

The onset of decomposition of iodide compounds [17] occurred at lower temperatures than compounds containing only chloride and bromide. This was attributed to reduction of the cation ($Co^{3+} \rightarrow Co^{2+}$) by the iodide ion, which is oxidized to I_2. In the other reactants, similar reduction is accompanied by oxidation of ammonia to N_2. The residual halides were CoY_2, where Y is the ionic constituent of the original reactant, except for the two compounds with Y = I^- in which I^- was partially oxidized.

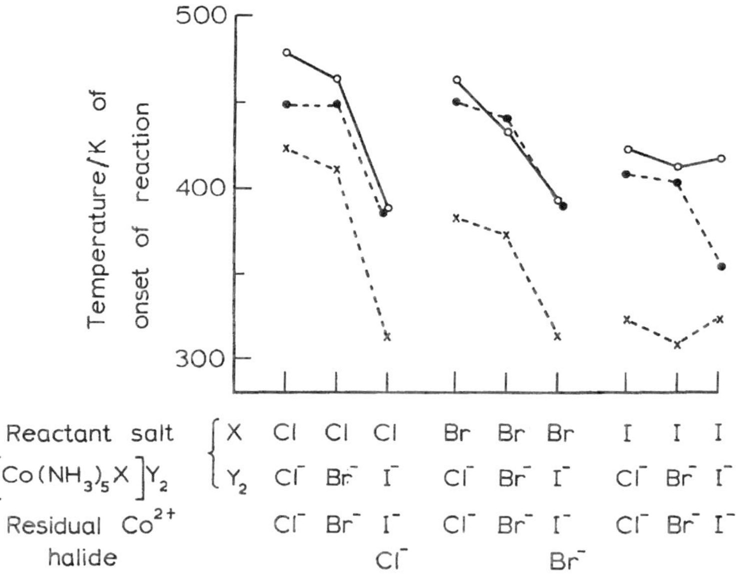

Figure 17.3.
Onset temperatures of decompositions of pentaamminehalocobalt(III) halides [17]. (o) DTA data in helium, (●) thermomagnetic data and (x) TG data.

Wendlandt and Smith [17] also studied the reactions of pentaamminehalocobalt(III) nitrates, $[Co(NH_3)_5X](NO_3)_2$ using TG between 400 and 450 K. They found that the stability sequence for the initial dissociation step was (X =) F^- = Cl^- > Br^- > I^-. Thereafter there was a rapid mass change between 488 and 560 K. DTA studies indicated that an initial endothermic reaction, attributed to dissociative evolution of ammonia, was followed by a strong exotherm having a maximum between 488 and 538 K, believed to result from an oxidation step. The evolved gases included [4] NH_3, N_2O, H_2O, NO and N_2 and the final residue was Co_3O_4. In discussion of the mechanisms of breakdown of the halides (X = F^-, Cl^-, Br^-), it was suggested that two nitrate groups may be accommodated within the coordination shell through stepwise substitution of ammonia.

Pentaammineaquacobalt(III) halides are relatively unstable. Substitution of coordinated water by halide (*anation* [19]) is detectable at about 360 K. Conductivity measurements have shown [19] that the water evolved may remain temporarily as a liquid before evaporation and this endothermic effect may mask the exothermic anation. This is consistent with the observation [4] that the minimum temperatures of cobalt reduction in the reactions of $[Co(NH_3)_5H_2O]X_3$ ($X = Cl^-$, Br^- and I^-) were very close to those of the corresponding halopentaammines.

Zheng *et al.* [20] showed that KI decreases the temperature and diminishes the activation energy for the decompositions of five cobalt(III) ammine chloride coordination compounds containing the cations $[Co(NH_3)_5X]$, where $X=N_3^-$, NO_2^-, Cl^-, CO_3^{2-} or NH_3. The electron transfer step to Co^{3+} is regarded as exerting an important control on the decomposition.

Wendlandt *et al.* [21,22] have investigated the decompositions of di- and triaquaamminecobalt(III) halides. The two molecules of water in the coordination shell of $[Co(NH_3)_4(H_2O)_2]Cl_3$ were replaced [21] by two chloride ions in distinct steps having endothermic peaks at 383 and 453 K. On completion of dehydration, there was an exothermic process (473 K), attributed to a phase transformation because this was not accompanied by gas evolution. Endotherms due to decomposition of the tetraamminedichloro compound were detected at 533, 563 and 623 K. The bromide analogue behaved similarly, although the water evolution steps occurred at lower temperatures 363 and 433 K and the anhydrous salt decomposed at 493 K. In vacuum, both reactants gave CoX_2 as residue, NH_4Cl sublimate and gaseous products NH_3, H_2O and N_2. Water is similarly lost [22] from the triamminetriaqua halides in two steps, the first between 333 and 373 K forming $[Co(NH_3)_3(H_2O)_2X]X_2$ where $X^- = Cl^-$, Br^- or I^-. The chloride and bromide reacted further above 373 K to give $[Co(NH_3)_3X_3]$, whereas the iodide evolved I_2 before the release of water was completed.

Azido coordination compounds. Kinetic studies [23] of the isothermal (370 to 420 K) decompositions of solid hexaammine-, azidopentaammine- and *cis*- and *trans*-diazidotetraamminecobalt(III) azides included investigations of the influences of ammonia and residual products on reaction rates and were supplemented by optical microscopy and X-ray identification of the phases present. Reactions were little influenced by the composition of the coordinated cation, the nature of the salt or the crystal structure. The decompositions of all four reactants were so similar that the operation of a common reaction mechanism was indicated. This similarity of behaviour was unexpected. Decompositions of simple *ionic* azides (Chapter 11) are believed to be initiated by an electron promotion step or exciton formation whereas the decompositions of covalent azides are envisaged as proceeding by N-N bond rupture in an asymmetric azide group. The hexaammine compound contains azide *ions* only, whereas the other reactants contain either one or two azide groups *covalently* attached to the cation. The possibility that all of these reactions proceed through progressive accommodation of the azide within the coordination shell, with expulsion of NH_3, is thought to be improbable and restricted, if it occurs at all, to the reaction interface. Optical and X-ray diffraction evidence provides little support for such a substitution. Such a reaction would be expected to be opposed by gaseous ammonia, whereas the observations show that ammonia either has no effect or accelerates reaction. There is some evidence that at low temperature in 100 Torr NH_3, *cis*-$[Co(NH_3)_4(N_3)_2]N_3$ yields some $[Co(NH_3)_3(N_3)_3]$.

These apparent inconsistencies have been resolved by the proposal of the following mechanism [23]. All the four reactants mentioned above decompose by the following alternative reactions, shown here for the hexaammine. Analogous reactions may be written for the penta- and tetraammines:

$$[Co(NH_3)_6](N_3)_3(s) \nearrow CoN(s) + 6NH_3(g) + 4N_2(g)$$

$$[Co(NH_3)_6](N_3)_3(s) \searrow Co(NH_3)_2(N_3)_2(s) + 4NH_3(g) + 1.5N_2(g)$$

These processes will be referred to as the *CoN* and *cobalt(II) reactions*, respectively (discussed individually below). Each sample of reactant decomposes to yield predominantly one or other product. The kinetics of *both* reactions for each of the four salts have been investigated, using small samples to minimize the possible danger of explosion. There were problems in identifying the products of the cobalt(II) reaction (which may include $[Co(NH_3)_4(N_3)_2]$), which were increased by the variations in behaviour of the reactant crystallites with particle size and the sensitivity of the kinetics to the presence of ammonia.

The CoN reaction. Following an induction period and a short acceleratory stage ($\alpha < 0.1$), the decomposition to yield CoN was rapid and predominantly deceleratory. The induction period increased in the presence of small quantities of ammonia, but was shortened by mixing some solid product with the reactant. The values of E_a for both the induction period and the main reaction were between 84 and 110 kJ mol^{-1}. In the presence of ammonia (product), E_a could be as high as 210 kJ mol^{-1}.

The first step in reaction (shown here for the hexaammine) is suggested [23] (as for some ionic azides) to be electron transfer:

$$[Co(NH_3)_6]^{3+} + e \rightarrow [Co(NH_3)_6]^{2+}$$

The divalent ion may then react with azide ions:

$$[Co(NH_3)_6]^{2+} + 2N_3^- \rightarrow \{Co(NH_3)_2(N_3)_2\} + 4NH_3$$

where the brackets {} signify an unstable intermediate, probably amorphous. This intermediate breaks down to yield CoN, through a sequence of unidentified steps.

The cobalt(II) reaction. The cobalt(II) reaction exhibited sigmoid α-time curves giving values of E_a between 170 and 230 kJ mol^{-1}, except for the hexaammine compound in ammonia where E_a was 132 kJ mol^{-1}. The mechanism proposed is initially identical to that detailed above for the CoN reaction, but the divalent intermediate is suggested to be stabilized within a crystalline product:

$$\{Co(NH_3)_2(N_3)_2\} \rightarrow [Co(NH_3)_2(N_3)_2](cryst)$$

The residue may also include some $[Co(NH_3)_4(N_3)_2]$, but identification and measurement of this constituent is difficult.

Reaction during the induction period. The conditions existing during the induction period are believed to determine which of the alternative reactions operates in any particular reactant. The factors preceeding the onset of the main reaction may be delicately balanced between the alternative possible breakdown paths. During initiation a small amount of the disorganized phase {} is formed by decomposition, which may either react further to CoN, or recrystallize and thus stabilize this compound as product, thereby inhibiting the tendency towards CoN production. On completion of this period of finely balanced alternatives, the main reaction is either the nucleation and growth process yielding $[Co(NH_3)_2(N_3)_2]$ through effective recrystallization, or reaction of the unstable phase to form CoN at a well-established reaction front. Reaction, once established, is relatively rapid and is insensitive to the presence of ammonia. The most significant feature of these comparative studies is the very similar pattern of reactions of salts with comparable cations but varied compositions and structures.

17.2.2. Cobalt(II) ammines

From TG, DTA and evolved gas analyses, Wendlandt and Smith [24] identified two intermediates in the dissociation of $[Co(NH_3)_6]Cl_2$:

$[Co(NH_3)_6]Cl_2 \rightarrow [Co(NH_3)_2Cl_2] + 4NH_3$
(TG 360 to 460 K, DTA endo maximum 416 K)

$\rightarrow [Co(NH_3)Cl_2] + NH_3$
(TG 460 to 493 K, DTA endo maximum 504 K)

$\rightarrow CoCl_2 + NH_3$
(TG 493 to 573 K, DTA endo maximum 559 K)

DTA maxima for the *cis* and *trans* forms of $[Co(NH_3)_2Cl_2]$ were at 507 and 500 K, respectively, and for the dissociation of the monoammine compound was 559 K. These observations confirm that the cobalt(II) compounds are less stable than those of cobalt(III), which is consistent with the postulated formation of cobalt(II) compounds as unstable intermediates in those decompositions of cobalt(III) compounds which are initiated by an electron transfer step.

The behaviour of the bromide was similar to that of the chloride, involving the formation of the intermediates $[Co(NH_3)_2Br_2]$ and $[Co(NH_3)Br_2]$. DTA peaks were detected at 441, 510 and 581 K, with evidence of melting of the monoammine compound above 523 K. Decomposition of the iodide yielded $[Co(NH_3)_2I_2]$ at about 450 K. This compound melted at about 480 K and decomposed at about 570 K to form CoI_2 (I^- was not oxidized).

$[Co(NH_3)_6](NO_3)_2$ lost ammonia at 410 K to produce the tetraammine compound which, after a phase transformation (453 K), exploded at about 480 K with the formation of Co_3O_4, NH_3, H_2O, N_2, and smaller amounts of NO and N_2O. The corresponding sulfate apparently lost two molecules of ammonia at 386 K, giving $[Co(NH_3)_4]SO_4$, and anhydrous cobalt(II) sulfate was obtained at 613 K. Again the deammination reactions of these salts are significantly influenced by the constituent anion.

Allan *et al.* [25] studied the decompositions of compounds of the type $[CoL_4]X_2$, where X = Cl^-, Br^- and I^- and L = pyridine; α-, β- and γ- picoline; 2,6-lutidine and 2,4,6-collidine. The changes occurring can be generally represented by the series of consecutive reactions:

$CoL_4X_2 \rightarrow CoL_2X_2$ (between 373 and 413 K) $\rightarrow CoLX_2$ (between 423 and 523 K) $\rightarrow CoL_{2/3}X_2$ (between 513 and 563 K) $\rightarrow CoX_2$ (above 563 K),

where all compounds were octahedral except for the tetrahedral CoL_2X_2. Differences of behaviour between individual salts were influenced both by L and by X. The decomposition [26] of $CoLCl_2$, where L is methylpyrazine or 2,5-dimethylpyrazine, proceeds to completion in two stages, also (see above) through the intermediate formation of $CoL_{2/3}X_2$. The reactions of the bromide analogues, believed to occur as dimers with tetrahedral structure ($Co_2L_2Br_4$), proceed to completion in a single stage. Because the peak of the endotherm for both bromides coincided at 613 K, it was concluded (from sterochemical considerations) that the cation-ligand bond occurs at the N-1 atom of the ligand.

17.2.3. Chromium(III) ammines

The thermal decompositions of the hexaamminechromium(III) compounds $[Cr(NH_3)_6]X_3$ (where X = Cl^-, Br^-, I^- and NO_3^-) have been studied by Wendlandt and Chou [27]. DTA curves for these compounds heated in helium are shown in Figure 17.4.

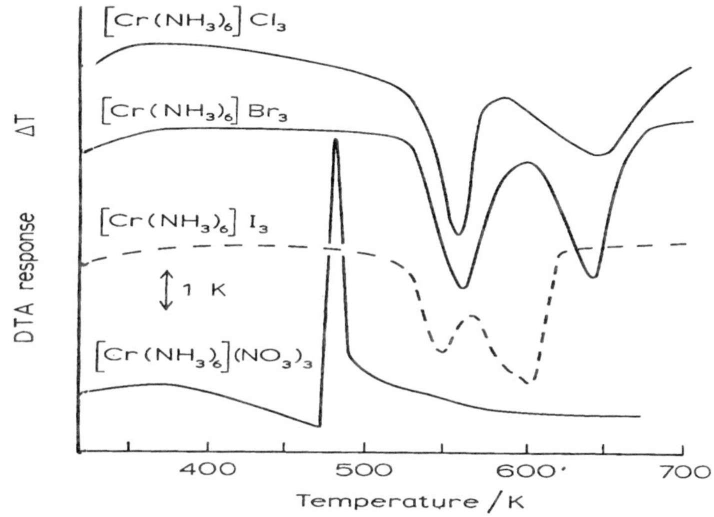

Figure 17.4. DTA curves for $[Cr(NH_3)_6]X_3$ (where X = Cl^-, Br^-, I^- and NO_3^-) [27].

During reactions of the chloride and bromide, $[Cr(NH_3)_5X]X_2$ and $[Cr(NH_3)_3X_3]$ were formed as intermediates. The latter reacted at about 700 K:

$$4[Cr(NH_3)_3X_3] \rightarrow 3CrX_3 + CrN + 8NH_3 + 3NH_4X$$

Problems of establishing the structure of this reactant have been discussed [4]. The first reaction of the iodide (DTA endotherm maximum at 553 K) was probably to form $[Cr(NH_3)_3I_3]$, but the final products were not characterized. $[Cr(NH_3)_6](NO_3)_3$ exploded at about 480 K, some 40 K above the temperature of onset of an endothermic process. This is reminiscent of the behaviour of the Co^{3+} analogue [11] and is attributed to the oxidizing power of the anion.

From DTA and TG studies [28] the steps during reaction of $[Cr(NH_3)_6]Cl_3$ in air were established as:

$$[Cr(NH_3)_6]Cl_3 \rightarrow [Cr(NH_3)_2]Cl_3 + 4NH_3 \text{ (519 to 578 K; } E_a = 20 \text{ kJ mol}^{-1})$$

$$[Cr(NH_3)_2]Cl_3 \rightarrow CrCl_3 + 2NH_3 \text{ (577 to 629 K; } E_a = 20 \text{ kJ mol}^{-1}).$$

$CrCl_3$ reacted with oxygen ($\rightarrow Cr_2O_3$) above 647 K.

Nagase and Tanaka [29] studied the decompositions of $[Cr(NH_3)_6]X_3$ compounds (where X = Cl^-, Br^-, I^- and NCS^-) using isothermal pressure-time measurements. The values of E_a (shown in brackets) for the stepwise substitution of the ammonia ligand by chloride were:

$$[Cr(NH_3)_6]Cl_3 \rightarrow [Cr(NH_3)_5Cl]Cl_2 \text{ (about 429 K, 125 kJ mol}^{-1}) \rightarrow$$

$$[Cr(NH_3)_4Cl_2]Cl \text{ (159 kJ mol}^{-1}) \rightarrow [Cr(NH_3)_3Cl_3] \text{ (130 kJ mol}^{-1})$$

The value of E_a for the first reaction of the bromide, to yield $[Cr(NH_3)_5Br]Br_2$, was 180 kJ mol^{-1}. The iodide decomposed differently:

$$[Cr(NH_3)_6]I_3 \rightarrow CrN + 3NH_4I + 2NH_3 \text{ (205 kJ mol}^{-1})$$

The reactions of the thiocyanates were:

$$[Cr(NH_3)_6](NCS)_3 \rightarrow [Cr(NH_3)_3(NCS)_3] + 3NH_3 \text{ (117 kJ mol}^{-1})$$

$$[Cr(NH_3)_5(NCS)](NCS)_2 \rightarrow [Cr(NH_3)_3(NCS)_3] + 2NH_3 \text{ (150 kJ mol}^{-1})$$

Stability, as judged by the temperature at which a defined rate of reaction was attained, was in the sequence:

NCS^- (350 K) < Cl^- (429 K) < Br^- (457 K) < I^- (467 K)

This sequence of reactivity is different from that found for the corresponding cobalt(III) compounds, in which electron transfer was identified as the initial step. The reactions of the chromium(III) salts (which are not reduced) proceed with direct substitution of the coordinated group and the following mechanisms are proposed: (i) an S_N1 mechamism, for which the intermediate is a 5-coordinated ion, e.g. $[Cr(NH_3)_5]^{3+}$, which possibly occurs in the reactions of the bromide and the iodide; or (ii) an S_N2 mechanism, which necessitates temporary 7-coordination, which possibly occurs during the decomposition of the chloride.

The decompositions of compounds containing the ions $[Cr(NH_3)_5(H_2O)]^{3+}$ or $[Cr(NH_3)_4(H_2O)_2]^{3+}$ were studied by Wendlandt *et al.* [30,31]. In subsequent investigations of the thermal deaquation and anation reactions of $[Cr(NH_3)_5(H_2O)]X_3$ (X = Cl^-, Br^-, I^- and NO_3^-), Nagase and Yokobayashi [32] supplemented TG and thermochemical studies with isothermal measurements of evolved gas pressure. α-time curves were sigmoid and values of E_a/kJ mol^{-1}, based on first-order rate constants, increased slightly in the sequence: NO_3^- (102) < Cl^- (110) < Br^- (124) < I^- (128). These values differ from earlier values estimated from non-isothermal data.

Kinetic data for the deaquation-anation reactions [33] of a range of mixed amine coordination compound salts of chromium(III) were discussed with a view to determining the mechanisms of the reactions observed. From the results it was concluded that behaviour is influenced by the available free space in the crystal structure and that dehydration, accompanied by isomerization, involves the simultaneous rupture of Cr-N (organic ligand) and Cr-O (water) bonds.

A kinetic study [34] of the decomposition of chromium(III) *tris*-N-benzoyl-N-phenylhydroxylamine between 433 and 453 K, showed that the stable products were benzanilide, chromium(III) benzoate and an unidentified chromium product. α-time curves were sigmoid and fitted the Avrami-Erofeev equation with $n = 3.9 \pm 0.1$, except for the lowest (and least accurate) temperature. Observation of partially reacted material identified the formation of liquid regions, characteristically above the melting point (434 K) of the major product, benzanilide. The autocatalytic process was considered to be initial decomposition of the solid, followed by more rapid breakdown of reactant dissolved in the melt. E_a was about 170 kJ mol^{-1}. Progressive melting is also consistent with the sigmoid α-time curve.

17.2.4. Platinum and palladium ammines

Wendlandt and Funes [35] characterized the dissociation of the coordination compounds $[Pd(NH_3)_4]X_2$ as:

$$[Pd(NH_3)_4]X_2 \rightarrow trans\text{-}[Pd(NH_3)_2X_2] + 2NH_3$$

For X = Cl^-, Br^- or I^-, these reactions (studied in helium by TG) were completed at 458, 453 and 388 K, respectively. Further ammonia loss from the Br^- and I^- salts yielded $PdBr_2$ and PdI_2 at 633 and 483 K, respectively. The reaction of the chloride proceeded to metal, NH_4Cl and other products. The diammine intermediate was not detected in the decompositions of the nitrate and sulfate compounds. Reactions to palladium metal residue were completed at about 530 and 600 K, respectively.

The DTA curves for both α- and β-picoline (pl) platinum(II) coordination compounds, *cis*-$[Pt(pl)_2Cl_2]$, showed [36] exothermic peaks at 478 K which are caused by isomerization to the more stable *trans* isomer. At higher temperatures the organic ligand was released to yield $PtCl_2$. The pyridine (py) compound *cis*-$[Pt(py)_2Br_2]$ behaves similarly undergoing an exothermic transformation to the *trans* form at 528 K, prior to dissociation at 553 K.

Möller and Meyer [37] used complementary experimental techniques capable of characterizing the phases and intermediates present, including X-ray powder diffractometry, to formulate the mechanisms of the complicated decompositions of ammonium containing platinum metal group coordination compounds. The proton in NH_4^+ may act as an acid or an oxidant and N^{3-} may behave as a reductant or a base. The decomposition of $(NH_4)_2[PtCl_6]$ is regarded as an internal redox reaction between Pt^{4+} and N^{3-} (from NH_4^+). X-ray studies detected $(NH_4)_2[PtCl_6]$ and, above 660 K, the product was finely divided Pt metal powder only. $PtCl_4$ decomposes above 760 K and is not, therefore, an intermediate. Similar behaviour was described for a number of comparable coordination compounds of Pt, Pd and Rh.

17.2.5. Nickel(II) ammines

The onset of deammination of the coordination compound $[Ni(NH_3)_6]Cl_2$ became appreciable [38] at 358 K, with the loss of $4NH_3$ and a further loss of $2NH_3$ at 488 K. No evidence of monoammine formation was obtained in these TG and DTA studies. $[Ni(NH_3)_6]SO_4$ gave some evidence of diammine production which further reacted to $NiSO_4$ by about 800 K. $[Ni(NH_3)_6](NO_3)_2$ lost 2 to 3 NH_3 prior to an exothermic oxidation reaction (533 K). Prost *et al.* [39] concluded that in this reaction NH_3 is replaced by NO_3^- during very slow oxidation of Ni^{2+} while maintaining the octahedral coordination of the central atom. The complexity of oxidation reactions involving nitrate decomposition in these coordination compounds is shown by a study [40] which included the analysis of the products of the reaction of $[Ni(aniline)_2](NO_3)_2.4H_2O$. In addition to H_2O and NH_3, some 44 different organic products were detected. This contrasts with the formation of aniline which was the only organic substance detected from breakdown of the corresponding chloride, bromide, iodide and sulfate.

The first step of the decomposition of hexaamminenickel(II) perchlorate [41] is the reversible dissociation:

$$[Ni(NH_3)_6](ClO_4)_2 \leftrightarrows [Ni(NH_3)_2](ClO_4)_2 + 4NH_3$$

Isothermal α-time curves (340 to 393 K) showed an induction period followed by a deceleratory reaction which could be described by the contracting volume equation with a discontinuity at $\alpha \approx 0.35$. The value of E_a (71 kJ mol^{-1}) for this step was close to the enthalpy of dissociation and the apparent energy barrier to the diffusion-controlled reammination reaction was small (about 8 kJ mol^{-1}). The energy barrier to deammination was identified [41] as rupture of the cation-ligand bond. The subsequent irreversible breakdown of the diammine (above 513 K) yielded residual NiO and the kinetics were comparable in some respects with the decomposition of NH_4ClO_4.

Jóna *et al.*, [42] in a similar study of the dissociation of various coordination compounds $[NiL_2(NCS)_2]$ (L = pyridine, α- or β-picoline, 2,6-lutidine or quinoline) showed that the nature of the ligand influenced whether the reactions occurred in one or in two steps, and that reaction also depended on the physical conditions. The same workers [43] have shown that the sequence of steps involved in the dissociation of $[Ni(NH_3)_4](NCS)_2$ was:

$$[Ni(NH_3)_4](NCS)_2 \rightarrow [Ni(NH_3)_3](NCS)_2 \rightarrow [Ni(NH_3)_2](NCS)_2 \rightarrow Ni(NCS)_2$$

Decomposition of a sample of the same salt prepared by a heterogeneous reaction differed in that the triammine intermediate was not produced.

17.2.6. Copper(II) ammines

Hexaamminecopper(II) chloride and bromide [44] undergo reversible [4] dissociation to yield the diammine:

$$[Cu(NH_3)_6]X_2 \rightleftarrows [Cu(NH_3)_2X_2] + 4NH_3$$

and the latter, at higher temperature, may react further with cation reduction:

$$6[Cu(NH_3)_2X_2] \rightarrow 6CuX + 6NH_4X + N_2 + 4NH_3$$

Intermediate monoammine formation was detected in reaction of the chloride, but not the bromide. The reaction of the iodide $[Cu(NH_3)_4]I_2$ first resulted in ammonia release ($-2NH_3$) and thereafter the anion participated in cation reduction:

$$2[Cu(NH_3)_2 I_2] \rightarrow 2CuI + I_2 + 4NH_3$$

The relative stabilities of these halides ($I^- < Br^- < Cl^-$) and the behaviour of the iodide are reminiscent of the reactivities of the corresponding hexaamminecobalt(III) salts.

Decomposition of $[Cu(NH_3)_4]SO_4.H_2O$ was initiated [45] by dehydration followed by ammonia dissociation ($-2NH_3$) to yield first the diammine and then the monoammine. The latter step was accompanied by some reduction, although Cu_2SO_4 formation is thought to be improbable. $[Cu(NH_3)_4]CrO_4$ decomposed [46] in two processes, at 420 and 480 K, each involving the release of $2NH_3$. Values of E_a for the consecutive isothermal processes were 75 and 163 kJ mol^{-1}, respectively. At 524 K the rates of the two reactions were equal and this behaviour suggested the direct production of $CuCrO_4$, because the quantity of the $[Cu(NH_3)_2]CrO_4$ intermediate present underwent no systematic variation with time.

Studies [47] of the dehydrations of two hydrated copper(II) chloride compounds, containing piperazinium or pyrimidinium ligands in which the planar $[CuCl_3H_2O]^-$ units are linked by longer semi-coordinate bonds (Cu - Cl) to form stacks, showed that, on heating, crystals of the pyrimidinium salt were dehydrated between 333 and 373 K. Following surface reorganization, there was a sudden production of liquid that burst from the particles undergoing crystallization at 393 K. This was ascribed to the formation of an impermeable layer of product that inhibited escape of evolved water. This pattern of behaviour contrasted with the smooth dehydration of powdered samples up to 413 K. Dehydration of the related piperazinium salt depended on reactant ageing. Low temperature dehydration (343 to 353 K) was absent from aged material. Reactants of both types showed a phase transition at about 413 K, before dehydration was completed, identified as resulting from the intervention of a lower hydrate. Again, a liquid phase was detected during the later stages of reaction. The pattern of behaviour has similarities with the dehydrations of the simple salts, $CuCl_2.2H_2O$ and $NiSO_4.6H_2O$, and mechanisms are discussed with reference to the layer structures of the present reactants.

The dissociation of $[Ag(NH_3)_2]_2SO_4$ is reversible [48] involving all the constituent ammonia. The rate of dissociation is initially at a maximum, diminishing thereafter. The contracting volume equation applied. Thermal pretreatment enhanced the rate of reaction.

17.2.7. Metal hexacyanoferrates

Raj and Danon [49] have used Mössbauer spectra to characterize changes in the valence state of iron during the decompositions of anhydrous alkali metal hexacyanoferrate(III)s and propose the following scheme:

$$2M_3[Fe^{3+}(CN)_6] \rightarrow 6MCN + 2Fe(CN)_2 + (CN)_2$$

$$4MCN + Fe(CN)_2 \rightarrow M_4[Fe^{2+}(CN)_6]$$

$$M_3[Fe^{3+}(CN)_6] + Fe(CN)_2 \rightarrow MFe^{3+}[Fe^{2+}(CN)_6] + 2MCN$$

$$3Fe(CN)_2 \rightarrow (Fe^{2+})_2[Fe^{2+}(CN)_6]$$

In the hydrated coordination compounds, one water molecule underwent an anation reaction to yield the aquapentacyano coordination compounds which were dehydrated between 370 and 470 K:

$$2M_2[Fe^{3+}(CN)_5H_2O] \rightarrow M_4[Fe^{2+}(CN)_6] + Fe(CN)_2 + 2H_2O + (CN)_2$$

Seifer [50] concluded that the stabilities of the alkali metal hexacyanoferrate(II)s increased in the sequence Cs < Rb < K < Na in:

$$M_4[Fe^{2+}(CN)_6] \rightarrow 4MCN + Fe(CN)_2$$

and below 720 K all of the iron remains coordinated. The cadmium and zinc hexacyanoferrate(II) monohydrates lost water [50] at 453 and 513 K, respectively. Subsequent decomposition of the zinc salt:

$$Zn_2[Fe(CN)_6] \rightarrow 2Zn(CN)_2 + Fe(CN)_2$$

occurred at 603 K, whereas the cadmium analogue was stable to almost 1000 K in argon.

17.2.8. Other metal cyano-compounds

The decomposition of $K_2[Pt(CN)_4Br_2]$ was studied [51] by TG, infrared analyses and X-ray diffraction. The first step was elimination of cyanogen with the formation of a Pt(III) compound including bridging cyanide groups. On further heating this dissociated giving a platinum compound which later decomposed with the release of platinum metal. The coordination compound $K_2[Pt(CN)_4].8H_2O$ was first dehydrated during progressive heating and later evolved cyanogen leaving the residue Pt+2KCN.

In a comparative study [52] of the temperatures and E_a values for decompositions of $Cu_3[Co(CN)_6]_n$, where n = 1 or 2, and the corresponding Hg^+ salts, the univalent cations were found to promote cleavage of the Co - CN bond with simultaneous transfer of electrons from ligand to the cation.

17.2.9. Dehydration/anation reactions yielding binuclear coordination compounds

Thermogravimetric studies of the solid state reactions:

$$[M(H_2O)(NH_3)_5][Cr(CN)_6] \rightarrow [(NH_3)_5MNCCr(CN)_5] \text{ (M=Rh,Ir) and}$$

$$[M(H_2O)(NH_3)_5][CrNO(CN)_5] \rightarrow [(NH_3)_5MNCCrNO(CN)_4] \text{ (M=Co,Rh,Ir)}$$

showed [53] that these were generally represented by the Avrami-Erofeev equation ($n = 2$) and values of E_a were between 104 and 135 kJ mol^{-1}. A dissociative ($-H_2O$) S_N1 reaction mechanism is proposed. Water escapes relatively easily from the Rh(III) and Ir(III) crystals because of the large cation/anion size ratio.

Kinetic studies [54] of the thermal deaquation-anation reactions of the isostructural coordination compounds $[M(H_2O)(NH_3)_5][Co(CN)_6]$ where M is Co^{3+}, Rh^{3+} or Ir^{3+}, showed that, following water loss, new compounds $[(NH_3)_5M\text{-}NC\text{-}Co(CN)_5]$ were formed, which were identified by spectroscopic measurements. Their stabilities decreased in the sequence Ir > Rh > Co. Magnitudes of activation energies showed the inverse order: Co = 130, Rh = 110 and Ir = 95 kJ mol^{-1}. The Avrami-Erofeev equation fitted the data ($0.2 < \alpha < 0.8$). It is concluded that the deaquation-anation reaction proceeds by a S_N1 dissociative mechanism with a square-based pyramidal activated coordination compound. Water is eliminated as Frenkel-type defects. Greater interlattice space is expected in the Rh and Ir compounds due to the greater volume of these coordinated cations, thus facilitating escape of water molecules.

17.3. MONONUCLEAR COORDINATION COMPOUNDS WITH POLYDENTATE LIGANDS

17.3.1. Ring formation

Compounds containing bidentate ligands, which form a chelate ring, are relatively more stable than analogous compounds with monodentate ligands. Such ring systems are generally least readily dissociated when comprised of five or six atoms (including the central atom). The coordination compounds derived from acetylacetone (acac) are stable non-electrolytes, e.g. $[Al(acac)_3]$ and $[Cr(acac)_3]$, containing the structure :

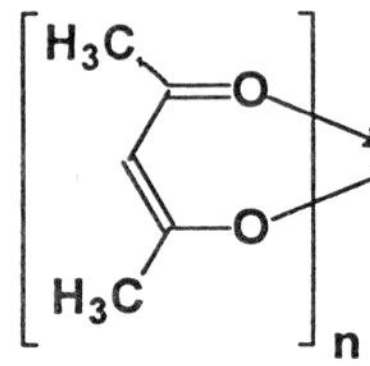

Some may be volatilized unchanged above 550 K. Salicylaldehyde, and ligands of related structure [17], form compounds containing the ring system:

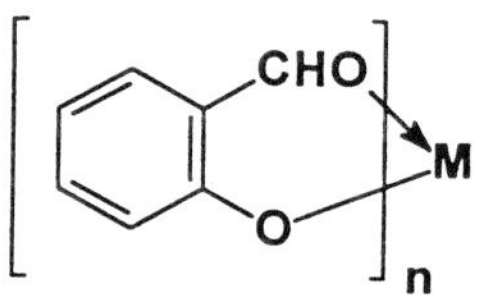

(MX_2 where M^{2+} is Cd, Cu, Co or Ni) which are more stable than coordination compounds containing phenol or benzaldehyde.

17.3.2. Ethylenediamine(en) cobalt(III) coordination compounds

Using several complementary experimental methods (TG, DTA, magnetic, spectroscopic measurements and mass spectra), Collins *et al.*, [55] concluded that, in nitrogen, the chloride $[Co(en)_3]Cl_3$ and the analogous bromide decompose as follows:

$$[Co(en)_3]Cl_3 \rightarrow trans\text{-}[Co(en)_2Cl_2]Cl \rightarrow CoCl_2 + (NH_4)_2CoCl_4 \rightarrow CoCl_2 + 2NH_4Cl$$

The ammonium ion is a product of breakdown of the organic ligand. In vacuum, the intermediate formation of the *trans* compound was not detected. The thiocyanate analogue of the *trans* diammine decomposed to yield $(NH_4)Co(SCN)_3$ and the sequence of steps involved in decomposition of $[Co(en)_3]I_3$ could not be established.

Zsakó *et al.* [56] concluded that the formation of $[Co(en)_2X_2]X$ is only possible when $X = Cl^-$, because the other ligands studied (Br^-, I^-, NO_3^-, CNO^-) are susceptible to reactions involving the Co^{3+} ion. These conclusions are not completely consistent with those of Collins *et al.* [55], although the two sets of observations refer to slightly different conditions.

17.3.3. Other cobalt(III) diamine coordination compounds

Le May [57] has studied the early stages of the decompositions of *trans*-$[Co(pn)_2Cl_2](H_5O_2)Cl_2$ (pn = 1,2-propanediamine) and *trans*-$[Co(en)_2Cl_2](H_5O_2)Cl_2$, which release H_2O and HCl. The reaction of the $(en)_2$ coordination compound was a nucleation and growth process. The initial advance of interface across surfaces was observed to be rapid and subsequent penetration of the crystal bulk proceeded more slowly. The contracting volume equation described the kinetics. Nucleation of the $(pn)_2$ compound was predominantly on the (100) faces and advance of the interface, viewed at right angles, appeared as a contracting area. The crystal planes of packed (pn) groups in this reactant were not penetrated by H_2O and HCl molecules. The rates of decomposition of both salts were, in part, controlled by water diffusion and values of E_a were about 75 kJ mol^{-1}, independent of particle size.

Chou and Olsen [58] found an unusual dependence of rate on particle size in the decomposition of isothiocyanatopentaammine cobalt(III) perchlorate. When α was less than 0.09, the larger particles decomposed relatively rapidly (E_a = 138 kJ mol^{-1}) and thereafter the reaction rate decreased and E_a was between 88 and 117 kJ mol^{-1}, depending on crystallite size. Initial internal stresses present in the larger particles are suggested to produce more reactive sites, but later ($\alpha > 0.09$) retention of product NH_3 in these largest crystals resulted in decreased reactivity.

In a review, discussing a wide range of previous studies of isomerization in coordination compounds of cobalt or chromium with several representative diamines, Yamaguchi *et al.* [59] concluded that, even in the solid state, any ligand in these metal coordination compounds was unexpectedly easy to move. Most of these isomerizations, including 22 reactions (*trans* to *cis* and *cis* to *trans* steps), occurred between about 400 and 520 K.

17.3.4. Ethylenediamine(en) chromium(III) coordination compounds

Zsakó *et al.* [56] found that values of E_a (obtained from non-isothermal data) for the reactions:

$$[Cr(en)_3]X_3 \rightarrow [Cr(en)_2X_2]X$$

increased in the sequence X = $NCS^- < Cl^- < Br^- < SbS_4^- < I^-$. Again the behaviour of Cr^{3+} contrasts with that of Co^{3+} in that the central cation was not reduced during decomposition.

Wendlandt and Stembridge [60] identified the following steps in the vacuum decomposition of $[Cr(en)_3]Cl_3$:

$$[Cr(en)_3]Cl_3 \rightarrow \textit{cis}\text{-}[Cr(en)_2Cl_2]Cl\ (488\ K) \rightarrow Cr_2(en)_3Cl_6\ (503\ K); \rightarrow CrCl_2\ (700\ K)$$

The reactions of the corresponding salts containing the anions SCN^-, Br^- and I^- showed similarities with the above sequence of changes, while the presence of oxidizing groups resulted in explosions: NO_3^- at about 500 K and $Cr_2O_7^{2-}$ at 420 K. Ammonium halides were identified [60] as catalysts for the decomposition of some $[Cr(en)_3]^{3+}$ compounds. The first step in such processes was suggested [61] to be the protonation of a nitrogen of the ethylenediamine followed by dissociation at the Cr-N coordinate linkage.

House and Bailar [62] showed that the solid state reaction:

$$[Cr(en)_3](NCS)_3 \rightarrow \textit{trans}\text{-}[Cr(en)_2(NCS)_2](NCS)$$

was catalyzed by NH_4NCS and was first-order in concentration of catalyst. Values of E_a for the catalyzed and uncatalyzed reactions were 198 and 138 kJ mol^{-1}, respectively, and corresponding values for reaction of the $(pn)_3$ analogue were 230 and 186 kJ mol^{-1}. The activity of the additive is similarly identified with the proton which, on attachment to the nitrogen, decreases the strength of the Cr-N bond, thus decreasing the energy barrier (E_a) for substitution of thiocyanate into the coordination sphere. The stereochemical requirements of this process were discussed, but the relative significances of protonation and bond rupture were not characterized in detail.

Hughes [63] concluded that the reaction of $[Cr(en)_3]Cl_3$ in an inert atmosphere was catalyzed by NH_4Cl but not by NH_4Br or NH_4I (as had been suggested elsewhere). The reaction rate was sensitive to the disposition of salt within the reaction zone and could be influenced by oxidizing atmospheres. The activity of NH_4Cl is believed to reside in HCl resulting from dissociation, values of E_a for reaction of the pure salt, salt + NH_4Cl and salt in HCl are reported as 175, 128 and 79 kJ mol^{-1}, respectively.

17.3.5. Ethylenediamine(en) nickel(II) coordination compounds

The dissociation reactions of ethylenediamine coordination compounds of nickel probably [4] involved the evolution of a single (en) molecule as the first step in salt breakdown:

$$[Ni(en)_3]Br_2 \rightarrow [Ni(en)_2]Br_2 + en \quad (570\ K)$$

White beam synchrotron radiation was used in a topographical study [64] of the phase transition in $[Ni(en)_2\ (NCS)_2]$ single crystals. The short exposure times possible using this intense radiation enabled the strained centre of the crystal to be identified as the most imperfect zone, within which the nucleation and growth process was initiated. The phase transition commenced at defect sites. The role of generation and relaxation of mechanical stresses in influencing kinetics of solid state transformations was later reviewed [65].

17.3.6. Metal oxalato coordination compounds

Nagase [66] has shown, from measurements of the temperature of onset of reaction, that the sequence of stability for potassium salts containing the ions $[M(C_2O_4)_n]^{m-}$ was:

(M=) Mn < Co < Pd < Pt < Rh < Fe < Cu < Cr < V < Al < Zn

This order is approximately the same as that of the electron affinity of the central atom, but is not that of the strength of the M-O bond, measured spectroscopically. Reactions of the Co, Pd, Pt, Rh, Fe and Cu salts resulted in reduction of the central atoms.

Tanaka and Nanjo [67] made a comparative study of oxalate and ammine ligands. The sequence of relative stability of the cobalt(III) coordination compounds is:

$$[Co(NH_3)_6]Cl_3 > [Co(NH_3)_4(C_2O_4)]Cl > NH_4[Co(NH_3)_2(C_2O_4)_2] > K_3[Co(C_2O_4)_3]$$

This order is explained by the assumption that the initial step in decomposition involves electron transfer and that the oxalate group donates an electron more readily than ammonia. The quite different order of stabilities of some corresponding chromium(III) compounds:

$K_3[Cr(C_2O_4)_3] > [Cr(NH_3)_6]Cl_3 > [Cr(NH_3)_4(C_2O_4)]Cl$

was interpreted as evidence of the occurrence of a different mechanism of reaction.

$[Co(NH_3)_6]_2(C_2O_4)_3.4H_2O$ loses all water and $2NH_3$ by an endothermic reaction [68]. The solid product decomposes further to form CoC_2O_4 (together with NH_3, H_2O, CO, CO_2 and small amounts of N_2). Decomposition in air gives Co_3O_4, CoO and Co, and in argon the products are Co and CoO. Values of E_a determined varied with the experimental conditions.

A kinetic study [69] of the isothermal decomposition of potassium trioxalatoferrate(III):

$$2K_3[Fe(C_2O_4)_3] \rightarrow 2FeC_2O_4 + 3K_2C_2O_4 + 2CO_2$$

was made in helium between 506 and 535 K using gas chromatography to measure α. The sigmoid α-time curves were fitted by the Prout-Tompkins equation and E_a was 192 kJ mol^{-1}. The addition of iron(II) oxalate decreased the acceleratory period but did not change the value of the rate coefficient. The activation step is suggested to be the promotion of an electron from the oxalate group to the antibonding orbitals of the iron(III) ion. Iron(II) participates in the conversion of the activated coordination compound to products, but is not regarded as a catalyst because the magnitude of E_a is not decreased by participation of Fe^{2+}.

Following dehydrations (below 450 K) the sequence of steps in the decompositions of strontium and barium oxalatoferrates(III) was [70]:

$$Ba_3[Fe(C_2O_4)_3]_2 \rightarrow Ba_3Fe_2(C_2O_4)_5 \rightarrow 3BaCO_3.Fe_2O_3 \rightarrow BaCO_3.2BaFeO_{2.8} \rightarrow Ba_3Fe_2O_6.$$

There is loss of CO_2 and CO at each stage, except the last in which oxygen is also evolved. Successive steps may overlap. The behaviour of the strontium compound was similar, except that reaction temperatures were slightly different. The iron(III) is reduced to iron(II) at 570 K, reverts to iron(III) by 670 K and a significant proportion is converted to the unusual valence state, iron(IV) by 970 K. This form is reduced to iron(III) at higher temperatures (about 1000 K) with oxygen release. These four valence changes were characterized and confirmed using Mössbauer spectra of the cooled salt. A physical mixture of oxalates, containing the same proportions of cations, showed only the behaviour of the individual components and, in particular, the formation of iron(IV) was not observed.

17.3.7. Metal oximato coordination compounds

Zsakó and Lungu [71] found compensation behaviour in kinetic data obtained from non-isothermal TG measurements for 134 decomposition reactions of bis(dioximato)cobalt(III) compounds. The slopes of the compensation trends are influenced by both the organic ligand and the constituent anion which enters the coordination shell during reaction.

17.4. BINUCLEAR COORDINATION COMPOUNDS WITH POLYDENTATE LIGANDS

The reaction of urea with cobalt sulfate in aqueous solution gives the coordination compound $[Co_2(NH_2)_2(SO_4)_2(H_2O)_2]$ which is binuclear and includes two bridging sulfate groups [72]:

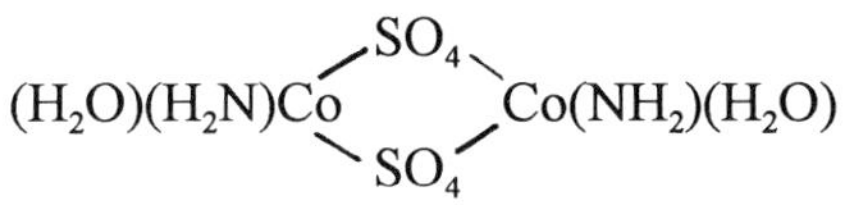

The first step in decomposition is the loss of $2H_2O$ (553 K, E_a = 97 kJ mol^{-1}). At 636 K an endothermic decomposition (E_a = 450 kJ mol^{-1}) yields $[Co_2(SO_4)_2]$ + NH_3 + $0.5N_2$ + $0.5H_2$. At 675 K (E_a = 15.5 kJ mol^{-1}) a reaction accompanied by no loss of mass was ascribed to breaking of the coordination bonds of the sulfato groups to give $CoSO_4$.

The thermal decompositions of the heterobinuclear coordination compounds M^{2+} : Cr^{3+} : diethylenetriaminepentaacetic acid (1:1:1) (M was Mn^{2+}, Fe^{2+}, Co^{2+}, Ni^{2+}, Cu^{2+} or Zn^{2+}) have been compared [73]. The complicated chemistry of these reactions is discussed only qualitatively but it is concluded that stability depends mainly on the chromium present and is less influenced by the M^{2+} ion.

17.5. FACTORS INFLUENCING THE THERMAL STABILITY OF COORDINATION COMPOUNDS

17.5.1. Effects of heating

The main changes which may occur on heating coordination compounds are removal (sometimes accompanied, or followed, by breakdown) of ligands, reaction of the ligand with either the central atom or the counterion, acceptance of an anion into the coordination sphere (*anation*) and/or isomerization. There is also the possibility of anion decomposition.

(i) *Removal of ligands*. Thermal decomposition usually involves a decrease in the ligand:central atom ratio. This process may occur in a stepwise manner, but each stage does not necessarily imply a decrease in the coordination number, because dimers and polymeric structures may be formed, or a site vacated on release of a ligand may be occupied by an anion (anation). Ligands may dissociate unchanged and these reactions may be reversible. The sequence of relative stabilities of groups within the coordination sphere [5] is given in Table 17.1.
(ii) *Reaction of the ligand*. The ligand group may react with either the central atom (e.g. coordinated NH_3 is converted to N_2 by reduction of $Co^{3+} \rightarrow Co^{2+}$) or the counterion (e.g. ligands are oxidized during decomposition of NO_3^- or ClO_4^- groups). Decompositions of coordination compounds may thus be accompanied by changes in oxidation number of the central atom. When iodide is present, this is usually oxidized to I_2. Oxalate ions readily yield electrons ($C_2O_4^{2-} \rightarrow 2CO_2 + 2e$). Copper(II) is readily reduced but chromium(III) frequently remains in the trivalent state.
(iii) *Reaction of the anion*. It has been suggested [74] that the stability of a coordination compound is greatest where the anion radius is largest. This is the opposite of the general trend shown in Table 17.1. and is clearly subject to additional considerations. On heating coordination compounds, the anion may be accepted into the coordination sphere (most easily studied by employing gentle conditions), or may initiate decomposition (e.g. nitrate, perchlorate, etc. are unstable and breakdown readily, while sulfate, chloride, etc. are less readily decomposed). The anion may also participate in, although not necessarily initiate, a chemical transformation (e.g. oxidation of I^- by cobalt(III) reduction), or remain as an anion on completion of decomposition of the compound.
(iv) *Isomerization*. The relative positions of ligands within the cooordination sphere may be interchanged without breakdown (e.g. *cis/trans* isomerization), or the bond structure within certain groups may be reorganized:
(e.g. $A\text{-}C{\equiv}N\text{-}B \rightarrow A\text{-}N{\equiv}C\text{-}B$).

A relative enhancement of stability is sometimes found when a 5- or 6-membered ring chelate is formed. The presence of a ring structure does not, however, invariably result in a decrease of reactivity. Other factors must also control or influence compound breakdown. The related compounds $[Cr(NH_3)_6]Cl_3$ and $[Cr(en)_3]Cl_3$ react at comparable temperatures, although the course of the changes at 553 K are different (yielding $3NH_3$ and (en) respectively [6]). The chemical properties of the ligand are important in controlling reactivity. The decrease in reaction temperature [67] as oxalate groups are substituted for ammonia in cobalt(III) coordination compounds is attributed to the greater ease of transfer of an electron from an oxalate group to the cation. In the comparable chromium(III) coordination compounds, where it is thought [67] that decomposition is not initiated by electron transfer, stability is not decreased by incorporation of oxalate into the coordination sphere.

Table 17.1.
Trends of stabilities of groups* inside the coordination sphere [5]

Decreasing stability →

$NH_3 > RNH_2 > R_2NH > R_3N$

$H_2O > ROH > R_2O > RCOR > RCHO$

$R_3As > R_3P > R_2S > PX_3$

$CO > NO$

$CN^- > SCN^- > F^- > OH^- > Cl^- > Br^- > I^-$

* R = alkyl or aryl radical
X = halogen

17.5.2. The influence of experimental conditions

The course of an observed change can also be sensitive to the experimental conditions. (i) The kinetics of reversible dissociations are sensitive to the prevailing pressure of product and to the presence of an inert gas which can inhibit the diffusive loss of a volatile product. (ii) Observed behaviour can be significantly influenced by the rate of sample heating. (iii) The chemical transformations occurring will be different when a reactive gas (e.g. O_2,H_2,Cl_2, etc.) is present which may interact directly with the reactant or the primary products. (iv) The significance of phase transformations in modifying reactivity and the influences of melting or aqueous fusion on the course of decomposition must be considered during the interpretation of data. (v) The possibility of reaction with, or catalytic changes on, the surfaces of the sample container must also be examined.

17.5.3. Rate limiting steps

The rate limiting step during thermal decomposition of a coordination compound may, in principle, be controlled by the chemical properties of any one of the constituents (the central atom(s), the ligands, or the counterions) or by interactions between some or all of these constituents. In general, the reactivities of crystalline coordination compounds are most influenced by the following features of the reactant components.

(i) *Chemical properties of the central atom*: the strength and number of sigma bonds present and which can be formed, together with the tendency of the central atom to change its oxidation state.

(ii) *Chemical properties of the ligands*: more than a single species may be present and behaviour will depend on whether the ligands are released unchanged or tend to oxidize/reduce the other groups present, and/or themselves to be oxidized/reduced by the other constituents.

(iii) *Disposition of ligands about the central atom*: bond arrangement (octahedral, tetrahedral, planar or other), stereochemistry [5], bridging by polydentate ligands which often imparts additional thermal stability to the coordination compound [5], bridging isomerism [75] resulting from coordination at different constituent atoms (e.g. Cd^{2+}-N≡C-Fe^{3+} → Cd^{2+}-C≡N-Fe^{3+}).

(iv) *Chemical properties of the counterion*: reactivity (oxidation/reduction) with or by other components of the compound or the products of its breakdown. Counterions are sometimes accepted into the coordination shell replacing another ligand.

(v) *Nature of the prevailing atmosphere*: this may sometimes oppose the diffusive escape of a reversibly dissociated molecule, or provide an alternative molecule that may be incorporated into the coordination shell. Any reactive molecule (O_2, Cl_2, etc.) may be capable of causing a different chemical transformation by reacting with the coordination compound under investigation, or any of its breakdown products.

17.6. CONCLUSIONS

Many of the experimental studies of the thermal decompositions of coordination compounds have been restricted to non-isothermal measurements primarily directed towards identification of the occurence of a reaction and the characterization of the major products of this change. The improved sensitivity of experimental methods (notably TG and DSC or DTA) has revealed the chemical complexity of the thermal reactions of solid coordination compounds. Much comparative information concerning the relative reactivities of related materials has been obtained. While some similarities of behaviour have been found for compounds containing comparable components, few general patterns of reactivity, related to either the physical or the chemical properties of this diverse collection of reactants, have been established, in spite of the hundreds of publications on the subject.

For many of the decompositions of coordination compounds which have been described in the literature, there is insufficient information concerning the following features of the rate processes under consideration. (i) The *reversibility* of dissociation. This is particularly important during dynamic kinetic experiments where local variations of pressure of the volatile product within the reactant mass can influence the overall rate of progress of breakdown. (ii) The occurrence of reactions in which an existing ligand is replaced by acceptance of an anion within the coordination sphere may be masked by the onset of decomposition at a slightly higher temperature. A particular chemical transformation may be a single reaction or it may be the result of concurrent or consecutive changes proceeding within the several different bonds that constitute the structure of the coordination compound. (iii) Few studies have been concerned with the *geometry* of interface advance during reaction. Kinetic studies of thermal breakdown of this type of reactant have rarely been extended to include microscopic examinations.

The unreliability of some of the reported kinetic parameters (A, E_a and the rate equation) through incomplete control of experimental conditions (e.g. allowing the reverse reaction to contribute), increases the difficulties of identifying reactivity controls and in formulating reaction mechanisms.

Joyner's work [23] on the reactions of the amminecobalt(III) azides showed that the presence of a reaction interface was apparently of greater significance than the internal structural features of the reactant. Despite the unusual feature of this group of reactants, the availiability of two distinct reaction paths, the behaviour of four different salts was largely independent of whether the initial azide groups were covalently bonded to the cation or present as the anion. Reactant restructuring preceded the main decomposition reaction.

Dollimore [5] has discussed some aspects of the influence of the central atom on the thermal stabilities of solid coordination compounds. The most fully characterized compounds are those of Co, Cr and Pt. Central atoms of relatively small radius but high oxidation number coordinate most effectively. The possible influences of melting and dependences upon reaction conditions increase the difficulties of identification of the factors which control reactivity. Stabilities are influenced [5] by the electronic structure of the coordinated ion, whether this is a normal or a penetration compound. In the latter species, covalent bonding involves $3d^24s4p^3$ orbitals, whereas in a normal coordination compound the $4s4p^34d^2$ orbitals participate. Bonding is predominantly ionic, or ion-dipole, and covalent links are weaker and more polar.

Corbella and Ribas [33], in a systematic study of the factors that influence the kinetics of dehydration and anation in solid coordination compounds, identified the availability of free space in the crystal structure as an important controlling factor in reactions proceeding by a dissociative S_N1 mechanism.

Tanaka *et al.* [76] have studied the decompositions of various solid hexaammine, halopentaammine and oxalatoammine cobalt(III) compounds. Kinetic data for these reactions (A and E_a) were determined from changes in the concentrations of Co^{3+} with time, and values of E_a were found to parallel the magnitudes of the orbital energies of the compounds.

Liptay *et al.* [77] have sought to identify the factors which determine the strengths of the metal - N(pyridine) bond in the compounds $[M(py)_n]X_z$, where M = Mn^{2+}, Co^{2+}, Ni^{2+}, Cu^{2+}, Zn^{2+} and Cd^{2+}, (py)= pyridine, α-, β- or γ-picoline, X = Cl^-, Br^-, I^-, OCN^-, SCN^-, NO_3^- or SO_4^{2-}, n = 2, 3, 4 or 6 and z = 1 or 2. The stabilities are shown to increase in the same sequence (Mn < Co < Ni < Cu) as that of the high spin transition metal coordination compounds.

From studies of isomerizations and racemizations of optically active octahedral coordination compounds, Fujiwara and Bailar [78] concluded that none of the several mechanisms proposed was capable of accounting for the observations for all of the reactions investigated.

There is a need for more critical reviews of this field and for studies in greater depth rather than greater breadth.

REFERENCES

1. G. Wilkinson, R.D. Gillard and J.A. McCleverty (Eds), Comprehensive Coordination Chemistry, Pergamon, Oxford, 1987, Vols 1-7.
2. B.F.G. Johnson, Comprehensive Inorganic Chemistry, (Eds J.C. Bailar, H.J. Emeleus, R. Nyholm and A.F. Trotman-Dickenson), Pergamon, Oxford, 1973, Vol.4, Chap.52.
3. S.J. Ashcroft and C.T. Mortimer, Thermochemistry of Transition Metal Complexes, Academic Press, London, 1970.
4. W.W. Wendlandt and J.P. Smith, The Thermal Properties of Transition Metal Ammine Complexes, Elsevier, Amsterdam, 1967.
5. D. Dollimore, Differential Thermal Analysis, (Ed. R.C. Mackenzie), Academic Press, London, Vol.1, 1970, Chap.14.
6. W.W. Wendlandt, Thermal Analysis, Wiley, New York, 3rd Edn, 1986.

7. O. Carp and E. Segal, Rev. Roum. Chim., 39 (1994) 1123.
8. J.O. Hill, Thermochim. Acta, 200 (1992) 187.
9. G. D'Ascenzo, R. Curini, G. De Angelis, E. Cardarelli, A.D. Magri, M. Tomassetti and A. Marino, Gazz. Chim. Ital., 113 (1983) 367.
10. G. Liptay, Thermochim. Acta, 15 (1976) 159.
11. W.W. Wendlandt, J. Inorg. Nucl. Chem., 25 (1963) 545.
12. E.L. Simmons and W.W. Wendlandt, Anal. Chim. Acta, 35 (1966) 461; J. Inorg. Nucl. Chem., 28 (1966) 2187.
13. G.W.Watt, Inorg. Chem., 3 (1964) 325.
14. N. Tanaka and M. Nanjo, Bull. Chem. Soc. Japan, 37 (1964) 1330.
15. L.W. Collins, W.W. Wendlandt and E.K. Gibson, Thermochim. Acta, 8 (1974) 303, 315.
16. Z.D. Zivkovic, J. Thermal Anal., 41 (1994) 99.
17. W.W. Wendlandt and J.P. Smith, J. Inorg. Nucl. Chem., 25 (1963) 843, 1267; 26 (1964) 445.
18. E. Ingier-Stocka, J. Thermal Anal., 37 (1991) 521, 769.
19. Z.Halmos and W.W. Wendlandt, Thermochim. Acta, 5 (1972) 165.
20. L.M. Zheng, W-H. Zhang and X-Q. Xin, Thermochim. Acta, 222 (1993) 273.
21. W.W. Wendlandt, W.R. Robinson and W.Y. Yang, J. Inorg. Nucl. Chem., 25 (1963) 1495.
22. W.W. Wendlandt and J.P. Smith, J. Inorg. Nucl. Chem., 26 (1964) 1619.
23. T.B. Joyner, J. Phys. Chem., 69 (1965) 1723; 71 (1967) 3431; 72 (1968) 703, 4386; 74 (1970) 1552, 1558, 1563.
24. W.W. Wendlandt and J.P. Smith, J. Inorg. Nucl. Chem., 25 (1963) 985.
25. J.R. Allan, D.H. Brown, R.H. Nuttall and D.W.A. Sharp, J. Inorg. Nucl. Chem., 26 (1964) 1895.
26. E. Koros, S.M. Nelson, F. Paulik, L. Erdey and F. Ruff, Magy. Kém. Foly., 70 (1964) 468.
27. W.W. Wendlandt and C.Y. Chou, J. Inorg. Nucl. Chem., 26 (1964) 943.
28. Z.D. Zivkovic, N. Milosavljevic, M. Grotowska and W. Wojciechowski, J. Thermal Anal., 36 (1990) 1411.
29. K. Nagase and N. Tanaka, Bull. Chem. Soc. Japan, 46 (1973) 2435.
30. W.W. Wendlandt and J.L. Bear, J. Inorg. Nucl. Chem., 22 (1961) 77.
31. W.W. Wendlandt and W.R. Robinson, J. Inorg. Nucl. Chem., 26 (1964) 531.
32. K. Nagase and H. Yokobayashi, Bull. Chem. Soc. Japan, 47 (1974) 2036.
33. M. Corbella and J. Ribas, Inorg.Chem., 26 (1987) 3589.

34. J.C. Machado, M.M. Braga, A.M.P.R. Da Luz and G. Duplatre, J. Chem. Soc., Faraday Trans. I, 76 (1980) 152.
35. W.W. Wendlandt and L.A. Funes, J. Inorg. Nucl. Chem., 26 (1964) 1879.
36. M.I. Ivanova, Zh. Neorg. Khim., 2 (1957) 1317, 1324, 1775.
37. A.Möller and G. Meyer, Thermochim. Acta, 210 (1992) 147.
38. T.D. George and W.W. Wendlandt, J. Inorg. Nucl. Chem., 25 (1963) 395.
39. M. Prost, É. Garbowski and M-V. Mathieu, J. Chim. Phys., 71 (1974) 1519.
40. P. Lumme and J.Korvola, Thermochim. Acta, 9 (1974) 109.
41. R.A.F. Sherriff and A.K. Galwey, J. Chem. Soc. A, (1967) 1705.
42. E. Jóna, T. Šramko and J. Gazo, J. Thermal Anal., 4 (1972) 61.
43. E. Jóna, T. Šramko and J. Gazo, J.Thermal Anal., 7 (1975) 551.
44. J.P. Smith and W.W. Wendlandt, J. Inorg. Nucl. Chem., 26 (1964) 1157.
45. W.W. Wendlandt and T.M. Southern, Anal. Chim. Acta, 32 (1965) 405.
46. C.S. Kelley, L.L. Pytlewski and L. Eng, J. Inorg. Nucl. Chem., 37 (1975) 1367.
47. T. Manfredini, G.C. Pellacani, A. Bonamarti-Corradi, L.P. Bataglia, G.G.T. Guarini, J.G. Giusti, G. Pon, R.D. Willett and D.X. West, Inorg. Chem., 29 (1990) 2221.
48. T.I. Torgonskaya and M.M. Pavlyuchenko, Vesti Akad. Navuk Belaruss. SSR, Ser.Khim. Navuk, (1970) 5.
49. D. Raj and J. Danon, J. Inorg. Nucl. Chem., 37 (1975) 2039.
50. G.B. Seifer, Zh. Neorg. Khim., 7 (1962) 1242; 8 (1963) 1172.
51. P.K. Gallagher and J.P. Luongo, Thermochim. Acta, 12 (1975) 159.
52. P. Tribunescu and G. Pirlea, Zh. Neorg. Khim., 13 (1968) 3256.
53. J. Ribas, A. Escuer and M. Monfort, Inorg. Chem., 24 (1985) 1874.
54. J. Ribas, A. Escuer and M. Monfort, Thermochim. Acta, 76 (1984) 201.
55. L.W. Collins, W.W. Wendlandt and E.K. Gibson, Thermochim. Acta, 8 (1974) 205, 307.
56. J.Zsakó, Cs. Várhelyi, G. Liptay and K. Szilágyi, J. Thermal Anal., 7 (1975) 41.
57. H.E. Le May, Inorg. Chem., 7 (1968) 2531.
58. C.J. Chou and F.A. Olson, Anal. Chem., 44 (1972) 1841.
59. S.Yamaguchi, Y. Hayashi, M.Suzuki and A.Uehara, Thermochim. Acta, 208 (1992) 191.
60. W.W. Wendlandt and C.H. Stembridge, J. Inorg. Nucl. Chem., 27 (1965) 129, 569, 575.
61. W.W. Wendlandt and L.V. Sveum, J. Inorg. Nucl. Chem., 28 (1966) 392.

62. J.E. House and J.C. Bailar, J. Amer. Chem. Soc., 91 (1969) 67.
63. M.A. Hughes, J. Thermal Anal., 8 (1975) 99.
64. T.P. Shahtshneider, E. Yu. Ivanov and V.V. Boldyrev, React. Solids, 2 (1987) 339.
65. A.P. Chupakhin, A.A. Sidelnikov and V.V. Boldyrev, React. Solids, 3 (1987) 1.
66. K. Nagase, Chem. Lett., (1972) 205; Bull. Soc. Chem. Japan, 45 (1972) 2166.
67. N. Tanaka and M. Nanjo, Bull. Chem. Soc. Japan, 40 (1967) 330.
68. E. Ingier-Stocka, J. Thermal Anal., 36 (1990) 2139; 40 (1993) 1357.
69. J.D. Danforth and J. Dix, Inorg. Chem., 10 (1971) 1623.
70. A.H. Verdonk, Thermochim. Acta, 4 (1972) 85.
71. J. Zsakó and M. Lungu, J. Thermal Anal., 5 (1973) 77.
72. S.M. Taleb, Thermochim.Acta, 228 (1993) 131.
73. R. Bucci, A.D. Magrì, A.L. Magrì and A. Napoli, Thermochim. Acta, 217 (1993) 213.
74. A.V. Nikolaev and A.M. Rubinshtein, Bull. Acad. Sci. URSS, Sci. Chim., (1940) 787.
75. R.J. Cosmano and J.E. House, Thermochim. Acta, 13 (1975) 127.
76. N. Tanaka, K. Nagase and S. Nagakura, Bull. Chem. Soc. Japan, 41 (1968) 1143.
77. G. Liptay, K. Burger, É. Mocsari-Fülöp and I. Porubszky, J. Thermal Anal., 2 (1970) 25.
78. T. Fujiwara and J.C. Bailar, Inorg. Chem., 25 (1986) 1806.

Chapter 18

THE PRESENT POSITION AND PROSPECTS FOR FUTURE PROGRESS IN STUDIES OF THERMAL DECOMPOSITIONS OF SOLIDS

18.1. OVERVIEW

Research in the field of kinetics of solid state reactions has been strongly influenced by the theories and practices of homogeneous kinetics. The widespread acceptance and application of theoretical concepts successfully used in discussions of rates of chemical changes in the gas or the liquid phase have probably hampered the development of insights into factors which control the reactivity of solids where mechanisms generally involve quite different steps. These include the growth of nuclei by interface advance, diffusion of species through immobile matrices, the influences of phase transformations, heat transfer and the previous history of the reactant, crystal imperfections, etc., all of which have no parallels in homogeneous kinetics.

The long term, and probably ambitious, aim of studying the decompositions of solids is the development of a theory that explains the relative thermal stabilities of different solids and permits the prediction of the thermal behaviours of untested compounds. Progress towards such a goal has been found to be more difficult than was perhaps appreciated by the authors contributing to Garner's influential book [1] in 1955. Some of the unresolved problems of the subject are discussed in this Chapter as a basis for possible future development. The term *crystolysis reactions* [2], proposed for distinguishing specifically decompositions proceeding in the crystalline solid state, has not received very wide acceptance, but, if adopted, would at least simplify literature searches.

As in the development of most theories, information obtained from experiments has to be examined and classified according to some logically acceptable scheme so that patterns of common behaviour begin to emerge. This process has proved to be far from straightforward. Some proposed classifications have shown promise for limited groups of compounds, but no general theory explaining the thermal stability of solids, let alone their wider reactivity, has been achieved.

Before describing some of the proposed classifications, it is useful to recognize that experimental investigations have ranged, in depth and purpose, from *empirical* studies, where information has been collected simply to characterize the behaviour of a particular sample heated under a set of conditions as closely related to an industrial-scale process as practicable, to *semi-theoretical* calculations of expected behaviour based on as few as possible physical properties of the compound (such as bond lengths, lattice energies, crystal structures, etc.).

At the empirical end of this range come those kinetic studies which concentrate on providing an accurate mathematical description of the experimental results (the *rate equation* or *kinetic model*), without providing a physical interpretation of the (mainly statistical) parameters used in this description. Provided that sufficient detail of the experimental conditions used in these studies is recorded, these data may eventually prove useful for reanalysis when (or if) behaviour patterns begin to emerge. If attempts are made to relate the mathematical description to the geometrical development of a reaction interface, confirmation of such topochemical conclusions by microscopic observation is required, or if a diffusion process is indicated, then information on the diffusing species should be sought.

The value of a study increases considerably when it reports the influences on the rate processes of changing the experimental conditions (also called procedural variables). Further value is added by attempts to relate these results to processes occurring at the molecular level, i.e. to formulate a *reaction mechanism*.

Throughout this book emphasis has been placed on proceeding beyond mathematical descriptions of reaction rates to *chemical* features of the transformations of solid reactants into one or more different products (possibly via identifiable, relatively stable intermediates).

18.2. THE CONTROLLING STEP IN SOLID STATE DECOMPOSITIONS

The processes which may control the rates of solid state decompositions are represented in Table 18.1. It should, in principle, be possible to estimate the thermal stability of any particular reactant from its chemical constituents together with the physical properties (atomic spacings, bond strengths, band structures, etc.) by assuming appropriate models for the "activated complex" and applying principles analogous to those familiar from the Transition State Theory of homogeneous reactions. It is probable, however, that the stereochemical restraints imposed by the crystal structure, immobilizing reactants, products and any intermediates, may invalidate the assumption (made in homogeneous kinetics) that a single activated state controls the rate of reaction. Concurrent or consecutive processes taking place in the reaction zone may result in influences by more than one type of energy barrier. The participation of crystal imperfections may also cause the numbers and the nature of these transition states to change considerably during the course of reaction. Under these circumstances Arrhenius parameters might be expected to vary with extent of decomposition. The fact that initial decomposition is localized at surfaces or, even more specifically, at certain surface sites, indicates that it is differences in the bonding configurations which lead to reaction. The consequences of initial reaction may then be either the replenishment of such favourable reaction situations, or their consumption prior to continued reaction at the next most favourable set of sites.

There are considerable difficulties in demonstrating dominant control by any one of the steps in Table 18.1. and in establishing the factors which determine the magnitude of the activation energy, E_a. The principal bond-reorganization step may be influenced by a variety of factors, including the following.

(i) The precursors to reaction are expected to interact with the reactant and/or the product phase. This interaction may be regarded as intercrystalline strain at the interface (energy band distortion) or appear, through bonding to the product surface, as an adsorbed species that undergoes catalytic breakdown. Solid reaction products can be finely divided and may not crystallize to stoichiometric phases at reaction temperature.

(ii) The kinetics of interface processes are very different from those of homogeneous reactions. Within an interface, constituents are largely immobilized and in some systems are probably oriented. Chemical change may arise through the cooperative movements of several atoms rather than a bimolecular collisional encounter, as assumed in Transition State Theory. The initial steps in reaction may lead to a sequence of changes involving low, or very low, total amounts of intermediates immobilized within the reaction zone. These small amounts of (locally available) participants would be expected to have little effect if dispersed in concentration-dependent homogeneous reactions in the liquid or gaseous states.

(iii) Interface structures can be expected to exert considerable control over chemical changes within a thin (perhaps of molecular dimensions) zone of enhanced reactivity. The interface can be regarded as an essentially two-dimensional extended imperfection. All constituents (reactants, products and intermediates) may be less stable and more mobile than when confined to regular crystal sites, so that reaction occurs preferentially within this zone.

Table 18.1.
Major bond redistribution steps during decompositions of solids (crystolysis reactions)
These may be rate controlling and may also be reversible

CRYSTAL BONDING	CHEMICAL STEP	REPRESENTATIVE REACTIONS ($\uparrow$ signifies subsequent decomposition)	SEE CHAPTER	NOTES
Ionic	Electron transfer	$N_3^- \rightarrow N_3 \uparrow + e$	11	1
(including covalent bonds within ions)		$2MnO_4^- \rightleftarrows MnO_4 \uparrow + MnO_4^{2-}$	14	
	Proton transfer	$Ca(OH)_2$: $2OH^- \rightarrow H_2O + O^{2-}$	8	
		NH_4^+ salts: $NH_4Cl \rightarrow NH_3 + HCl$	15	
		Malonates $\rightarrow$ acetate intermediate	16	
	Coordination bond rupture	Dehydrations	7	2
		Deamminations	17	
	Homolytic covalent bond rupture	Transition metal carboxylates	16	3
		Oxalates: C - C bond rupture	16	
	Heterolytic covalent bond rupture	Carbonates: $CO_3^{2-} \rightarrow CO_2 + O^{2-}$	12	
		Other oxyanions: SO_4^{2-}, CrO_4^{2-}	14	
	Product phase recrystallization	Dehydrations:	7	4
		$Ca(OH)_2$, $NiSO_4.6H_2O$, alums	7,8	

(continued)

Table 18.1.(continued)
Major bond redistribution steps during decompositions of solids (crystolysis reactions)
These may be rate controlling and may also be reversible

CRYSTAL BONDING	CHEMICAL STEP	REPRESENTATIVE REACTIONS ($\uparrow$ signifies subsequent decomposition)	SEE CHAPTER	NOTES
Covalent (extended anion)	as for ionic solids	Mineral dehydrations (diffusive escape of product H_2O)	7,8	5
Metallic	Product desorption Product precipitation Interstitial diffusion	PdH, Fe_4N Ni_3C Intermetallic compounds	10	6
Molecular (van der Waals)	Breakdown of constituent groups before melting	Explosives: organic azides, nitrates, etc.		7
All	Recrystallization Melting Sublimation	Many solids undergo recrystallization and/or melting without chemical changes of the constituents		8

(continued)

Table 18.1. Notes

1. Boldyrev *et al.* [3] have classified decompositions of inorganic solids as follows:

FIRST GROUP	No electron transfer	1. Irreversible	: Alkali and alkaline earth metal formates, oxalates.
		2. Reversible	: Ammines, hydrates, carbonates, alkali and alkaline earth metal hydroxides.
SECOND GROUP	Cleavage of bonds accompanied by electron transfer	1.	Transfer within ionic structure: permanganates, alkali metal perchlorates.
		2.	Anionic radicals formed: $Ag_2C_2O_4$, AgN_3. Cationic radicals formed: no examples known.

2. The removal of ligands (including H_2O) from coordination compounds may be accompanied by recrystallization and either a change in coordination number, structure or ligands present through formation of new bonds with the anions, anation, etc. Constituent water in crystalline hydrates may be as coordinated ligands and/or structural water. A scheme for the classification of dehydration reactions is described in Table 7.3.

3. Boldyrev [4] suggested that the first step in oxalate decomposition is rupture of the C - C bond:

$$\begin{array}{lll} C_2O_4^{2-} \rightarrow 2CO_2^- & \rightarrow \text{(i) recombine} \\ \quad\text{(radicals)} & \rightarrow \text{(ii) } CO_3^{2-} + CO \\ & \rightarrow \text{(iii) electron transfer} \rightarrow 2CO_2 \end{array}$$

Acheson and Galwey [5] identify electron transfer as the controlling step in the decompositions of mellitates and oxalates from correlations between E_a and the strength of the M - O bond.

4. Structural reorganization to the stable solid product may occur at a different interface from the reaction zone, as discussed for the dehydrations of $Ca(OH)_2$, $NiSO_4.6H_2O$ and alums.

5. Few decompositions of covalent compounds have been studied in detail. In the dehydrations of some minerals, the chemical step is rapid and the rate is controlled by the diffusive escape of water from between layers of extended anions.

6. Few reactions of these solids have been studied in detail. Rate may be controlled by removal of product (by combination of atoms to form gaseous molecules, by precipitation of a solid product, or by diffusion).

7. Decomposition of molecular crystals below their melting points requires that the stability of the molecules be lower than the intermolecular forces in the crystals.

8. The most probable consequence of heating crystalline solids of all types is melting without chemical change.

18.3. THE STRUCTURAL COMPLEXITY OF THE INTERFACE

The forms and functions of some interface models that have been proposed to explain observations are described and discussed in Section 6.8. Recent investigations, using microscopy, have shown that their structures and properties are even more complicated than was formerly accepted. Classification of reaction mechanisms and predictions of reactivities in untested systems may thus also be more difficult than has been generally recognized.

The following examples illustrate difficulties that have been encountered for specific reactions where the simplest model has been found, on closer examination, to require modification.

(i) Dehydration reactions involve elimination of water from the reactant structure. This step may be followed by recrystallization. The two processes may be appreciably separated [6,7] in space and/or time. In reactants consisting of extended stable constituents, such as layer-type minerals, the recrystallization step may be absent and the reaction kinetics are then controlled by the rate of diffusive escape of water, or any other mobile molecular component.

(ii) The dehydration of α-$NiSO_4.6H_2O$ at low temperature in vacuum yields the monohydrate by a nucleation and growth process, in which the elimination of water does not necessarily coincide with the recrystallization step [8]. Above about 370 K and in the presence of water vapour, however, the di- and tetrahydrates are formed as an impervious surface layer that changes the kinetics of dehydration [8,9]. This pattern contrasts with the more usual behaviour of other hydrates that progressively lose water as the temperature is raised, e.g. $MgSO_4.7H_2O$ [10], which yields successive intermediates containing 6, 4, 2.5, 2 and $1H_2O$, before forming anhydrous $MgSO_4$.

(iii) Herley *et al.* [11] used scanning electron microscopy to identify the textural changes during interface advance in the decomposition of aluminium hydride ($2A\ell H_3 \rightarrow 2A\ell + 3H_2$). On completion of the induction period, nuclei appeared in cracks as clusters of short, thick and randomly oriented needles [12]. As reaction progressed, more nuclei appeared together with the formation of voids and internal fragmentation.

(iv) The initial solid product of $CaCO_3$ dissociation was first identified as a metastable phase before it was shown to be finely divided CaO [13-15].

(v) In other rate processes, e.g. the dehydrations of $CaSO_3.\frac{1}{2}H_2O$ [16] and of $MgC\ell_2.2H_2O$ [17], the particle sizes were too small and/or the textural changes too indistinct to permit the disposition or structure of the interface to be investigated. Electron microscopy confirmed that the reactants underwent no comprehensive melting. Similar examinations of reaction zones in the decompositions of nickel squarate [18] and of nickel formate [19] showed that the relatively dense metallic nickel product was formed as finely divided particles located in active areas (nuclei) at cracks and pits. Again the interface textures were well below the limits of resolution by the microscope.

The information available demonstrates that interface structures are more complicated than can be represented by an advancing monomolecular layer within which the overall change is completed. Different controlling factors and textures may apply to different reactants. When formulating chemical mechanisms, it is necessary to include results of detailed examinations of interface textures. A promising approach is the determination of crystallographic structures at a sequence of sites that extend across the interface by synchrotron diffraction methods [6]. This provides information concerning all phases participating and their positions. Spectral studies of energy levels associated with reaction zones could be used to complement the more conventional evidence from kinetic data, including the magnitude of E_a and the stoichiometric requirements.

18.4. REACTANT SURFACE CHEMISTRY AND NUCLEATION

The theory of solid state reaction kinetics includes no consideration of surface properties other than the recognition that crystal faces are the most probable location for nucleation in many reactions. Dehydration studies have provided evidence that, in many such processes, *all* surfaces are modified soon after the onset of chemical change [20,21]. This is ascribed to a surface reaction that is limited in extent and can continue only at local sites of special reactivity where the recrystallization required for nucleation is possible. In other decompositions there is evidence that the modified reactivity associated with surfaces may influence the overall reaction [11] and may also preserve the identity of crystals. This, incidentally, masks the occurrence of melting during decomposition [22].

The changes occurring at crystal boundaries before or early in reaction have been reviewed recently [23]. These changes are particularly important in discussing the mechanisms of nucleation. The new product phase may be generated in the modified material, in the unchanged reactant beneath, or in the transition zone [7]. The occurrence of a surface reaction may explain the difficulties encountered [24] in associating growth nuclei with the surface terminations of line dislocations. The theory of nucleation is, as yet, incomplete and much further work is required to identify the factors that control the initiation of product formation. The destruction of precursor states by the nucleation step increases the problems of experimental investigation.

18.5. EXPERIMENTAL EVIDENCE IN FORMULATING REACTION MECHANISMS

18.5.1. The crystolysis reaction

The steps which may contribute to a crystolysis reaction are summarized in Table 18.2. Formulation of a mechanism requires reliable experimental observations and as many complementary types of measurements as can be obtained. The contributions from precursor and secondary reactions (Table 18.2.) must also be considered. Some comments about the relevance of different types of information are given below.

18.5.2. Reaction stoichiometry : analytical data

For any kinetic study, the chemical change should be clearly specified, though this is not always achieved [25], and the identities of all reactants, intermediates and products must be established using suitable analytical techniques. Unlike homogeneous reactions, the crystallographic structures, topotactic relationships and states of sub-division of all participating solids should also be described.

Many studies have been intentionally concerned with relatively simple rate processes to minimize stoichiometric problems. However, even relatively simple reactions do not always give a single product, for example barium azide has been reported to give about 70% Ba_3N_2 together with the metal [26]. The extensively studied decompositions of oxalates require that the identity of the residual solid be identified, for each constituent cation, as either carbonate, oxide or metal, together with any change of cation valency. The chemistry of oxalate breakdown can, however, be much more complicated, as has been shown for the Y, Eu and Yb salts [27].

More complicated reactions, involving intermediates and consecutive or concurrent rate processes, require separate investigations of the contributing chemical steps. For example, the decomposition of $(NH_4)_2Cr_2O_7$ yields NH_3, N_2, N_2O and H_2O as volatile products by kinetically distinct rate processes [28]. The residual solid was not readily identified as a simple compound and nitrate and nitrite were detected as intermediates. In other reactions, the residual products may not crystallize to form recognizable stoichiometric phases, for example, in the decomposition of $KMnO_4$ [29].

Rate measurements are also influenced by the atmosphere and the environment. An inert static atmosphere may cause the retention of volatile products in the vicinity of the solid reactant and, hence, control the rate and nature of the chemical change occurring, particularly for reversible reactions [8,9]. The availability of

Table 18.2.
Steps contributing to a crystolysis reaction

REACTANT AS PREPARED → REACTION PRECURSOR

During heating to reaction temperature, one or more of the following steps may change the structure, texture and reactivity of the reactant.

(i) Dehydration (→ lower hydrate) → anhydrous salt
(ii) Amorphous phase → recrystallized phase
(iii) Textural changes → disintegration
→ imperfection redistribution

Decomposition
Nucleation

↓

Interface (growth)
Diffusion processes
Retexturing
Recrystallization

↓

PRIMARY PRODUCTS

↓ ↓

Residual Volatile
Secondary reactions

↓ ↓

Solid products Evolved gases

(a) Crystal phases in the residual products should be characterized and the possibility of a topotactic relationship with the reactant considered.
(b) The roles of any reversible reactions and any condition-dependendent kinetic behaviour should be examined.
(c) The identities of the products of ligand dissociation and decomposition / reaction should be determined.

gaseous impurities to the reactant (O_2, H_2O, CO_2, etc.) may change the reaction being studied. The walls of a reaction vessel may also exhibit a catalytic influence, or a glass container may react with an alkaline solid product.

18.5.3. Diffraction measurements

Probably the most powerful tool for the exploration of phase relationships at interfaces is the use of synchrotron radiation to determine structures. This can be undertaken for a sequence of small volumes across the reaction zone [6]. The potential of the method is considerable, but, as yet, the number of applications has been small because there are few of these expensive facilities available. X-ray diffraction measurements are, however, widely used to confirm or identify the structure of both reactants and products, and can be extended to detect any topotactic relationship between them. Assumptions do have to be made concerning the absence of structural changes on cooling from the reaction temperature to the temperature of structure determination, if these temperatures are different.

18.5.4. Microscopic observations

Many of the earliest studies of decompositions of solids made use of optical microscopy as a main experimental technique [1]. Observations of interest were often at or near the resolution limits, but such information was of value in developing the theory of interface reactions. Later workers have often omitted microscopic observations and drawn conclusions about the geometry of interface development from the fit of yield-time data to a rate equation. The greater resolving power of scanning electron microscopy has been exploited in characterizing the textural changes that accompany reaction. Replication techniques [30] can be used for solids that are unstable to direct examination in the vacuum chamber of the microscope. This approach is also valuable in determining intranuclear crack structures.

Observations of samples of partially decomposed solid (at an appropriate series of α values) and across crystals sectioned after reaction, can enable reactant and product phases to be distinguished. The occurrence of melting, sintering, cracking and retexturing, knowledge of which is essential in formulating reaction mechanisms, may also be revealed. The detailed (molecular level) interface structures may, however, remain below the limits of present resolution. Atomic force microscopy may be able to reveal greater detail within the reaction zone.

Sesták [31] has warned of some of the limitations of the techniques used and of interpretations based on microscopic studies. Only a small portion of the sample is observed and there is a need to ensure that the observed behaviour is representative. The possible destabilizing effects of conditions in the electron microscope have already been mentioned. The effects of temperature changes and possible further reaction of partially decomposed samples before or during examination have also to be taken into account in interpreting observations.

18.5.5. Spectroscopic observations

Information about the distribution of energy amongst chemical species at the interface (which is required in a fundamental (transition state) theoretical treatment of interface processes [32,33]) can (in principle) be obtained from spectral measurements (see Chapter 6). Vyazovkin and Linert [34], for example, identify the rate limiting step in the decomposition of polyethylene terephthalate as the rupture of a C - O bond, through evidence that the isokinetic temperature (622 K) is consistent with the energy associated with a far infrared absorption band. Comparable methods might be useful in the investigation of selected crystolysis reactions. The interface levels described by Galwey and Brown [35] should also be detectable by spectroscopic measurements. The changes in the vibrational spectrum of the nitrate ion in the temperature range close to the $Ba(NO_3)_2$ phase transition at 573 K have been investigated in detail and the stability changes have been associated with structural reorganization.

Infrared and other spectral measurements have been widely used [25] in experimental studies of crystolysis reactions, usually as an analytical tool for the identifications of reactants, products and, sometimes, intermediates [28]. There are difficulties in associating spectral features with interface species and structures because these may be present at very low concentrations, or their absorptions may obscured by other stronger absorptions, or by the opacity of a solid product (perhaps a metal or carbon). Nevertheless, the approach offers the possibility, in favourable systems, of identifying or confirming steps controlling, or participating in, interface reactions.

18.5.6. Thermochemical measurements

Thermochemical measurements, using DTA or DSC, can provide evidence of the occurrence of phase transitions such as melting and crystallographic transformations, in addition to chemical changes, together with their onset temperatures and enthalpy changes. Contributing reactions that may occur concurrently under isothermal conditions may sometimes be distinguished. Such information is of considerable value in planning kinetic investigations of complex reactions. Modulated temperature DSC [36] has the advantage of separating reversible and irreversible contributions to the heat flow processes. Comparison of the rate of heat evolution with other measurements of reaction rate, such as the rate of loss of mass by the sample, under either programmed or constant temperature conditions [37], can reveal possible complexities of the processes being studied.

Evolved gas analysis (EGA) can provide information on the rates of formation of individual gaseous products.

18.6. KINETIC PARAMETERS

18.6.1. Reliability

The most reliable measured kinetic parameters, adjudged by the agreement of values reported by different workers, have been obtained for *irreversible decompositions*. The well-known sigmoid α-time curve for the $KMnO_4$ decomposition [38] has recently been shown to be well represented by the Avrami-Erofeev equation [29] and problems in identifying the equation giving the best fit were discussed. Distinguishability within the sigmoid group of conversion functions (Avrami-Erofeev equation with $n = 2$, 3 or 4 and the Prout-Tompkins equation) and within the deceleratory group (contracting volume, contracting area and first-order) is notoriously difficult [39,40]. The analysis may depend greatly on the accuracy of measurement of the final product yield [37], which involves the assumption that reaction has gone to completion. Interpretation of kinetic data in terms of observed nucleation and growth behavior is also not straightforward [29].

In spite of differences in the conversion function chosen, values of E_a for the decompositions of $KMnO_4$, $RbMnO_4$ and $CsMnO_4$ were usually within the range 160 to 170 kJ mol^{-1} [29,41]. Different kinetic studies of NH_4ClO_4 decomposition have shown that the sigmoid α-time curves also fit the Avrami-Erofeev equation and many values of E_a were in the range 120 ± 20 kJ mol^{-1} [42]. While agreement between different workers is not perfect, there is evidence that substantial kinetic agreement can be achieved. The probable reason is that irreversible reactions are relatively insensitive to locally prevailing conditions.

The kinetics of *reversible decompositions* are often highly sensitive to reaction conditions [43]. For example, the values of A and E_a for the decomposition of $CaCO_3$ show unusually wide variations, owing to the sensitivity of reaction rate to the availability of CO_2 [44,45]. The spread of apparent E_a values is considerable [46] and some values are close to the dissociation enthalpy [1]. However, Beruto and Searcy [47] concluded that, under high vacuum conditions, the constant rate of interface advance in large crystals was probably controlled by the dissociation step in the absence of a perceptible contribution from the reverse process. The decomposition activation energy (205 kJ mol^{-1}) was appreciably larger than the dissociation enthalpy (178 kJ mol^{-1}). This is probably the most precise kinetic measurement for the calcite decomposition [48].

Studies of the influences of procedural variables (sample mass, partial pressure of CO_2, heating rate, etc.) on the kinetics of calcite powder dissociation continue to be published [49-51]. Such kinetic observations are the sum of different reaction rates within an inhomogeneous reactant assemblage and hence have limited fundamental significance. In general, much of the published kinetic information concerned with reversible reactions probably refers to rates measured only under the particular conditions of that experiment.

For programmed temperature experiments at high heating rates (and in a flow of an inert gas) there will be temperature variations within the mass of a reactant undergoing an endothermic reaction and the pressure of product gas (e.g. CO_2) will be greater towards the centre of the particle assemblages. The work of Beruto and Searcy on the decomposition of $CaCO_3$ [47,48], together with that of Flanagan *et al.* [52] on the dehydration of $NiC_2O_4.2H_2O$, demonstrates clearly the considerable effort and careful design of experiments that are required for the measurement of the forward rates of reversible reactions. Kinetic information of great fundamental value would be obtained by extending single crystal studies, of the type referred to above, to quantitative determinations of the influences on reaction rate of constant, known product gas pressures [48], at known product layer thicknesses, and on the influence of changing reaction temperatures, etc. Fundamental aspects of the effects of gaseous products on endothermic decompositions have been discussed by Beruto and Searcy [47]. The kinetics of reactions of powders, even though reaction conditions are generally ill-defined, are, however, of great industrial importance.

Values of E_a measured for reversible reactions under conditions permitting absorption equilibration between gas and solid products are often comparable with the reaction enthalpy [1,47]. Under such conditions the identification of the rate expression, $g(\alpha) = kt$, on the single criterion that this equation gives the most acceptable correlation coefficient is not a sufficient foundation to characterize a geometric reaction model. The use of additional information, for example microscopy, can provide confirmatory evidence concerning interface development. Similarly, the value of studies which conclude that kinetic results are satisfactorily described by equations based on diffusion models is increased considerably if the identity of the migrating species is established [53].

Kinetic parameters measured by rising temperature techniques are considered by some workers to be less reliable than those calculated from isothermal data because of the additional assumptions required by the theory. Use of different applicable equations can give different values of A, E_a and $g(\alpha)$ [25,54,55]. Ingier-Stocka [55] concludes that "it is impossible to choose one kinetic model from dynamic measurements." Levchik *et al.* [56] emphasize the value of complementary isothermal and non-isothermal experiments in the determination of details of complicated kinetic behaviour where different phases may be involved.

The ever-increasing computing power available has enabled kinetic analysis to become much more sophisticated. Large sets of data can be handled readily and standard statistical analyses can be applied to results.

18.6.2. Compensation behaviour

Reports of compensation behaviour appear to have been mainly empirical findings which have not contributed much to the theory of solid state reactions. Identification of the effect as a direct consequence of isokinetic behaviour [57,58] does, however, offer a theoretical foundation for interpretation of the observed correlated changes in apparent Arrhenius parameters. The fit of Arrhenius parameters for crystolysis reactions to the expression:

$$\ln A = b\,E_a + c$$

is most frequently described for data sets that refer to: (i) the same reaction studied under different conditions; or (ii) the decompositions of a group of similar reactants that undergo comparable chemical changes.

An explanation [58] of compensation behaviour in heterogeneous catalytic reactions could be applicable here. "Compensation may (therefore) be explained as a set of reactions exhibiting common reactivity (isokinetic behaviour) because of a common mechanism whereby similar adsorbed intermediates undergo comparable changes in a dynamic active surface equilibrium in the same temperature interval." Because similar (or identical) bonds are redistributed during the steps involved in the overall chemical change, all reactions of a particular (compensation) set will become detectable at similar (or identical) temperatures (isokinetic behaviour) [58]. The concentrations of species participating in these different contributory processes can, however, be expected to vary with temperature across the reaction interval because of variations of the prevailing controls or conditions (i.e. within different reactants, or with different equilibrations of reaction precursors depending on the availability of volatile product). A consequence of such temperature dependent changes of reactant concentrations is the appearance of linked variations of $\ln A$ and E_a [59]. The occurrence of reactions in the same temperature range, but with different temperature coefficients of reaction rate, results in compensation behaviour [59].

Values of A and E_a that depend upon experimental conditions are not of fundamental significance and cannot be reliably associated with a particular rate limiting step. The compensation pattern observed may, however, indicate the conditions under which the reaction rate becomes significant, but the apparent kinetic parameters may vary with sample mass, particle sizes, thermal contact and pressures of product gases (which may be inhomogeneous within the sample assemblage). These are the procedural variables [49]. Changes in the availability of reactant at the reaction site have been shown [59] to result in compensation behaviour. Quantitative identification of the effects of conditions on reaction rates is essential in the design of industrial processes and also contributes to understanding how factors other than a chemical step may determine the kinetics of solid state reactions.

The observation of compensation behaviour for the decompositions of a group of comparable reactants under similar conditions is ambiguous because such behaviour could be explained either (following the approach used in with heterogeneous catalysis [59]) as the result of relatively small crystal structural differences in the rate determining bond rupture step, or (as above) by differences in the physical conditions existing within each individual reactant sample.

Identification of a compensation trend may thus be an indication of reaction reversibility. Magnitudes of A and E_a cannot be regarded as providing fundamental mechanistic information without deeper exploration of the source of the compensation behaviour.

18.7. REACTION MECHANISMS

18.7.1. Thermal stability

The development of a theory capable of explaining the observed differences in thermal stabilities of different solid compounds and of predicting the behaviour of untested compounds, must start from recognizing similarities of chemical changes shown by a series of well-defined reactants. (Formulation of such a theory might be a step towards a wider theory of solid state reactivity.)

Decompositions of chemically dissimilar substances (including NiC_2O_4, PbC_2O_4, KN_3, $(NH_4)_2Cr_2O_7$) may exhibit similar kinetic behaviour, whereas the decompositions of similar reactants (e.g. oxides) may exhibit apparently unrelated rate characteristics [60]. Decomposition data for very different reactants may be described satisfactorily by the same rate expression, for example the Avrami-Erofeev equation, with $n = 2$, has been reported as representing the decompositions of FeC_2O_4 [61], $KMnO_4$ [29], silver malonate [62], nickel squarate [18], lead citrate [63] and *d* - LiK tartrate [64] (no common constituents in these six reactants).

Decompositions of reactants containing similar or related components are discussed below.

18.7.2. Decompositions of metal malonates

Comparisons of observations for the decompositions of five metal malonates (Cu(II) [22], Ag [62], Ni [65], Co [66] and Ca [67]) illustrate the behavioural idiosyncrasies and individualities of these closely related compounds. Anion breakdown was accompanied by hydrogen transfer so that four of the reactants gave appreciable yields of acetate or acetic acid. (No analytical measurement was made for the cobalt salt.) The principal kinetic results (Table 18.3.) show the remarkably contrasting behaviours of copper(II) malonate, where decomposition occurs in a melt [22], and the nucleation and growth process observed for silver malonate [62]. These differences occurred in reactions studied in overlapping temperature ranges and with comparable E_a values. The essential chemical differences in the reactions are the stepwise cation reduction in the Cu^{2+}($\rightarrow$ Cu^{+} $\rightarrow$ Cu^{o}) salt together with formation of the molten acetate phase. Copper metal, in contrast with the Ni and Ag salts, appears not to promote anion breakdown. While the reactions of both the Ag and Ni salts are believed to proceed by interface advance, the reactivity of the nickel metal is attenuated by carbon deposition and carbide (Ni_3C) formation. A quite different mechanism must apply in the decomposition of calcium malonate [67] because cation reduction to the metal cannot be expected. Proton transfer was proposed as the probable controlling step in anion breakdown to yield the principal residual product, $CaCO_3$. Cation properties are, therefore, identified as strongly influencing malonate anion breakdown. The reaction temperature rises and the mechanism changes with increase in the electropositive character of the constituent metal, Table 18.3. This is consistent with the possibility that electron transfer controls the reactivities of these compounds.

Table 18.3.
Kinetic characteristics of metal malonate decompositions

CATION	RANGE T/K	INITIAL REACTION	MODEL (MAIN)	MECHANISM	E_a/kJ mol^{-1}	ε^o /V*
Ag^{+}[62]	453 - 485	4%	Avrami-Erofeev $n = 2$	Nucleus and growth	170	0.799
Cu^{2+}[22]	468 - 505	3.5%	Exponential $\ln \alpha = kt$	Melting	200	0.344 ($Cu^{2+} + 2e \rightarrow Cu$) 0.158 ($Cu^{2+} + e \rightarrow Cu^{+}$)
Ni^{2+}[65]	543 - 613	12%	Zero order	Nucleus and growth	176	-0.250
Co^{2+}[66]	572 - 600	-	Zero order	Nucleus and growth	181	-0.277
Ca^{2+}[67]	612 - 653	15%	Mainly deceleratory	Did not melt	465	-2.87

* Standard reduction potential

18.7.3. Decompositions of metal oxalates

The initial or rate limiting step for anion breakdown in metal oxalate decompositions has been identified as either the rupture of the C - C bond [4], or electron transfer at a M - O bond [5]. This may be an oversimplification, because different controls may operate for different constituent cations. The decomposition of nickel oxalate is probably promoted by the metallic product [68] (the activity of which may be decreased by deposited carbon, compare with nickel malonate mentioned above [65]). No catalytically-active metal product is formed on breakdown of oxalates of the more electropositive elements. Instead, they yield oxide or carbonate and reactions may include secondary processes [27]. There is, however, evidence that the decompositions of transition metal oxalates may be accompanied by electron transfers. The decomposition of copper(II) oxalate [69] ($Cu^{2+} \rightarrow Cu^{+} \rightarrow Cu^{o}$) was not catalytically promoted by the metal and the acceleratory behaviour was ascribed to progressive melting. Similarly, iron(III) oxalate decomposition [61,70] was accompanied by cation reduction ($Fe^{3+} \rightarrow Fe^{2+}$). In contrast, evidence was obtained that the reaction of MnC_2O_4 was accompanied by the intervention of Mn^{3+}, believed to be active in anion breakdown [71]. These observations confirm the participation of electron transfer steps in breakdown of the oxalate ion, but other controls influence the overall behaviour. Dollimore has discussed [72] the literature concerned with oxalate pyrolyses, including possible bond rupture steps involved in the decomposition mechanisms.

18.7.4. Decompositions of metal formates

Reports of investigations of metal-catalyzed decompositions of formic acid have discussed the participation of the metal formate as a reaction intermediate, for example in the decomposition of $Ni(OOCH)_2$ [73]. This indicates the possible value of complementary investigations of both classes of heterogeneous rate processes: crystolysis and metal surface chemistry. There is an extensive literature concerned with the decompositions of adsorbed species, including formic acid, at low coverages on almost perfect metal crystal faces. Results of such work may not, however, be directly applicable to the more crowded states existing at solid state reaction interfaces.

For the decompositions of the metal carboxylates, alternative mechanisms involving breaking of each of the different principal linkages in the R-CO-O-M group have been proposed as rate controlling (reference [40] p.210), in addition to mechanisms based on charge transfer, interface strain and catalysis by an active product.

18.7.5. Decompositions of copper(II) salts

Stepwise cation reduction has been recognised as a common mechanistic feature in the decompositions of a variety of copper(II) salts. This was reported by Lomovsky *et al.* [74] for copper(II) hypophosphite ($Cu(H_2PO_2)_2 \rightarrow CuH_2PO_2 \rightarrow Cu$). Later work on the thermal reactions of the copper(II) salts of a range of organic acids provided evidence that decompositions proceeded with $Cu^{2+} \rightarrow Cu^{+} \rightarrow Cu^{o}$. The mechanisms of these reactions, containing this common constituent, have been reviewed [75].

18.7.6. Decompositions of cobalt(III) ammine azides

Joyner [76], reviewing a series of comparable kinetic studies of four solid cobalt(III) ammine azides, discussed the factors that determine the two alternative decomposition processes that yield either product CoN or cobalt(II) complexes. He concluded that the mechanisms are "independent of the nature of the cation, the nature of the salt and properties dependent on crystal structure." From a very careful appraisal of the evidence, he concluded that the course of the reaction is determined by "the very early interplay between the CoN and cobalt(II) reactions." This conclusion emphasizes the subtlety of the factors controlling decomposition behaviour.

18.7.7. Decomposition of ammonium salts containing oxidizing anions

Dodé [77] proposed that NH_4NO_3 should be the expected final residual product from the decomposition of any ammonium salt containing a sufficiently oxygenated anion, e.g., $(NH_4)_2Cr_2O_7$ [28], NH_4MnO_4 and NH_4ClO_3. This rule appeared not to apply to NH_4ClO_4 [78], which underwent incomplete decomposition at low temperatures and sublimed at higher temperatures [42]. This apparent inconsistency has since been shown to be due to the ability of nitrate to catalyze NH_4ClO_4 breakdown through formation of the intermediate NO_2ClO_4. Earlier investigation of the $NH_4ClO_4 + NH_4NO_3$ reaction [79] could have assisted understanding of the chemical controls of this intensively studied decomposition.

18.8. SOME PATTERNS OF KINETIC BEHAVIOUR

18.8.1. Basic assumption

Instances of different reactions that proceed at similar rates with similar kinetic characteristics (data fit the same rate expression and give comparable A and E_a values) *may* be regarded as evidence of similar chemical control. The observations available may, however, not be sufficient to characterize unambiguously a common rate limiting step. The kinetics of the following sets of reactions (Sections 18.8.2 - 5.) appear to be in this category.

18.8.2. Dehydrations of $CaSO_3.½H_2O$ and of $Ca(OH)_2$

The dehydration of $CaSO_3.½H_2O$ was studied [16] between 573 and 673 K (an unusually high temperature for this type of reaction) and the value of E_a for the loss of water of crystallization was 173 kJ mol^{-1}. The rate of the established reaction (but not the rate equation) was very close to that measured under similar conditions for dehydration of $Ca(OH)_2$ (with E_a = 174 kJ mol^{-1}) [80]. It was concluded that both reactions proceeded in a temperature range where proton transfer occurs readily [81]. The bond structure of the water in calcium sulphite hemihydrate may contain features similar to those in $Ca(OH)_2$ at reaction temperatures [80].

18.8.3. Dehydration of $MgCl_2.2H_2O$ and oxidation of $MgCl_2$

The dehydration of $MgCl_2$. $2H_2O$ [17] and the reaction of $MgCl_2$ with oxygen [82] both yield product MgO and proceed at similar rates between about 620 and 700 K [17]. The precise controlling step was not identified, but an interaction between $MgCl_2$ and an oxygen-containing molecule is a probable common kinetic feature.

18.8.4. Dehydrations of lignite and of alums

The rate of evaporation of water from the carbonaceous residue (to which it is not chemically bonded) during lignite dehydration [83] is only slightly greater than that from alum nuclei [20]. This supports the view that intranuclear water in alums, temporarily retained [43] and believed to promote product recrystallization, is not strongly bonded to the residual sulfates and further dehydration rapidly replaces evaporative loss.

Although the structures and compositions of the isomorphic alums are identical, the dehydration steps vary from one compound to another [20,43]. $KAl(SO_4)_2.12H_2O$ loses about $9H_2O$ in a single reaction below 370 K (E_a =108 kJ mol^{-1}), whereas $KCr(SO_4)_2.12H_2O$ evolves about $5H_2O$ below 335 K and a further $4H_2O$ below 390 K (E_a = 70 kJ mol^{-1}). The shapes of nuclei and the kinetics of nucleation differ, showing that the properties of the reactant constituents influence the solid state reactions occurring.

18.8.5. Decompositions of $KMnO_4$, $RbMnO_4$ and $CsMnO_4$

The closely similar reactivities and E_a values measured for the decompositions of $KMnO_4$, $RbMnO_4$ and $CsMnO_4$ [29,41] are consistent with control by the same step (probably electron transfer in the anionic sub-lattice) within very similar reactant structures. The influence of the non-volatile products is difficult to assess because reorganization does not yield stoichiometric and well-defined crystalline solid products at reaction temperature.

18.8.6. Rate equations not characteristic of solid state reactions

The kinetics of some decompositions are found not to be adequately described by the usual set of rate equations (Table 3.3.) characteristic of solid state reactions. This suggests that behaviour is more complicated than assumed in the derivation of these models. Interpretation of such kinetic results, supported by microscopic observations, has lead to the recognition of melting during the decompositions of copper(II) malonate [22], ammonium dichromate [84] and the dehydration of *dl*-$LiKC_4H_4O_6.H_2O$ [85]. During some such decompositions, e.g. those of copper(II) malonate and $(NH_4)_2Cr_2O_7$, several concurrent rate processes contribute to the overall chemical change.

18.8.7. Dehydrations of lithium potassium tartrate hydrates

The influence of structure on reactivity was shown in the dehydrations of *d*- [7], *dl* -[85] and *meso* - [86] lithium potassium tartrate hydrates. Significant differences in kinetic behaviour were found for reactants that contained identical components packed in different arrays. The dehydrations and decompositions [64] were later compared and discussed further [87].

18.8.8. Product texture : CaO from $CaCO_3$ decomposition

Beruto *et al.* [15] have investigated, in unusual detail, the texture of the CaO formed in the decomposition of $CaCO_3$, to determine the role of CO_2 in promoting diminution of the surface area of the initially highly divided oxide product. At low pressures, CO_2 catalyses the decrease in CaO surface area to about 90 $m^2\ g^{-1}$. Further coarsening proceeds by grain boundary and bulk diffusion processes.

18.8.9. Conclusion

The most obvious conclusion from the above (admittedly selective) survey is that solid state reactions are sensitive to the chemical properties of *all* the constituents and the crystal structure of the reactant. The trends in behaviour discerned, however, provide only limited insights into the factors controlling thermal stability. The observations are difficult to interpret at the levels of molecular, or ionic, interactions and rate determining bond redistributions. More precise, detailed and systematic comparative studies are required to advance understanding of the interface chemistry.

18.9. THE LITERATURE CONCERNED WITH CRYSTOLYSIS REACTIONS

18.9.1. Literature reviews

Since the appearance in 1955 of the seminal reviews in *"Chemistry of the Solid State"*, edited by Garner [1], contributions to the subject have been dispersed in a wide cross-section of often interdisciplinary chemical journals, with a high proportion in *Thermochimica Acta* and *Journal of Thermal Analysis*. There have been few reviews compared with the considerable publication output [25].

Many typical research papers do not include introductory or discussion material that places the contribution in its widest possible literature context. Attempts to classify the vast amounts of available material have produced few generally useful patterns or correlations, so that the continued accumulation of often widely dispersed individual reports does little to advance the subject as a whole. The following topics are in particular need of critical review.

(i) Roles of different types of crystal imperfections in crystolysis reactions.

(ii) Methods of kinetic analysis of non-isothermal data with assessment of the strengths and weaknesses of the different approaches that are employed.

(iii) Comparisons of reactivities of different compounds with similar crystal structures.

(iv) Comparisons of reactivities of similar compounds with different crystal structures.

(v) Studies of the solid state decompositions of coordination compounds.

(vi) Comparison of the kinetics and mechanisms of solid state decompositions with related chemical changes in a melt or solution phase.

(vii) Comparison of the kinetics and mechanisms of crystolysis reactions with related reactions on catalyst surfaces (e.g., metal + carboxylate).

(viii) Comparison of the kinetics and mechanisms of crystolysis reactions with pyrolysis reactions of polymers.

(ix) Compilations and correlations of the kinetic characteristics (f(α) or g(α), A and E_a) for decompositions of selected groups of related reactions (e.g. carbonates, carboxylates, salts of oxyacids, etc.).

(x) Comparative studies of selected crystolysis reactions, e.g. the decomposition of $CaCO_3$, under different experimental conditions.

(xi) Surveys of the chemical mechanisms of crystolysis reactions (to extend, develop, apply and discuss alternative interpretations to those presented in this book).

Dollimore [72], specifically referring to the decompositions of oxalates, states that "the existing literature ..... is found to often be repetitive in nature." New research should develop and extend identified previous work for clearly stated reasons, which should be more positive than that a particular reactant "has not previously been studied". Imaginative research is needed in this field where so many patterns of reactivity remain unexplained.

18.9.2. A survey of the content of the recent literature

Thermochimica Acta was identified in a previous literature review [25] as the international journal containing the largest number of articles concerned with crystolysis reactions. Seventy-one relevant entries were found in an extended search that located altogether 368 contributions to the topic dated 1981. A more recent search of this same journal for the interval, January 1994 to December 1995, Volumes 232-270, resulted in the identification of 262 articles on subjects falling within the scope of the present book. These research reports were concerned with decompositions of diverse compounds stated, or presumed, to proceed in the solid state, together with aspects of the relevant instrumentation and theory, particularly theoretical aspects of reaction kinetics. This selected set of relevant articles is sufficiently large to highlight trends and topics of particular interest and importance. In the analysis that follows, bracketed numbers indicate the number of articles principally concerned with each topic discussed for the (**262** entries) that constitute the source set.

Simple salt reactants (**131** entries). Articles concerned with decompositions of simple salts were often concerned with kinetic characteristics, many used non-isothermal data, and stoichiometric information was provided for some of these chemical changes. Several of these studies were concerned with determining trends of behaviour through comparisons between related salts. Detailed descriptions of the chemical steps and identifications of the rate-controlling processes in the mechanisms were less frequently provided. A small proportion of the papers was concerned with previously well-studied reactions such as the dissociations of carbonates (**13** entries), including the effects of procedural variables on the decompositions of $CaCO_3$ (**4** entries) and of dolomite (**5** entries).

Although about a quarter of these papers discussed or reported aspects of dehydration (**33** entries), few described the use of reaction conditions known to eliminate the influence of the reverse process on the measured kinetic characteristics or attempted to establish detailed mechanisms. Almost every dehydration study in the set was concerned with a different reactant and the set included various inorganic salts (**14** entries), a range of salts of organic acids (**8** entries) and diverse hydroxides and basic salts (**11** entries).

Decompositions of simple inorganic (**37** entries) and organic (**16** entries) compounds were concerned with such a wide range of substances that there were remarkably few instances of reactant duplication. A few reports (**6** entries) were concerned with decompositions of mixtures, some of which may involve melting. The remaining articles were concerned with phase transitions (**5** entries), thermal reactions of various minerals (**10** entries), zeolites (**3** entries), or the properties of industrial products (**9** entries). The inorganic reactants included some of those compounds that have been the subject of continuous investigation throughout the development of the subject (permanganates, carbonates, etc.). There were three studies of various vanadates and two each of decompositions of different iodates and tungstates, otherwise no obvious common features of the reactants could be discerned. Similarly the organic reactants contained no common constituents, other than three studies of the reactions of metal acetates. Six comparative studies of less simple reactants were based on examinations of 4 to 14 compounds in each article (2 were concerned with sequences of lanthanides).

Coordination compounds (**72** entries). The diversity of reactants investigated across this set of articles offered no obvious basis for classification. Half of the studies were concerned with comparisons of behaviour within groups of five or more (up to 27) related reactants. A small proportion of the papers was concerned with the properties of a single compound. There was an inverse trend between the number of reactants mentioned and the detail in which each individual substance was described. Reaction stoichiometries were often based simply on the reactant mass loss.

No sets of related studies could be correlated or compared through common reactant features. The collection of studies mentioned numerous complexed metal atoms, ligands, cations and anions. Compounds containing nickel were mentioned in about a third of the papers examined (**25** entries), closely followed by compounds of Cu, Co, Fe, Zn and the lanthanides. Many other constituents were investigated including cations of all types (NH_4^+, alkali and alkaline earth metals, other transition metals, etc). Anions were similarly diverse, including halides, oxyanions, carboxylates, etc. Very many different types of ligand were included in these researches ranging from simple to remarkably complicated molecules. The reasons for the selection of specific ligands for study were not always adequately explained and the chemistry of their dissociation or breakdown on heating the complex was not usually determined.

The survey demonstrates the diversity of research activities in this field where no compound or reaction type can currently be identified as a focus of attention.

Instrumentation (**21** entries). The number of papers describing developments in instrumentation was smaller than anticipated (**16** entries), but this probably reflects the maturity of the subject and the wide availability of accurate and reliable commercial instruments. Nevertheless several advances in balance design were described. A small number (**5** entries) of papers were concerned with calibration.

Theory and kinetic analysis (**38** entries). Many aspects of the theory of kinetic analysis were discussed (**27** entries). Some papers were specifically concerned with discrimination of fit of data between alternative kinetic expressions or with constant reaction rate thermal analysis. Other articles (**11** entries) were concerned with aspects of the fundamental theory of the subject and with the compensation effect. The content of papers concerned with kinetic analyses appeared to accept the common basis of the applicability of the rate equations listed in Table 3.3.

Comments on the survey. The main conclusion that emerges from this limited (but probably representative) survey is that, although there is a high level of interest and activity in the field, no particular aspects are currently attracting particular attention. This may indicate a maturity of the subject, in which established experimental methods and agreed interpretative theory are used to obtain accepted types of results. Alternatively, it may indicate some stagnation with the need for new ideas and techniques. The conflicting ideas that were a feature of the early development of the subject have now largely been resolved or bypassed.

18.10. RELATIONSHIPS WITH NEIGHBOURING SUBJECT AREAS

This book, together with much of the literature devoted to crystolysis reactions, predominantly seeks correlations *within* the confines of the topic. (The Arrhenius equation, and attendant theory, is an obvious exception). Cross-fertilization of ideas with neighbouring fields may be hampered by the jargon existing within each subject area. Reviews addressed to wider readerships, cross-disciplinary conferences and plenary lecturers capable of communication beyond the narrow confines of their speciality are required.

Interpretations of reaction mechanisms must be based on up-to-date reactant models, including crystal properties (bonding, band structures, energy transfer, etc.) together with quantitative knowledge of all relevant features of any defective structures represented as participating in each step contributing to the overall change.

The subject area of decompositions of ionic solids would benefit from cross-fertilization of ideas from work done on polymers. Whereas the structures of polymers contain (by definition) an extended, often directed, network of strong linkages, together with much weaker, van der Waals intermolecular links, bonds in ionic solids tend to be of intermediate strength and are not directed. Ionic crystals are often more ordered and undergo a wider variety of types of chemical changes. Comparisons of the influences of reactant structures and bonding from both subject areas *may* yield valuable insights into the controls of reactivity in variously ordered condensed phases. Polymer chemistry may also provide insights into the role of melting or dissolution on the kinetics of decomposition.

Studies of the thermal reactions of those solids which are believed to involve catalysis by a solid product, would clearly benefit from progress in developing the theory of heterogeneous catalysis. These studies range from experiments in high vacuum with clean, almost perfect, single crystal faces maintained at low reactant coverage, to more realistic industrial-type working conditions of maximum activity, at high temperatures in the presence of excess quantities of reactants and products [73].

18.11. SOME GENERAL COMMENTS

18.11.1. Introduction

Predictions of directions for profitable future development need to build on the strengths and achievements of the subject and to avoid or overcome the weaknesses identified. Those model reactions that have contributed to the advance of the subject will, no doubt, be re-examined in the future. No reactant should be excluded from examination on the grounds that it has been investigated previously, provided that novel aspects of the reaction are addressed, or that new experimental techniques have become available. Progress through persistence is demonstrated in the thought-provoking review by Herbstein *et al.* [88] entitled "Old and new studies of the thermal decomposition of potassium permanganate" and a subsequent conference presentation [89] "What almost two hundred years of research has taught us about the thermal decomposition of potassium permanganate."

A major weakness is the lack of development or application of new experimental techniques which can give more direct chemical information about the processes occurring within a relatively inaccessible and mobile reaction zone, possibly of molecular dimensions. An ideal probe capable of providing such information would have to operate at the reaction temperature without causing disturbance within the reaction zone, but have high spatial and time resolutions. The most promising, but not generally accessible, technique has been diffraction studies using synchrotron radiation [6]. Hot-stage microscopy continues to provide useful geometrical information as well as phase identification. There is a need for development of reaction chambers for scanning electron microscopy so that observations are not limited to high vacuum and ambient temperature conditions. Atomic force microscopy shows promise as a technique for providing insights into surface processes.

Experimental studies, which may prove to be of value in advancing the theory of the subject, include investigations to establish the role of crystal structure in crystolysis reactions. One possible approach would be by comparing reactivities, through kinetic parameters for the decompositions of a single suitable compound in the crystalline state, in the melt [90] and in an inert solvent. There is a need for the list of rate equations given in Table 3.3. to be extended to include model systems based on allowances for partial and complete melting and for such rate equations to be tested for real reactions. It is uncertain whether the participation of melting could be recognized on kinetic evidence alone, but restriction of consideration to solid state models only does bias an investigation.

Progress requires both precise and imaginative approaches. The check list in Table 18.4. summarizes techniques and concepts that are currently widely applied. This is a starting point and may be profitably extended. Investigation of any crystolysis reaction should include determination of the role (if any) of precursor changes (e.g., recrystallization and/or a previous reaction such as dehydration, together with reactant cracking or diffusive loss of constituents). When an interface can be recognized, the dominant process in this zone may be identifiable as recrystallization accompanying or following a catalytic or strain-promoted chemical change. Retention of crystallographic structural features (topotaxy) or loss of local order (melting) must also be investigated. Changes which may occur behind the reaction front include secondary reactions between primary products, volatile product retention (including reversible rate processes) and retexturing of the residual solid. The rate processes upon which kinetic analyses are based need to be clearly defined, including, where appropriate, the reaction stoichiometry. Much also remains to be established concerning the chemical steps in nucleation processes and structures of reaction interfaces.

Of all crystolysis reactions, understanding of dehydrations appears to be most advanced and it remains to be seen whether the mechanistic patterns of behaviour recently proposed (Chapter 7) are of wider applicability. Systematic comparisons

Table 18.4.
Solid state decompositions: Interpretation of observations and formulation of mechanisms (both reaction geometry and interface chemistry). A check list.

EXPERIMENTAL TECHNIQUES	ANALYSIS OF OBSERVATIONS	MECHANISTIC CONCLUSIONS
Reactant and product compositions confirmed. Where possible, intermediates and impurities are measured.	Overall chemical change and intermediates identified. Identities of primary and secondary products distinguished.	Retexturing and/or recrystallization, dehydration.
		Surface reactions Chemistry, kinetic model and energetics of any initial reaction identified.
X-ray diffraction All participating phases identified including possible topotaxy.	**Textural changes** Interface geometry, melting, sintering, particle disintegration.	**Nucleation** Mechanism, precursor and final structures.
Microscopy Examine reactants, products and partially decomposed samples. Determine interface geometry, interface texture, particle sizes and textural changes.	**Rate equation** Interface geometry. Reversibility, diffusion control. Observations should apply to clearly defined rate processes.	**Interface chemistry** Discussion of overall reaction, rate limiting step, topochemistry, diffusion, stereochemistry, strain, catalysis, melting.
Thermochemistry Phase changes, enthalpy of reaction.	**Arrhenius parameters** Chemical steps. Compensation effects.	**Post-interface changes** Reversibility, sintering, recrystallization, secondary reactions.
Kinetic measurements Isothermal and non-isothermal experiments. Reproducibility, reversibility.		**Reaction geometry** Correlation of kinetic and microscopic data. Consistency of reaction models between geometric and textural observations.
Spectroscopy Identify characteristic bonds or species present in all participants.		**Comparative chemistry** Compare results with previous studies of the same reaction, with related reactions and with relevant theory.

with the thermal behaviour of other groups of related solid coordination compounds could yield results of theoretical value.

A representative selection of decompositions has been reviewed in Chapters 7 to 17. The subject area, however, lacks a unifying theory and accepted criteria for classification of decomposition reactions into groups on the basis of similarities or of differences in properties, mechanisms or behaviour. This topic is considered further below.

18.11.2. The thermal stability of solids

Prediction of behaviour. Prediction of the behaviour to be expected on heating a solid must include consideration of both thermodynamic and kinetic aspects. Thermodynamics will determine the temperature ranges over which endothermic decompositions are feasible, because entropies of decomposition are invariably positive. Thus explanations of differences in behaviour during exothermic decompositions will involve control by kinetic factors. Because most decompositions occur under conditions far from equilibrium, with non-homogeneous distributions of reaction zones, predictions based on equilibrium thermodynamics should generally be replaced by treatments using irreversible thermodynamics.

Crystal structure. The differences between heterogeneous reactions proceeding in the solid state and homogeneous reactions, in the gaseous state or in solution, have been discussed in Chapter 1. During chemical changes within crystals, the precursor species are relatively immobilized and may be subjected to stereochemical constraints by the ordered reactant structure and/or product phases, together with any (possibly transitory) topotactic relationships between them. Therefore, one factor which can be expected to influence the thermal stability of a solid, to a greater or lesser extent, is its crystal structure.

Attempts at identifying the influence of structure on stability have generally been inconclusive. For example, some alkali metal permanganates with comparable structures show similarities of decomposition behaviour [29], while, in contrast, the decompositions of several cobalt(III) ammine azides show little evidence of structural influences [76]. Significant differences in behaviour were found for the various crystal forms of the LiK tartrate hydrates [87] and, also, for the dehydrations of the isomorphic alums [20,43]. However, some reactants, for example those prepared by the dehydration of hydrated metal carboxylates [5], may be amorphous to X-rays, thus preventing recognition of any control of stability by crystal structure.

Each crystalline substance has a unique structure. Groups of compounds classified as *isomorphous* have similarities of lattice symmetry, but dimensions, and hence interionic forces, are different. Moreover, a particular substance can adopt alternative structures under changed conditions of temperature, pressure, crystallization conditions, presence of impurities, etc. Ordered packing, with symmetrical intracrystalline forces, appears to confer enhanced stability within the bulk solid so that decomposition processes usually occur at surfaces within a restricted reaction zone. Interfaces can be regarded variously as complex imperfections, zones of destabilizing strain, or (product) sites of catalytic activity.

Although a particular compound, in crystallizing, will usually adopt the same overall lattice, each real sample of the compound will have a unique set and distribution of imperfections of various kinds. Different specimens of a particular solid may be closely similar, but no two will be identical. Imperfections are relatively energy-rich sites of local distortion and so are important participants in solid state reactions. These deviations from the ideal lattice may exert a greater influence on the thermal stability of a solid than the crystal structure. This is evident from the differences in kinetic behaviour often observed between single crystals and powdered samples of the same reactant. (The reaction interface is itself an atypical distortion of the crystal structure.) Traces of impurities, or radiation damage, can also significantly enhance reactivity.

Reproducibility. An expected consequence of the types of reactant heterogeneity described above, would be irreproducibility of kinetic behaviour in which every specimen studied exhibited different kinetic characteristics (see, for example, [7] p.171-2). From these differences in behaviour, kinetic analyses might, in principle, provide information on the numbers, types and reactivities of the imperfections present in each individual specimen studied. (An assumption would have to be made that irreproducibility stems only from the characteristics of the samples and not from slight variations in the experimental conditions, e.g. temperature control and/or the efficiency of removal of gaseous products.) In practice, the main techniques used for measurement of the extent of reaction, α, (e.g. thermogravimetry or evolved gas pressures, see Chapter 6) are all *averaging* techniques and are thus insensitive to variations in behaviour at a molecular level. For example, loss of a minimum detectable mass of 1 μg from a 10 mg sample of $CaCO_3$ corresponds to the formation of 10^{-7} mol of CO_2 and the decomposition of $6x10^{16}$ CO_3^{2-} ions.

Differences in kinetic behaviour have been reported for relatively large single crystals [7,91] and these differences have been correlated with the dimensions and shapes of the crystals. In contrast, reasonably reproducible behaviour is often found from one portion to another for a single sample powder preparation of fairly uniform particle size. Reproducibility often becomes less satisfactory as the range of particle sizes increases. It is also common for samples of different preparations (by precipitation or crystallization from solution) of a reactant to show decreased reproducibility [63] compared to that for different portions from one preparation.

Sample heterogeneity is not often specifically investigated or discussed in formulating decomposition mechanisms. Many studies deal only with the behaviour of one preparation of the reactant, usually sufficiently finely ground to provide powder samples which yield acceptably reproducible results.

If acceptable reproducibility is obtained, it is tempting to argue, by analogy with homogeneous reactions [35], that a single specific process controls the rate of reaction. The concept of a rate determining step in homogeneous reactions is based upon the assumption that there is a single path for converting reactants to products and that movement along the reaction coordinate involves surmounting the energy barrier at its minimum. This analogy is strengthened by the observed applicability of the Arrhenius equation (Chapter 4) to solid state reactions. A close to linear Arrhenius plot over a wide temperature interval would support the suggestion that most (if not all) of the chemical change proceeds via a single type of bond redistribution step (characterized by a single value of E_a). Arrhenius plots for many solid state decompositions are, however, only approximately linear and the uncertainties in the estimated values of E_a may indicate participation of structural imperfections in providing alternative reaction paths with a range of E_a values.

The advance of a reaction interface, once established, into regions of almost perfect crystal, in preference to the appearance of new centres of decomposition in the bulk material, indicates that the reaction contributes to further decomposition by generating conditions that were not present originally. This is a form of autocatalysis. The crystal structure may, however, influence the direction and rate of advance, e.g. along layers in layer-type structures. Distortion of the structure adjacent to the reaction zone may lead to destabilization, or reaction may produce surfaces which act as a catalyst for further reaction, e.g. the particles of metal product formed during decompositions of metal carboxylates [18,59,73 and Chapter 16] provide active surfaces for anion breakdown. If a decomposition takes place in several stages, the stability of an intermediate may determine the overall reaction rate [22,79].

Initial processes, completed at small values of α, and with relatively low E_a values are observed for many decompositions. Such processes are often associated with surfaces or other imperfections [18]. Annealing or sintering of samples can decrease their imperfection contents. Some decompositions occur at temperatures close to the melting point of the sample. The reactant constituents, including imperfections, will have relatively high mobilities and mechanisms of crystolysis may begin to approach those of homogeneous processes. Further complexity develops when the initial product forms a molten eutectic mixture [22].

Many decompositions are preceded by a dehydration step (see Chapter 7) which has a disrupting effect on the original crystal structure and often yields an amorphous product as the reactant in the decomposition step. Surface textures are observed to change considerably during dehydration. The kinetics of the subsequent decomposition may be very dependent upon the dehydration conditions. If, however, the anhydrous sample recrystallizes before decomposing, the defect content will again be modified. Retention of water may promote recrystallization [20] and rates of decomposition of recrystallized and of amorphous material may be very different.

18.12. AN AGENDA FOR THE FUTURE

The formulation of mechanisms of individual crystolysis reactions is more difficult than might be expected. Investigations invariably require detailed interpretations of the many observations obtained from complementary experimental techniques. Little is known, at present, about the controls and chemistry of processes at interfaces. The subject lacks theoretical principles and models capable of accounting systematically for the diversity of behaviour of the many thermal decompositions described in the literature. In the absence of such models, classification of behaviour has not made the necessary progress.

18.12.1. Classification of solid state reactions

The recognition of behaviour patterns, both similarities and differences, amongst crystolysis reactions is necessary in any attempt to identify the factors that control reactivity. Establishment of such patterns requires detailed reviews of the vast literature (of which a small but representative part has been surveyed in Chapters 7 to 17). As an agenda for the future, some approaches which might provide a basis for a system of classification are discussed below.

The character and role of reaction interfaces. This approach was introduced in Chapter 6 and developed for dehydrations in Chapter 7. Such a classification scheme could be extended to a wider range of crystolysis reactions. In interpreting interfacial phenomena, cooperative interactions between species within the reaction zone need to be considered, including catalytic processes such as occur on the nickel metal product during decompositions of some nickel carboxylates [65].

Kinetic parameters. The literature contains numerous reports of the rate equations identified for particular crystolysis reactions, together with the calculated Arrhenius parameters. However, reproducible values of E_a (Section 4.1.) have been reported by independent researchers for relatively few solid state decompositions. Reversible reactions often yield Arrhenius parameters that are sensitive to reaction conditions and can show compensation effects (Section 4.9.4.). Often the influences of procedural variables have not been carefully identified. Thus, before the magnitudes of apparent activation energies can be compared, attempts have to be made to relate these values to particular reaction steps.

The distributed reactivity models used by Burnham and Braun [92] in the kinetic analysis of complex materials (see Section 5.5.12.) deserve further consideration, particularly in view of the results obtained by Christy *et al.*, [93] for the kinetics of dehydration of calcium oxalate monohydrate. Water loss proceeds at different rates from different lattice sites in this monohydrate.

A theoretical explanation for the applicability of the Arrhenius equation to solid state reactions has recently been provided [35] (see also Chapter 4.). Determinations of the available energy levels associated with the reaction zone and their occupancy, perhaps by spectroscopic methods, may help in establishing the steps which control bond redistributions under these conditions.

Two interesting current developments which may disturb some of the stagnation which has affected the subject are described below.

The L'vov sublimation model. The applications (outlined in Sections 2.4.5. and 2.4.6.) of the Hertz-Knudsen-Langmuir vaporization model to decompositions by L'vov *et al.* [94] are based on the assumption that decomposition involves an initial sublimation step, followed by condensation of the less volatile products. Because sublimation is an endothermic process, the condensation process would need to make a significant energetic contribution for the decomposition to be exothermic overall.

This theoretical model has been applied, mainly in very recent work [94], to correlating the behaviours of diverse reactants. It is too early to assess the impact that this approach may have on the development of the subject, but the work offers positive prospects for stimulating debate. The model has been applied to reversible and irreversible, endothermic and exothermic decompositions and is an alternative to the Polanyi-Wigner treatment (Section 4.5.) with which it shares some very general features. L'vov's papers include critical analyses of published results. The behaviours of individual single reactants and the patterns discerned within groups of different compounds, in which a common constituent (the anion) undergoes breakdown, are considered. The influences on the decomposition kinetics of inhomogeneous effects such as the reaction enthalpy (which may alter the reactant temperature) and the prevailing pressures of volatile product within the reactant mass are discussed. For example, the temperature and water vapour distributions developed within $Li_2SO_4.H_2O$ during dehydration [95] have been suggested to be capable of explaining the Smith-Topley effect (Section 7.7.).

An interesting extension of L'vov's theory could be towards establishing correlations with other rate processes involving solids. This may lead to identification of parallels with heterogeneous catalysis, where some mechanistic features, such as common intermediates in a few particular reactions, have already been described (Chapter 16.).

It is also necessary, however, to recognize that the validity of the underlying assumptions, particularly for processes occurring within the reaction interface, has to be confirmed. The model will need to be developed to account for complicating factors, such as the formation of reaction intermediates, the occurrence of melting, the role of impurities, and the observed topotactic nature of some decompositions.

The mathematical kinetic models proposed by Korobov. Korobov [96] has discussed the limitations of the traditional geometric-probabilistic approach to describing solid state reaction kinetics. He proposes that some of the more recently developed mathematical techniques (see Section 6.8.5.) should be used to provide improved descriptions of the advance of the reaction front within the structural symmetry of an individual reactant.

18.12.2. Conclusion

This survey of the decompositions of ionic solids records many of the achievements of the past and outlines some of the considerable problems and challenges to continued progress. The formulation of chemical mechanisms for reactions of solids has turned out to be far more difficult than was foreseen in early work, but considerable practical and theoretical rewards and intellectual satisfaction await those contributing to the future development of this important subject.

REFERENCES

1. W.E. Garner, Chemistry of the Solid State, (Ed. W.E. Garner), Butterworths, London, 1955, Chap.8.
2. A.K. Galwey and M.A. Mohamed, J. Chem. Soc., Faraday Trans. I, 81 (1985) 2503.
3. V.V. Boldyrev, I.S. Nev'yantsev, Yu.I. Mikhailov and E.F. Khairetdinov, Kinet. Katal., 11 (1970) 367.
4. V.V. Boldyrev, Kinet. Katal., 6 (1965) 934.
5. R.J. Acheson and A.K. Galwey, J. Chem. Soc. A, (1967) 1167.
6. V.V. Boldyrev, Y.A. Gapanov, N.Z. Lyakhov, A.A. Politov, B.P. Tolochko, T.P. Shakhtshneider and M.A. Sheromov, Nucl. Inst. Method. Phys. Res., A261 (1987) 192.
7. A.K. Galwey, G.M. Laverty, N.A. Baranov and V.B. Okhotnikov, Phil. Trans. R. Soc. (London), A347 (1994) 139, 157.
8. G.G.T. Guarini, J. Thermal Anal., 41 (1994) 287.
9. N. Koga and H. Tanaka, J. Phys. Chem., 98 (1994) 10521.
10. M. Lallemant and G. Watelle-Marion, C. R. Acad. Sci. Paris, 264 (1967) 2030; 265 (1967) 627.
11. P.J. Herley, O. Christofferson and R. Irwin, J. Phys. Chem., 85 (1981) 1874, 1882, 1887.
12. P.J. Herley and W. Jones, J. Mater. Sci. Letters, 1 (1982) 163.
13. D. Beruto and A.W. Searcy, Nature (London), 263 (1976) 221.
14. K.M. Towe, Nature (London), 274 (1978) 239.
15. D. Beruto, L. Barco and A.W. Searcy, J. Amer. Ceram. Soc., 67 (1984) 512; 62 (1979) 580; 65 (1982) C42; 66 (1983) 893.
16. C. Anderson and A.K. Galwey, Canad. J. Chem., 70 (1992) 2468.
17. A.K. Galwey and G.M. Laverty, Thermochim. Acta, 138 (1989) 115.
18. A.K. Galwey and M.E. Brown, J. Chem. Soc., Faraday Trans. I, 78 (1982) 411.
19. M.E. Brown, B. Delmon, A.K. Galwey and M.J. McGinn, J. Chim. Phys., 75 (1978) 147.
20. A.K. Galwey, R. Spinicci and G.G.T. Guarini, Proc. R. Soc. (London), A378 (1981) 477.
21. G.G.T. Guarini and L. Dei, J. Chem. Soc., Faraday Trans. I, 79 (1983) 1599.
22. N.J. Carr and A.K. Galwey, Proc. R. Soc. (London), A404 (1986) 101.

23. M.E. Brown, A.K. Galwey and G.G.T. Guarini, J. Thermal Anal., 49 (1997) 1135.
24. J.M. Thomas and G.D. Renshaw, J. Chem. Soc. A, (1967) 2058; (1969) 2753.
25. N.J. Carr and A.K. Galwey, Thermochim. Acta, 79 (1984) 323.
26. K. Torkar and H.T. Spath, Monat. Chem., 98 (1967) 2020.
27. A. Glasner, E. Levy and M. Steinberg, J. Inorg. Nucl. Chem., 25 (1963) 1119, 1415; 26 (1964) 1143.
28. J. Simpson, D. Taylor and D.M.W. Anderson, J. Chem. Soc., (1958) 2378.
29. M.E. Brown, A.K. Galwey, M.A. Mohamed and H. Tanaka, Thermochim. Acta, 235 (1994) 255.
30. A.K. Galwey, R. Reed and G.G.T. Guarini, Nature (London), 283 (1980) 52.
31. J. Sesták, J. Thermal Anal., 36 (1990) 1997.
32. R.D. Shannon, Trans. Faraday Soc., 60 (1964) 1902.
33. H.F. Cordes, J. Phys. Chem., 72 (1968) 2185.
34. S. Vyazovkin and W. Linert, J. Solid State Chem., 114 (1995) 392.
35. A.K. Galwey and M.E. Brown, Proc. R. Soc. (London), A450 (1995) 501.
36. M. Reading, Thermochim. Acta, 292 (1997) 179.
37. M.E. Brown, A.K. Galwey and A. Li Wan Po, Thermochim. Acta, 220 (1993) 131; 203 (1992) 221.
38. E.G. Prout and F.C. Tompkins, Trans.Faraday Soc., 40 (1944) 488.
39. A.K. Galwey and M.E. Brown, Thermochim. Acta, 269/270 (1995) 1.
40. M.E. Brown, D. Dollimore and A.K. Galwey, Reactions in the Solid State, Comprehensive Chemical Kinetics, (Eds C.H. Bamford and C.F.H. Tipper), Vol.22, Elsevier, Amsterdam, 1980.
41. A.K. Galwey, S.A. Lyle and S.A.A. Mansour, Thermochim. Acta, 235 (1994) 239; 228 (1993) 379.
42. P.W.M. Jacobs and H.M. Whitehead, Chem. Rev., 69 (1969) 551.
43. A.K. Galwey and G.G.T. Guarini, Proc. R. Soc. (London), A441 (1993) 313.
44. J. Zsakó and H.E. Arz, J. Thermal Anal., 6 (1974) 651.
45. P.K. Gallagher and D.W. Johnson, Thermochim. Acta, 14 (1976) 255.
46. M. Maciejewski and A. Reller, Thermochim. Acta, 110 (1987) 145.
47. D. Beruto and A.W. Searcy, J. Chem. Soc., Faraday Trans. I, 70 (1974) 2145; A.W. Searcy and D. Beruto, J. Phys. Chem., 82 (1978) 163.
48. T. Darroudl and A.W. Searcy, J. Phys. Chem., 85 (1981) 3971.

49. F.W. Wilburn, J.H. Sharp, D.M. Tinsley and R.M. McIntosh, J. Thermal Anal., 37 (1991) 2003.
50. H. Wei and Y. Luo, J. Thermal Anal., 45 (1995) 303.
51. G.N. Karagiannis, T.C. Vaimakis and A.T. Sdoukos, Thermochim. Acta, 262 (1995) 129.
52. T.B. Flanagan, J.W. Simons and P.M. Fichte, Chem. Comm., (1971) 370.
53. T.B. Grimley, Chemistry of the Solid State, (Ed. W.E. Garner), Butterworths, London, 1955, Chap.14.
54. A.A. El-Bellihi, J. Thermal Anal., 41 (1994) 191.
55. E. Ingier-Stocka, J. Thermal Anal., 37 (1991) 769.
56. S.V. Levchik, E.E. Bolvanovich, A.I. Lesnikovich, O.A. Ivashkevich, P.N. Gaponik and S.V. Vyazovkin, Thermochim. Acta, 168 (1990) 211.
57. A.K. Galwey and M.E. Brown, Thermochim. Acta, 300 (1997) 107.
58. A.K. Galwey, Thermochim. Acta, 294 (1997) 205.
59. A.K. Galwey, Adv. Catal., 26 (1977) 247 (see Appendix I, p.311).
60. D.A. Young, Decomposition of Solids, Pergamon, Oxford, 1966.
61. M.A. Mohamed and A.K. Galwey, Thermochim. Acta, 213 (1993) 269.
62. M.A. Mohamed and A.K. Galwey, J. Chem. Soc., Faraday Trans. I, 81 (1985) 2503.
63. M.E. Brown, J. Chem. Soc., Faraday Trans. I, 69 (1973) 1202.
64. A.K. Galwey and G.M. Laverty, Proc. R. Soc. (London), A440 (1993) 77.
65. A.K. Galwey, S.G. McKee, T.R.B. Mitchell, M.A. Mohamed, M.E. Brown and A.F. Bean, React. Solids, 6 (1988) 173,187.
66. A.K. Galwey, D.M. Jamieson, M. Le Van and C. Berro, Thermochim. Acta, 10 (1974) 161.
67. A.K. Galwey and M.A. Mohamed, Solid State Ionics, 42 (1990) 135.
68. P.W.M. Jacobs and A.R. Tariq Kureishy, Trans. Faraday Soc., 58 (1962) 551; Reactivity of Solids, (Eds J.H. de Boer *et al.*), Elsevier, Amsterdam, 1961, p.353.
69. M.A. Mohamed and A.K. Galwey, Thermochim. Acta, 217 (1993) 263.
70. A.K. Galwey and M.A. Mohamed, Thermochim. Acta, 213 (1993) 279.
71. M.E. Brown, D. Dollimore and A.K. Galwey, J. Chem. Soc., Faraday Trans. I, 70 (1974) 1316.
72. D. Dollimore, Thermochim. Acta, 117 (1987) 331.
73. J. Fahrenfort, L.L. Van Reyen and W.M.H. Sachtler, The Mechanism of Heterogeneous Catalysis, (Ed. J.H. de Boer), Elsevier, Amsterdam, 1960, p.23.

74. O.I. Lomovsky, Yu.I. Mikhalov, V.V. Boldyrev and V.M. Mastikhin, Thermochim. Acta, 43 (1981) 135.
75. A.K. Galwey and M.A. Mohamed, Thermochim. Acta, 239 (1994) 211.
76. T.B. Joyner, J.Phys.Chem., 74 (1970) 1552, 1558, 1563.
77. M. Dodé, C.R. Acad. Sci. Paris, 200 (1935) 63; Bull. Soc. Chim. Fr., 5 (1938) 170.
78. L.L. Bircumshaw and B.H. Newman, Proc. R. Soc. (London), A227 (1954/5) 115, 228.
79. A.K. Galwey and M.A. Mohamed, Proc. R. Soc. (London), A396 (1984) 425.
80. A.K. Galwey and G.M. Laverty, Thermochim. Acta, 228 (1993) 359.
81. O. Chaix-Pluchery and J-C. Niepce, React. Solids, 5 (1988) 69.
82. M.C. Ball, Thermochim. Acta, 21 (1977) 349.
83. M.E. Brady, M.G. Burnett and A.K. Galwey, J. Chem. Soc., Faraday Trans.I, 86 (1990) 1573.
84. A.K. Galwey, L. Pöppl and S. Rajam, J. Chem. Soc., Faraday Trans.I, 79 (1983) 2143.
85. S.D. Bhattamisra, G.M. Laverty, N.A. Baranov, V.B. Okhotnikov and A.K. Galwey, Phil. Trans. R. Soc. (London), A341 (1992) 479.
86. A.K. Galwey, G.M. Laverty, V.B. Okhotnikov and J. O'Neill, J. Thermal Anal., 38 (1992) 421.
87. A.K. Galwey, Drug Stability, 1 (1995) 3.
88. F.H. Herbstein, M. Kapon and A. Weissman, J. Thermal Anal., 41 (1994) 303.
89. F.H. Herbstein, Proc.Israel-Hungary Binational Conference on Thermal Analysis and Calorimetry of Materials, Ein-Bokek, Israel, 1996, p.6.
90. C.E.H. Bawn, Chemistry of the Solid State, (Ed. W.E. Garner), Butterworths, London, 1955, Chap.10.
91. J.M. Thomas, E.L. Evans and T.A. Clarke, J. Chem. Soc. A, (1971) 2338.
92. A.K. Burnham and R.L. Braun, Energy and Fuels, submitted; R.L. Braun and A.K. Burnham, Energy and Fuels, 1 (1987) 153; A.K. Burnham, R.L. Braun and T.T. Coburn, Energy and Fuels, 10 (1996) 49.
93. A.A. Christy, E. Nodland, A.K. Burnham, O.M. Kvalheim and B. Dahl, Appl. Spectrosc., 48 (1994) 561.

94. B.L. L'vov, Spectrochim. Acta, B52 (1997) 1; Zh. Anal. Khim., 45 (1990) 2144; Microchim. Acta, Wein II, (1991) 299; Thermochim. Acta, 291 (1997) 179; 303 (1997) 161; B.L. L'vov and A.V. Novichikhin, Spectrochim. Acta, B50 (1995) 1427, 1459; Thermochim. Acta, 290 (1997) 239; B.L. L'vov, A.V. Novichikhin and A.O. Dyakov, Thermochim. Acta, 315 (1998) 135,169.
95. B.L. L'vov, Thermochim. Acta, 315 (1998) 145.
96. A. Korobov, Heterogeneous Chemistry Reviews, 3 (1996) 477.

GLOSSARY

Short explanations of a limited number of selected terms, or groups of terms, in the subject area are given below. These are less formal than rigid or legal-type definitions and are presented here as an aid to understanding.

Arrhenius equation, Activation energy(E_a), Frequency (or Pre-exponential) factor (A)

The increases with temperature of the rates of most reactions, including the majority of decompositions of solids, are well represented by the Arrhenius equation, $k = A \exp(-E_a/RT)$. Application of this equation to rate processes in solids does, however, require reconsideration (see Chapter 4) of the theory developed for homogeneous kinetics, where the magnitude of E_a is regarded as providing a measure of the energy barrier to reaction and A is identified as the frequency of occurrence of the reaction situation. Regardless of theoretical shortcomings, the values of A and E_a can provide a convenient method of quantitatively reporting and comparing reactivities. Limits of error reported for these parameters for many solid state rate processes are often greater than the limits achieved routinely for homogeneous reactions. The meaning to be attached to the term 'mole' for E_a values (units: kJ mol^{-1}) remains enigmatic (Chapter 4).

Rate equation, Kinetic model, Reaction mechanism

These three terms are sometimes treated as being synonymous. The distinctions are, however, important.

The **rate equation** specifies the mathematical function ($g(\alpha) = kt$ or $d\alpha/dt = k\,f(\alpha)$) that represents (with greatest statistical accuracy, Chapter 3) the isothermal yield (α) - time data for the reaction. For reactions of solids these equations are derived from geometric **kinetic models** (Chapter 3) involving processes such as nucleation and growth, advance of an interface and/or diffusion. $f(\alpha)$ and $g(\alpha)$ are known as **conversion functions** and some of these may resemble the concentration functions in homogeneous kinetics which give rise to the definition of order of reaction.

The term **reaction mechanism** specifies the sequence of chemical steps through which reactants are transformed into products. In the collision model of homogeneous reactions the steps are described in terms of their **molecularity**. However, the sequence of bond redistributions and other processes (diffusion, recrystallization, etc.) by which a solid reactant is converted into products will generally be far more complex (see Chapter 18) and the information required to characterize contributing steps is far less accessible. Description of these steps, including the formation and role of intermediates, requires both chemical and structural information.

A kinetic/geometric analysis **and** a description of the bond redistribution steps, which are consistent with each other, are required to provide a complete description of the decomposition of a solid.

Reversible reactions, Dissociations

Reversible reaction is a more general term than **dissociation**. **Dissociation** implies a single reactant and hence for a one-step decomposition of a single solid the terms are synonymous. Reversible solid-gas reactions proceed until the equilibrium pressures (activities) of the volatile participants are reached. Some dissociations of solids are completely reversible, i.e. the reaction direction is determined only by the deviation of the prevailing pressure from the equilibrium value. Reactions are subject to kinetic controls that may include geometric factors and diffusion limitations across a product barrier layer.

Measured rates of reactions may include contributions from reverse processes which must be investigated and allowed for in the interpretation of kinetic observations (see Chapter 12).

Concentrations, Reaction orders

Concentrations of crystalline solids are constants detemined by their crystal structures and such values have no useful applications in the kinetic description of decompositions that are completed within a narrow reaction interface that progressively advances into the unchanged reactant. Concentrations within the solid may change through diffusion and/or structural change as the migrating zone of chemical transformation arrives. The reactant concentration then rapidly drops to zero and that of the product becomes large. Measurements of concentration variations across such an interface might give mechanistic information, but we are aware of no successful measurements of this type. The experimental difficulties would be considerable because interface thicknesses are small, perhaps of molecular dimensions.

Rate expressions based on **reaction orders** (Chapter 3) have found applications in solid state decompositions due to the important distinction that rate may be determined by the total amount of reactant present rather than its concentration. Several explanations of the fit of data for solid state decompositions to kinetic equations expressed as reaction order (usually first-order) have been given, including the following. (i) The fit may be empirical, perhaps approximate, without mechanistic significance. (ii) The assumption that reaction proceeds in the solid state may be incorrect in complex rate processes where there is reactant melting. (iii) Commercial software for kinetic analysis sometimes restricts coverage to reaction orders, and the wider range of rate equations (Table 3.3.) is simply not considered. (iv) Some specific reaction models have the same form as reaction orders. For example, random nucleation within a large number of small crystals can be regarded as formally identical with a first-order reaction.

Compensation effect, Isokinetic temperature, θ rule

A set of related rate processes in which the decrease in rate expected to result from an increase in activation energy is offset by a corresponding increase in the pre-exponential term is referred to as a **compensation effect** and is expressed:

$$\ln A = a + b\,E_a$$

At the **isokinetic temperature** (often referred to as θ, hence the **θ rule**) all reactions of the set have rate constants of the same magnitude (k_θ).

Sets of reactions that have been identified as exhibiting compensation behaviour incorporate common chemical features, such as reactants containing one or more of the same constituents, similar ligands in coordination compounds, etc.

A number of instances of compensation behaviour that have been reported for studies of (partially or completely reversible) decompositions of a particular solid reactant carried out under different reaction conditions. The significance of such observations is discussed in Chapter 4.

Thermal decomposition (of solids), Dissociation, Pyrolysis, Crystolysis

These terms are used to describe the major changes in bond distributions that occur on heating a solid to a sufficiently high temperature. These changes involve the bonding *within* individual constituents, in contrast to the bonds *between* constituents that are reorganized in phase transformations or recrystallisations. Volatile products may be formed. **Thermal decomposition** and **pyrolysis** are non-specific terms that can be applied to all such reactions. Pyrolysis has probably been used most frequently for the breakdown of organic reactants. **Dissociation** refers specifically to reversible processes. **Crystolysis** has been proposed to identify decompositions proceeding in the (initially crystalline) solid state only.

Crystal surface, Crystal bulk

Processes occurring at a **crystal surface** refer to any boundary phenomena within a zone of unspecified and often indeterminate thickness. Many chemical changes, including nucleation, are regarded as surface processes because these are initiated at, or near, crystal faces. There have been few measurements or theoretical assessments made of the variation in reactivity with depth below a crystal surface. Surface reactivities may vary at different crystallographic faces. Surfaces are susceptible to damage and may retain impurities which influence reactivities. Cleaved surfaces are sometimes used in kinetic investigations of nucleation and growth processes but few quantitative comparisons of properties have been made between freshly exposed and as-grown crystal faces.

By contrast, the **crystal bulk** is the intracrystalline material wherein there is no significant influence from a surface. Reactivity is regarded as constant and the bulk material is usually more stable than that in the surface.

Fractional reaction, Product yield, α

These are equivalent measures of the fractional amount of a defined reaction that has taken place. In kinetic studies, measurements of reactant mass loss, gas released, heat evolved (or absorbed), etc., are used to calculate the magnitudes of α at known values of time, t, and temperature, T (Chapter 2).

Non-isothermal kinetics

This term refers to the determination of kinetic parameters (f(α) or g(α), A and E_a) for a reactant subjected to a predetermined heating programme, usually, but not necessarily, a constant rate of temperature increase ($dT/dt = \beta$) (Chapter 5). Isothermal data may provide the more sensitive tests for distinguishing the 'best fit' rate equations ($g(\alpha) = kt$), whereas rising temperature observations may be preferred for the determination of Arrhenius parameters (A and E_a). Reasons for any differences noted in the results of the alternative treatments should be investigated.

Reaction interfaces, Nucleation

The essential feature of the advancing reaction interface model is that the chemical changes of crystal constituents occur, and are completed, in the immediate vicinity of the zone of contact between reactant and product, with only minor displacements of the constituents of both the reactant and the product phases. Progress of the reaction zone is regarded as formally analogous to an energetic wave wherein chemical bonds are redistributed but solid materials are not transported with it. The Polanyi-Wigner model (Chapter 3) regards the reaction zone as being of approximately molecular thickness, or the minimum layer within which chemical change can occur.

The steps that culminate in generation of an interface, or formation of a nucleus, have been discussed in Chapter 3. Little progress has been made in establishing detailed mechanisms for specific reactants, because formation of product necessarily destroys any precursor states together with any intermediates. **Nucleation** may occur at surfaces, in the vicinity of a crystal imperfection (Chapter 1), but the demonstration that the onset of reaction, at least for some hydrates (Chapter 7), is accompanied by modifications to all reactant surfaces, indicates that the processes occurring are more complicated than was formerly recognized.

Similar conclusions apply to interface structures (Chapter 6). Textural examinations of reactant-product contact zones have revealed greater structural complexities than were recognised in earlier work. The interface model leads to the kinetic representation that the rate of product formation is directly proportional to the area of reactant-product contact and its geometric pattern of development (Chapter 3).

Annealing

Reorganization within the crystalline material, below its melting point, that relaxes disorder. Dislocations may be expelled or merged and impurities expelled or accommodated at imperfections, thus decreasing their bulk concentrations. Surfaces may be stabilized.

Sintering

Localised fusion at points of contact between particles, resulting in an effective increase of particle size and decrease in specific surface area. Sintering may be accompanied by annealing.

INDEX

Note: Entries in **Bold** face, marked with a **G**, refer to the **Glossary**, p.566-571.

N

O

P

T